CONCRETE FORMWORK

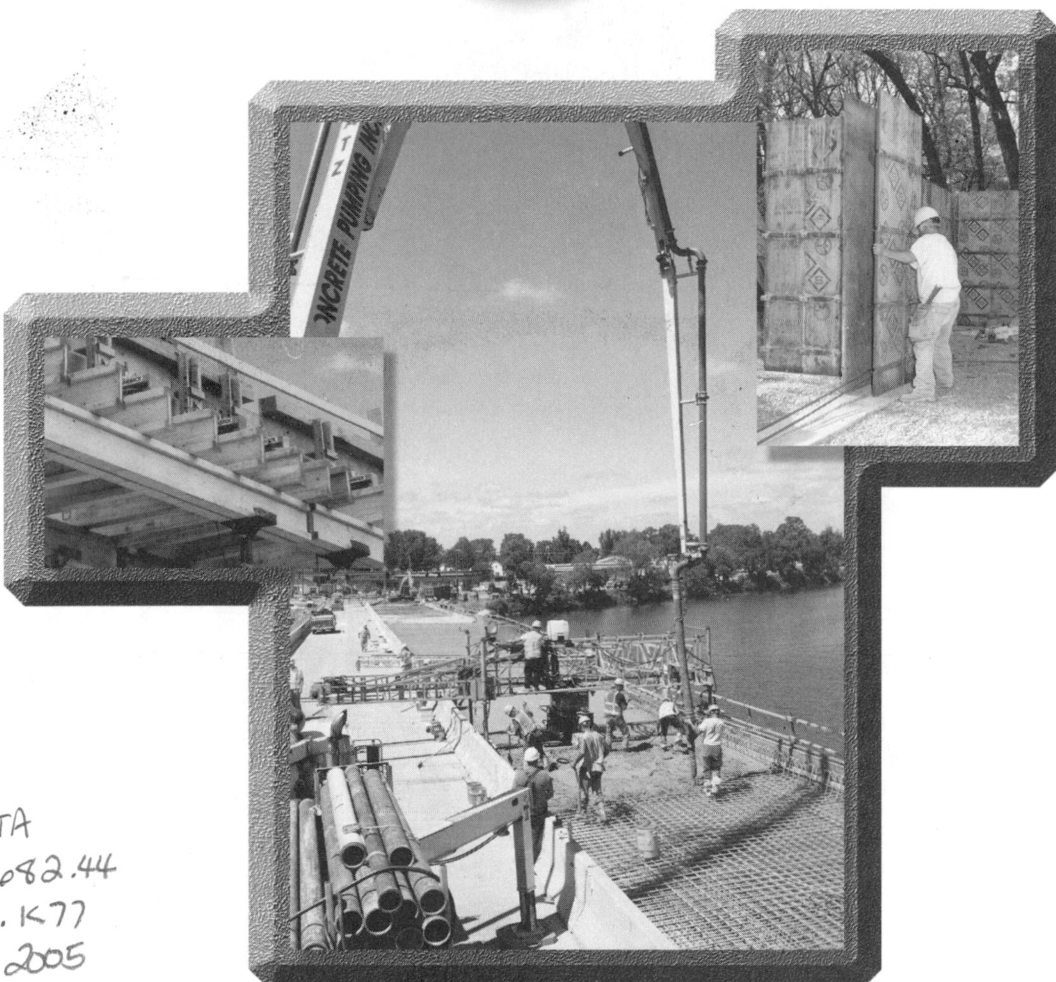

AMERICAN TECHNICAL PUBLISHERS, INC.
HOMEWOOD, ILLINOIS 60430-4600

Leonard Koel

Concrete Formwork, Third Edition, contains procedures commonly practiced in industry and the trade. Specific procedures vary with each task and must be performed by a qualified person. For maximum safety, always refer to specific manufacturer recommendations, insurance regulations, specific job site and plant procedures, applicable federal, state, and local regulations, and any authority having jurisdiction. The material contained is intended to be an educational resource for the user. American Technical Publishers, Inc. assumes no responsibility or liability in connection with this material or its use by any individual or organization.

B.V. Plyform is a registered trademark of APA—The Engineered Wood Association. The National Electrical Code and NEC are registered trademarks of the National Fire Protection Association. Steel-ply is a registered trademark of Symons Corporation. Styrofoam is a registered trademark of Dow Chemical Company. Velcro is a trademark of Velcro Industries. International Building Code is a registered trademark of the International Code Council. International Residential Code is a trademark of the International Code Council.

3 4 5 6 7 8 9 – 05 – 9 8 7 6 5 4 3 2

Printed in the United States of America

ISBN 0-8269-0708-3

Contents

Acknowledgments

The author and publisher are grateful for the technical information and assistance provided by the following companies and organizations:

American Concrete Institute
American PolySteel
American Society for Testing and Materials
APA—The Engineered Wood Association
Barclay and Associates
The Burke Company
Case Foundation
Chris P. Stefanos Associates, Inc.
David White Instruments
ECO-Block, LLC
ELE International, Inc.
The Euclid Chemical Company
The Garlinghouse Company
Gates & Sons, Inc.
Gomaco Corporation
Hilti, Inc.
Increte Systems
John Deere Construction & Forestry Company
Laser Alignment, Inc.
Leica Geosystems
Meadow Burke Products, Inc.
Metal Forms Corporation
MFG Corporation
Occupational Safety and Health Administration
Patent Construction Systems
Portland Cement Association
RJD Industries, Inc.
Simpson Strong Tie Company
Stanley Tools
Symons Corporation
Wacker Corporation

Introduction

Concrete Formwork, Third Edition, presents information on the safe construction of formwork for residential, light commercial, and heavy construction projects. This new edition features new and expanded information covering safety, excavating and trenching, scaffolding, post-tensioning of concrete, anchoring systems, tilt-up construction, insulating concrete forms (ICFs), total stations, and other topics. Reference is made throughout the book to International Building Code® (IBC) and International Residential Code™ (IRC) standards. Also incorporated are the latest American Concrete Institute (ACI) recommendations and Occupational Safety and Health Administration (OSHA) regulations.

Chapters 1 and 2 provide information about building site safety and the materials and methods used to form walls. New information addresses excavating and trenching, scaffolding, ICFs, and shielding. Chapters 2 through 5 cover residential foundation construction, flatwork, and heavy construction with new anchoring device information and an expanded welded wire reinforcement section. Other new information presented includes slip forms, flying forms, post-tensioning concrete, shoring, and reshores. Chapter 6 covers precast concrete construction and tilt-up construction including information on bracing procedures for wall panels and tilt-up sandwich panels. Chapter 7 covers concrete mix and placement, concrete curing, and form stripping. Chapter 8 covers fundamental skills in three sections including Math Fundamentals, Printreading, and Form Materials and Concrete Quantity Takeoff.

The Appendices and Glossary provide valuable information to supplement the chapter material. Appendix A includes general formulas and tables used by a form builder or estimator. Appendix B provides information on lumber dimensions, reinforcement, and quantities of concrete required for concrete walls, footings, and floor slabs. Appendices C and D include ACI Recommended Practices and OSHA Concrete and Shoring Regulations. Appendix E presents use of the builder's level, transit-level, laser transit-level, and total station instrument. The comprehensive Glossary includes a listing of terms commonly used in industry.

Review Questions throughout the book test for understanding of chapter material. Answers to Review Questions involving math are rounded to two places after the decimal point. Printreading Exercises test for basic knowledge of measurement, print elements, and interpretation. The question types used in the Review Questions and Printreading Exercises are completion, multiple choice, and identification. Answers to the questions should be recorded in the space provided. Answers for all questions are listed in the *Concrete Formwork Answer Key.*

Concrete Formwork is one of many high-quality training products available from American Technical Publishers, Inc. To obtain information about related training products, visit the American Tech web site at www.go2atp.com.

The Publisher

CHAPTER 1

The Building Site

The building site is the area in which construction occurs. A building site may be a small residential lot or a large area used for a heavy commercial construction project. Property lines indicate the boundaries of a building site and are a reference for groundwork and building layout.

The foundation design of a structure and the amount of groundwork required are based on the soil conditions of the building site. The soil characteristics determine the bearing capacity of the soil and the amount of settlement that can be expected. Sandy soil contains larger soil particles than soil with a higher percentage of clay. Sandy soil has a greater bearing capacity and therefore less soil and foundation settlement.

Preliminary groundwork is completed before foundation construction. A set of prints is used to determine the location of a structure and the amount of groundwork to be completed. Tradesworkers establish building lines based on the prints. Building lines indicate the location of the foundation of the structure. Operating engineers perform the groundwork using earth-moving equipment such as backhoes, bulldozers, motor graders, etc. Groundwork may range from a small amount of grading to massive trenching and excavating. Carpenters verify grade levels during groundwork and shore earth walls of heavy excavations.

The foundation of a structure is constructed after the groundwork is completed. The size and shape of a foundation are based on the prints and local building codes and zoning ordinances.

SOIL CONDITIONS

Soil conditions determine the type of foundation design required for a building. One of the most important factors influencing foundation design is the type of soil found beneath the structure. Some soils have a higher bearing capacity than others. *Bearing capacity* is the ability of soil to support weight. A number of different earth layers (strata) often exist below the ground surface of a building site. The bearing capacity varies for each layer of soil. Excavation (removal) of some of the earth layers may be required so the foundation footings rest on soil of adequate bearing capacity. **See Figure 1-1.**

All foundations settle over time. The soil conditions present determine the amount of settlement. Excessive or uneven settlement can cause cracks in the foundation, resulting in structural damage to the building. Foundations built in proper soil conditions can reduce settlement problems.

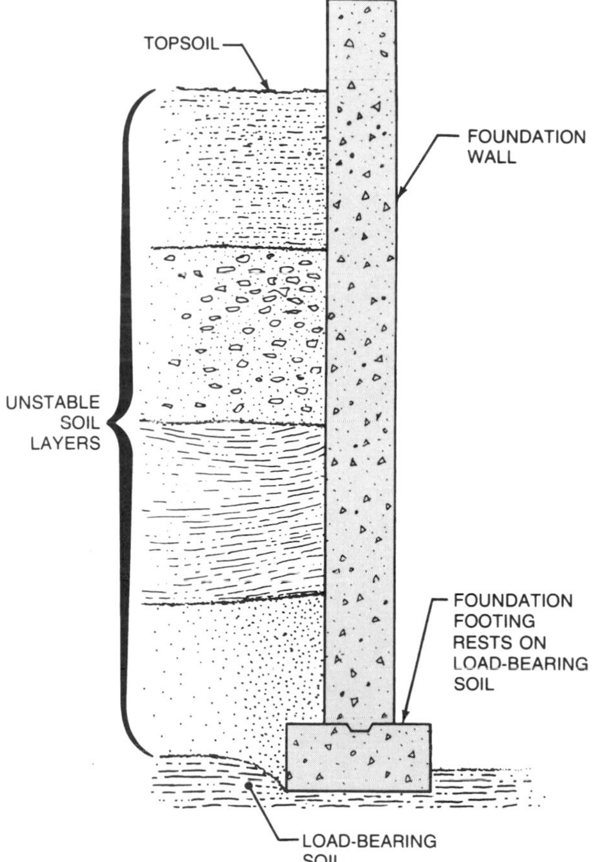

Figure 1-1. Excavation for a foundation footing requires that the foundation footing rest on soil with adequate bearing capacity.

Soil Mechanics

Soil mechanics is the study of soil types and the resulting effect upon the behavior of the soil. Two important factors in soil mechanics are the type of soil and the stress factors that have a weakening effect on the stability of a trench or excavation. One of the more frequent stress factors is tension cracks. A *tension crack* is a crack in soil caused by increases and decreases in moisture content and by earth movement. Tension cracks contribute to sliding and toppling. Other factors that cause trench disturbances are subsidence and bulging, heaving or squeezing, and boiling. **See Figure 1-2.**

Soil Composition

All soils consist of particles that originate from the breakdown and decomposition of solid rock. The major factor in differentiating soils is the size of the soil particles. Soils are generally classified as granular or cohesive. A *granular (coarse-grained) soil* is a soil that consists mostly of sand and gravel with large, visible particles. A *cohesive (fine-grained) soil* is a soil that consists mostly of silt and clay with particles that usually can be seen only with a microscope. **See Figure 1-3.**

Samples taken from building sites often show mixed soils. The predominant soils found in the mixture determine the final analysis of the sample and its expected bearing capacity. Classification charts are available that give the approximate bearing capacities of various soil mixtures common to the location. These soil classifications apply to normal conditions. Specific job sites may require further testing for an accurate analysis. **See Figure 1-4.**

Soil Compressibility. Compressibility of the soil below the foundation determines the amount a foundation will settle. The amount of compressibility is determined by the reduction of the spaces (voids) between the soil particles that contain air and/or water. Sandy soil contains larger soil particles than silt or clay, which results in less compression and foundation settlement. Silt or clay containing small soil particles compresses more, resulting in more foundation settlement.

Soil Type. The Occupational Safety and Health Administration (OSHA) classifies soil and rock deposits into four types: stable rock, Type A, Type B, Type C, and layered geological strata. The type of soil has a strong bearing on what protective measures must be taken in the excavation or trench. The soil at building sites requiring deep excavations should be inspected and approved by a qualified engineer. Deep excavations in Type A, Type B, Type C, and layered geological strata always require shoring and other protective measures.

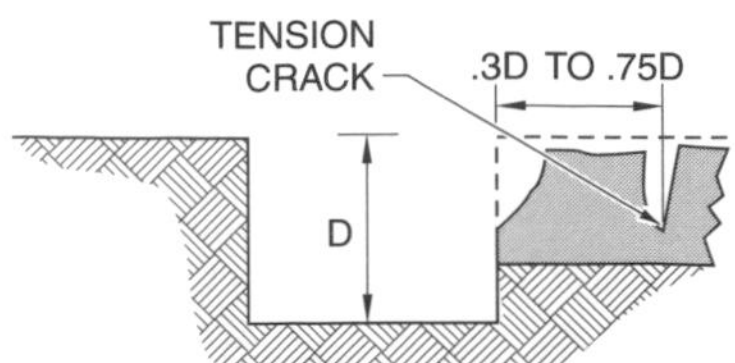

Tension cracks typically occur at a horizontal distance of .3 times to .75 times the depth of trench

TENSION CRACK

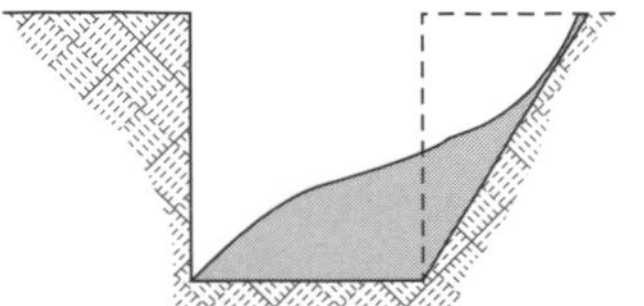

Sliding, also called sluffing, can occur as a result of tension cracks

SLIDING

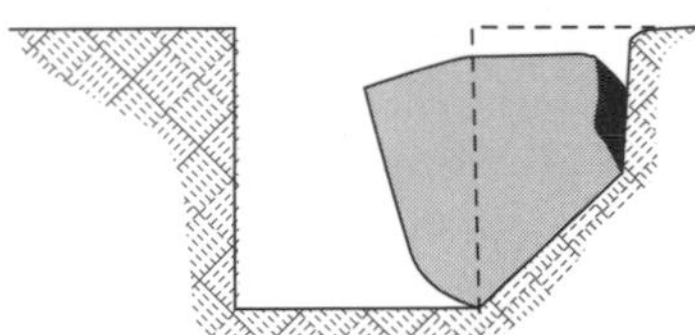

Toppling may also be a result of tension cracks

TOPPLING

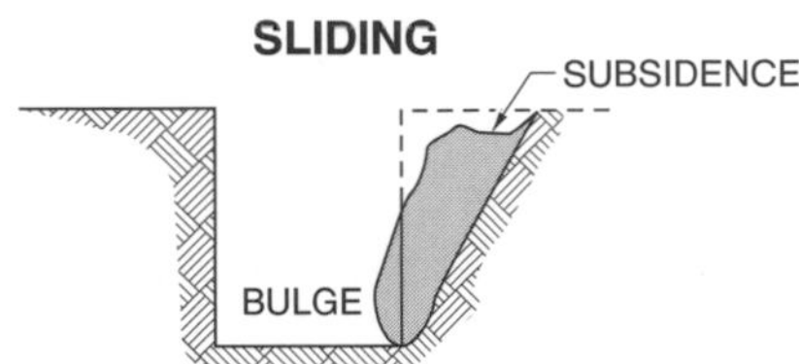

Subsidence and bulging are caused by stress which causes subsidence at the surface and bulging in the vertical face of the trench

SUBSIDENCE AND BULGING

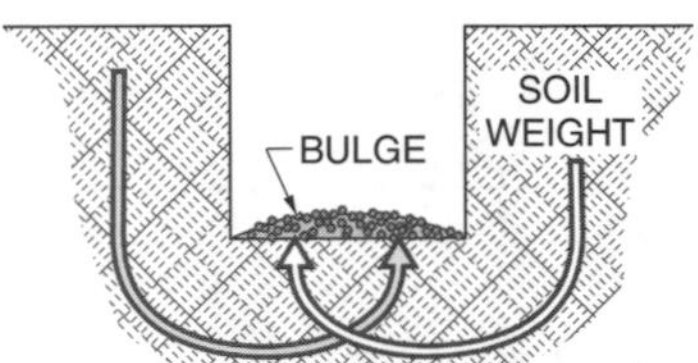

Heaving or squeezing is caused by downward pressure from the adjoining soils, causing a bulge in the bottom of the trench

HEAVING OR SQUEEZING

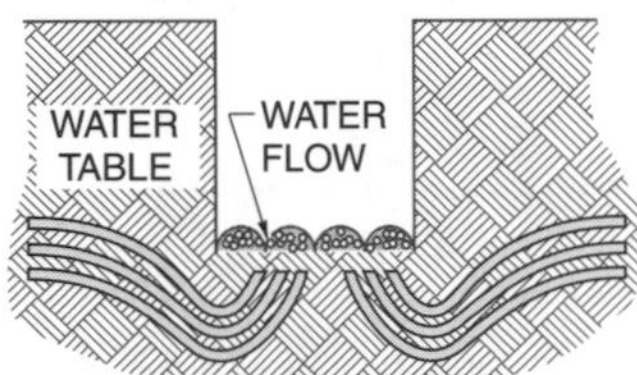

Boiling is an upward water flow into the bottom of the trench, often caused by a high water table in the surrounding area

BOILING

Figure-1-2. Frequent stress factors in excavations are tension cracks, sliding, toppling, subsidence, bulging, heaving or squeezing, and boiling.

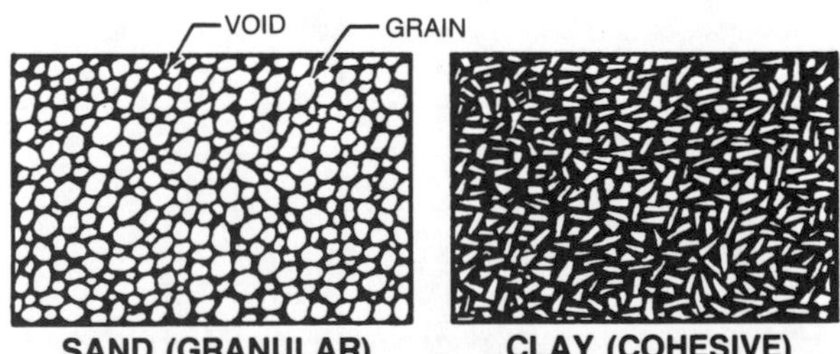

SOIL TYPES	APPROXIMATE SIZE LIMITS OF SOIL PARTICLES
Boulders	Larger than 3″ diameter
Gravel	Smaller than 3″ diameter but larger than #4 sieve
Sand	Smaller than #4 sieve* but larger than #200 sieve†
Silts	Smaller than 0.02 mm diameter but larger than 0.022 mm diameter
Clays	Smaller than 0.022 mm diameter

* approximately ¼″ in diameter
† particles less than #200 sieve not visible to the naked eye

Figure 1-3. A major factor in soil classification is the size of the soil particles. Sand grains are larger than clay grains and have fewer air and water voids between grains.

Stable rock is usually identified by a rock name such as granite or sandstone. Because stable rock is a solid mineral it can be excavated with exposed vertical sides and requires no shoring or minimum shoring.

Type A soils are considered cohesive soils. Examples of Type A soils are clay, silicate clay, sandy clay, and clay loam. Type A soil is subject to movement and vibration and will require shoring.

Also cohesive soils are Type B soils made up of angular gravel, silt, and silt loam. Type B soils may also include dry, unstable rock. Type B soils are subject to disturbance by movement and vibration and require shoring.

Type C soils are also cohesive soils. Type C soils include granular soils such as gravel, sand, and loamy sand. Type C soils may also include soils from which water is freely seeping and submerged rock that is not stable.

Layered geological strata refers to a condition that defines layers of different soils in the excavation. The protective methods used are based on the requirements of the weakest soil in the strata.

Soils can also be classified as rock, virgin soil, and fill. Rock has the greatest soil-bearing capacity if it is level and free of faults (cracks). Virgin soils include, in the order of their strengths, gravel, sand, silt, and clay. Most construction takes place in virgin soils. Fill consists of soil brought from some other location and deposited at the building site. Fill does not provide as dependable a foundation base as virgin soil. When constructing on a site with fill, foundation footings must rest on ground excavated to firm, undisturbed virgin soil, or additional support must be provided.

Organic material makes up approximately 2% to 5% of the topsoil in humid regions and less than .5% of the topsoil in arid regions.

Soil Moisture

Soil moisture is water in the soil that affects soil conditions. The type and amount of soil moisture must be considered in the design and construction of a foundation. If water collects in an enclosed space, such as the area beneath the floor of a residential crawl space foundation, it can cause odors, mold, and wood decay. Water penetrating the walls of a full basement foundation can make the basement area unsuitable as a storage, work, or living space. Methods commonly used to prevent potential surface and groundwater problems include proper grading around the perimeter of the building, drain tile, and vapor barriers. **See Figure 1-5.**

SOIL CLASSIFICATION*

Major Division			Typical Names	Presumptive Bearing Capacity†,‡ Tons Per Sq Ft
Coarse-Grained Soils (More than half of material is larger than the smallest particle visible to the naked eye)	Gravels (more than half of coarse fraction is larger than ¼″)	Gravels with fines (appreciable amount of fines)	Well-graded gravel, gravel-sand mixtures, little or no fines	5
			Poorly graded gravel or gravel-sand mixtures, little or no fines	5
		Clean sands (little or no fines)	Silty gravel, gravel-sand-silt mixtures	2.5
			Clayey gravel, gravel-sand-clay mixtures	2
	Sands (more than half of coarse fraction is smaller than ¼″)	Clean gravels (little or no fines)	Well-graded sand, gravelly sand, little or no fines	3.75
			Poorly graded sand or gravelly sand, little or no fines	3
		Sands with fines (appreciable amount of fines)	Silty sand, sand-silt mixtures	2
			Clayey sand, sand-clay mixtures	2
Fine-Grained Soils (More than half of material is smaller than the smallest particle visible to the naked eye)	Silts and clays (liquid limit is less than 50)		Inorganic silt, very fine sand, rock flour, silty or clayey fine sand, or clayey silt with slight plasticity	1
			Inorganic clay of low to medium plasticity, gravelly clay, sandy clay, silty clay, lean clay	1
			Organic silt, and organic silty clay of low plasticity	
	Silts and clays (liquid limit is greater than 50)		Inorganic silt, micaceous or diatomaceous fine sandy or silty soil, elastic silt	1
			Inorganic clay of high plasticity, fat clay	1
			Organic clay of medium to high plasticity, organic silt	
Highly Organic Soils			Peat and other highly organic soils	

* based on ASTM D2487—*Classification of Soils for Engineering Purposes*
† International Building Code, 2003
‡ in tons/sq ft

Figure 1-4. Bearing capacity of soil is expressed in tons per square foot for common soil mixtures.

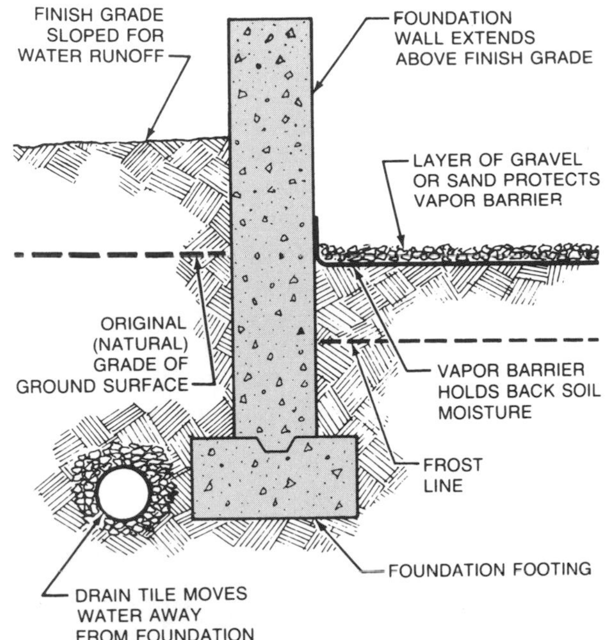

Figure 1-5. Methods to control surface water and groundwater include sloping the finish grade outside the building and placement of drain tile and a vapor barrier.

Soil moisture is caused by surface water, groundwater, and capillary action. Surface water results from rain, downspout discharge, and melting snow. By sloping the finish grade away from the foundation walls, surface water can be directed away from the building. For nonpaved areas, a recommended slope is 6″ in 10′ (5%) away from foundation walls. If the area around the foundation is paved, a slope of ⅛″ in 1′ (1%) is usually adequate.

The water table is the highest point below the surface of the ground that is normally saturated with water. Water table levels vary in different geographical areas. In addition, water tables tend to rise during wet seasons because of water penetration from rain and melting snow. During dry seasons, water tables subside to their normal levels.

The presence of groundwater in a deep excavation can be a serious problem and may hold back the progress of construction work. Water may also temporarily collect because of rain or melting snow. However, more serious problems are caused by underground streams or high water tables in the area. Mechanical pumps are commonly used to remove groundwater accumulating in the excavation. Groundwater in the excavation can also be controlled by lowering the water table in the excavation area. This requires sinking a series of well points and removing the water with a suction pump.

The amount of moisture in the soil is also affected by capillary action of the soil. *Capillary action* is a physical process in soil that causes water and vapor to rise from the water table and move up toward the surface of the ground. Capillary action occurs in all types of soil. However, water and vapor rise higher in porous, fine-grained soils such as silt and clay than in coarse-grained soils such as gravel and sand. **See Figure 1-6.**

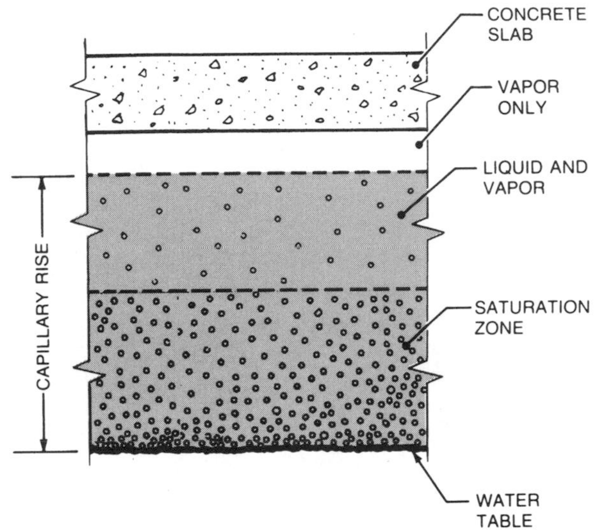

CAPILLARY RISE	SOIL TYPE	SATURATION ZONE
11.5′	Clay	5.7′
11.5′	Silt	5.7′
7.5′	Fine sand	4.5′
2.6′	Coarse sand	2.2′
0.0′	Gravel	0.0′

Figure 1-6. Capillary action of water from the water table is greater in silt and clay than in sand and gravel.

Capillary action can create dampness at the surface of basement floor slabs and slabs that rest directly on the ground. Moisture accumulation must also be avoided in the enclosed areas of crawl space foundations.

Vapor barriers (ground covers) are widely used to control surface moisture caused by capillary action. A *vapor barrier* is a waterproof membrane placed under slabs-on-grade to contain ground surface dampness. Four-mil polyethylene film is commonly used as a vapor barrier because of its resistance to decay and insect attack. Concrete may be placed directly on the polyethylene film to form a slab. When concrete is not placed directly on the polyethylene film, such as in a crawl space, a covering layer of pea gravel or sand is recommended to protect the film from damage and retain the desired position. The amount of capillary action can also be controlled by removing the more porous soil adjoining the foundation walls and replacing it with gravel.

Drain tile or drain pipe around the foundation provides the best means of directing groundwater away from the foundation. Some of the more traditional drain tile methods use round clay or concrete sections of pipe with a ¼″ space between each section. The groundwater enters the tile through this open space and then flows along the tiles. A strip of asphalt-saturated paper is placed over the top of the space between the tiles to prevent soil from falling inside the tile. Another type of drain tile has holes at the bottom of the tile sections. The groundwater seeps into the holes and flows through the inside of the tiles. **See Figure 1-7.**

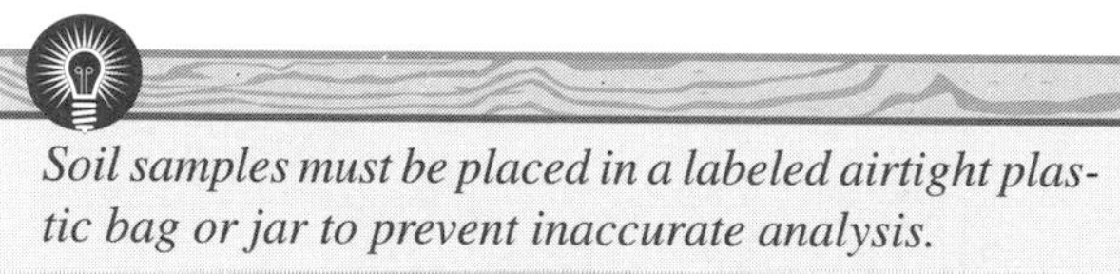

Soil samples must be placed in a labeled airtight plastic bag or jar to prevent inaccurate analysis.

Soil Compaction

Soil compaction is the process of applying energy to loose soil to increase its density and load-bearing capacity through consolidation and removal of voids. Soil compaction is commonly specified for building foundations, trench backfills, curbs and gutters, bridge supports, slab work, driveways, sidewalks, cemeteries, and other confined work areas.

During excavating, grading, and/or trenching, air infiltrating the soil causes air voids and an increase in volume. Air voids in the soil cause weakness and inability to carry heavy loads. Air voids in uncompacted soil can also cause undesirable settlement of a structure, resulting in cracks or complete failure. Compacted soil reduces water penetration and related problems. Water penetration swells the soil during the wet season. Water removal contracts the soil during the dry season. In addition, water expands when frozen in the soil, causing heaving and cracking of walls and floor slabs.

Soil Compaction Methods. Soil compaction methods include impact force, vibration, and static force. *Impact force compaction* is compaction using a machine that alternately strikes and leaves the ground at high speed to increase soil density. *Vibration compaction* is compaction using a machine to apply high-frequency vibration to the soil to increase soil density. *Static force compaction* is compaction using a heavy machine that squeezes soil particles together without vibratory motion to increase soil density.

Cohesive (fine-grained) soils are best compacted by impact force compaction. Cohesive soils do not settle under vibration due to natural binding between small soil particles. They tend to lump, forming continuous laminations with air pockets between the laminations because of their light weight and "pancake shape," which prevents them from dropping into voids under vibration. Impact force compaction on cohesive soils produces a shearing effect that squeezes air pockets and excess water to the surface and moves soil particles closer together.

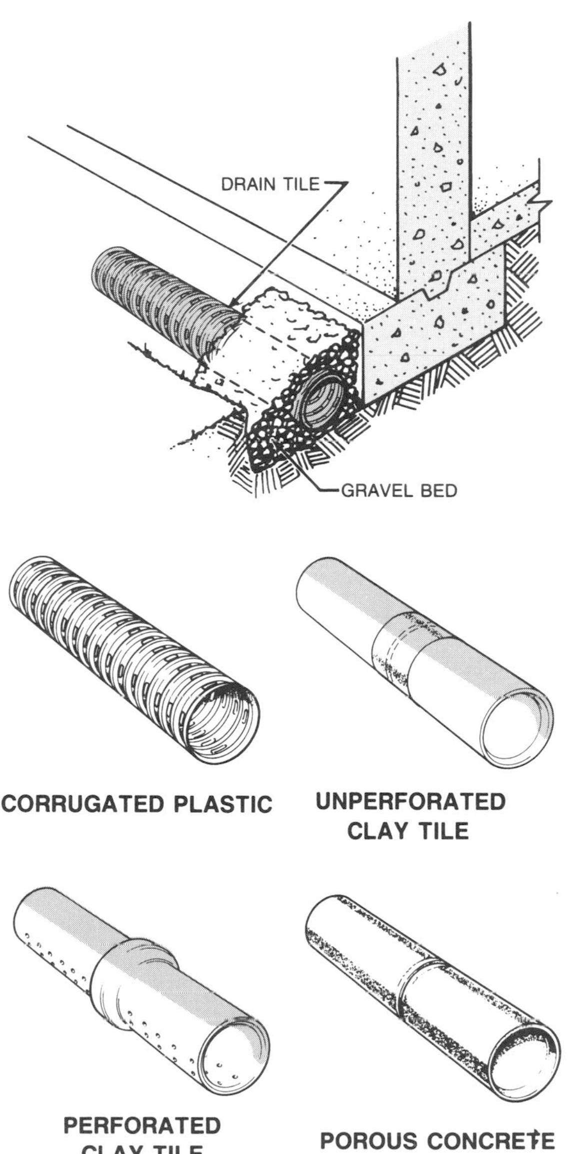

Figure 1-7. Drain tile systems move groundwater away from the foundation.

Granular (coarse-grained) soils are best compacted by vibration compaction. Vibration reduces the frictional forces at the contact surfaces, allowing the soil particles to fall freely under their own weight. This occurs as the soil particles are set in vibration and become separated from one another. Vibration motion allows soil particles to twist and turn into voids that limit the movement, resulting in compaction. A combination of impact force compaction and vibration compaction can be used to achieve compaction of granular soils.

Soil Compaction Equipment. Soil compaction equipment use is based on the type of soil to be compacted. Common soil compaction equipment includes the rammer, vibratory plate, and roller. **See Figure 1-8.** A *rammer* is a soil compaction tool that alternately strikes and leaves the ground at high speed to increase soil density. It is most effective on cohesive soil composed of very small particles such as silt and clay. A *vibratory plate* is a soil compaction tool that applies high-frequency vibrations to the ground to increase the soil density. Vibratory plates are commonly used for compacting granular soils such as sand and gravel, and mixes of granular and cohesive soils. A *roller* is a soil compaction tool that uses weight, or weight and vibration, to increase soil density.

Frost Line

The *frost line* is the depth to which soil freezes in a particular area. Soil freezing is caused by temperature drop and surface water penetrating the soil. The frost line varies with different climate areas, ranging from 0″ to 5″ on the West Coast to 72″ to 108″ in some of the far northern sections of the country. **See Figure 1-9.** In addition, the amount of frost penetration is affected by the type of soil underlying the building site. For example, clay and silt tend to absorb and hold moisture, allowing for deeper frost penetration. Coarse sands and gravels drain well, resulting in shallower frost penetration. Increasing water drainage through the use of gravel and drain tile helps limit the depth of frost penetration in more porous soils.

The frost line must be considered in excavation work and the design and construction of a foundation. The footings of any foundation system should always be placed below the frost line. Footings placed above the frost line are subject to soil-heaving action. Soil heaving occurs as the soil freezes and expands, resulting in the building structure moving upward. When the soil thaws, the structure drops again. This up-and-down movement can result in structural damage to the foundation.

Soil is excavated below the frost line before being compacted to lessen the amount of movement during freezing and thawing.

RAMMER

Portland Cement Association

VIBRATORY PLATE

ROLLER

Figure 1-8. Common soil compaction equipment includes the rammer, vibratory plate, and roller.

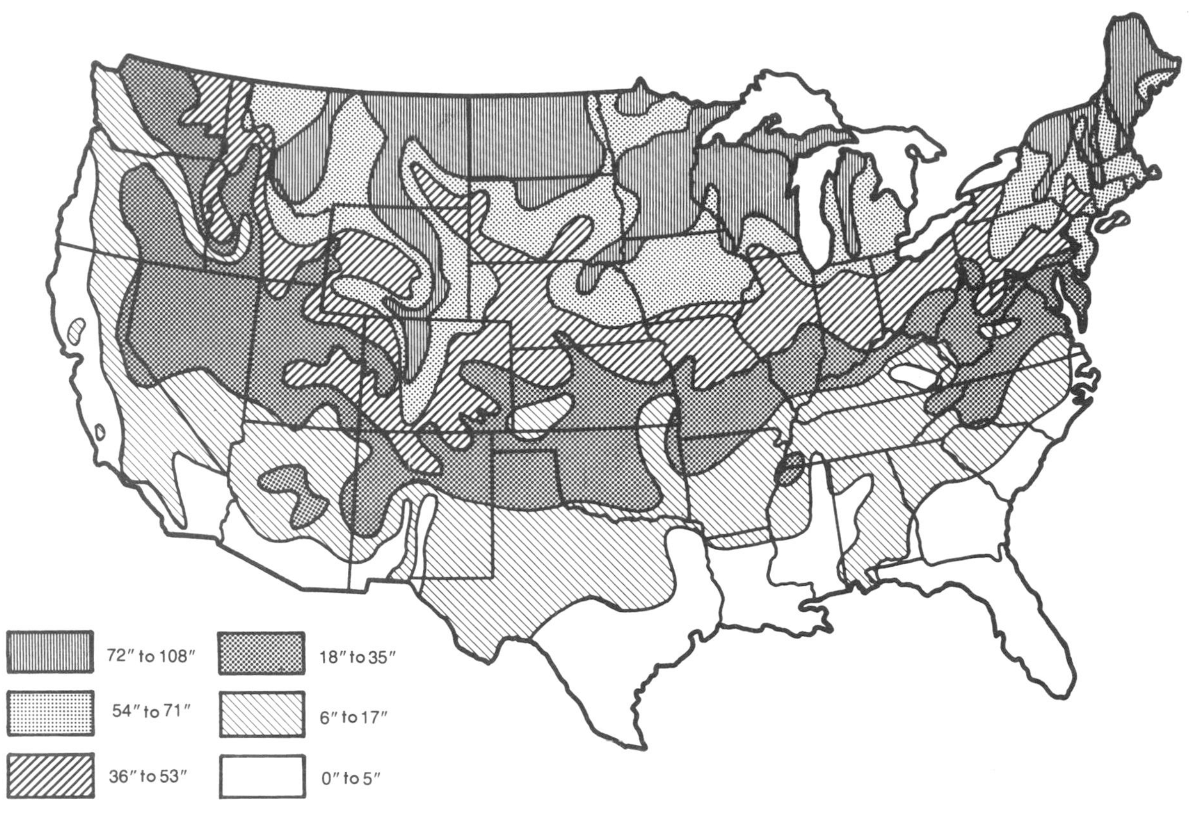

Figure 1-9. The frost line varies according to climate.

SITE LAYOUT

Layout is completed before foundation construction begins. The initial layout for the building is determined by information provided on the plot plan of the prints. Additional data may be contained on the foundation plan and the written specifications of the prints. All this information must conform to the local building codes and zoning regulations.

The Plot Plan

The *plot plan* (site plan) is a drawing that provides information regarding preliminary sitework. Before drawing a plot plan, the architect has a survey performed on the lot to establish and mark the legal boundaries of the property and record the existing grades. Additional information provided on the plot plan includes the setback, compass direction, benchmark and elevations, utility hookups, roads, sidewalks, terraces, tree locations, and easements. **See Figure 1-10.**

Property Lines. A *property line* is a line that defines the boundaries of a building lot. Building lots are usually mapped and recorded by local building or zoning authorities. The lot surveyor studies the zoning maps and records to determine the exact boundaries of the property. The surveyor establishes the two front corners of the lot by measuring from existing reference points, which have been established in previous surveys by city or county agencies. In some localities, street curbs are used as reference points. In other localities, property corners are found by measuring from the center of a road or from markers placed at intervals in sidewalks. The two rear lot corners are normally laid out with a transit-level by sighting at a right angle from the two established front corner stakes.

To identify the lot corners, the surveyor places corner stakes (hubs) at each corner. Traditionally, the hubs were wooden stakes driven so that their tops were flush with the surface of the ground. Small nails driven into the top of each corner stake marked the exact corners of the property. Today, hubs are often a piece of rebar with a brightly colored plastic cap placed over the top end. Another method of marking a hub is to use a piece of pipe with a cork or lead plug. The property lines and measurements shown on the plot plan represent the lines extending from the four corners of the lot. For example, a lot may measure 140′-0″ × 80′-0″.

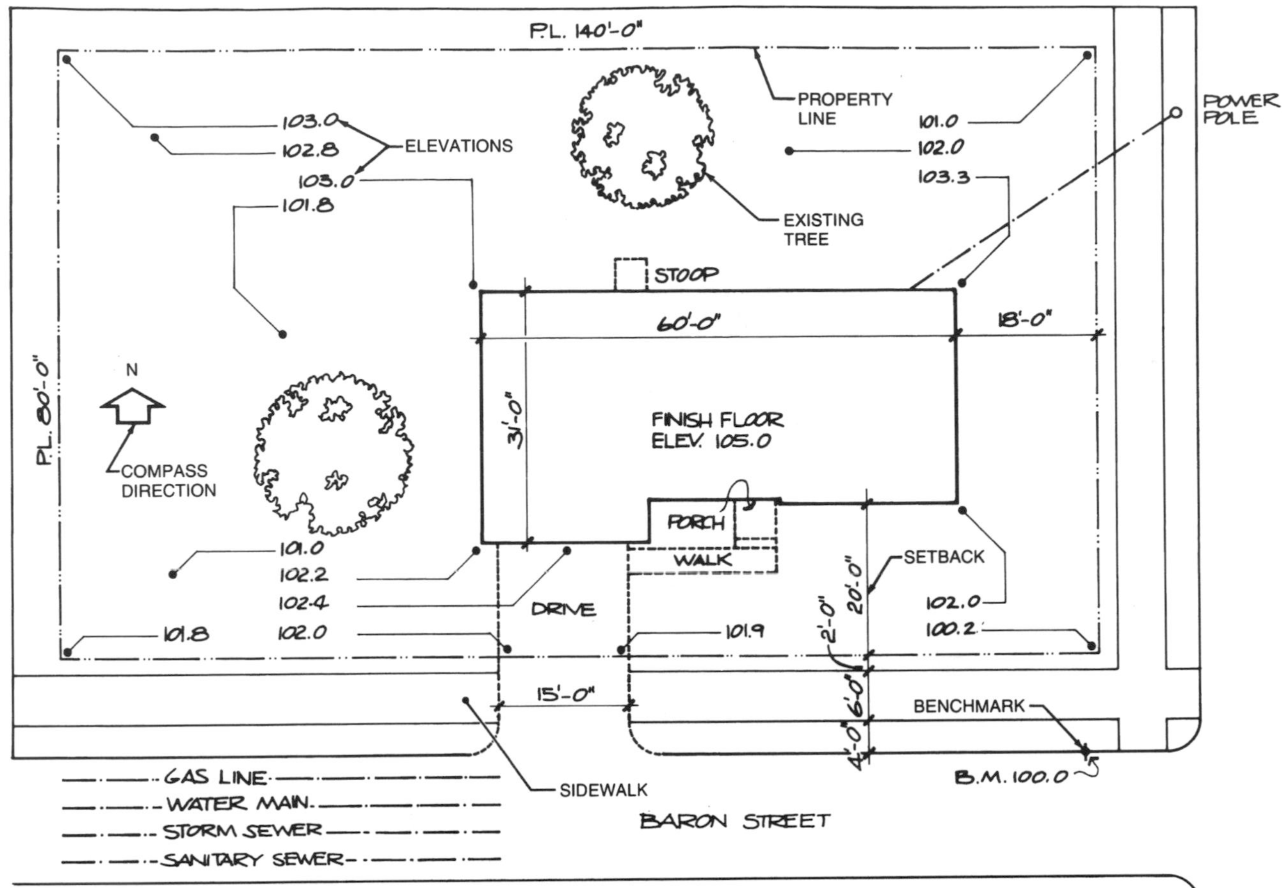

Figure 1-10. The plot plan provides information related to the preliminary site work.

Grades and Elevations. Grades and elevations for the job site are specified on the plot plan. The plot plan indicates the grade levels at all four corners of the lot. Grade levels may be shown at the building corners as well as at different points within the interior of the lot. Grade levels marked on the plot plan are usually noted in engineering measurements of feet and tenths or hundredths of a foot. Carpenters and other construction workers use rules with feet and inch measurements; therefore, it is often necessary to convert engineering measurements to inch and fractional equivalents. A table can be used for this purpose, or the conversions can be performed mathematically. (See Chapter 8.) The grade levels show how the ground surface will be sloped and contoured. Some plot plans include contour lines to help clarify the desired slopes of the ground. A *contour line* is a line that shows the slope of the ground with lines extending from identified grade levels. **See Figure 1 -11.**

A plot plan typically shows a benchmark (job datum) established by the surveyor at a convenient spot close to the property. A *benchmark* is a point of reference for grades and elevations on a construction site. The benchmark may be a mark chiseled at the street curb, a plugged pipe driven into the ground, a brass marker, or a wood stake placed near one corner of the lot. In addition, benchmarks may be established on adjoining buildings or power poles.

The benchmark shown on a plot plan may be identified as the number of feet the ground is above sea level at that point, or it may be shown as 100.0′. The benchmark establishes a reference for all other elevations on the plot plan. For example, the elevation along the south end of the building is 104.8′; therefore, the surface at that point is 4.8′ (104.8′–100.0′=4.8′) higher than the benchmark. The grade level at the southwest edge of the lot is 103′; therefore, the elevation is 3′ (103.0′–100.0′=3′) higher at that point.

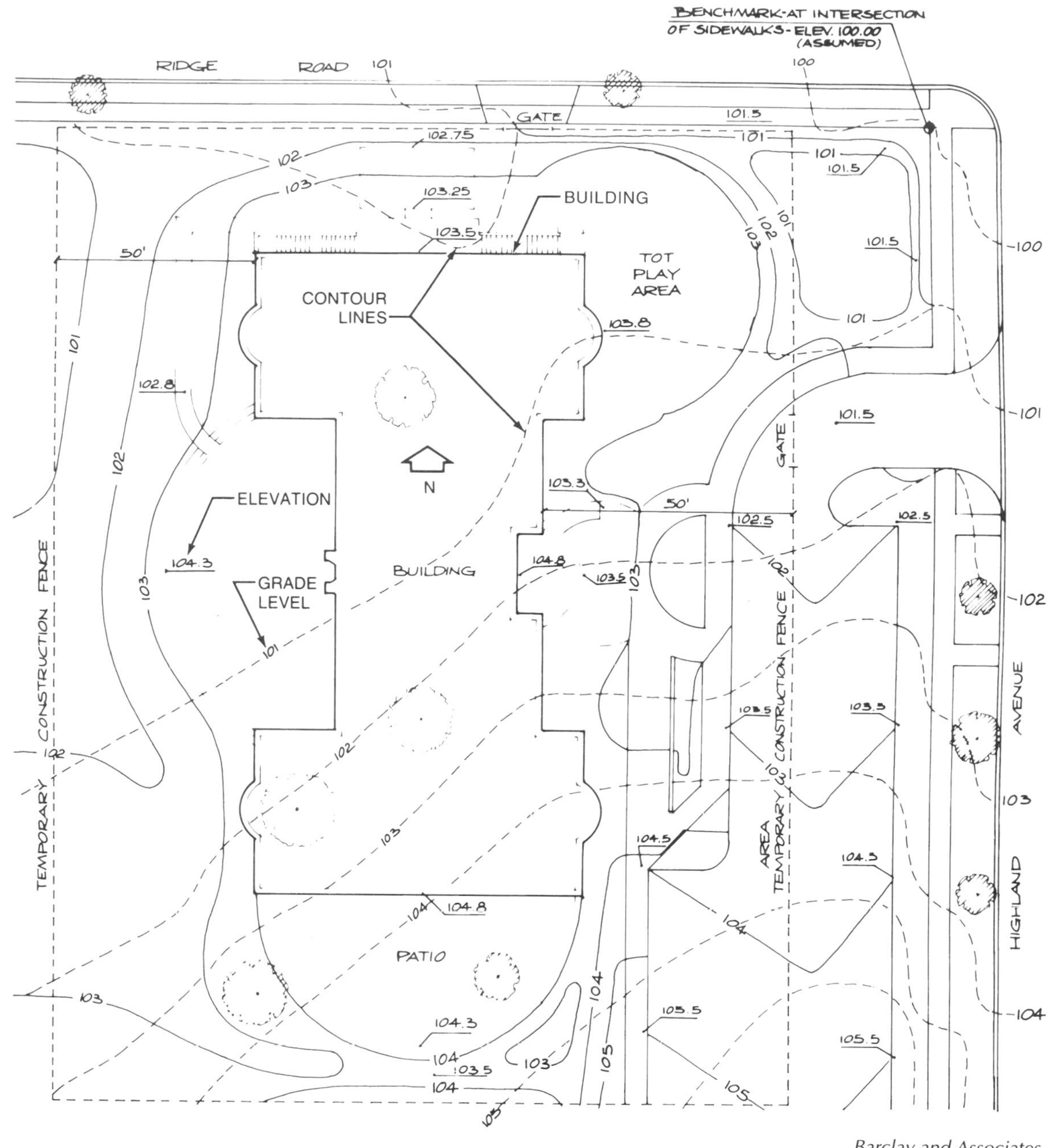

Barclay and Associates

Figure 1-11. Contour lines or grade levels are used on the plot plan to indicate grades or elevations. The dashed contour lines represent the natural grades of the lot in its original state. The solid contour lines represent the finish grades after groundwork has been completed.

GROUNDWORK

Groundwork is the preliminary grading, excavating, trenching, and backfilling required at a job site. Groundwork is completed before the foundation construction begins, although additional grading and backfilling are often required after the foundation work has been completed. The amount of groundwork required depends on the existing contours of the lot and the depth of the foundation footings. Most groundwork is done with earth-moving equipment such as bulldozers, motor graders, power shovels, backhoes, and other heavy equipment. **See Figure 1-12.** Operating engineers run heavy equipment. Other tradesworkers may work with the operating engineers by setting up lines to guide the earth-moving work and assisting in checking grade levels and depth of excavations.

BULLDOZER USED TO START EXCAVATIONS AND STRIP ROCKS AND TOPSOIL AT SURFACE SITE

HYDRAULIC POWERED EXCAVATOR USED FOR DEEPER EXCAVATIONS, TRENCHING, AND LOADING

BACKHOE LOADERS USED FOR TRENCHING, BACKFILLING, AND SMALL LOADING OPERATIONS

MOTOR GRADERS USED FOR GRADING OPERATIONS

John Deere Construction & Forestry Company

Figure 1-12. Heavy earth-moving equipment is used for grading, excavation, and trenching.

Grading

The amount of grading required is determined by the condition of the ground. Lots that are fairly flat and level or only slightly sloped may require very little grading. Lots that are steeply sloped or have very uneven surfaces require considerable grading. In order for the surface to conform to the grade levels shown on the plot plan, soil may have to be removed from or added to different areas of the lot. The finished surface should be sloped away from the building so that surface water caused by rain and melting snow will flow away from the building.

During grading operations, the levels of grading can be measured and checked by using a rod and leveling instrument such as a laser transit-level, builder's level, or transit-level. (See Appendix E.)

Trenching

Trenches 5′ or more in depth, in hard, compact soils are commonly shored by placing vertical timbers on opposite sides of the trench. Timber shoring may consist of 2 × 4 wooden struts placed between vertical timbers. These timbers are held in place by wood cross braces or screw jacks. The wood cross braces or screw jacks and upright timbers should be spaced a maximum of 5′-0″ OC. One cross brace is required for every 4′ of the trench depth, with no less than two braces. Trenches dug in loose and unstable soil should be supported with wood panels reinforced with 4 × 4 stringers and cross braces. Maximum spacing between stringers is 5′-0″ OC. **See Figure 1-13.**

Backfilling

Backfilling is the replacing of soil around the outside foundation walls after the walls have been completed. Backfilling is performed after the forms have been stripped (removed) from the walls and the waterproofing and drain pipe work have been finished. The soil used for backfill should be free of wood scraps and any other type of waste material. Backfill must be placed carefully against the foundation wall and well compacted to avoid future shrinkage. Gravel is trucked in and used for backfill on many jobs since it allows better water drainage around the building. **See Figure 1-14.**

Figure 1-14. Gravel backfill is placed around the outside of a completed foundation wall.

To prevent a collapse when backfilling around a foundation, ensure that heavy equipment does not come too close to the edge of the excavation.

SCHEDULE FOR WOOD HORIZONTAL BRACES FOR HARD, COMPACT SOIL TRENCHES	
Trench Width	**Minimum Timber Size**
1′-0″	4″ × 4″
3′-0″	4″ × 6″
6′-0″	6″ × 6″
8′-0″	Increases proportionately

SIZE OF WOOD SHEET PILING FOR LOOSE-SOIL TRENCHES	
Trench Depth	**Minimum Thickness**
4′-0″ TO 8′-0′	2″
OVER 8′-0′	3″

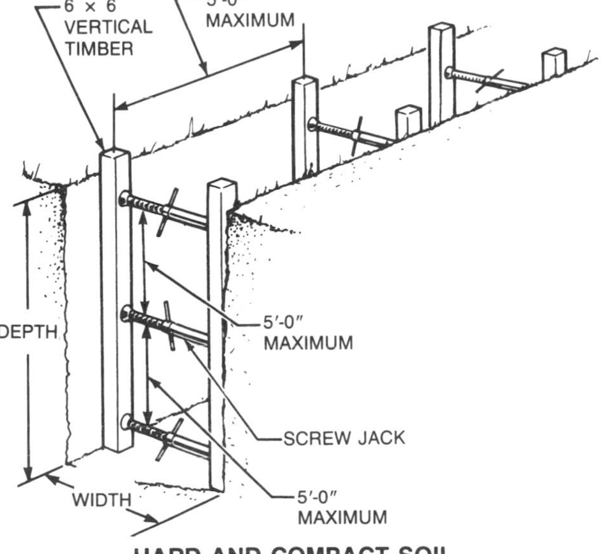

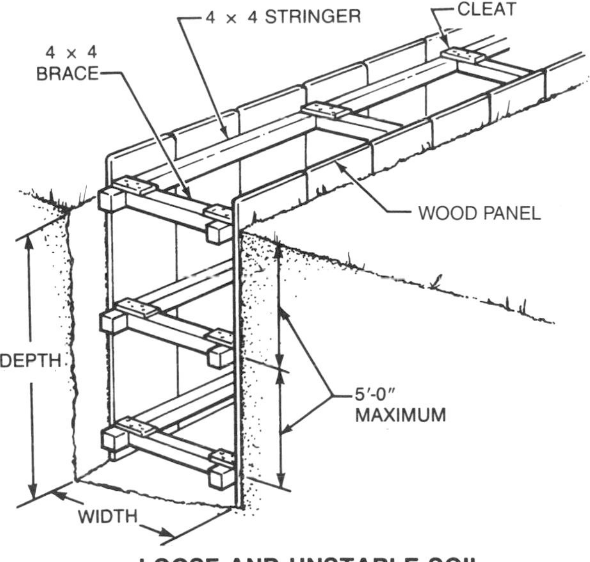

Figure 1-13. Shoring for trenches in hard, compact soil requires vertical timbers and cross bracing. Shoring for trenches in loose, unstable soil requires wood panels in addition to stringers and cross bracing.

EXCAVATIONS

The term excavation applies to any cut, depression, or trench in a ground surface. A *trench* is a narrow excavation that is deeper than it is wide. A trench can be up to 15′ wide in its bottom width. Concrete structures often require deep excavations or trenches; therefore, concrete formwork normally begins at the bottom of the excavation. Excavation methods used to prevent collapse when working below ground level include sloping and benching, shoring, and shielding.

Sloping and Benching

Sloping is an excavation method that slants the sides away from the bottom of the excavation. The sides of shallow excavations can be sloped or benched to prevent collapse of the sides. If there is sufficient space around the construction site, sloping the earth banks may be all that is necessary to prevent collapse. Sloping cannot be used if buildings or streets are immediately adjacent to the excavation. The soil conditions present on the job site determine the angle of slope. *Benching* is an excavation method in which a series of steps is carved with vertical surfaces between the levels. Benching requires more room around the excavation than sloping. **See Figure 1-15.**

OSHA regulations for the construction industry recommend a 45° slope for excavations with average soil conditions. Solid rock, shale, or cemented sand and gravels may require less slope. Compacted sharp sand or well-rounded loose sand may require more than a 45° slope.

Temporary spoil is the soil material and other waste that is dug out of the excavation. Temporary spoil should be placed a minimum of 2′ from its base to the side of the excavation to prevent loose rock or soil in the spoil pile from falling back into the excavation. Spoil should also be placed so that it channels rainwater and other runoff water away from the trench. **See Figure 1-16.**

Benching is commonly used in areas where the soil is well compacted and will not collapse during construction.

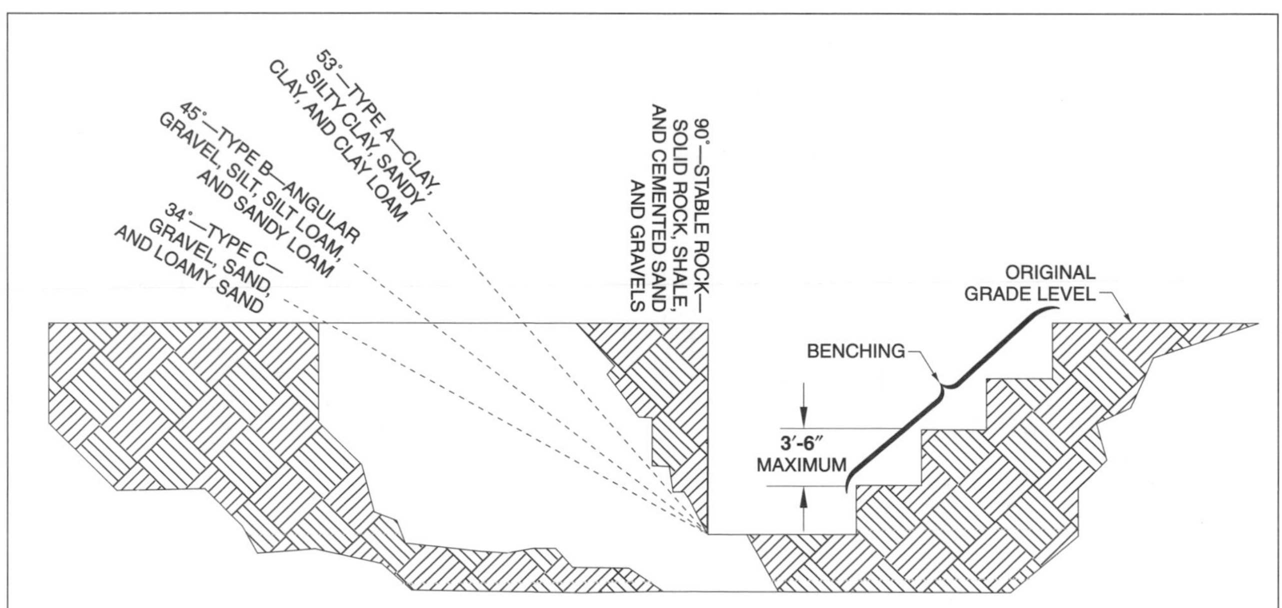

Figure 1-15. Sloping and benching are commonly used methods to prevent the collapse of earth banks in deep excavations.

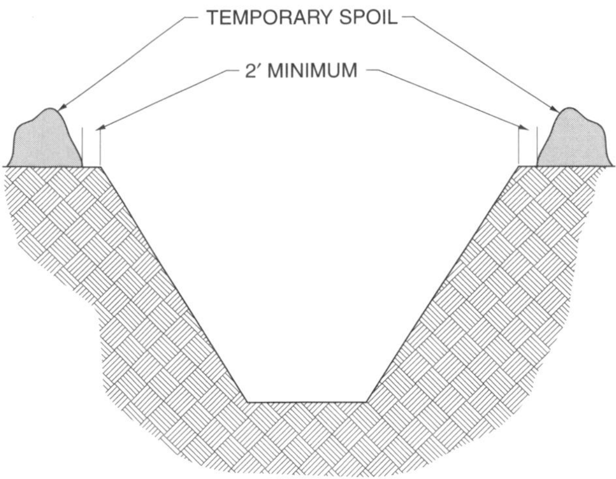

Figure 1-16. Temporary spoil should be placed a minimum of 2′ from its base to the side of the excavation, and so that it channels rainwater and other runoff water away from the trench.

Shoring Open Excavations

Shoring (supporting) the vertical walls of an open excavation is required when sloping or benching is considered unsafe or inadequate. Soil types such as clays, silts, loams, or nonhomogenous soils usually require shoring. Shoring may also be required where there is insufficient room for sloped banks; particularly in downtown urban areas where new construction is immediately adjoining existing buildings. In addition to preventing injury from collapse of excavation banks, the stability of the foundation walls of adjoining buildings must be protected. Shoring for high vertical walls in open excavations is designed by a civil engineer and the installation is supervised by qualified personnel. The shoring system should not be removed until the construction in the excavated area is completed and all necessary steps are taken to safeguard workers. Two methods commonly used to shore open excavations are interlocking sheet piling and soldier piles. **See Figure 1-17.**

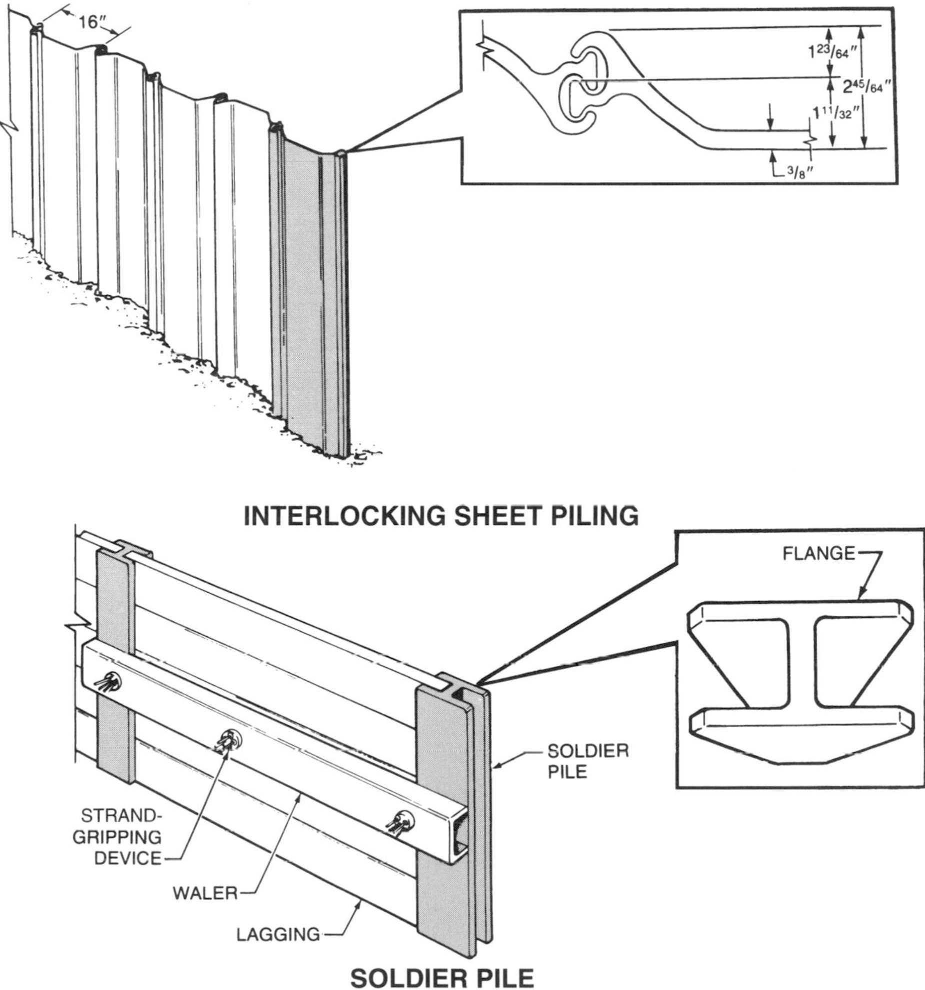

Figure 1-17. Interlocking sheet piling and soldier piles are used to shore the walls of deep excavations for heavy construction projects.

Interlocking sheet piling consists of steel piles that can be reused many times and offers the additional advantage of being watertight. Each individual sheet piling is lowered by crane into a template that holds it in position. The piling is then driven into place with a pile-driving rig. Braces may also be installed to help support the metal sheets.

Soldier piles, also called soldier beams, are H-shaped piles that are driven into the ground with a pile-driving rig and are spaced approximately 8′ apart. Three-inch thick wood planks called lagging are placed between the flanges, or directly against the front of the piles. Soil conditions and the depth of the excavation may require tie-backs, which consist of steel strand cables placed in holes drilled horizontally into the banks of the excavation. The holes are drilled with a power auger and are often 50′ or more in length. The tie-back cables are inserted through an opening in the pile and are secured in the earth by power-grouting the hole.

After the grout has set up, a strand-gripping device consisting of a gripper and gripper casing is placed over the cables. A hydraulic tensioning jack is used to tighten the cables. When the jack releases the cables, the gripping device holds them and maintains the required tension against the pile. The number of tie-backs required should be determined by an engineer whose decision will be based on soil conditions and the depth of the excavation. Some soldier pile systems may also include a heavy horizontal steel waler held in place with tie-backs.

Shoring Deep Trenches

Shoring deep trenches is necessary when the depth of the trench makes it impractical to slope back the maximum allowable angle. Shoring systems consist of posts, walers, struts, and sheeting. The two prevalent types of trench shoring are hydraulic and timber.

Hydraulic shoring is the preferred shoring method. Hydraulic shoring consists of a prefabricated strut and/or waler system made of aluminum or steel. Hydraulic shoring offers an important safety advantage over timber shoring because workers do not have to be in the trenches to place or remove the hydraulic shoring. It can also be easily adapted to various trench widths and depths. Aluminum hydraulic shoring systems are commonly used. **See Figure 1-18.**

Shielding

Trench shields, or trench boxes, are made of steel. Trench boxes consist of two heavy steel plate sides separated by steel pipe struts. Trench boxes are not designed to give total support to the trench face, but are intended primarily to protect workers from cave-ins or objects falling into the trench. The clearance between the outside of the trench box and the face of the trench should be as small as possible. Trench boxes are moved along as the work progresses. Trench boxes can be stacked in even layers or can accommodate stepped trenches. They are also placed in partially sloped trenches. **See Figure 1-19.**

Inspection

Inspections shall be in accordance with OSHA recommendations (29 CFR Part 1926 Subpart P). OSHA recommendations call for inspections to be performed by a qualified person with training, experience, and knowledge of soil analysis and use of protective systems. The inspector must have the ability to detect conditions that could result in cave-in, failure of protective systems, or hazardous atmospheres. Such inspections must take place on a daily basis, or more often if conditions require.

Excavations must be inspected after every rainstorm and other natural events such as windstorms, snowstorms, thaws, and earthquakes. Inspection is also required when tension cracks, fissures, water seepage, undercutting, and bulging occur at the bottom of a trench. Any change in movement of adjacent structures, or change in the size, placement, and location of spoil piles also warrants immediate inspection.

Deep excavations that cannot be benched or sloped require shoring before workers are allowed to enter the excavation.

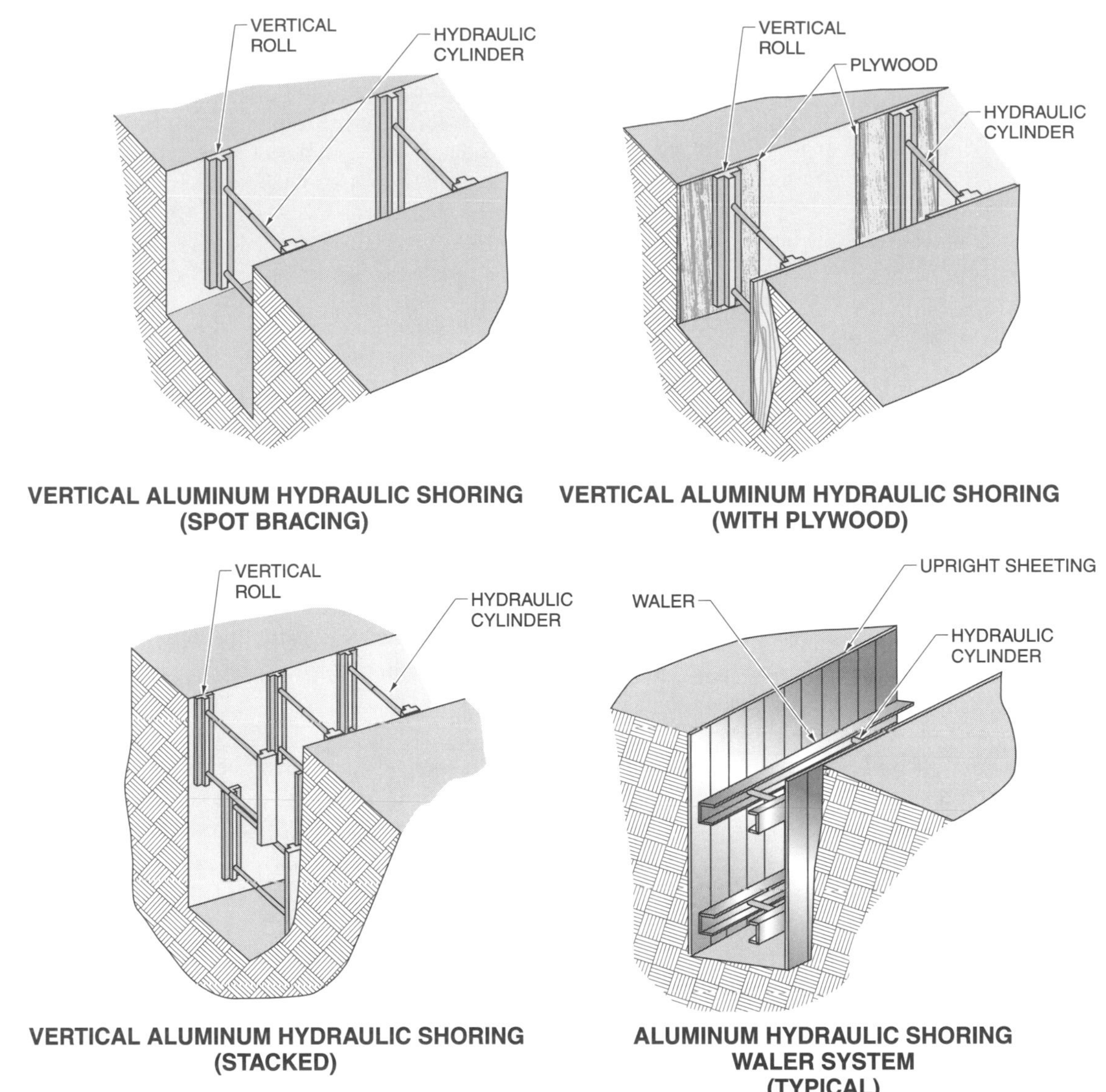

Figure 1-18. Hydraulic shoring consists of a prefabricated strut and/or waler system made of aluminum or steel.

Excavation Safety

Safety is a prime concern when working in excavations. The fatality rate for excavation work is 112% higher than the rate for general construction.

OSHA 29 CFR 1926 Subpart P, *Excavations* provides specific guidelines for excavation work and safety. Safety in excavation work requires knowledge of soil mechanics and proper sloping and benching. Knowledge of general health measures and safety considerations is also required.

Ingress and Egress. Safe and immediate access to or exit from a trench must be available at all times to persons working in an excavation. Trenches that are 4′ deep or more must have a fixed means of exit. Ladders should be placed so that a worker will not have to travel more than 25′ in a lateral direction to climb out of the trench. Ladders must be stabilized and must extend at least 36″ above the ground surface. Wooden ladders are preferred, particularly when electrical utilities are in the vicinity.

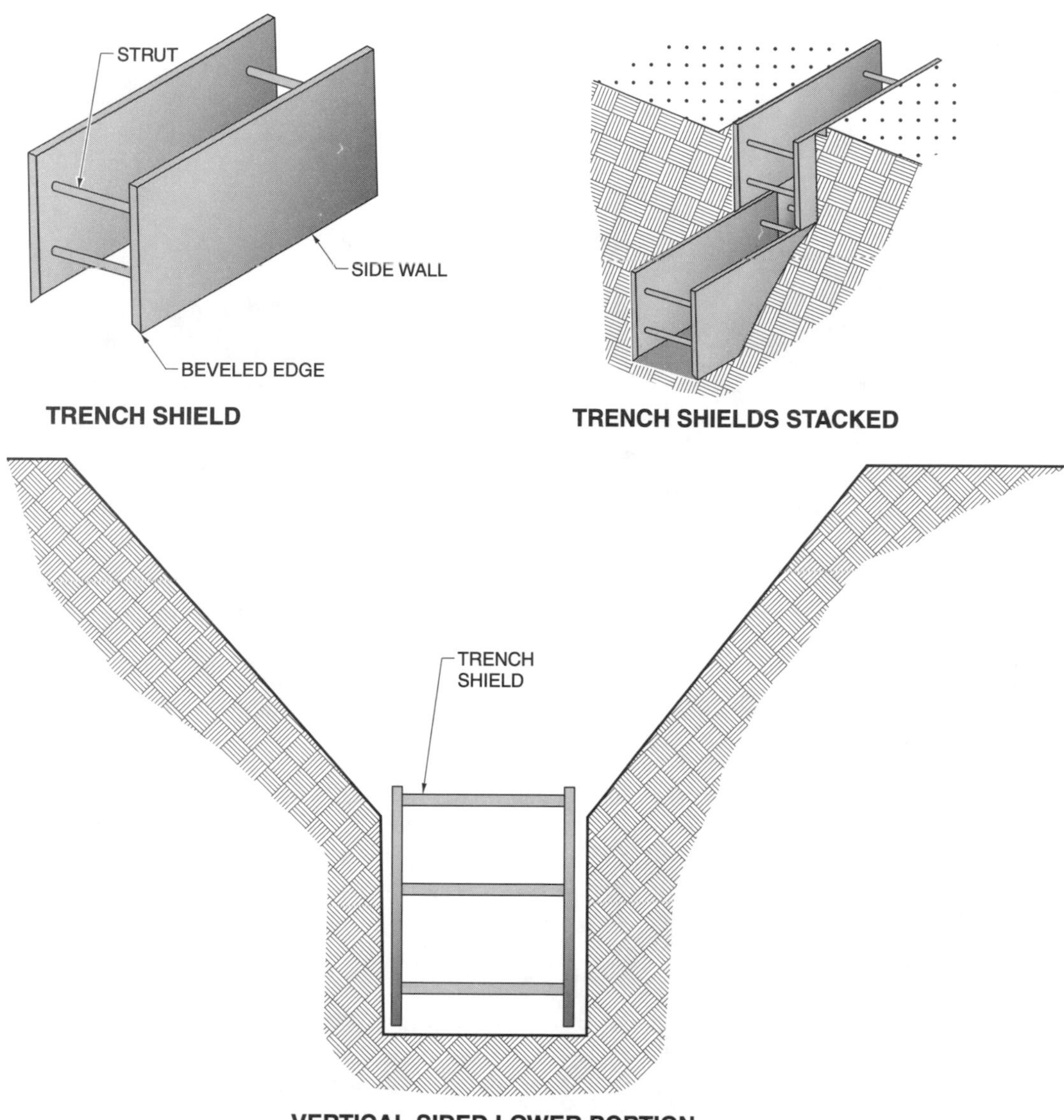

Figure 1-19. Trench shields are designed to protect workers from cave-ins or objects falling into the trench.

Some construction jobs may require ramps or temporary stairs to provide a means for workers to move about on the job and for materials to be transported. Ramps and runways may also be constructed for the movement of wheelbarrows and power buggies used to transport concrete and other materials. **See Figure 1-20.**

Such locations might be landfills, hazardous substance storage areas, and gas pipelines. Low oxygen levels and inflammable gases are two of the main hazardous atmospheres encountered. Such circumstances require respiratory equipment and/or forced ventilation.

Water Accumulation. Excavation in areas subject to runoff from heavy rain or in close proximity to natural water conditions must have a special support or shield system. Water removal equipment must be on hand to control water accumulation. Ditches and dikes can be constructed to prevent water from entering the excavation site. Steps should be taken to provide adequate drainage of the areas adjacent to the excavation site.

Hazardous Atmospheres. Atmospheric testing must be conducted in excavations over 4′ deep where hazardous atmospheres can reasonably be expected to exist.

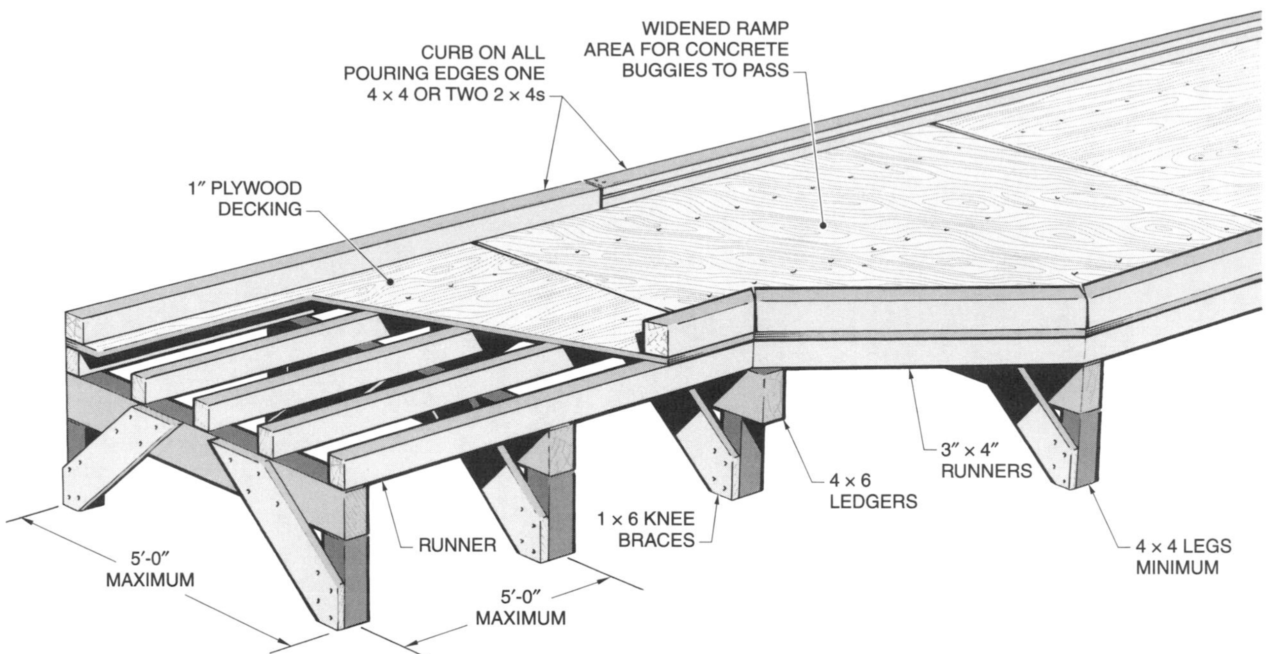

Figure 1-20. Construction jobs often require ramps or temporary stairs to provide a way for workers to move about on the job and for materials to be transported.

When equipment must enter an excavation, such as when pouring concrete, workers must wear safety vests so they are visible to vehicle drivers.

Vehicular Traffic. Moving vehicles operating in and around excavations pose a danger to workers. When pedestrian or vehicular traffic is present on construction jobs, carpenters erect barricades around the job site to prevent unauthorized persons from entering the construction area. In many cases, barricades provide overhead protection to prevent objects from falling in the traffic area. Workers must be provided with warning vests or other garments made of reflectorized or high-visibility materials so that they are visible to vehicular traffic. When deemed necessary, there should be a trained flag person present to direct traffic in addition to signs, signals, and barricades.

Surface crossings over trenches are generally discouraged. However, if trenches must be crossed using walkways or bridges, the walkways or bridges must be designed and installed under the supervision of a professional engineer. Walkways and bridges should have a safety factor of 4, meaning that they must be able to support four times any expected load. The crossing must extend a minimum of 24″ over each edge of the excavation.

Guardrails protect the safety of workers on the job. Guardrail requirements are detailed in OSHA 29 CFR 1926.502, *Fall Protection Systems Criteria and Practices.* Working and walking surfaces 6′-0″ or more above a lower surface must be guarded by guardrails. Guardrails are placed across openings for exterior doors if there is a drop of more than 4′. Guardrails are also required if the bottom of a window opening is less than 39″ above the working surface. Guardrails must be constructed on all unprotected sides or edges of floor openings such as openings for stairways and skylights. If the openings are used for worker access, such as ladderways, the openings must be protected with a gate so workers cannot walk directly into the opening. Guardrails may also be installed on ramps and runways, elevator shafts, balconies, and other parts of the building under construction.

Danger from Falling Loads. Because of the presence of lifting and digging equipment on a job site, there is the danger of falling loads and objects. Workers must never work or walk under raised loads and must stand away from equipment as it is being loaded and unloaded.

Danger from Mobile Equipment. Mobile equipment includes cranes, bulldozers, motor graders, and excavators. There is always a danger of moving machinery falling into trenches. Grading the soil away from an excavation helps prevent vehicles from coming near the trench. A safe distance from the edge of the excavation, determined by soil conditions, should be maintained at all times. Only front-end loaders that are backfilling should come near an excavation, and only after the operator has been signaled that all workers have exited the excavation.

Hand or mechanical signals should be used to direct operating mobile equipment. Barricades may also be required. Stop logs can also be placed near trenches.

Heavy Construction Excavation

Heavy construction excavation is required in the construction of large buildings. When deep and extensive trench and excavation work is necessary, safety becomes a major factor. Heavy construction excavation work is subject to strict local, state, and national code regulations. Precautions must be taken to ensure that the excavation banks do not collapse and cause injury or death to persons working in the excavation. The method used to protect excavation banks from collapsing depends on the type of soil in the area, depth of the excavation, type of foundation being built, and space around the excavation.

Before beginning an excavation, the builder must secure all possible information regarding any underground utility installations in the area including sewer, water, fuel, and electrical lines. Precautions must be taken not to disturb or damage any existing utility while digging, and to provide adequate protection after they have been exposed. All owners of underground facilities in the area of the excavation should be advised of the work prior to the start of the excavation.

Many safety codes also require that a qualified person inspect the excavation after a rainstorm or any other hazardous natural occurrence. Earth bank cave-ins or landslides may be averted by increasing the amount of shoring and other means of protection.

Some soil types pose greater problems than others during excavation. Sandy soil is always considered dangerous even when allowed to stand for a period of time after a vertical cut. The instability can be caused by moisture changes in the surrounding air or changes in the water table. Vibration from blasting, traffic movement, and material loads near the cut can also cause earth to collapse in sandy soil.

Clay soils present less risk than sand; however, clay can also be dangerous if it is soft. A simple test of clay conditions can be accomplished by pushing a 2 × 4 into the soil. If the 2 × 4 can be easily pushed into the ground, the clay is soft and may collapse. Silty soils are also unreliable and require the same precautions as sand.

Residential Excavation and Trenching

Crawl space foundations require trenches or shallow excavations dug down to the proper footing level. The trenches for a crawl space foundation must be deep enough to place the footing below the frost line and at depths prescribed by local building codes. Residential buildings or other types of small buildings may require a large excavation for a below-grade full basement foundation. **See Figure 1-21.**

If a footing form is required, the trench must be wide enough to allow room for the construction of the form. For crawl space or below-grade full basement foundations, lines must be set up showing the boundaries of the building in order to establish the perimeters of the excavation. The walls of the excavation should be at least 2′ outside the building lines to allow enough room for the formwork. However, if the soil is loose and unstable, the excavation for the foundation walls should extend farther back.

John Deere Construction & Forestry Company

Safe and immediate access to or exit from a trench must be available at all times for people working in it. Ladders must extend at least 36″ above the ground surface.

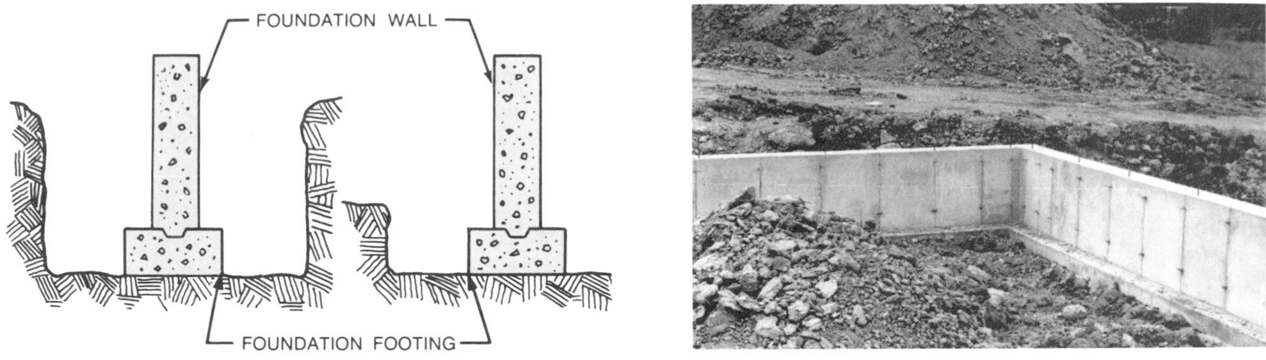

CRAWL SPACE FOUNDATION

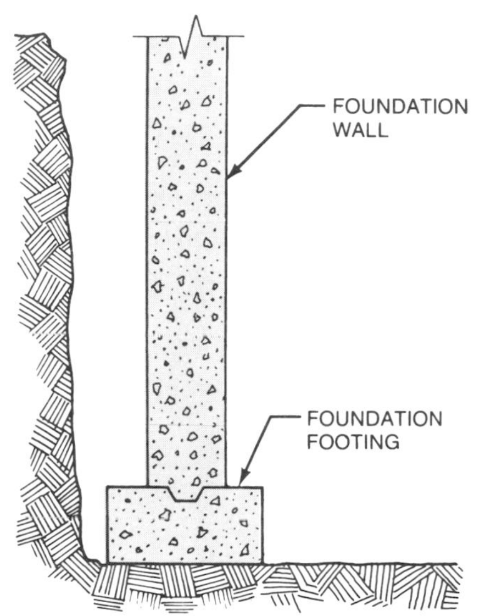

FULL BASEMENT FOUNDATION

Figure 1-21. Crawl space foundations require trenches dug down to the proper footing level. Full basement foundations require complete excavation of the area required by the basement.

Crawl space and basement foundations are dug below ground and are typically open and well ventilated. Although crawl space and basement foundations are typically considered open work areas, air quality should still be checked before entering. If a permit for entering the area is authorized, at least one attendant must be provided per OSHA 29 CFR 1910.146, Permit-Required Confined Spaces.

The depth of an excavation must extend to firm and stable soil and be below the frost line. A complete set of prints usually provides section view drawings that provide the information needed to determine the depth of the excavation. For a full basement, the excavation depth can be calculated by adding the slab-to-joist height, slab thickness, and footing height, and subtracting the distance the wall projects above the finish grade. **See Figure 1-22.** In some warm climate areas where frost lines are not a factor, many codes only require that the topsoil (a soft surface layer of earth in which vegetation grows) be removed and the trenches dug a minimum of 6″ into natural undisturbed soil.

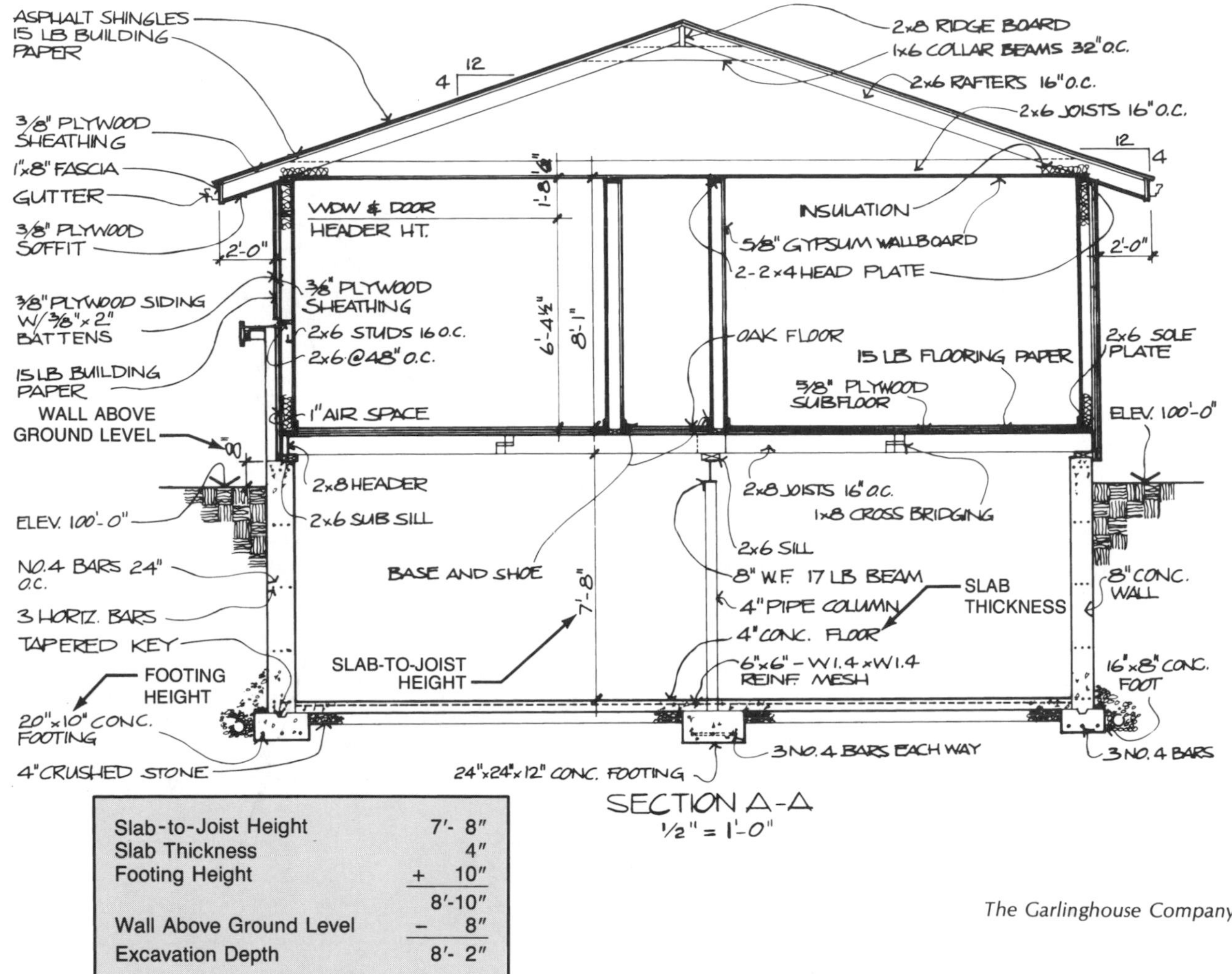

The Garlinghouse Company

Figure 1-22. The depth of excavation for a full basement foundation is calculated by adding the footing height and the foundation wall height, and then subtracting the wall projection above the finish grade.

WORKING SCAFFOLDS

A *working scaffold* is a temporary elevated structure providing a working platform supporting workers and materials. Scaffolds are commonly made of steel, aluminum, or fiberglass. The height of scaffolds may range from a few feet to hundreds of feet above floor or ground level. Older, traditional scaffolds were constructed of wood.

Steel Scaffolds

Steel scaffolds are the most common type of scaffold in construction work. Qualified workers who have received special training erect scaffolds. Steel scaffolds may be stationary or mobile.

Sectional Metal-Framed Scaffolds. Sectional metal-framed scaffolds are a common type of stationary steel scaffold. Sectional metal-framed scaffolds are placed inside and around the outside of a structure. They are designed and produced by different manufacturers; therefore, the manufacturer's instructions should be strictly followed in the erection of the scaffold. The main component of a metal-framed scaffold is the vertical frame unit. The two types of frames commonly used are the walk-through and open-end frames. **See Figure 1-23.** The frames are joined together with horizontal and diagonal braces. Vertical coupling pins are used to attach one frame on top of a lower frame.

Mobile Steel Scaffolds. Mobile steel scaffolds are commonly used for higher interior work. Mobile steel scaffolds (rolling scaffolds) are similar to stationary metal-framed scaffolds. They are equipped with casters (wheels) and can be moved about by hand on level paved or concrete surfaces. **See Figure 1-24.** Some mobile scaffolds include a power drive system for movement. Outriggers can be installed to help maintain the balance of taller mobile scaffolds.

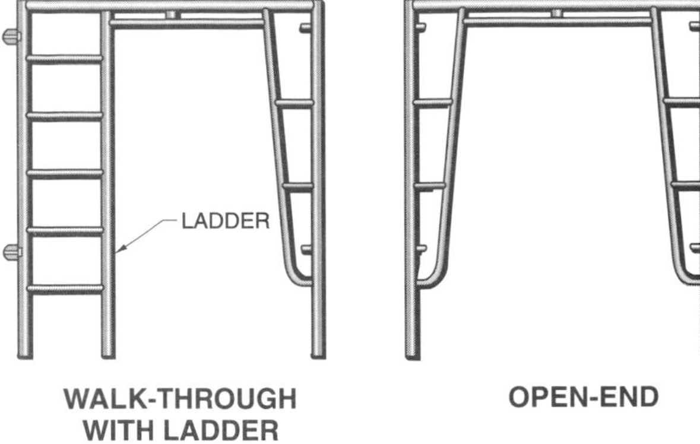

Figure 1-23. Walk-through and open-end frames are commonly used in the construction of steel scaffolds.

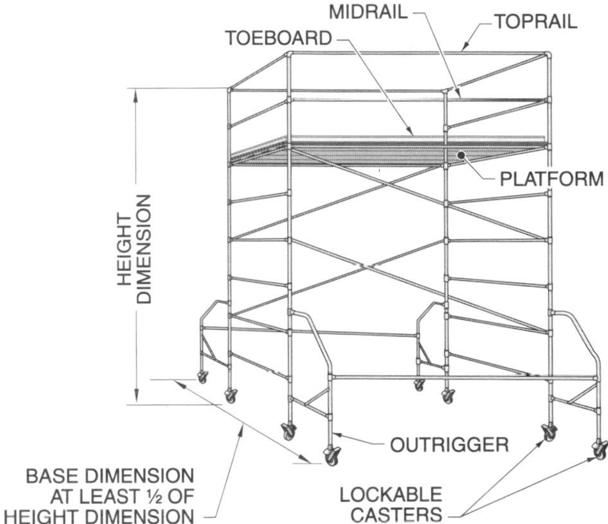

Figure 1-24. Casters at the base of mobile scaffolds allow movement of the scaffold. Outriggers help maintain the balance of taller mobile scaffolds.

Other Scaffold Systems

There are a variety of working platform designs that are included in the scaffold category. Some of these are form bracket scaffolds, suspension scaffolds, and outrigger scaffolds. When working at heights, workers must use fall arrest systems to ensure safe working conditions.

Form Bracket Scaffolds. Form bracket scaffolds are built into large panel and ganged form units. **See Figure 1-25.** Metal brackets are hooked over the walers or bolted to the walers and/or uprights. The platforms consist of planks placed over the brackets and enclosed by preinstalled guardrails.

Patent Construction Systems

Figure 1-25. Form bracket scaffolds are incorporated into the construction of large panel and ganged form units. The brackets of form scaffolds are hooked over walers or bolted to the walers or uprights.

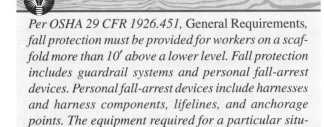

Per OSHA 29 CFR 1926.451, General Requirements, *fall protection must be provided for workers on a scaffold more than 10′ above a lower level. Fall protection includes guardrail systems and personal fall-arrest devices. Personal fall-arrest devices include harnesses and harness components, lifelines, and anchorage points. The equipment required for a particular situation depends on the scaffold being used.*

Suspension Scaffolds. Suspension scaffolds are used for heavy construction work in limited areas that do not require the use of scaffolds resting on and extending from the ground level. They are supported by overhead wire ropes that must be capable of supporting at least six times the maximum intended load applied or transmitted to the rope.

The major types of suspension scaffolds are the swinging platform, two-point suspension, and multiple-point suspension scaffolds. **See Figure 1-26.** The components of a swinging platform scaffold are a metal grid base supporting a wood platform. A steel stirrup to which the lower block of a block and tackle is attached supports each end of the platform. The wire ropes are tied to hooks or anchors on the roof of the building.

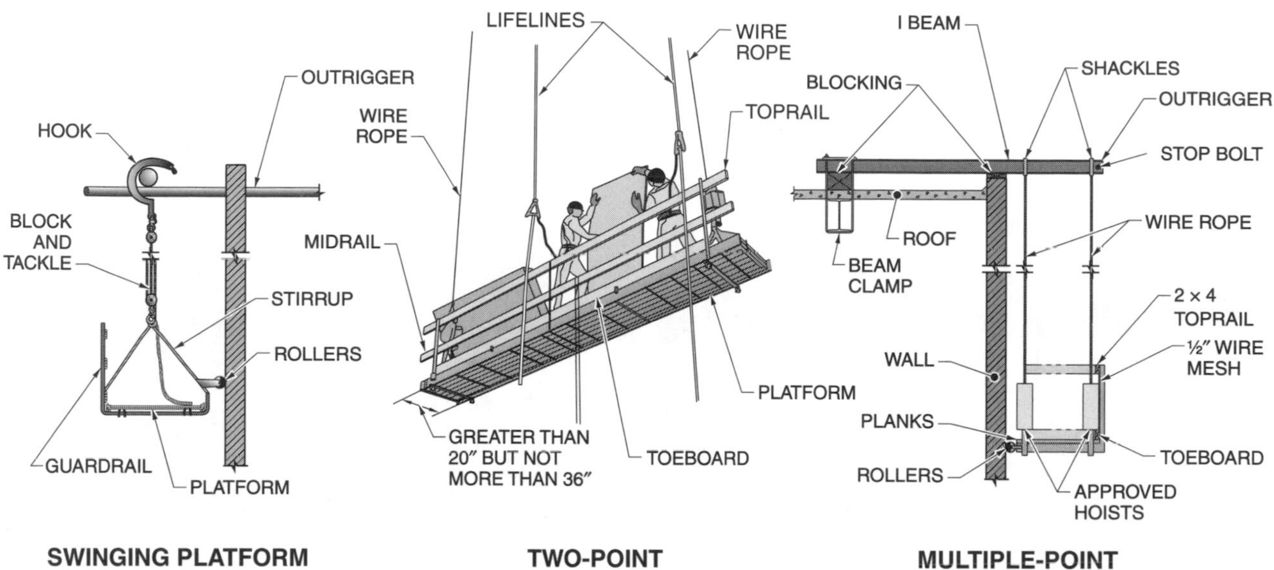

Figure 1-26. Suspension scaffolds are used to perform work in limited areas. They are supported by wire rope attached to hooks or outriggers.

Two-point suspension scaffolds are suspended from two points whereas multiple-point scaffolds are suspended from more than two points. Both types are heavier than swinging scaffolds and are used for heavier operations and materials. These scaffolds are available in sizes up to 6′ wide and 12′ long. Suspension scaffolds, supported by outriggers on the roof, are raised or lowered by power hoists.

Outrigger Scaffolds. Beams protruding out of a wall opening (usually a window) hold up outrigger scaffolds. The interior end of the beam is securely anchored to the floor. Planks are placed on top of the outside extension of the beams, and guardrails are added. **See Figure 1-27.**

Fall Arrest Systems. Fall arrest equipment must be worn when working at heights in excess of 10′. These life-saving devices include lifelines, harnesses, lanyards, and rope grabs. **See Figure 1-28.** Harnesses, when worn correctly, distribute the shock of an arrested fall, thus protecting the internal body organs, the spine, and other bones from injury. Lifelines are securely anchored to a fixed end above the work area and must never support more than one worker. Lanyards secure the worker's harness to the lifeline. Lifelines can be hooked to a metal ring attached to a rope grab.

Construction Safety and Regulations

The U.S. Occupational Safety and Health Administration (OSHA) has established strict guidelines and regulations pertaining to the construction of scaffolds and the safety of personnel working on scaffolds. These regulations are found in OSHA 29 CFR 1926 Subpart L, *Scaffolds*; OSHA 29 CFR 1910.28, *Safety Requirements for Scaffolding*; and OSHA 29 CFR 1910.29, *Manually Propelled Mobile Ladder Stands and Scaffolds (Towers)*.

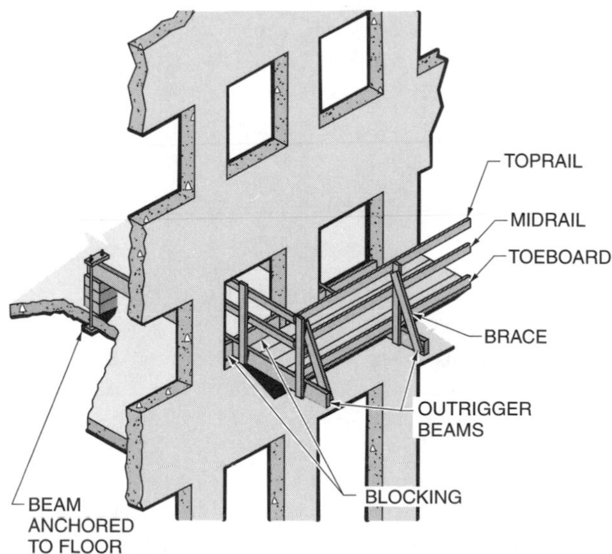

Figure 1-27. Beams extending out a wall opening (usually a window opening) support outrigger scaffolds.

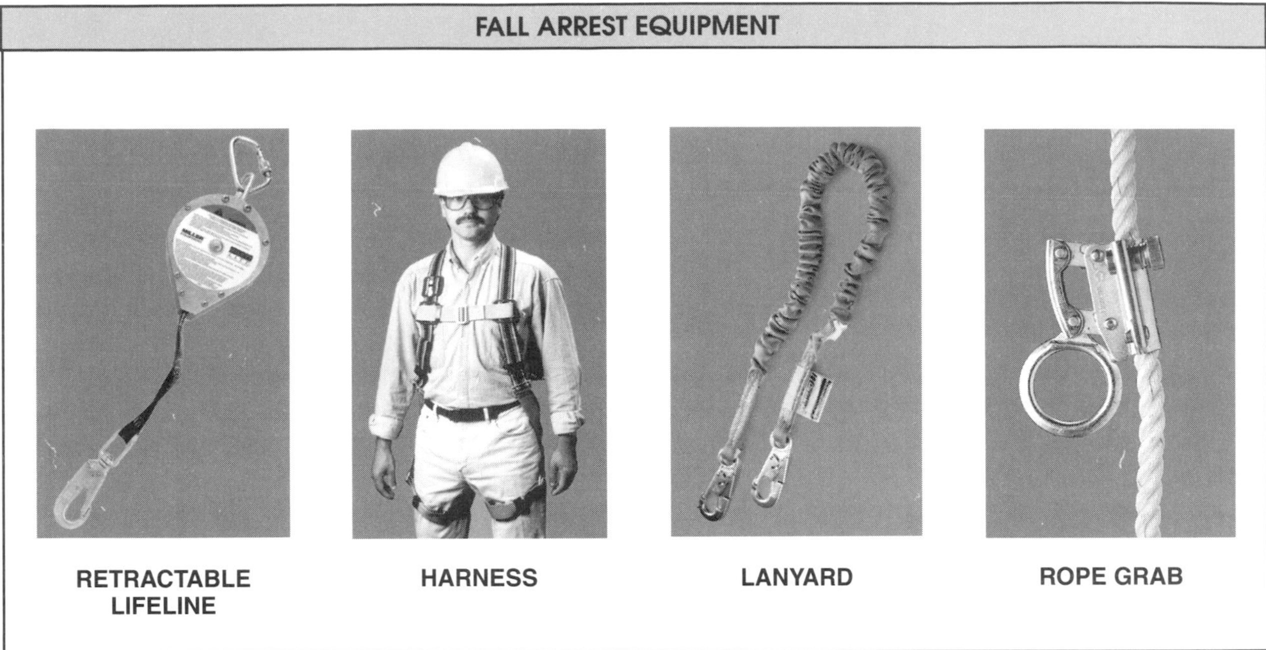

Figure 1-28. A typical fall arrest system includes a retractable lifeline, harness, lanyard, and rope grab.

JOB SITE SAFETY

Workers are exposed to dangerous situations on the job site every day. Some of the hazards are obvious, such as an open trench. Some hazards are not obvious, such as the dangers of silicosis. OSHA requires all employers to provide a safe environment for employees. Employers are responsible for safety training and for ensuring that employees follow OSHA regulations. Many large contractors have a comprehensive orientation program to familiarize employees with applicable safety regulations and company standards. Safety meetings and/or "toolbox talks" should be conducted frequently to discuss current safety topics and address employee safety concerns.

A safe environment is free of hazards and has precautions in place to protect employees. However safe the work environment, all workers must be aware of potentially dangerous situations and the proper ways to protect themselves. Proper techniques, habits, and personal protective equipment (PPE) must be used to protect workers from possible injury. Workers must be protected from direct contact with materials, falling objects, loud noises, lifting heavy objects, and falls.

Companies also establish substance abuse policies that maintain a safe working environment and promote high work standards. Substance abuse policies prohibit workers from working while under the influence of drugs, alcohol, or controlled substances. Failure to comply with an established substance abuse policy may result in serious injury and/or employment termination.

Many accidents can be prevented by safe work habits including the following:

- When working on an elevated surface, do not place tools where they may fall and injure another worker below.
- Always watch where you step.
- Look out for the safety of other workers as well as your own safety.
- Do not engage in horseplay on the job.

Good Housekeeping

Hazards on a job site may be caused by general sloppiness, poor organization, and careless storage of materials. OSHA 29 CFR 1926.25, *Housekeeping,* details good housekeeping practices on a job site. Rules for good housekeeping are as follows:

- Keep scrap lumber cleared from work areas, passageways, and stairs. Clinch or pull out protruding nails.

- Maintain well-defined passageways and walkways on a job site. Keep passageways and walkways well lit and free of tripping hazards. Provide walkways at convenient places to bridge ditches.
- Ensure that areas within 6′ of a building under construction are reasonably level.
- Keep material storage areas free of obstructions and debris.
- Stack materials in such a way that they will not fall, slip, or collapse. Lumber piles should not exceed 8′ in height if the lumber is to be handled manually, or 20′ if it is to be handled with equipment such as a forklift or material handler.
- Use cleanup crews to periodically remove waste materials from the job site.
- Store tools and equipment not being used in chests or tool sheds.
- Remove slush or snow from work areas or walkways before it turns into ice. Slipping hazards can be reduced by spreading sand, gravel, cinders, or other gritty material over the work areas.

Accident Reports

An *accident report* is a document that details facts about an accident. Accidents and safety hazards must be reported to the employer or supervisor regardless of their nature. If injuries occur, medical assistance should be obtained immediately.

Accident report forms commonly include the name of the injured person; date, time, and place of accident; immediate supervisor; and circumstances surrounding the accident. Accident reports are required for insurance claims and become a permanent part of company records. Information about causes of the accident can be used to prevent future injuries. Experience has proven that most accidents can be prevented by proper safety practices. Various studies stress the following facts:

- Strain or overexertion is the most common injury suffered by construction workers.
- Slips or falls from elevated work surfaces and ladders account for nearly one-third of construction injuries.
- Worker injuries resulting from use of machines and tools, or from a worker being struck by a machine or tool, account for one-fourth of injuries reported.
- The most common cause of death from job site accidents involves construction equipment such as loaders or material lifts.

Workers climbing and descending scaffold ladders must have both hands free for climbing and should remove foreign substances such as dirt and grease from shoes and hands.

Personal Protective Equipment

Personal protective equipment (PPE) is safety equipment worn by a construction worker for protection against safety hazards on a job site. PPE includes protective clothing, head protection, eye and face protection, hearing protection, respiratory protection, back protection, hand and foot protection, and knee protection.

Protective Clothing. Protective clothing is worn to prevent injury. Protective clothing made of durable material such as denim provides protection from contact with sharp objects, hot equipment, and harmful materials. Protective clothing should be snug, yet provide ample movement. Pockets should allow convenient access, but should not snag on tools or equipment. Loose-fitting clothing should not be worn, long hair must be secured, and jewelry must be removed to prevent its getting caught in rotating equipment. Metallic watches and rings should not be worn since serious injury may result if contact is made with an energized electrical circuit.

Head Protection. OSHA 29 CFR 1926.100, *Head Protection,* mandates that workers wear protective helmets (hard hats) in areas where there is the potential for head injury from impact, falling and flying objects, and electrical shock. **See Figure 1-29.** Protective helmets protect workers by resisting penetration and absorbing the blow of impacts. The shell of the protective helmet is made of a durable, lightweight material. A shock-absorbing lining consisting of crown straps and a headband keeps the shell of the protective helmet away from the head and allows for ventilation. Many contractors require protective helmets to be worn at all times on a job site.

Eye and Face Protection. Eye protection is required by OSHA 29 CFR 1926.102, *Eye and Face Protection,* when there is a reasonable probability of preventing injury to eyes or face from flying particles, molten metal, chemical liquids or gases, radiant energy, or any combination. Proper eye and face protection must be worn to prevent eye or face injuries caused by flying particles such as wood chips, metal particles, and chemicals.

ECO-Block, LLC

Figure 1-29. OSHA mandates that workers wear protective helmets in areas where there is the potential for head injury on a job site.

Personal protective equipment required is based on the job requirements and includes hard hats, gloves, and rubber boots.

Eye and face protection includes safety glasses, face shields, and goggles. **See Figure 1-30.** *Safety glasses* are glasses with impact-resistant lenses, reinforced frames, and side shields. Frames are designed to keep the lenses secured in the frame if an impact occurs and are used for protection from flying objects or splashing liquids. *Goggles* are a form of eye protection that have a flexible frame and are secured on the face with an elastic headband. Goggles must fit snugly against the face to seal the areas around the eyes, and may be used over prescription glasses. For general operations, goggles with clear lenses are typically worn. For welding and cutting operations, tinted lenses are required to protect against ultraviolet (UV) rays.

Hearing Protection. Operating equipment and power tools can produce high noise levels. Construction workers subjected to excessive noise levels may develop hearing loss over a period of time. The severity of hearing loss depends on the noise intensity and the duration of exposure. Noise intensity is expressed in decibels. A *decibel (dB)* is a unit used to express the relative intensity of sound. **See Figure 1-31.** Per OSHA 29 CFR 1926.101, *Hearing Protection,* ear protection devices must be worn when it is not feasible to reduce the noise intensity or duration of exposure. Ear protection devices inserted in the ears, such as earplugs, must be fitted or determined individually by a competent person. Cotton is not acceptable hearing protection.

Ear protection devices can be broadly classified as earplugs and ear muffs. An *earplug* is an ear protection device inserted into the ear canal and made of moldable rubber, foam, or plastic. An *ear muff* is an ear protection device worn over the ears. A tight seal around the ear muff is required for proper protection. Ear protection devices are assigned a noise reduction rating (NRR) number based on the noise level reduced. For example, an NRR of 27 means that the noise level is reduced by 27 dB when tested at the factory. To determine approximate noise reduction in the field, 7 dB is subtracted from the NRR. For example, an NRR of 27 provides a noise reduction of approximately 20 dB in the field.

Portland Cement Association

Figure 1-30. Safety glasses or goggles are commonly required for eye protection.

SOUND LEVELS		
Decibel (dB)	**Loudness**	**Examples**
140	Deafening	Jet plane taking off, air raid siren, locomotive horn
130	Pain threshold	
120	Feeling threshold	
110	Uncomfortable	
100	Very loud	Chain saw
90	Noisy	Shouting, auto horn
80	Moderately loud	Vacuum cleaner
70	Loud	Telephone ringing, loud talking
60	Moderate	Normal conversation
50	Quiet	Hair dryer
40	Moderately quiet	Refrigerator running
30	Very quiet	Quiet conversation, broadcast studio
20	Faint	Whispering
10	Barely audible	Rustling leaves, soundproof room, human breathing
0	Hearing threshold	Intolerably quiet

Figure 1-31. Noise intensity is expressed in decibels.

Hand and Foot Protection. Hand protection is required to prevent injuries to hands from cuts or chemical absorption. The appropriate hand protection is determined by the duration, frequency, and degree of the hazard to the hands. Gloves are recommended for many construction tasks, including handling and installing sharp and abrasive materials. Gloves should be snug-fitting. Gloves that are too large can pose a safety hazard when working around moving parts and equipment. **See Figure 1-32.** OSHA 29 CFR 1910.138, *Hand Protection,* details the use of hand protection.

Foot injuries are typically caused by objects falling less than 4′ and weighing an average of 65 lb. Safety shoes with reinforced steel toes protect against injuries caused by compression and impact. Some safety shoes have protective metal insoles and metatarsal guards for additional protection.

Oil-resistant soles and heels are not affected by petroleum-based products and provide improved traction. Synthetic rubber boots are commonly used by concrete workers during concrete placement to protect feet or footwear from exposure to concrete. Protective footwear must comply with ANSI Z41, *Personal Protection—Protective Footwear.*

Back Protection. Back injury is one of the most common injuries resulting in lost time. Most back injuries are the result of improper lifting techniques. Back injuries can be prevented through proper planning and lifting techniques. Assistance should be sought when moving heavy objects. When lifting objects from the ground, the path should be clear of obstacles and free of hazards. The knees should be bent and the object firmly grasped. The legs should be straightened and the back should be kept as straight as possible. The load should be kept as close to the body as possible when moving forward.

Figure 1-32. Gloves are recommended for many construction tasks.

Knee Protection. Workers who spend considerable time working on their knees may require knee pads. *Knee pads* are rubber, plastic, or leather pads strapped onto the knees for protection. Buckle straps or Velcro™ closures secure knee pads in position. **See Figure 1-33.**

Respiratory Protection. Respiratory protection is required to protect against airborne hazards. OSHA 29 CFR 1926.103, *Respiratory Protection,* details respiratory protection requirements. When harmful airborne substances are present on a job site, engineering control measures, such as enclosure of the operation, ventilation, and substitution of less toxic substances, should be undertaken. If proper engineering control measures are not feasible, appropriate respiratory protection must be used. The respiratory protection required is determined by the type of airborne hazard. The degree of risk from exposure to any given substance depends on the nature and potency of the substance and the duration of exposure.

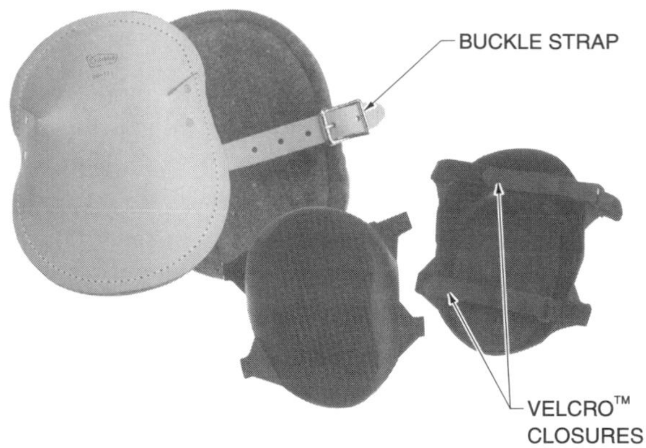

Stanley Tools

Figure 1-33. Workers who spend considerable time working on their knees may require knee pads.

Airborne chemical hazards in the form of mists, vapors, gases, fumes, or solids may be encountered on a job site due to the use of cleaners and solvent cements or to sawing or drilling operations performed by workers. Chemical hazards vary according to how they are introduced into the body. For example, some chemicals are toxic through inhalation. Others are toxic by absorption through the skin or through ingestion. Some chemicals are toxic by all three routes and may be flammable as well. Workers may be subjected to hazards such as vapors from aerosol cleaning solvents or dust from crystalline silica particles which can result in silicosis.

Silicosis is a disease of the lungs caused by inhaling dust containing crystalline silica particles. *Crystalline silica (quartz)* is a natural compound found in the crust of the Earth and is a basic component of sand and granite. As dust containing crystalline silica particles is inhaled, scar tissue forms in the lungs and reduces the ability of the lungs to extract oxygen from the air. Since there is no cure for silicosis, prevention is the only means of control.

Early stages of silicosis may go unnoticed. Continued exposure to dust can result in shortness of breath when exercising, fever, and possible bluish skin along the ear lobes or lips. Advanced silicosis leads to fatigue, extreme shortness of breath, loss of appetite, chest pain, and respiratory failure, which may cause death. Chronic silicosis usually occurs after 10 or more years of exposure to crystalline silica at relatively low concentrations. Accelerated silicosis results from exposure to high concentrations of crystalline silica and develops five to 10 years after the initial exposure. Acute silicosis occurs where exposure concentrations are the highest and can cause symptoms to develop within a few weeks to four or five years after the initial exposure.

Concrete Dust. Workers must be aware of dust containing crystalline silica particles when working with concrete. Concrete sawing and grinding, abrasive blasting of concrete, demolition of concrete structures, dry sweeping or pressurized air blowing of concrete, and concrete mixing produce dust containing crystalline silica particles. **See Figure 1-34.** Disposable particulates masks are required when crystalline silica particles are present in the air.

Figure 1-34. Concrete sawing produces dust containing silica particles. Disposable particulates masks should be worn when concrete sawing is taking place.

Exposure to dust containing crystalline silica should be limited to reduce the possibility of contracting silicosis. Workers should be aware of operations where exposure to crystalline silica may occur and use the appropriate personal protective equipment when applicable. Type CE positive-pressure abrasive blasting respirators should be used for sandblasting operations. For other operations where respirators may be required, a respirator approved for protection against crystalline silica-containing dust must be used.

The respirator must not be altered or modified in any way. When using tight-fitting respirators, workers cannot have beards or mustaches that interfere with the respirator seal to the face. If possible, workers should change into disposable or washable work clothes at the job site and, after showering, change into clean clothing before leaving the job site.

Tobacco products or cosmetics should not be used in areas where dust containing crystalline silica is present. In addition, food and drink should not be ingested in these areas. After leaving areas containing crystalline silica dust, hands and faces should be thoroughly washed before eating, drinking, smoking, or applying cosmetics.

Material Safety Data Sheets

A *material safety data sheet (MSDS)* is a written document used to relay hazardous material information from the manufacturer, importer, or distributor to the worker. The information is listed in English, provides precautionary information regarding proper handling, and includes emergency and first-aid procedures. **See Figure 1-35.** Chemical manufacturers, distributors, and importers must develop an MSDS for each hazardous material. If an MSDS is not provided, the employer must write to the manufacturer, distributor, or importer to obtain the missing MSDS. MSDS files must be kept up to date and made readily available to employees.

Information may be filed according to product name, manufacturer, or a company-assigned number. If two or more MSDSs on the same material are found, the latest version is used. An MSDS has no prescribed format. Formats provided in ANSI Z400.1, *Material Safety Data Sheet Preparation,* may be used. Employees should become familiar with the MSDS of any hazardous materials to which they are exposed. An MSDS includes the following information:

- manufacturer information
- product information
- hazardous ingredients and identity information
- physical and chemical characteristics
- fire and explosion hazard data
- reactivity data
- health hazard data
- spill or leak procedures
- safe handling and use information
- special precautions

Confined Spaces

A *confined space* is a space large enough and configured so an employee can physically enter and perform assigned work, has limited or restricted means for entry and exit, and is not designed for continuous employee occupancy. Confined spaces have a limited means of egress and are subject to the accumulation of toxic or flammable contaminants or an oxygen-deficient atmosphere. Oxygen deficiency is caused by the displacement of oxygen by leaking gases or vapors, combustion or oxidation processes, oxygen being absorbed by the vessel or product stored, and/or oxygen being consumed by bacterial action. Breathing oxygen-deficient air can result in injury or death.

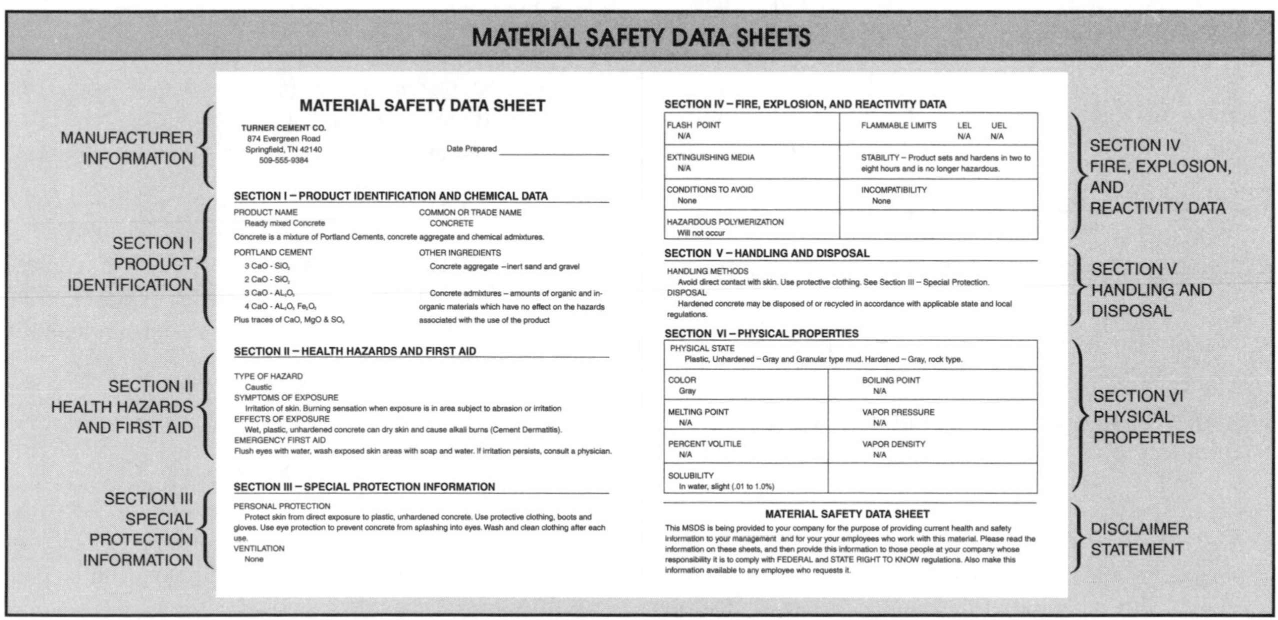

MATERIAL SAFETY DATA SHEET

TURNER CEMENT CO.
874 Evergreen Road
Springfield, TN 42140
509-555-9384

Date Prepared ____________

SECTION I – PRODUCT IDENTIFICATION AND CHEMICAL DATA

PRODUCT NAME
Ready mixed Concrete

COMMON OR TRADE NAME
CONCRETE

Concrete is a mixture of Portland Cements, concrete aggregate and chemical admixtures.

PORTLAND CEMENT
$3\ CaO - SiO_2$
$2\ CaO - SiO_2$
$3\ CaO - AL_2O_3$
$4\ CaO - AL_2O_3\ Fe_2O_3$
Plus traces of CaO, MgO & SO_3

OTHER INGREDIENTS
Concrete aggregate –inert sand and gravel

Concrete admixtures – amounts of organic and inorganic materials which have no effect on the hazards associated with the use of the product

SECTION II – HEALTH HAZARDS AND FIRST AID

TYPE OF HAZARD
Caustic
SYMPTOMS OF EXPOSURE
Irritation of skin. Burning sensation when exposure is in area subject to abrasion or irritation
EFFECTS OF EXPOSURE
Wet, plastic, unhardened concrete can dry skin and cause alkali burns (Cement Dermatitis).
EMERGENCY FIRST AID
Flush eyes with water, wash exposed skin areas with soap and water. If irritation persists, consult a physician.

SECTION III – SPECIAL PROTECTION INFORMATION

PERSONAL PROTECTION
Protect skin from direct exposure to plastic, unhardened concrete. Use protective clothing, boots and gloves. Use eye protection to prevent concrete from splashing into eyes. Wash and clean clothing after each use.
VENTILATION
None

SECTION IV – FIRE, EXPLOSION, AND REACTIVITY DATA

FLASH POINT N/A	FLAMMABLE LIMITS LEL N/A UEL N/A
EXTINGUISHING MEDIA N/A	STABILITY – Product sets and hardens in two to eight hours and is no longer hazardous.
CONDITIONS TO AVOID None	INCOMPATIBILITY None
HAZARDOUS POLYMERIZATION Will not occur	

SECTION V – HANDLING AND DISPOSAL

HANDLING METHODS
Avoid direct contact with skin. Use protective clothing. See Section III – Special Protection.
DISPOSAL
Hardened concrete may be disposed of or recycled in accordance with applicable state and local regulations.

SECTION VI – PHYSICAL PROPERTIES

PHYSICAL STATE Plastic, Unhardened – Gray and Granular type mud. Hardened – Gray, rock type.	
COLOR Gray	BOILING POINT N/A
MELTING POINT N/A	VAPOR PRESSURE N/A
PERCENT VOLITILE N/A	VAPOR DENSITY N/A
SOLUBILITY In water, slight (.01 to 1.0%)	

MATERIAL SAFETY DATA SHEET

This MSDS is being provided to your company for the purpose of providing current health and safety information to your management and for your your employees who work with this material. Please read the information on these sheets, and then provide this information to those people at your company whose responsibility it is to comply with FEDERAL and STATE RIGHT TO KNOW regulations. Also make this information available to any employee who requests it.

Figure 1-35. A material safety data sheet (MSDS) is used to relay hazardous material information from the manufacturer, importer, or distributor to the worker.

Confined spaces include, but are not limited to, storage tanks, process vessels, bins, boilers, ventilation or exhaust ducts, sewers, underground utility vaults, tunnels, pipelines, and spaces more than 4′ in depth with open tops such as trenches, pits, tubes, ditches, and vaults.

Hand Tool Safety

Accidents with hand tools may be reduced by following hand tool safety rules. Hand tool safety rules include the following:

- Point cutting tools away from body during use.
- Organize tools to protect and conceal sharp cutting surfaces.
- Transport sharp tools in a holder or with the blade pointed down.
- Keep tools sharp and in proper working order.

Power Tool and Equipment Safety

Accidents when using power tools and equipment may be reduced by following power tool and equipment safety rules. Power tool and equipment safety rules include the following:

- Follow all manufacturer operating instructions.
- Use UL- or CSA®-approved power tools that are installed in compliance with the National Electrical Code® (NEC®).
- Use power tools that are double-insulated or have a third conductor grounding terminal to provide a path for fault current.
- Ensure the power switch is in OFF position before connecting a tool to a power source.
- Ensure that all safety guards are in place before starting.
- Arrange cords and hoses to prevent accidental tripping.
- Stand clear of operating power tools. Keep hands and arms away from moving parts.
- Shut OFF, lock out, and tag out disconnect switches of power tools requiring service.

Pneumatic Tool Safety

When pneumatic tools are used, the safety of the operator and of other people in the immediate area must be considered. For example, before panels are fastened to studs or joists, ensure that no one is on the other side of the wall or below the joists. If the nail or staple misses the stud or joist, it could go through the panel and severely injure someone. Other essential safety precautions that must be observed when using pneumatic nailers and staplers are as follows:

- Follow proper operating procedures as outlined in the operator manual for the particular tool.
- Inspect pneumatic nailers and staplers daily to ensure that the firing mechanism and safety features are operating properly. Test-fire the tool into a block of wood before using the nailer or stapler for the desired application.
- Point the tip of a pneumatic nailer or stapler away from your body and other workers.
- Wear safety glasses or a face shield. Wood chips, concrete, or a deflected nail can cause serious eye injury.
- Use the right nailer or stapler for the job, and use the correct size of nail or staple. Refer to manufacturer recommendations.
- When using a portable air compressor, ensure that exposed belts have guards on both sides to reduce the possibility of injuries to fingers and hands.
- Ensure that air hoses and connections are in good operating condition. Air hoses that are larger than ½″ diameter must have a safety device at the air compressor or branch line to reduce air pressure in case of hose failure.
- Do not lift or carry a pneumatic nailer or stapler by the air hose.
- Disconnect pneumatic tools from the air supply when the tools are not in use. Pneumatic tools should be equipped with a fitting that releases air pressure from the tools when disconnected.
- Per OSHA 29 CFR 1926.302, *Power-Operated Hand Tools,* pneumatic nailers and staplers that operate at more than 100 psi air pressure must have a safety device on the tip to prevent the nailer or stapler from ejecting fasteners unless the tip is in contact with the work surface.

Powder-Actuated Tool Safety

Per OSHA 29 CFR 1926.302, *Power-Operated Hand Tools,* only workers who have been trained in the operation of a powder-actuated tool and have obtained an operator license can operate powder-actuated tools. The operator must take and satisfactorily pass a quiz on the safe operation of powder-actuated tools.

Powder-actuated tool operators must wear proper PPE such as safety goggles and earplugs or earmuffs. Powder-actuated tools should not be used in flammable or explosive environments since the ignition of the charge produces a spark. There must be adequate ventilation when the tool is used in confined spaces. Other workers should not stand close by or to one side of the operator.

Powder-actuated tools must be inspected and tested each day to ensure that safety devices are in proper operating condition. Tools found not to be operating properly should be immediately removed from service and repaired.

Powder-actuated tools should not be loaded until immediately prior to firing. The tool should be brought into the intended firing position before pulling the trigger. Powder-actuated tool operators should not drive fasteners into easily penetrated materials unless the materials are backed by a substance that will prevent the drive pin or threaded stud from passing completely through.

If a powder-actuated tool misfires, hold the tool in place for at least 30 seconds and then try to operate the tool a second time. If the tool does not fire again, hold the tool in the operating position for another 30 seconds and then proceed to remove the load following the manufacturer instructions. If the problem is due to a tool defect, remove the tool from service. Additional safety precautions to be observed when using powder-actuated tools include the following:

- Never point a powder-actuated tool at any other person, whether the tool is loaded or unloaded.
- Post warning signs stating that powder-actuated tools are in use within 50′ of the area where the tool is being used.
- Keep hands away from the open barrel end.
- Before firing a powder-actuated tool, inspect the chamber to ensure the barrel is clean and there are no obstructions.
- Be aware of electrical circuits and use caution when driving fasteners near electrical circuits.
- Do not leave a powder-actuated tool unattended.
- Do not carry a loaded tool from the job.
- Store tools and cartridges in a locked container when they are not in use. Be sure the tool is unloaded before storing it.
- Never carry fasteners or other hard objects in the same container or pocket as powder loads.
- Use a spall guard whenever possible.

ELECTRICAL SAFETY

Improper procedures with electrical tools and equipment can result in an electrical shock. *Electrical shock* is a condition that results when a body becomes part of an electrical circuit. Safe work habits and use of proper PPE are required to prevent electrical shock when working with or in proximity to electrical devices.

The severity of electrical shock depends on the amount of electrical current, measured in milliamps (mA), that flows through the body; the length of time the body is exposed to the current flow; the path the current takes through the body; and the physical size and condition of the body through which the current passes. **See Figure 1-36.**

ELECTRICAL SHOCK EFFECTS	
Approximate Current*	**Effect on Body†**
Over 20	Causes severe muscular contractions, paralysis of breathing, heart convulsions
15–20	Painful shock May be frozen or locked to point of electrical contact until circuit is de-energized
8–15	Painful shock Removal from contact point by natural reflexes
8 or less	Sensation of shock but probably not painful

*in mA
†effects vary depending on time, path, amount of exposure, and condition of body

Figure 1-36. Safe work habits and use of proper PPE are required to prevent electrical shock when working with or in proximity to electrical devices.

Extension Cords

An extension cord is used to supply power to portable electric tools and equipment. Heavy-duty, three-wire extension cords must be used for tools and equipment. Extension cords should be visually inspected each day for external damage such as deformed or missing prongs, damaged insulation, or indications of possible internal damage. OSHA 29 CFR 1926.405, *Wiring Methods, Components, and Equipment for General Use,* details safe work practices related to extension cords. Safety precautions to observe when using extension cords include the following:

- Do not use frayed extension cords and do not use electrical tape to make repairs.

- Extension cords must be of the three-wire type, which have a grounding conductor and ground prong.
- Protect extension cords that pass through doorways or other pinch points from damage.
- Extension cords should not be run through holes in walls, ceilings, or floors.
- Do not conceal extension cords behind walls, ceilings, or floors.
- Extension cords should not be hung from nails or wire, or fastened with staples.

FIRE PREVENTION

A serious concern for all construction workers is the danger of fire on a job site. Workers must be aware of potential fire hazards and understand what creates a fire hazard and how this danger can be reduced.

Fire Extinguishers

The National Fire Protection Association (NFPA) classifies fires as Class A, B, C, D, and K, based upon the combustible material. The appropriate fire extinguisher must be used on a fire to safely and quickly extinguish the fire. The types of fires and appropriate fire extinguishers are as follows:

- Class A fires occur with wood, paper, textiles, and similar materials. Class A fires are extinguished with water and other water-based agents.
- Class B fires occur with flammable liquids such as grease or solvent cements. Class B fires are extinguished with smothering agents such as carbon dioxide and chemical foams.
- Class C fires occur with live electrical equipment and are extinguished with nonconductive dry chemical agents.
- Class D fires occur with combustible metals such as magnesium, sodium, and potassium. Class D fires are extinguished with a coarse powder agent that seals the burning surface and smothers the fire.
- Class K fires occur with grease in commercial cooking equipment. Class K extinguishers coat the fuel with wet- or dry-base chemicals.

OSHA 29 CFR 1926.150, *Fire Protection,* details the fire protection requirements for a construction job site. Fire extinguishers must be periodically inspected to ensure proper operation. Fire extinguishers are placed around the job site so the travel distance from any point to the nearest fire extinguisher does not exceed 100′. At least one fire extinguisher must be provided on each floor of a building. In multistory buildings, at least one fire extinguisher shall be located adjacent to a stairway. Fire extinguishers should be located in clear view and should not be obstructed by building materials. **See Figure 1-37.**

If a fire occurs on the job site, an alarm should be sounded and the fire department called. However, small fires can be quickly extinguished if the proper fire extinguisher is present on the job.

Figure 1-37. Fire extinguishers are placed around the job site so the travel distance from any point to the nearest fire extinguisher does not exceed 100′.

Preventive Measures

The following preventive measures should be observed to reduce the threat of fire on a job site:

- Do not allow rubbish and combustible material to accumulate on a job site. Periodically clean the job site and place wood scraps and other rubbish in appropriate disposal containers.
- Keep volatile and flammable materials stored away from the immediate job site.
- Do not smoke near volatile materials such as solvent cements, as these materials readily evaporate at normal temperatures and pressures and produce flammable vapors.
- Keep all flammable liquids such as gasoline, paint thinner, oil, grease, and paint in tightly plugged or capped containers.

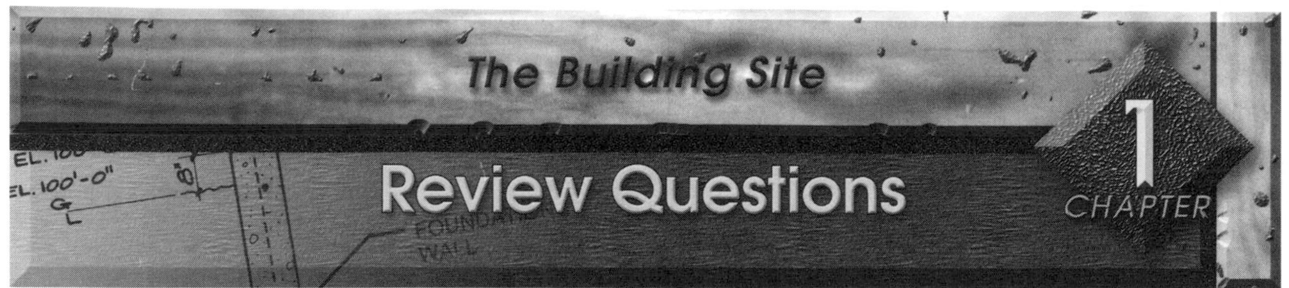

Name ______________________________ Date ______________

Completion

__________ **1.** The size of the soil ___ is the major factor differentiating types of soil.

__________ **2.** Soils are generally classified as granular or ___.

__________ **3.** Sand and gravel are referred to as ___-grained soils.

__________ **4.** Silt and clay are referred to as ___-grained soils.

__________ **5.** ___ is a disease of the lungs caused by inhaling dust containing crystalline silica particles.

__________ **6.** ___ is one of the most common injuries resulting in lost time.

__________ **7.** When constructing walkways and bridges, the crossing must extend a minimum of ___″ over each edge of the excavation.

__________ **8.** The recommended slope for finish grade around foundation walls is ___″ in ___′ for paved areas.

__________ **9.** The recommended slope for finish grade around foundation walls is ___″ in ___′ for nonpaved areas.

__________ **10.** The water ___ is the highest point below the surface of the ground that is normally saturated with water.

__________ **11.** ___ action is the physical process occurring in soil that causes water to rise from the water table.

__________ **12.** Vapor ___ are used for ground covers to control ground surface dampness.

__________ **13.** ___ tile is placed around foundation footings to move groundwater away from the foundation.

__________ **14.** The ___ line is the depth to which soil freezes in a particular area.

__________ **15.** ___ occurs in soil when it freezes and expands.

Multiple Choice

__________ **1.** Grade levels on a plot plan are usually expressed in ___.
A. feet and inches
B. fractions of a foot
C. feet and tenths or hundredths of a foot
D. a percentage of a foot

__________ **2.** Contour lines show the ___ of the ground.
A. slope
B. shape
C. location
D. none of the above

__________ **3.** A(n) ___ is established by a surveyor to locate various elevations.
A. hub
B. easement
C. setback
D. benchmark

__________ **4.** When grading a lot, the finished surface should ___.
A. be perfectly level
B. slope from front to back
C. slope away from the building
D. slope toward the center

__________ **5.** The walls of a foundation excavation should be ___ the building lines.
A. 1′ inside of
B. even with
C. at least 2′ outside of
D. 5′ outside of

__________ **6.** Print information regarding the depth of excavation is given on the ___ .
A. plot plan
B. foundation plan views
C. foundation section views
D. floor plan

__________ **7.** Accidents on the job site can be prevented by safe work habits that include ___.
A. always watching where you step
B. not engaging in horseplay on the job
C. not placing tools where they may fall and injure other workers
D. all of the above

__________ **8.** ___ is replacing the soil around the outside of a completed foundation wall.
A. Trenching
B. Excavating
C. Grading
D. Backfilling

__________ **9.** The recommended slope for excavations under average soil conditions is ___°.
A. 33
B. 45
C. 70
D. 90

__________ **10.** Soldier piles are used to ___.
A. support the floor of a building
B. shore vertical walls of excavations
C. stiffen formwork
D. shore trenches

__________ **11.** ___ is a disease of the lungs caused by inhaling dust containing crystalline silica particles.
A. Repsiratory illness
B. Asbestosis
C. Silicosis
D. Carcinogen poisoning

__________ **12.** A(n) ___ is any cut, depression, or trench in a ground surface.
A. foundation
B. benchmark
C. slope
D. excavation

__________ **13.** Ladders should be placed so that a worker will not have to travel more than ___′ in a lateral direction to climb out of the trench.
A. 5
B. 10
C. 15
D. 25

__________ **14.** Guardrails are placed across openings for exterior doors if there is a drop of more than ___′.
A. 3
B. 4
C. 6
D. 7

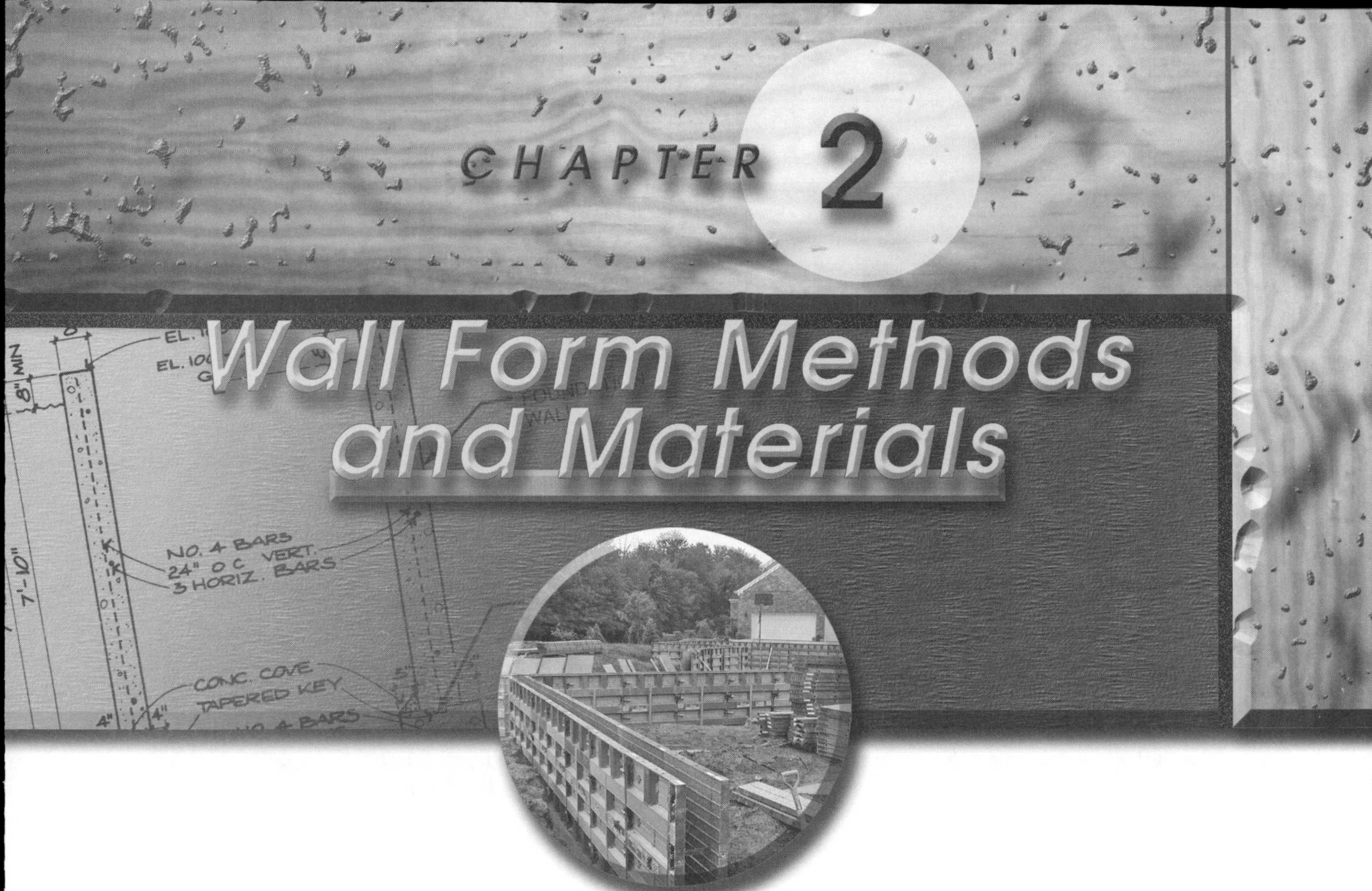

CHAPTER 2
Wall Form Methods and Materials

Concrete is used for various types of walls, including foundation walls of residential and heavy concrete structures. A foundation wall rests on a foundation footing, and ranges in height from low walls for crawl space foundations to higher and thicker walls for high-rises. The type of foundation footing used depends on the bearing capacity of the soil and the type of structure to be constructed.

A concrete wall is formed by placing concrete in a plastic state into wall forms and allowing it to set. Built-in-place wall forms and panel forms are used for residential and light commercial construction. Panel forms and ganged panel forms are commonly used for heavy construction projects. A built-in-place form is constructed in its final position. Panel forms are prebuilt panel sections constructed with studs and a top and bottom plate. Ganged panel forms are large forms constructed by bolting prefabricated panel forms together.

Materials for traditional forming methods are plywood panels for sheathing, and dimension lumber to stiffen and brace the form walls. There has been a major increase in the use of prefabricated form systems in recent years. One system consists of plywood panels set in a metal frame with horizontal metal stiffeners. All-metal systems of aluminum or steel are also widely used.

Safe and established construction procedures must be followed when constructing wall forms. Consult the American Concrete Institute (ACI) or Occupational Safety and Health Administration (OSHA) for information regarding safe construction procedures.

WALL FORM CONSTRUCTION

Wall forms are constructed in various shapes, heights, and thicknesses. Two methods are commonly used for high walls. One method consists of sheathing stiffened by studs and double walers. A second method consists of sheathing stiffened by single walers and strongbacks. **See Figure 2-1.** Both systems require braces to hold the walls in position and devices to tie the opposite walls together. Variations of these two methods are used in the construction of low wall forms.

Wall Form Systems

Wall form systems must be constructed to the same shape, width, and height as the foundation walls. The wall form must be supported and braced so that it is correctly aligned and adequately supports all vertical and lateral loads imposed on it. The form must be designed so that it can be conveniently stripped (removed) from the wall after the concrete has set. The form must be tight enough to prevent excessive leakage at the time the concrete is placed. Leakage can result in unsightly surface ridges, honeycombs, and sand streaks after the concrete has set. The wall form must be able to safely withstand the pressure of concrete at the time it is placed. Short cuts taken, lack of materials, and inadequate bracing can cause the form to collapse or move during concrete placement.

Symons Corporation

Wall form systems are designed to be flexible to fit various shapes of foundations and foundation walls.

For residential and other light construction projects, built-in-place or panel form systems are commonly used. Built-in-place forms are constructed in place on the job site. Panel systems consist of prefabricated panel sections framed with studs and a top and bottom plate. Heavy construction projects usually require panel forms or ganged panel forms. A ganged panel form consists of a number of prefabricated panel forms tied together to create a much larger single panel.

Wall forms may deflect when sheathing of inadequate thickness is used, or when studs or walers are spaced too far apart. Wall forms must be constructed so that they can contain the weight of the concrete without deflection or collapse.

Built-in-Place Forms. Built-in-place forms are constructed in place over a footing or concrete slab that acts as a platform for the wall form. The most common procedure is to fasten a base plate to the footing or concrete slab for the outside wall with powder-actuated fasteners or concrete nails. Wood studs are nailed to the base plate and tied together with a temporary ribbon board. The sheathing is then nailed to the studs and the walers are placed. **See Figure 2-2.** After securing the walers, wall ties are inserted through predrilled holes in the sheathing. Wall ties extending from the outside walls are inserted through holes in the inside wall panel. The inside form walls are constructed in a manner similar to the outside form walls. The inside walls are then aligned and braced.

A simplified built-in-place plank system for forming low walls is popular in many areas. The forms are constructed of 2″ thick planks (actual size 1½″) and tied together with 2 × 4 stakes or uprights. A flat form tie secured with tapered wedges spaces and holds the opposite walls together. **See Figure 2-3.** Standard-sized ties can be obtained for wall thicknesses ranging from 6″ to 16″. The simplified plank system is commonly used for low foundation walls and footings not exceeding 4′ in height. However, forms for higher walls may also be constructed using this method.

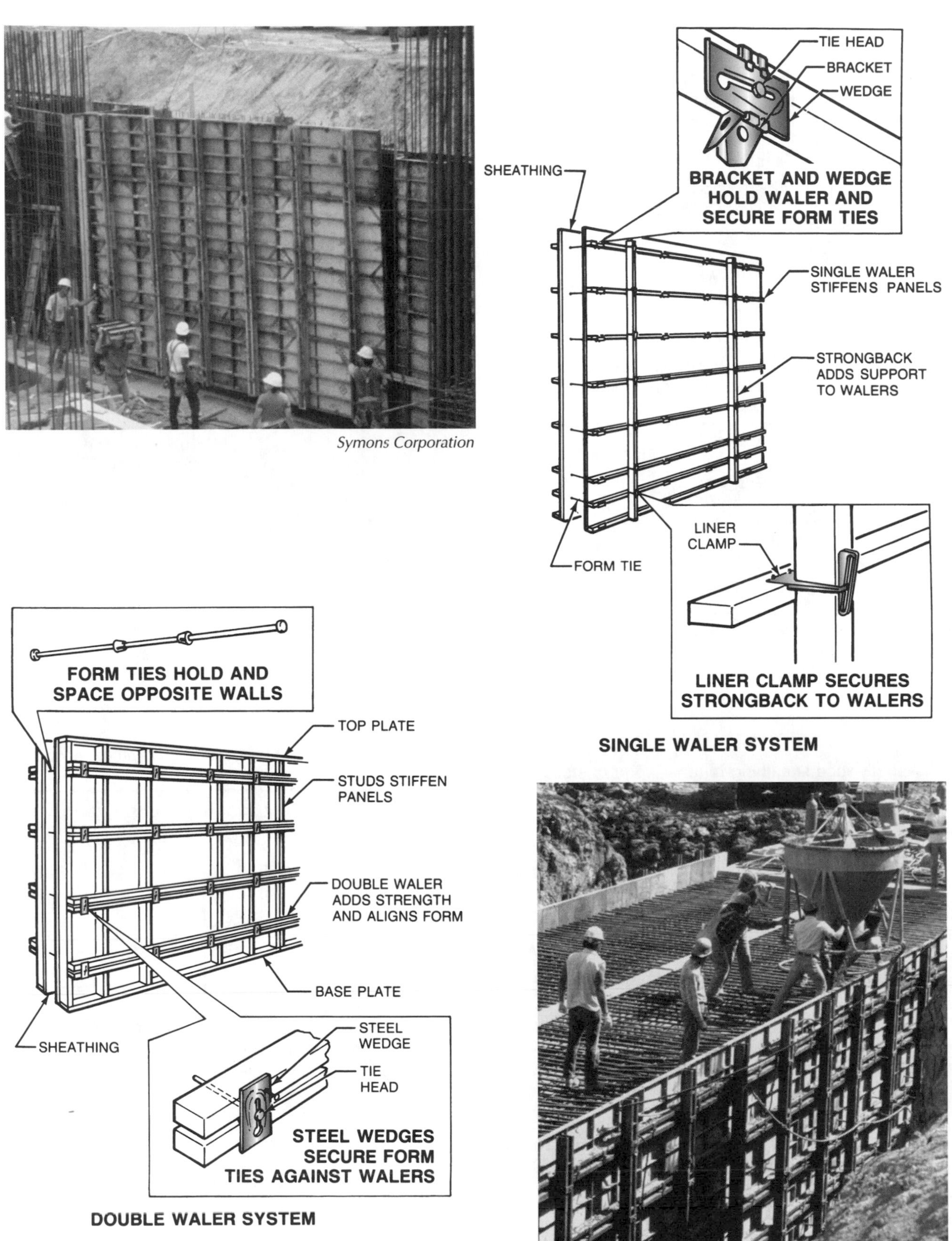

Figure 2-1. Single or double waler systems may be used to form high foundation walls.

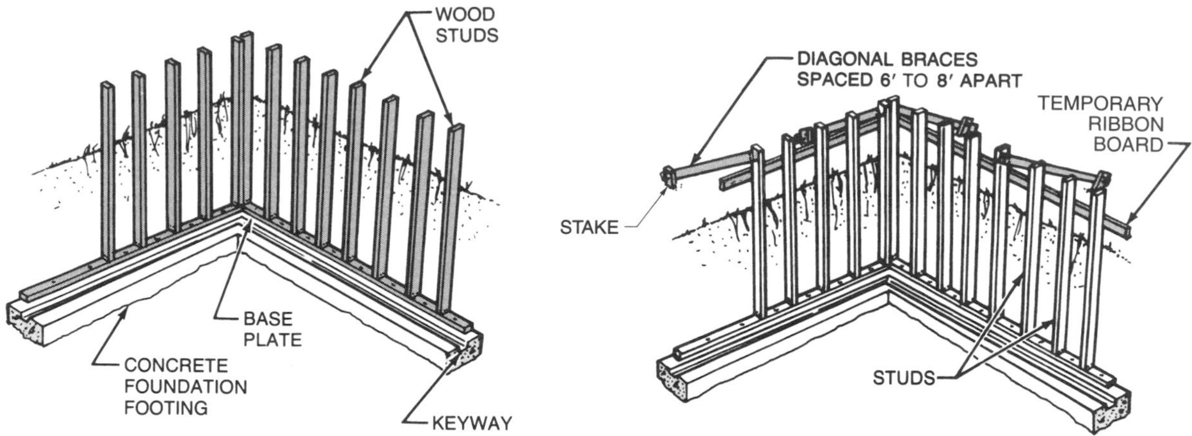

1. Fasten base plate for outside form wall to foundation footing with concrete nails or powder-actuated fasteners. Toenail studs to base plate.

2. Tie studs together with temporary ribbon board. Secure studs and ribbons with diagonal braces every 6′ to 8′.

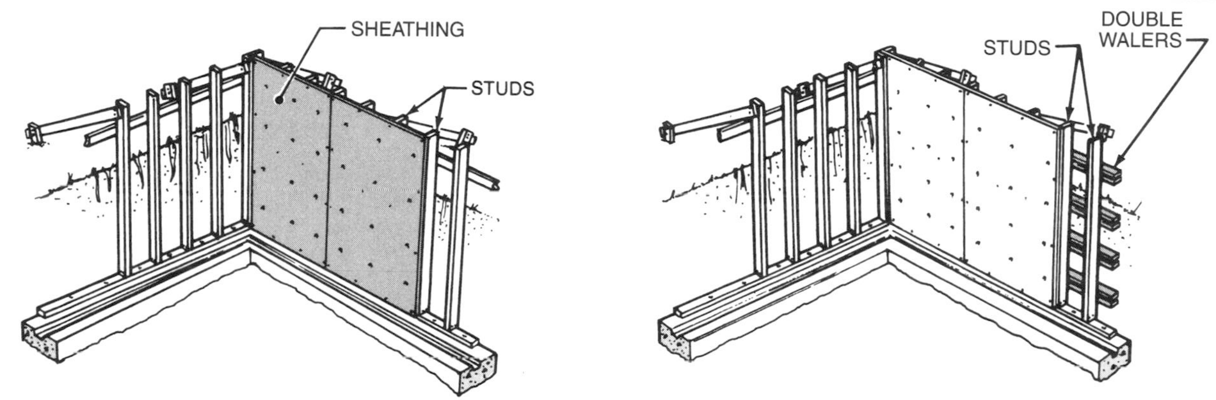

3. Nail plywood sheathing to inner edge of studs.

4. Attach double walers to outer edge of studs. Align and brace wall form.

Figure 2-2. Built-in-place wall forms are constructed over a foundation footing or floor slab. Plywood sheathing is reinforced with studs, walers, and braces.

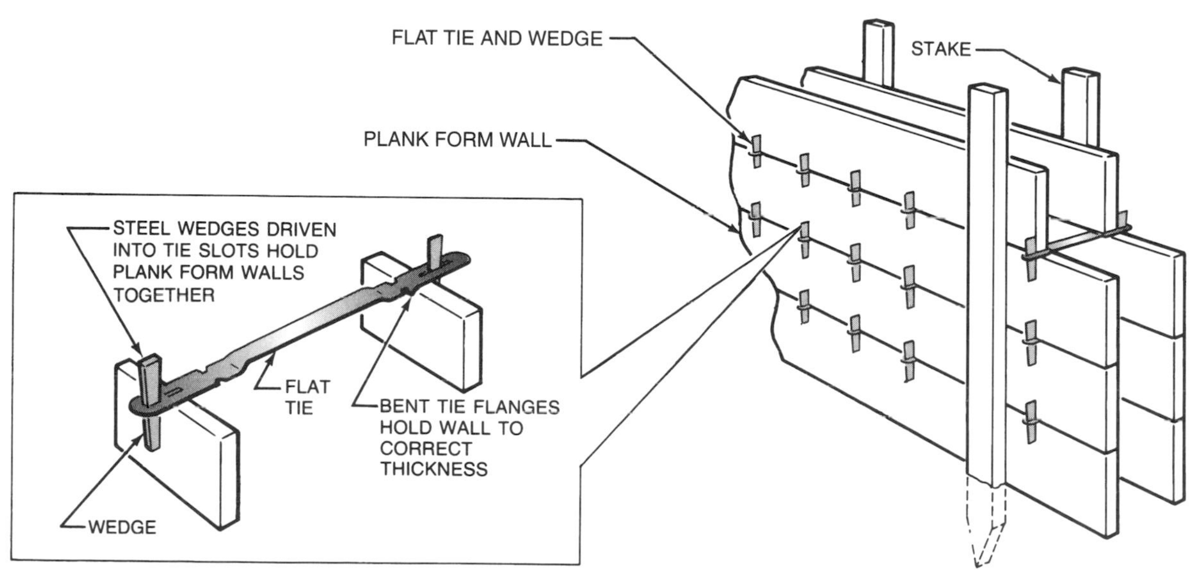

Figure 2-3. The built-in-place plank method is used to build low wall forms.

Panel Forms. Panel form systems consist of prebuilt panel sections framed with studs and a top and bottom plate. Panel form systems increase the speed and efficiency of construction. The panel sections can be built in the shop or on the job, and with proper care can be reused many times. If many similar panel sections are required, a template table facilitates the construction of the panels. A template table is constructed of plywood nailed to a frame and supported by legs. The studs are placed between cleats laid out to the correct spacing of the studs, and the sheathing is nailed to the studs. **See Figure 2-4.**

After the panel sections are prefabricated, they are placed into position. Smaller panels can be placed by hand; larger panels require lifting equipment. Panel sections are fastened to the footings by driving concrete nails or powder-actuated fasteners through the base plate of the panel section. The panels are fastened to each other with bolts or 16d duplex nails. A filler panel is placed in any leftover space that is less than a full panel width. Walers are then nailed to the panel studs to keep the panels aligned. Vertical strongbacks, if required, are then fastened to the walers. After securing the walers or strongbacks, wall ties are inserted through predrilled holes in the panels. The inside walls are constructed in a manner similar to the outside walls. As the inside panels are tilted into place, the wall ties are inserted through predrilled holes. Braces are attached to the walers or strongbacks and the wall is aligned and braced. **See Figure 2-5.**

Symons Corporation

Multiple prefabricated panel forms are used on construction sites for high walls. Forms can be constructed to the required dimensions and reused if cared for properly.

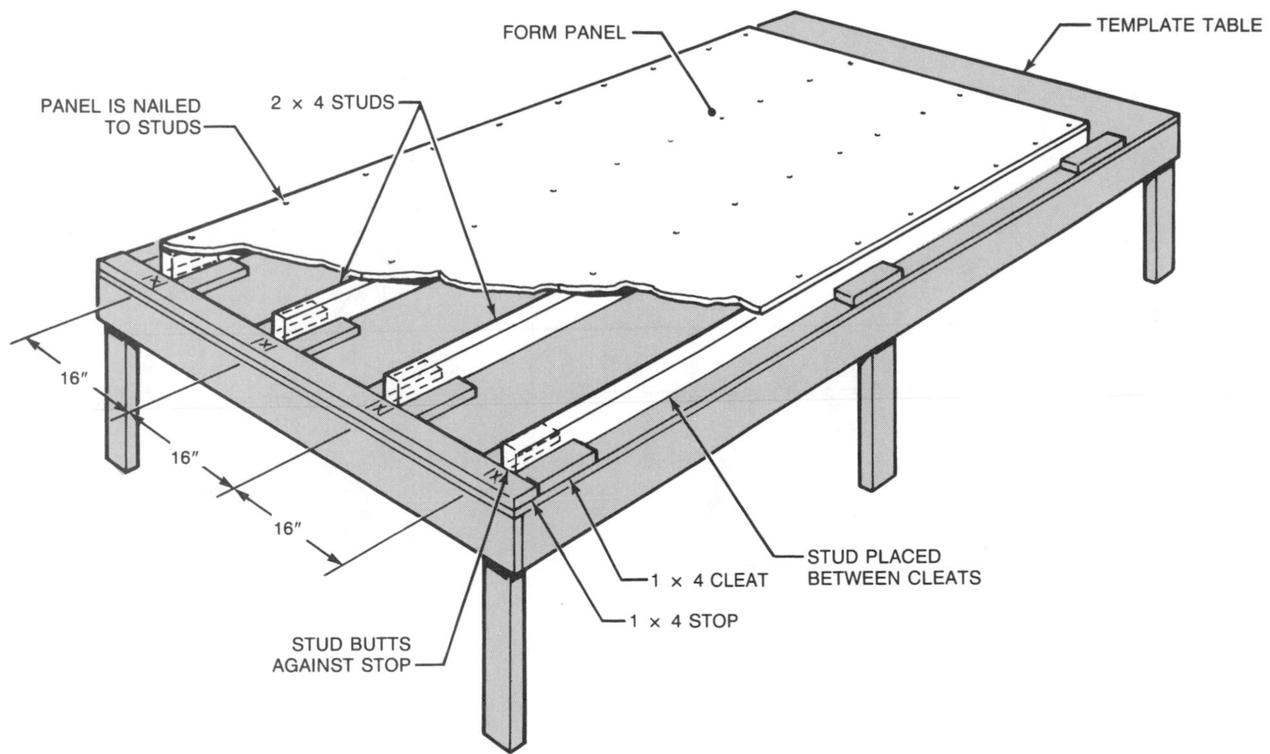

Figure 2-4. A template table facilitates construction of prefabricated panel form sections.

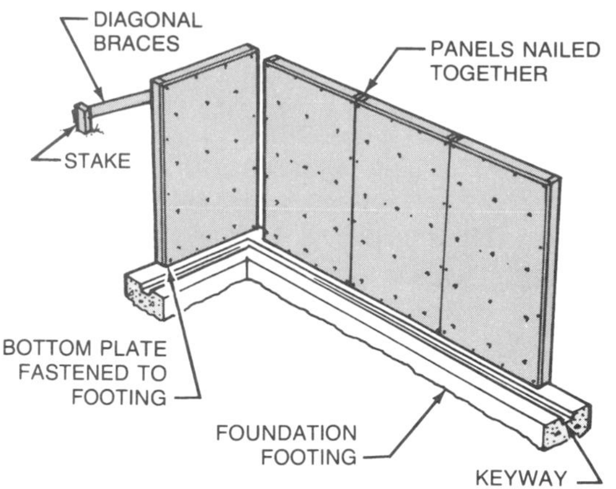

1. Fasten bottom plate of panel to foundation footing with concrete nails or powder-actuated fasteners. Secure panels in position with diagonal braces. Fasten panels together with 16d duplex nails or bolts.

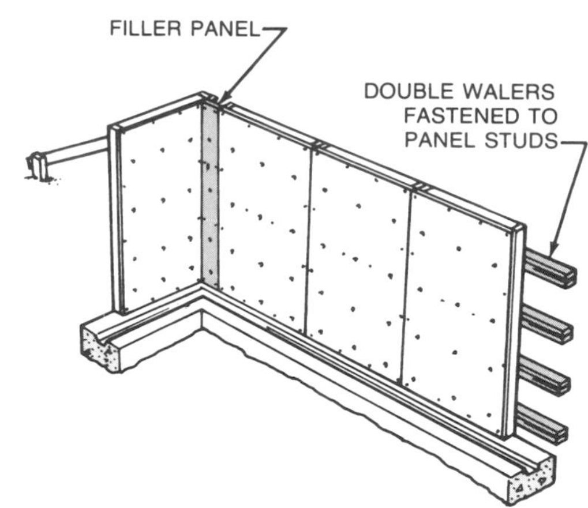

2. Place filler panel in any openings that are not a full panel width. Fasten double walers to panel studs. Align and brace wall form.

Figure 2-5. A panel form system is a cost-effective method of constructing wall forms.

Single Wall Forms. Single wall forms are built when conditions make it difficult to remove one side of a form after the concrete has set. This might occur where a foundation wall is to be constructed very close to or against the existing foundation of an adjoining building, or where a wall is constructed next to solid rock or extremely hard soil. Single wall forms consist of plywood sheathing, studs, walers, strongbacks, and braces. Single wall forms must be heavily braced to hold them in position. **See Figure 2-6.** Single wall forms may be built-in-place, or prefabricated panels can be used. Prefabricated steel-framed plywood panels are also available.

When an addition is constructed adjacent to an existing building, single wall forms allow the new foundation to be constructed abutting the existing foundation.

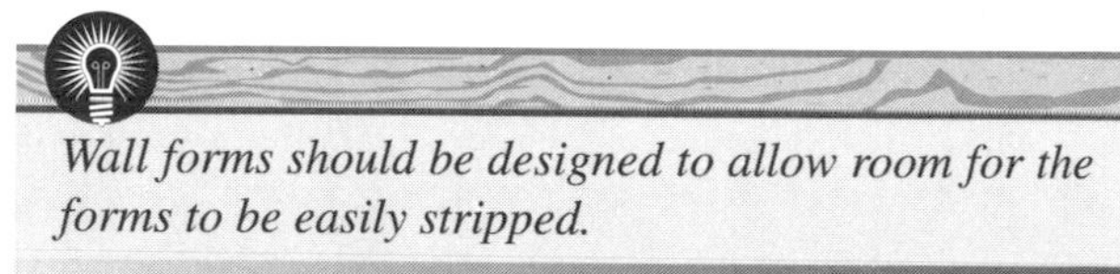

Wall forms should be designed to allow room for the forms to be easily stripped.

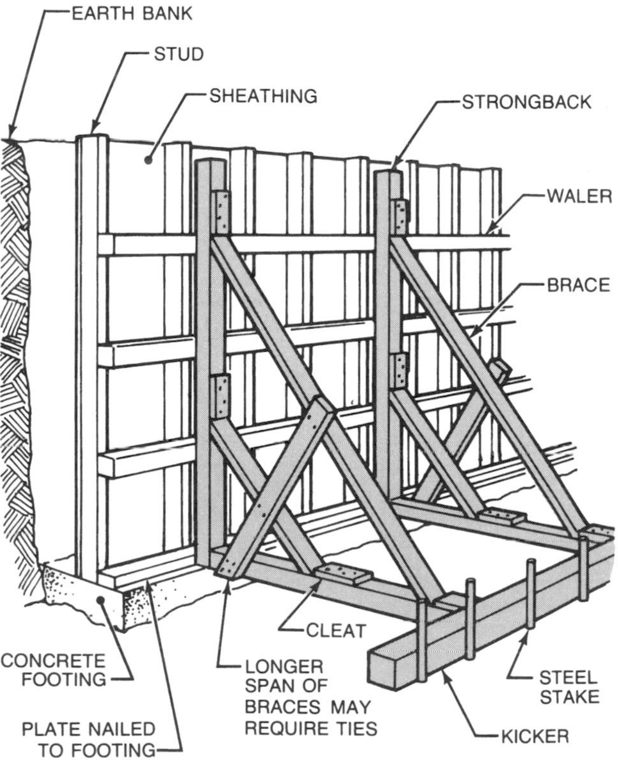

Figure 2-6. Single wall forms must be heavily braced to secure them in position.

Pilasters. A pilaster is a rectangular column incorporated with a concrete wall and used to strengthen the wall at areas of concentrated structural loads. Pilasters are also constructed to support the ends of beams or girders spanning concrete walls. Pilaster forms constructed of plywood sheathing, cleats, studs, and walers are erected along with the wall form. Kickers or patented ties are used to strengthen the corners of the form. **See Figure 2-7.** The footing should project out to form a base for the pilasters wherever they are placed in foundation walls.

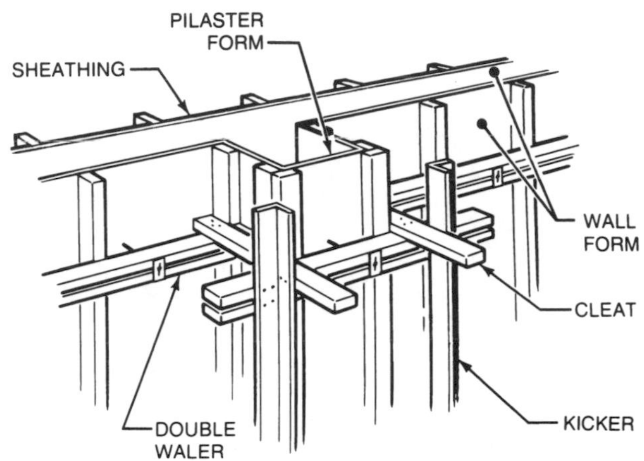

Figure 2-7. A pilaster is a rectangular column joined to a wall.

Wall Openings

When door and/or window openings occur in a cast-in-place concrete wall, preparations must be made inside the forms for these openings. This is done by placing and attaching reinforced metal, plastic, or wooden frames to the outside form walls before the inside form walls are erected.

Metal and plastic frames are commonly used for openings. Wood frames (bucks) were traditionally used for openings, and are still used, although not as often.

Metal Frames. Metal frames are typically made of steel or aluminum. Plastic frames are typically high-density polyethylene (HDPE). The frame is fastened to the outside form wall after the wall has been leveled and braced. A commonly used type of frame has brackets extending from the frame. The brackets are embedded in the concrete when the concrete is placed within the form walls, thus permanently securing the frame that remains in place after the form walls are removed. **See Figure 2-8.**

Western Forms

Figure 2-8. Permanent metal or plastic frames placed in the wall forms must be carefully plumbed, aligned, and braced.

Wood Frames. A typical wood frame (buck) design consists of an outside frame reinforced by 2 × 4s placed on edge, and horizontal cross braces. The frame may also be built of 2″ thick members or ¾″ plywood. More bracing is required with thinner material. Arches are formed over door and window openings by fastening an arched section at the top of the door or window buck. The arched section is laid out on two pieces of plywood that are then secured to the top of the buck. The arched form is then enclosed with ¼″ thick plywood. **See Figure 2-9.** If a recess is required to receive the frame, a recess strip is nailed to the outside of the buck's frame. If a wood nailing strip is to be provided in the concrete, a wedged piece is attached to the frame. An inspection pocket may be cut out at the bottom of a window buck to observe the flow and consolidation of the concrete beneath the buck. When the concrete reaches the bottom of the buck, the inspection pocket is replaced and cleated down.

Door and window bucks are made to accommodate any size and shape of door or window. Bucks should be prefabricated to rough opening size using dimensional lumber or a metal framing system. When installing door and window bucks, the openings needed for utilities should also be installed.

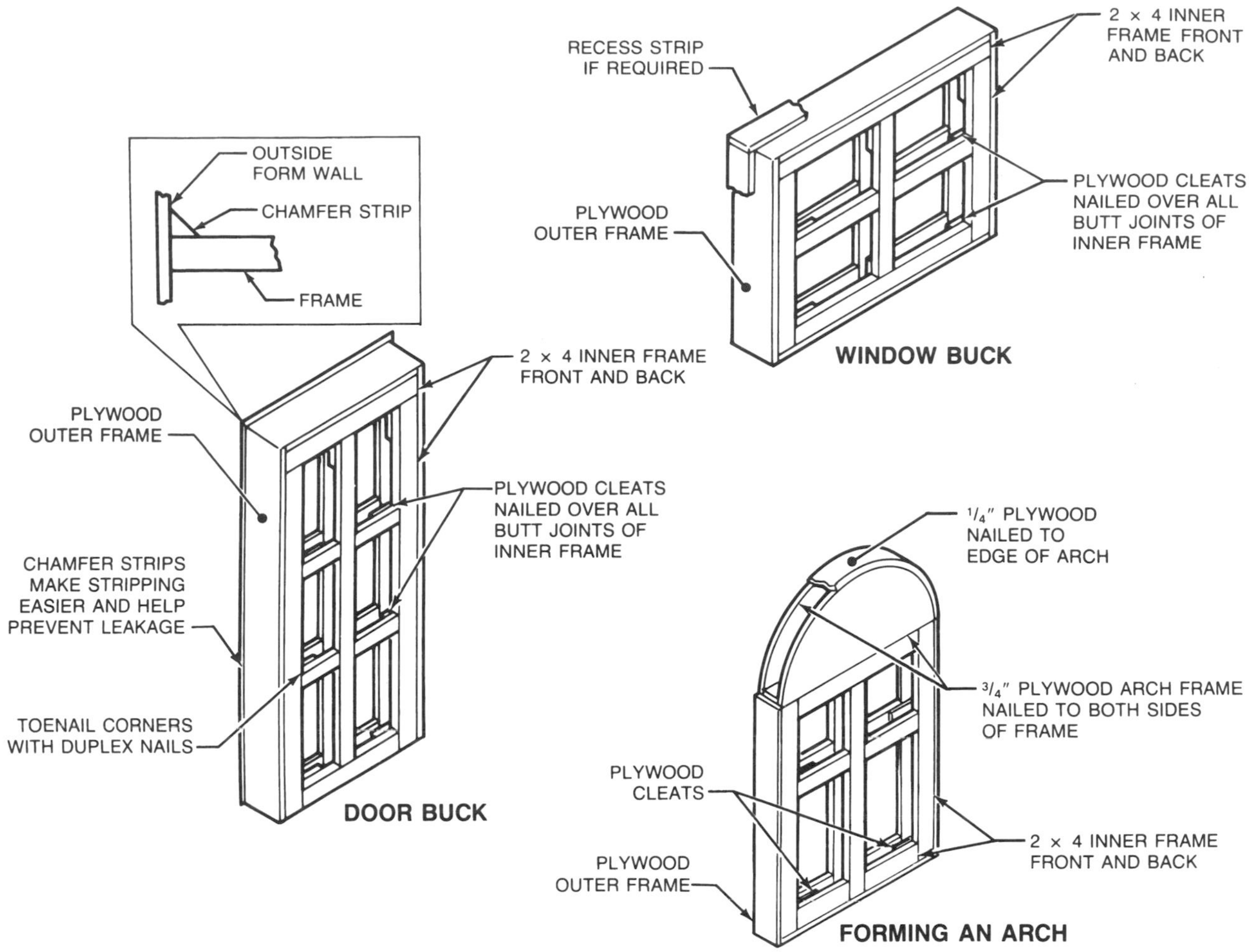

Figure 2-9. Bucks are heavily braced frames placed inside the wall form to provide door and window openings in concrete walls.

Door and window bucks are usually placed after the outside form walls have been erected. The bucks can be laid out and built against the wall. **See Figure 2-10.** Many form builders, however, prefer to prefabricate the bucks and then attach them to the form wall.

A wooden door or window buck may be designed so that a nailing strip will remain in the concrete after the buck is removed. This will provide a nailing surface for the finish jamb or window to be attached to in the opening. Another common procedure is to create an opening in which the metal door or window frame can be bolted to expansion bolts placed in the concrete.

Small openings for large-diameter pipes, vents, heating and ventilating ducts, and utilities may also be required. Small vent, duct, and pipe openings can be formed by constructing wood box frames to the size of the openings. Cellular plastic blockouts and fiber and sheet metal sleeves are also used.

MATERIALS AND COMPONENTS

Sheathing (usually plywood) and lumber (2 × 4s, 2 × 6s, 4 × 4s, etc.) used for structural form components such as studs, walers, strongbacks, and bracing are used to construct wall forms. The thickness of the sheathing and the size and spacing of the structural components are determined by the anticipated loads and pressures exerted on the form. Forms for small and light wall forms are designed based on established procedures. However, wall forms for larger and more complicated structures may be designed by a structural engineer.

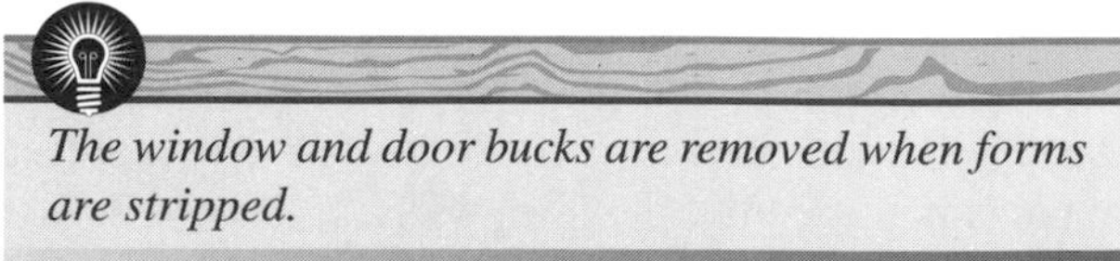

The window and door bucks are removed when forms are stripped.

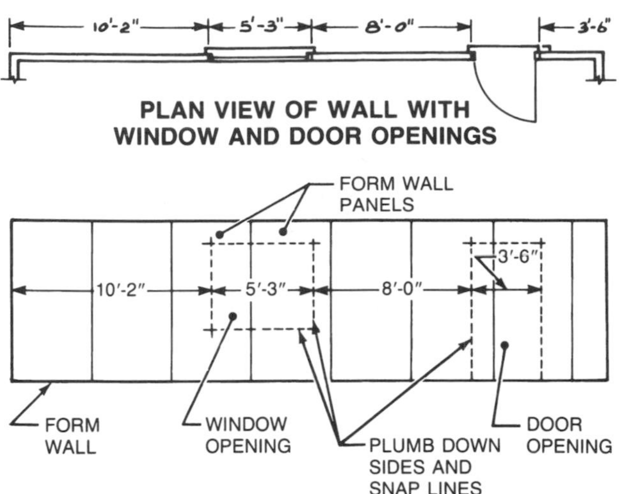

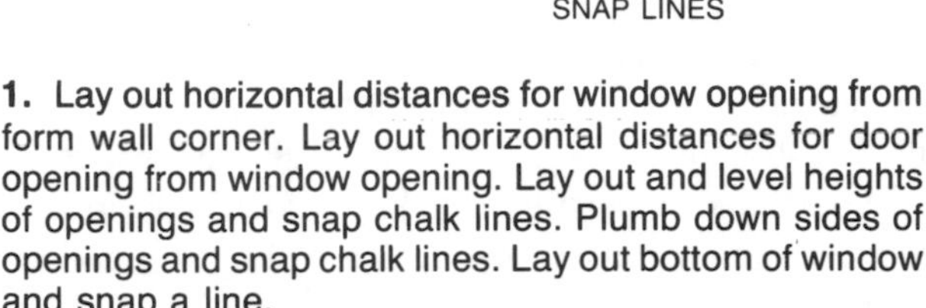

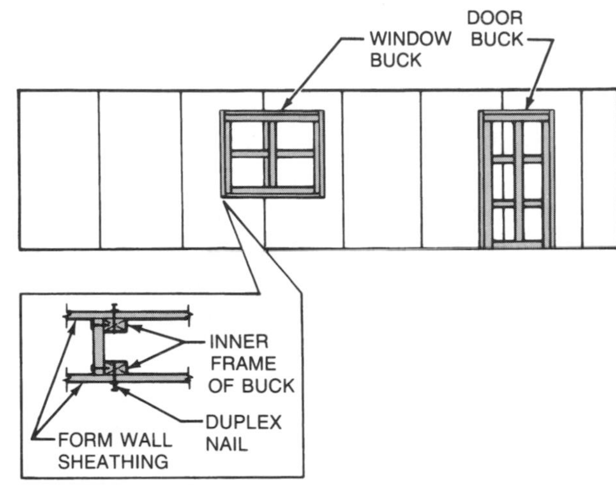

1. Lay out horizontal distances for window opening from form wall corner. Lay out horizontal distances for door opening from window opening. Lay out and level heights of openings and snap chalk lines. Plumb down sides of openings and snap chalk lines. Lay out bottom of window and snap a line.

2. Set door and window bucks to snapped lines. Drive duplex nails through the form wall sheathing into the inner frames of the bucks.

Figure 2-10. Door and window buck placement is determined using the plan view.

Plywood

Plywood sheathing is most commonly used to surface wall forms. Plywood is available in large panels, which reduces labor costs involved in constructing and stripping forms. In addition, plywood produces fewer joint impressions on the finished concrete walls, thus reducing the cost of finishing and rubbing the surface.

Plywood is a manufactured panel product made up of veneers. A *veneer* is one of an odd number of thin layers of wood that are glued together under intense heat and pressure. The odd number of layers allows the front and back veneer grain to run in the same direction. Each layer is placed with the grain running at a right angle to the adjacent layer. This process, known as cross-lamination, increases the strength of the panel and minimizes shrinking and warpage. The outside surfaces of a plywood panel are the face veneer and back veneer. The inner veneers are the crossbands and the center layer is the core. **See Figure 2-11.**

Most plywood panels used in formwork are manufactured from softwood lumber species such as Douglas fir. Other woods used are pine, spruce, larch, and hemlock. The most common form panel thicknesses are ½", ⅝", and ¾". Standard size 4' × 8' sheets are used most often; however, sheets 5' wide, and lengths ranging from 5' to 10', are also available. Some form builders prefer 5' × 10' panels because they can be cut to 2'-6" × 10' sections and can be handled more easily by one person.

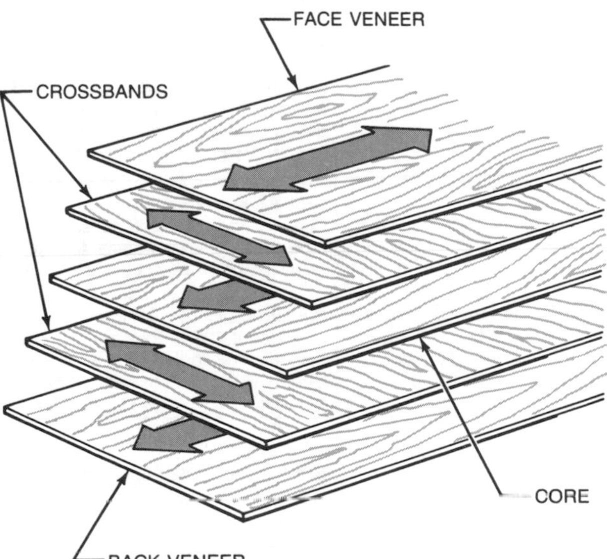

Figure 2-11. Plywood consists of an odd number of veneers. The grain direction of one layer runs perpendicular to the direction of the adjacent layer.

The manufacture of plywood is governed by strict product standards established by the U.S. Department of Commerce in cooperation with industry associations such as the APA—The Engineered Wood Association. The two major types of plywood are exterior and interior. Exterior plywood is bonded with waterproof glue and interior plywood is bonded with water-resistant glue. Each type is further broken down into grades that are based on the condition and appearance of the outer veneers. Grade-use guides that provide information about plywood grades are published by various industry associations.

Plyform®. *Plyform®* is a plywood product specifically designed for concrete formwork. Plyform® is recommended for most forming operations. The *Grade-Use Guide for Concrete Forms* identifies three primary grades of Plyform as B-B Plyform®, High Density Overlaid (HDO) Plyform®, and Structural 1 Plyform®. Additional information regarding Plyform, such as typical trademarks and recommendations for usage, may also be found in the guide. **See Figure 2-12.**

High Density Overlaid (HDO) Plyform is considered the "workhorse" of the concrete industry. With proper care, HDO can be reused more than 20 times. HDO produces uniform, smooth, glossy surfaces. Its tough resin overlay resists the alkali elements of concrete and the scouring action of the aggregates. Little surface discoloration occurs because the overlaid veneer is waterproof and nonabsorbent, thus keeping the hydration process uniform. *Hydration* is a chemical reaction between cement and water that produces hardened concrete.

The smooth face of HDO Plyform® aids the removal of air during the consolidation process, reducing bugholes and honeycombing. Stripping and cleaning of the panels are much easier. HDO Plyform® is nonabsorbent so form-release agents stay on the surface of the panels.

Textured Plywood. Textured plywood is used to produce special surface effects such as wood grain and boards. **See Figure 2-13.** A textured wall surface is produced by using ¼″ textured plywood as a liner for a thicker structural panel, or a thicker textured form panel without a liner. By using a thicker textured form panel, labor is minimized. However, the textured panels can only be reused a limited number of times because the textured surfaces are easily damaged when the panels are stripped from the concrete walls.

GRADE-USE GUIDE FOR CONCRETE FORMS*

Use These Terms When Specifying Plywood	Description	Typical Trademarks	Veneer Grade		
			Faces	Inner Plies	Backs
APA B-B Plyform Class I & II†	Specifically manufactured for concrete forms. Many reuses. Smooth, solid surfaces. Mill-oiled unless otherwise specified.	APA THE ENGINEERED WOOD ASSOCIATION PLYFORM B-B CLASS 1 EXTERIOR 000 PS 1-95	B	C	B
APA High Density Overlaid Plyform Class I & II†	Hard, semi-opaque resin-fiber overlay, heat-refused to panel faces. Smooth surface resists abrasion. Up to 200 reuses. Light application of releasing agents recommended between pours.	HDO · B-B · PLYFORM I · 60/60 · EXT · APA · 000 · PS 1-95	B	C-Plugged	B
APA Structural 1 Plyform†	Especially designed for engineered applications. All Group 1 species. Stronger and stiffer than Plyform Class I and II. Recommended for high pressures where face grain is parallel to supports. Also available with High Density Overlay faces.	APA THE ENGINEERED WOOD ASSOCIATION STRUCTURAL I PLYFORM B-B CLASS 1 EXTERIOR 000 PS 1-95	B	C or C-Plugged	B
Special Overlays, Proprietary panels and Medium Density Overlaid plywood specifically designed for concrete forming†	Produces a smooth uniform concrete surface. Generally mill treated with form release agent. Check with manufacturer for specifications, proper use, and surface treatment recommendations for greatest number of reuses.				
APA B-C EXT	Sanded panel often used for concrete forming where only one smooth, solid side is required.	APA THE ENGINEERED WOOD ASSOCIATION B-C GROUP 1 EXTERIOR 000 PS 1-95	B	C	C

*Commonly available in 19/32″, 5/8″, 23/32″, and 3/4″ panel thicknesses (4′ × 8′).
†Check dealer for availability in your area.

APA—The Engineered Wood Association

Figure 2-12. The grade-use guide is used to determine the type of plywood to use in the construction of forms. The trademarks are usually stamped on the back veneer.

Symons Corporation

Figure 2-13. Special surface designs, such as a brick pattern, are produced with textured plywood forms.

Deflection and Panel Strength. Deflection (bending) of a wall form occurs when panels of inadequate thickness are used, or when the stiffeners (studs or walers) are spaced too far apart. A wall form must be constructed to hold the concrete to a straight and true surface. The required panel thickness for a particular form design must take into consideration the maximum concrete pressure anticipated for that form, as well as the spacing and type of stiffeners used.

An important factor when positioning plywood in the wall form is the direction of the grain in relation to the panel stiffeners. Plywood has much greater strength when the face grain is placed perpendicular rather than parallel to the stiffeners. **See Figure 2-14.**

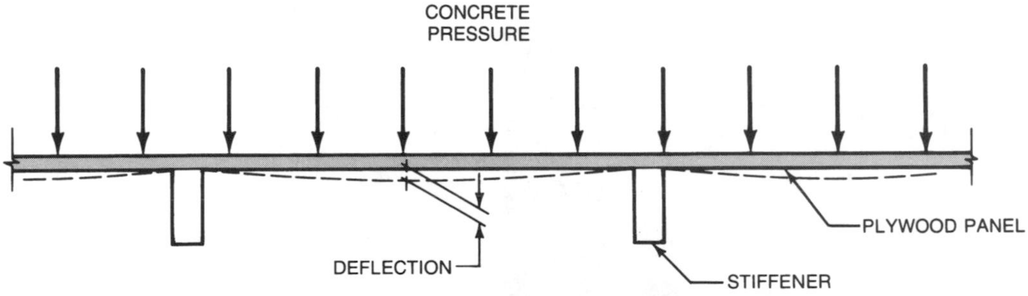

CONCRETE PRESSURE

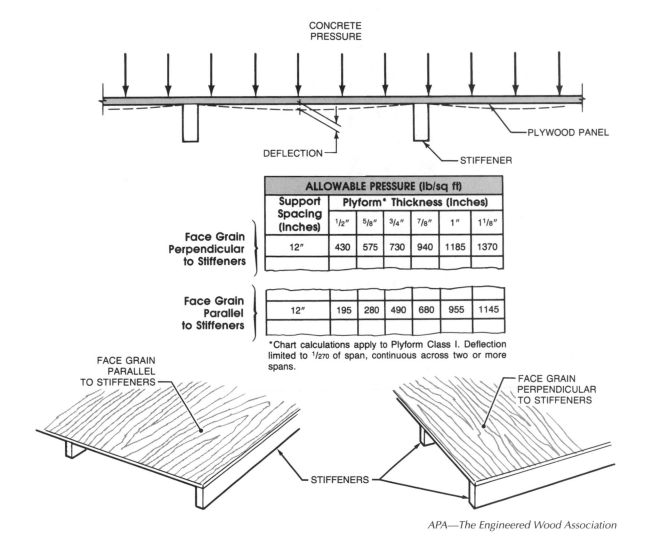

PLYWOOD PANEL

DEFLECTION

STIFFENER

ALLOWABLE PRESSURE (lb/sq ft)							
	Support Spacing (Inches)	Plyform* Thickness (Inches)					
		1/2"	5/8"	3/4"	7/8"	1"	1 1/8"
Face Grain Perpendicular to Stiffeners	12"	430	575	730	940	1185	1370
Face Grain Parallel to Stiffeners	12"	195	280	490	680	955	1145

*Chart calculations apply to Plyform Class I. Deflection limited to 1/270 of span, continuous across two or more spans.

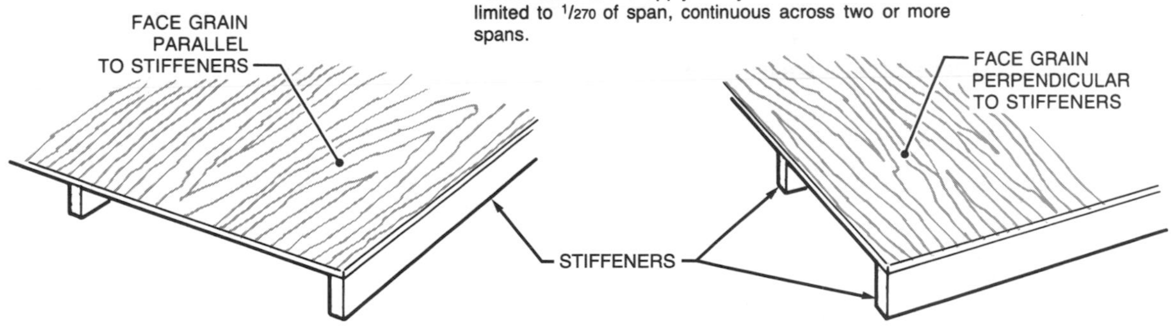

FACE GRAIN PARALLEL TO STIFFENERS

FACE GRAIN PERPENDICULAR TO STIFFENERS

STIFFENERS

APA—The Engineered Wood Association

Figure 2-14. Plywood has much greater strength when the face grain is placed perpendicular to the stiffeners rather than parallel.

Curved Forms. Plywood and Plyform® are also used to sheathe curved forms. The bending capacity of dry plywood is based on the minimum bending radius. The *minimum bending radius* is the smallest bending radius that the plywood can be subjected to without structural damage. **See Figure 2-15.** A 1/4″ thick panel bent across the grain has a 2′ bending radius and can be shaped to a curve having a 2′ radius. Shorter radius curves can be produced by bending plywood across the grain rather than with the grain.

Gates & Sons, Inc.

MINIMUM BENDING RADII		
Plywood Thickness (in.)	**Across the Grain (ft)**	**Parallel to Grain (ft)**
1/4	2	5
5/16	2	6
3/8	3	8
1/2	6	12
5/8	8	16
3/4	12	20

APA-The Engineered Wood Association

Figure 2-15. Plywood is bent to form curves. Shorter radius curves are produced by bending plywood across the grain rather than with the grain.

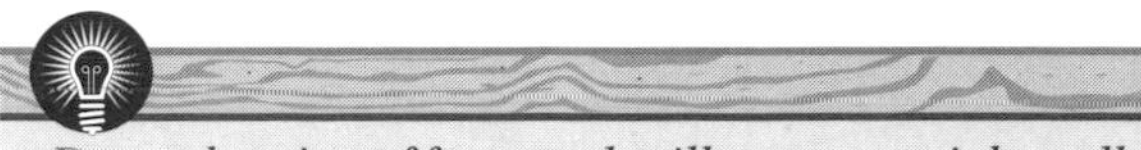

Proper bracing of formwork will ensure straight walls. A variety of wood, steel, and aluminum bracing systems are available. All window and door openings are braced and uninterrupted sections of wall are braced every 8′ to 10′.

Gradual curves can be made with dry panels ranging in thickness from 1/4″ to 3/4″. Shorter radius curves can also be constructed by wetting or steaming the panel, or cutting spaced kerfs in the panel. Another curving method is to use two thinner panels instead of a single thick panel.

Preparation and Maintenance. Proper preparation and maintenance allows plywood panels to be used several times. Oiling the faces of plywood panels before use reduces moisture penetration that may damage the panel. A liberal amount of oil should be applied a few days before the plywood panel is to be used, and then wiped clean so only a thin layer remains. The oil also acts as a release agent, making it easier to strip the form panels from the concrete walls. Other release agents including waxes, oil emulsions, cream emulsions, and water emulsions can also be sprayed on the panel.

After a panel has been stripped from the concrete wall, it should be inspected for wear and cleaned with a fiber brush. Necessary repairs such as patching holes with patching plaster or plastic wood should be made. Tie holes may be patched on the inside of the forms with metal plates. The panel should then be lightly oiled. Panels should be laid flat and face-to-face when stored, and kept out of the sun and rain. If this is not possible, the panels should be protected by a loose cover that allows air circulation without heat buildup.

Framework, Stiffeners, and Bracing

Framework, stiffeners, and bracing lumber for concrete forms should be straight and structurally sound. Partially seasoned (dry) stock is recommended because fully dried lumber tends to swell when it becomes wet and creates distortions when aligning the forms. Completely unseasoned (green) lumber dries out and warps during hot weather, which also causes distortions.

Softwood lumber species such as Douglas fir, southern pine, spruce, and hemlock are generally used for structural purposes. When ordering stock, the lumber estimate should be based on minimal waste because much of this material is reused after the forms are stripped from the finished concrete walls.

The most frequently used lumber stock is 2 × 4; however, 2 × 6s, 4 × 4s, and 2″ planks of various widths are also common. Actual lumber sizes differ from nominal lumber sizes; for example, a 2 × 4 is actually 1 1/2″ × 3 1/2″. This difference is a result of the amount of waste removed from the rough lumber during surfacing in the planing mill, and the anticipated shrinkage while it is drying. (See Standard Lumber Size table in Appendix B.)

The key factor in determining the size and spacing of stiffeners for a form wall is the pressure exerted when the concrete is being placed into the form. Other factors that determine the size and spacing of stiffeners include the rate of placement, ambient temperature, and consistency of the concrete.

Various methods are used to stiffen and brace low wall forms. One method requires studs or stakes, a single waler placed toward the top edge of the form, wood ties, and braces. Another low wall forming method requires 2″ thick planks, stakes, wood ties, and braces. **See Figure 2-16.**

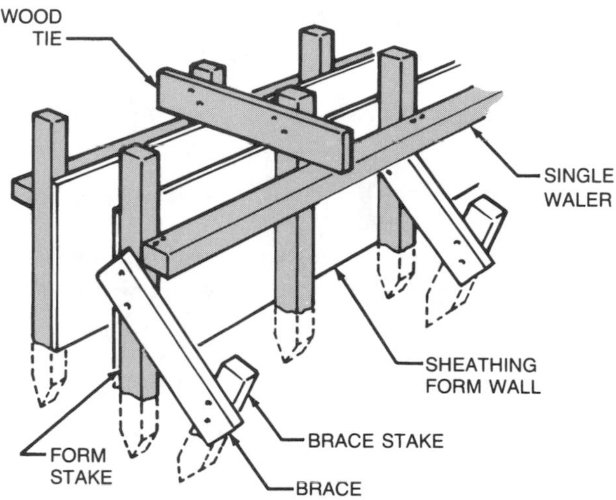

PLYWOOD STIFFENED BY FORM STAKES, A SINGLE WALER, AND WOOD TIES

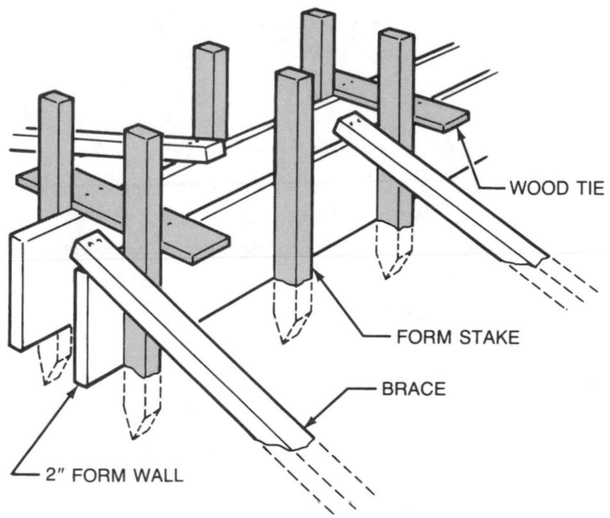

2″ THICK PLANKS STIFFENED BY FORM STAKES AND WOOD TIES

Figure 2-16. Low wall forms are stiffened by using stakes, studs, or wood ties.

Studs, Walers, and Strongbacks. Most high-wall forming systems are constructed and stiffened with a combination of studs and walers, or walers and strongbacks. An older but still widely used method for stiffening high form walls consists of vertical studs and double walers (wales). The studs give rigidity to the form panel and are spaced 12″, 16″, or 24″ apart, depending on the anticipated concrete pressure. The double walers (usually 2 × 4s) are positioned horizontally and are fastened to the studs with nails, clips, or brackets. The double walers reinforce the studs, align and tie together the form panels, and provide a wedging surface for patented wall ties. Typical spacing for double walers is to place walers within 12″ of the top and bottom of the wall, and no more than 24″ OC for the intervening rows. However, the spacing may vary, depending on concrete pressure and the size and spacing of other materials. For walls where precise alignment is required, vertical strongbacks may be fastened behind the double walers.

A newer forming method eliminates the studs and uses only single walers strengthened by vertical strongbacks. Eliminating the studs requires the walers to be spaced closer together. Single walers are placed within 8″ of the top and bottom of the wall, and no more than 16″ OC for the intervening rows. **See Figure 2-17.** The single walers may be spaced more closely to accept greater concrete pressure. Under normal circumstances, vertical strongbacks placed over single walers are spaced every 6′, or closer if conditions require.

Simpson Strong Tie Company

Single walers are a commonly used forming method. Since single walers are used without studs, walers must be spaced no more than 16″ OC.

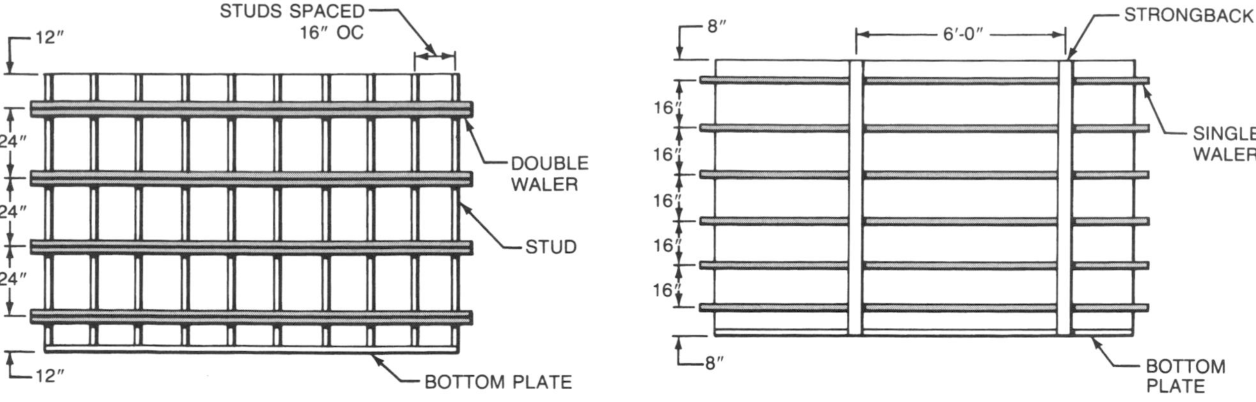

Figure 2-17. Minimum waler spacing must be maintained to accept the pressure of the concrete during placement.

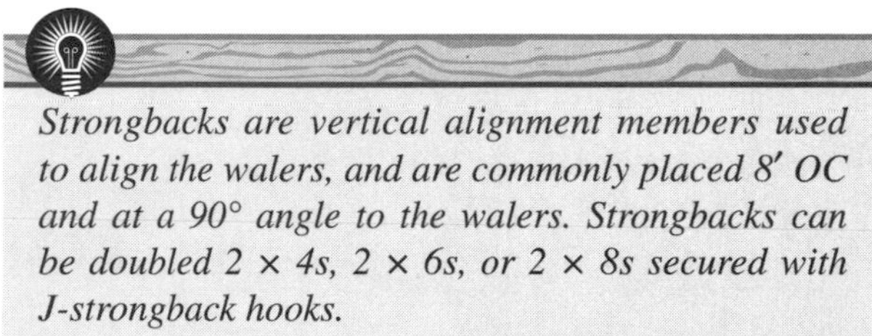

Strongbacks are vertical alignment members used to align the walers, and are commonly placed 8′ OC and at a 90° angle to the walers. Strongbacks can be doubled 2 × 4s, 2 × 6s, or 2 × 8s secured with J-strongback hooks.

Corner Ties. The corners of wall forms are subjected to extreme pressure during concrete placement. Corners must be tied and braced so that a form failure does not occur. An effective method to tie corners together is to overlap the walers at the corners and nail kickers against the walers. Patented metal devices are also available for tying the corners. **See Figure 2-18.**

Bracing Wall Forms. Lateral pressure against a wall form is caused by the movement and force exerted when concrete is placed, wind load, and pressures resulting from the weight of workers and materials on scaffolds attached to the forms. Adequate lateral bracing must be provided to keep the walls straight and to prevent collapse.

The two forces acting on a form brace are tension (pulling away) and compression (pushing against). Some braces are designed to handle both tension and compression and are only required on one side of the form. Other braces are only effective for compression and may be required on the two opposite sides of the form.

Wood form braces are usually 2 × 4s set at approximately a 45° angle. Braces set at less than a 45° angle are subject to greater force. Braces placed at more than a 45° angle are adequate for walls 6′ or less in height. With higher walls, when braces are placed at more than a 45° angle there is a danger of the braces bending and becoming ineffective. Long braces (12′ or more) should be strengthened at the midpoint by nailing stiffeners across the braces. The stiffeners should also be braced to the ground to resist bending. **See Figure 2-19.** For very high walls with long braces, two rows of stiffeners, one high and one low, may be required. Typical spacing for braces is every 8′ for form walls up to 8′ in height, and every 6′ for walls up to 10′ in height.

Braces on higher walls should be fastened to a waler or strongback. Braces should be attached at the lower end to the top stakes or pads. The method used depends on the type of surface present. **See Figure 2-20.**

Patented metal devices are also available to facilitate bracing operations. One commonly used device features a turnbuckle that is fastened at either end of a 2 × 4 wood brace by driving nails or screws through a metal angle bracket. When placed at the upper end of the wood brace, a metal brace plate fastens the turnbuckle to a waler or strongback. When placed at the lower end of a wood brace, a metal anchor bracket is secured to the ground with a steel stake. A turnbuckle allows easy adjustment for plumbing and aligning the wall form. **See Figure 2-21.**

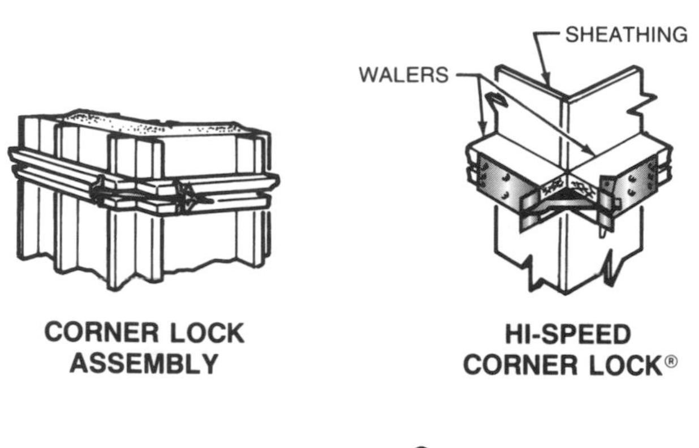

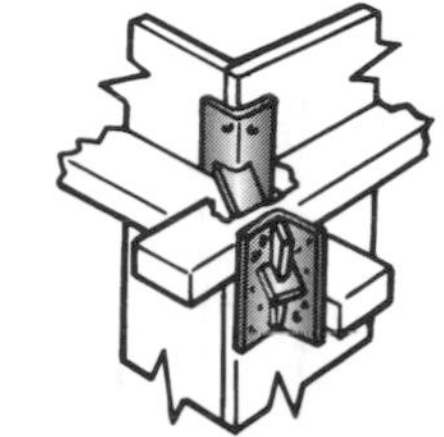

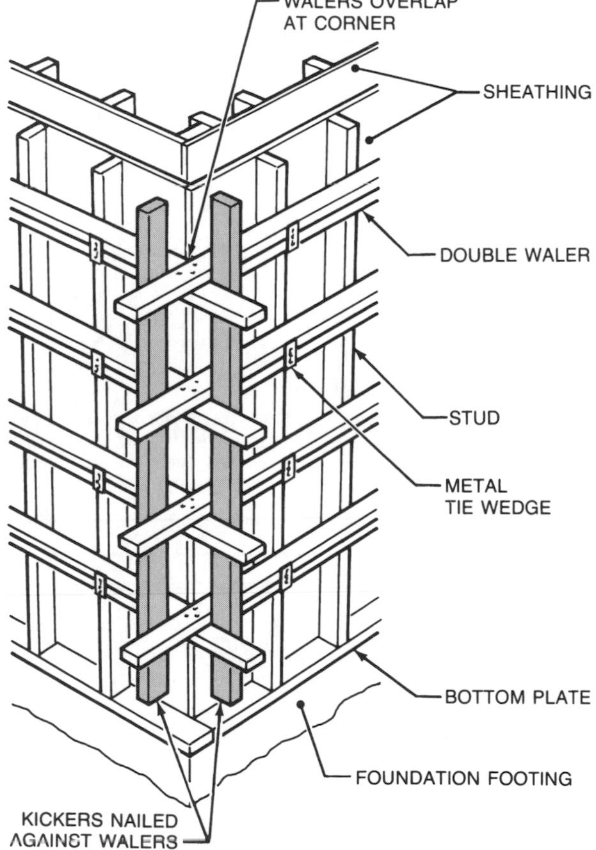

Figure 2-18. Wall form corners must be tied together to withstand the extreme pressure of concrete.

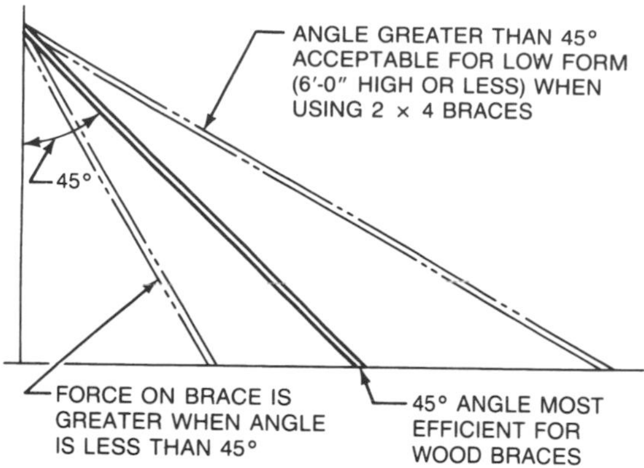

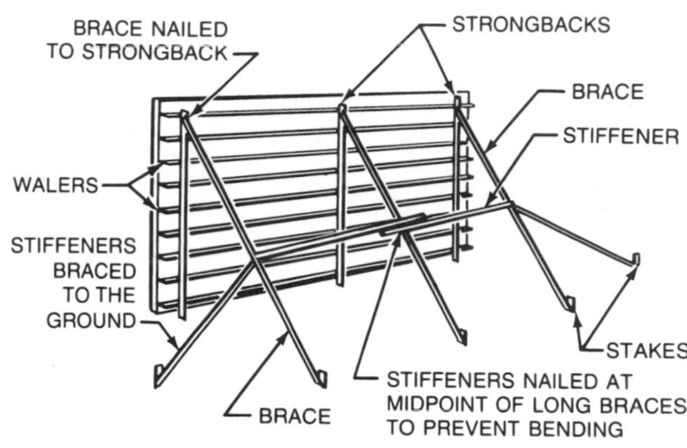

Figure 2-19. Forces affecting wall forms must be counteracted by using proper bracing angles and methods.

Gates & Sons, Inc.

Wall forms must be properly braced and stiffened when erected.

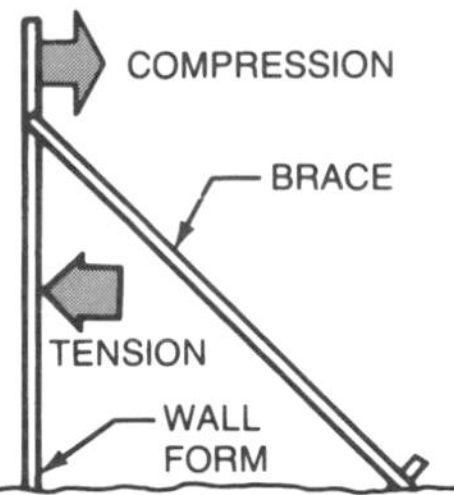

Braces designed for tension and compression strength are attached to one wall.

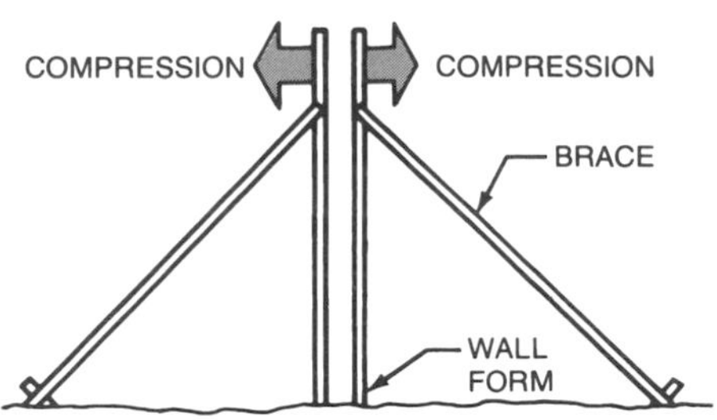

Braces designed for compression strength only are attached to both walls.

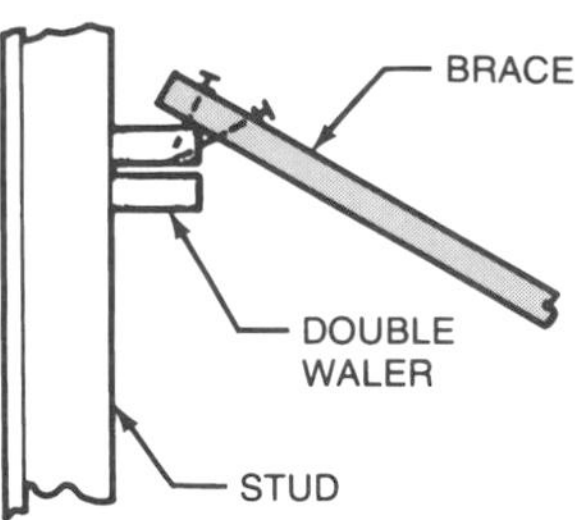

Brace attached to waler when strongback is not used. Brace can be attached to stud on low wall forms.

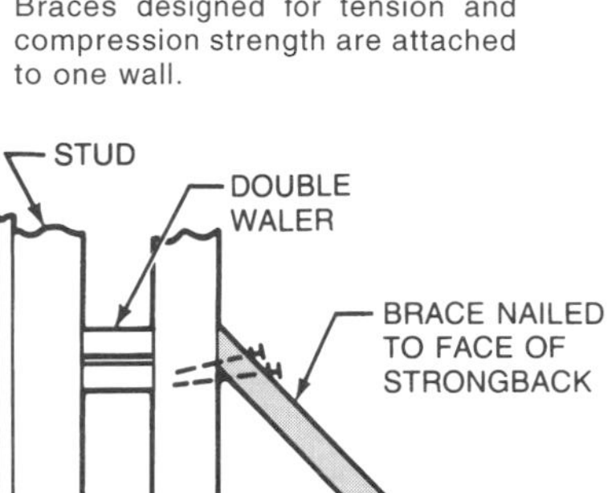

Brace nailed to face of strongback. Good compression strength.

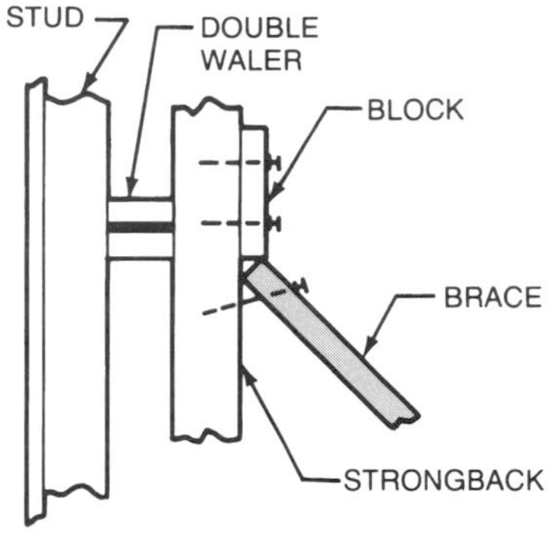

Brace nailed to strongback and block. Good compresion strength.

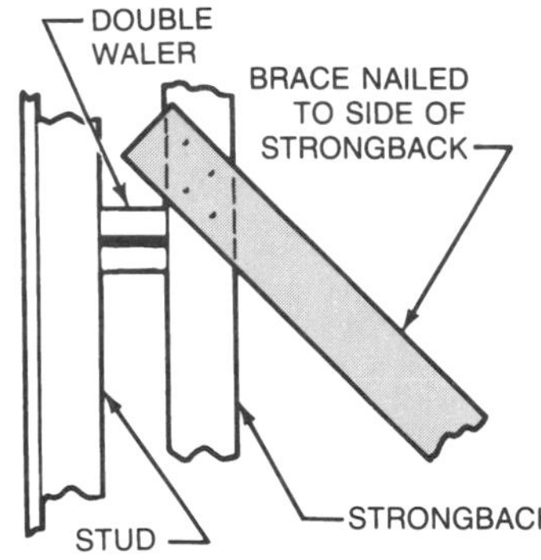

Brace nailed to side of strongback. Good tension and compression strength.

ATTACHING BRACES TO WALL FORMS

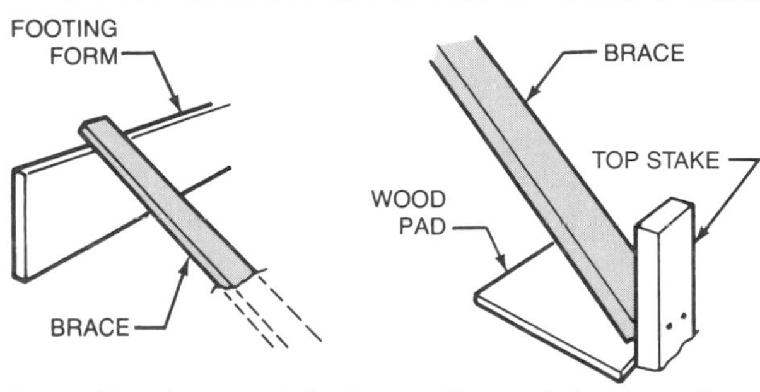

Brace driven into ground for low footing form. Good compression strength.

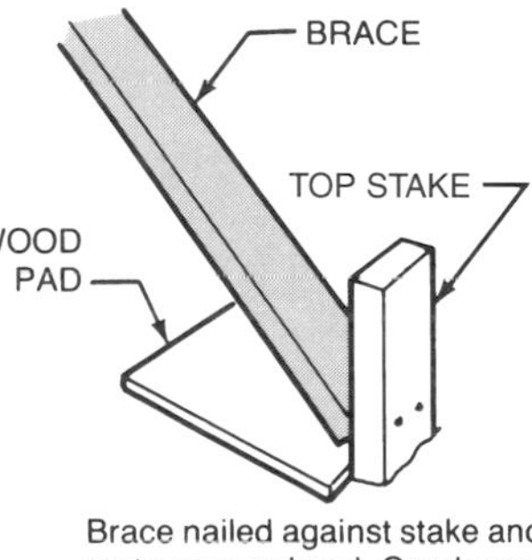

Brace nailed against stake and rests on wood pad. Good compression strength.

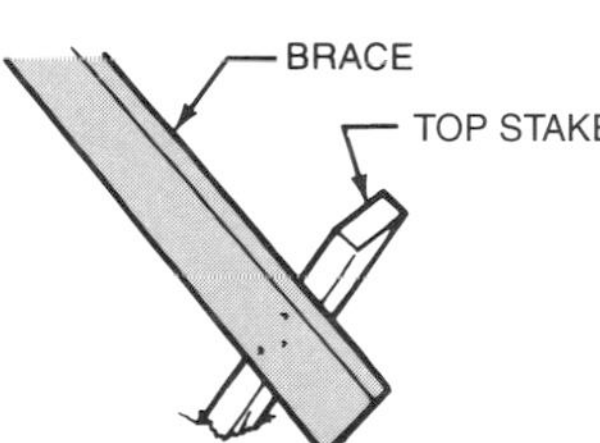

Brace nailed against edge of stake. Use for loose soil. Good compression and tension strength.

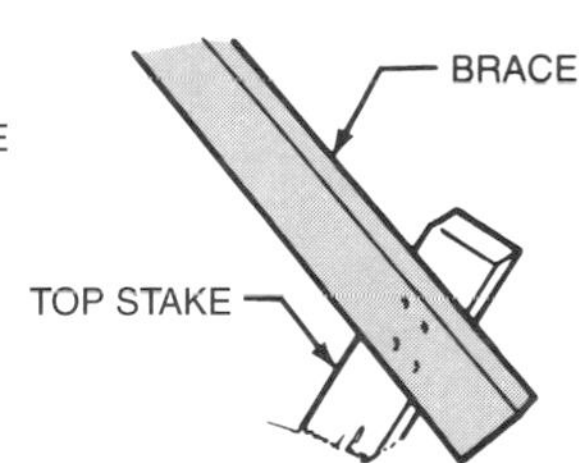

Brace nailed against face of stake. Use for firm soil. Good compression and tension strength.

ANCHORING BRACES TO THE GROUND

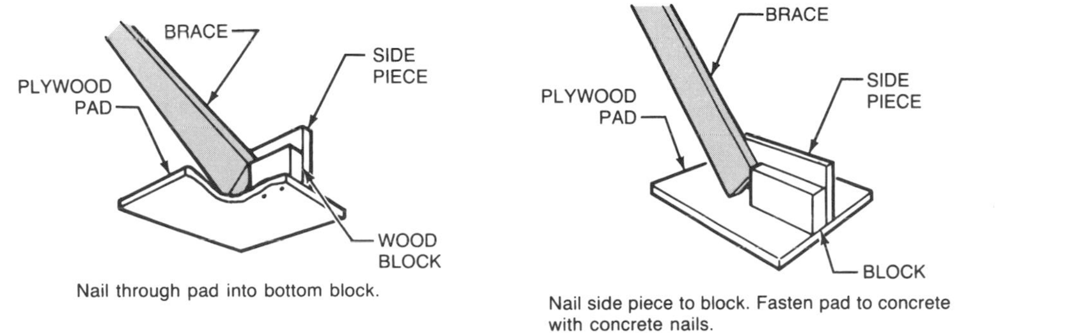

Nail through pad into bottom block.

Nail side piece to block. Fasten pad to concrete with concrete nails.

ANCHORING BRACES TO CONCRETE SLAB

Figure 2-20. Braces are attached to strongbacks or walers for maximum strength.

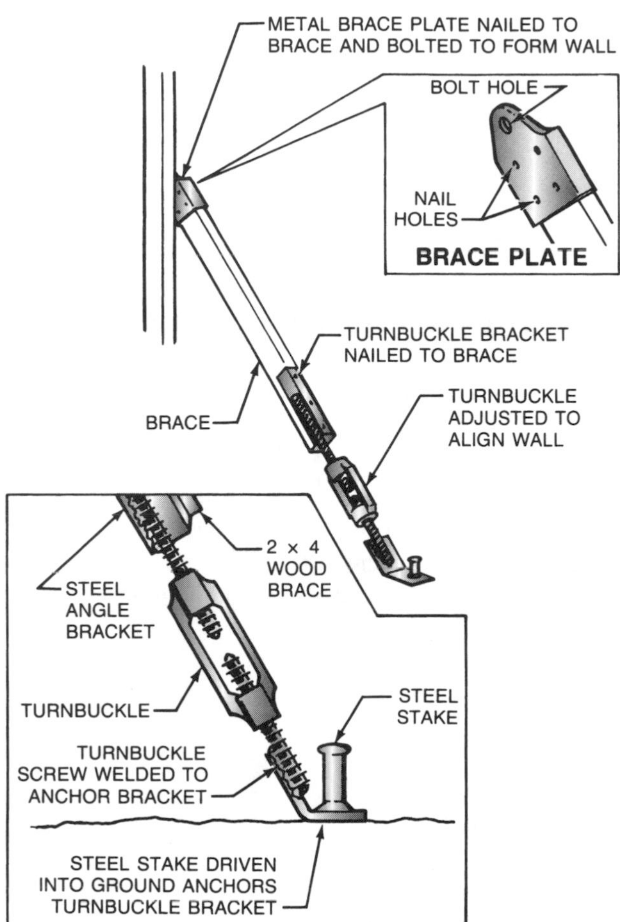

Figure 2-21. Patented metal devices are used with wood form braces.

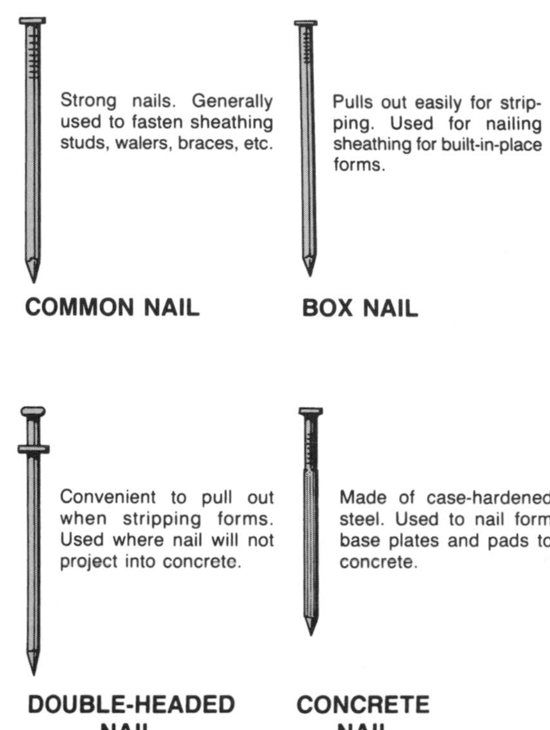

Figure 2-22. Many types of nails are commonly used in form construction. The type of nail used depends on the application.

Fastening Devices and Procedures

Nails are the most common fastener used in light form construction. Bolts, lag screws, and other devices are used in the construction of heavier forms. Proper nailing procedures must be used to ensure the strength and durability of the forms. However, because the forms must be stripped from the hardened concrete walls, the use of too many nails or nails that are too large should be avoided.

Nails are available in a variety of types and sizes. Nail lengths are designated by a number and the letter *d,* which is the designation for penny. Some of the nail lengths commonly used in formwork are 6d (2″), 8d (2½″), 10d (3″), and 16d (3½″). Spikes are 16d and larger nails. Common, box, double-headed, and concrete nails are commonly used in form construction. **See Figure 2-22.**

Common nails are used to fasten sheathing to the studs or walers and to fasten form frames, stiffeners, and bracing together. Plywood ⅝″ or thicker should be fastened with 6d common nails.

Box nails are primarily used to nail sheathing or liner material for built-in-place forms. Box nails have thinner shanks and flatter heads than common nails. The thinner shank is an advantage when stripping built-in-place forms because it is easier to pull from the forms. The flat head of a box nail also leaves less of an impression on the finished concrete wall.

Double-headed (duplex) nails are used extensively to nail walers, strongbacks, kickers, blocks, and other form components. The second head makes it convenient to pull the nail with a claw hammer or wrecking bar during stripping operations. As a result, there is less bruising and damage to reusable form lumber. Double-headed nails are recommended anywhere they do not protrude into the concrete at the time of the concrete placement.

Concrete (masonry) nails are used to fasten form base plates to concrete slabs or footings. Concrete nails are made of special case-hardened steel to resist bending. They must be driven straight to avoid chipping or breaking out the concrete.

Nailing Methods. Proper nailing methods are essential to good form construction. The strength of a nailed joint depends on the lateral and/or withdrawal resistance of the nail and the nailing procedure used. A few general rules should be followed for maximum withdrawal resistance. **See Figure 2-23.**

- The withdrawal resistance of a nail is much greater when driven into the edge grain rather than the end grain of wood.
- When fastening pieces of wood of different thicknesses together, nail through the thinner piece into the thicker piece.
- When nailing plywood to solid wood, drive the nail through the plywood and into the solid wood.
- Toenailing is often a better alternative to nailing into end grain. For best results, toenails should be driven at a 30° angle with one-third of the nail penetrating the piece being fastened.
- Maximum withdrawal resistance can be accomplished by clinching a nail across the grain.
- Drive nails straight into the pieces being joined. This facilitates removal when stripping forms.

Form Ties

A *form tie* is a device used to space and tie opposite form walls and prevent them from spreading or shifting while concrete is being placed. Low form walls are tied together with wood cleats nailed to the tops of the walls or to stakes extending above the walls. High form walls are held together with metal ties. A variety of patented ties are available for securing and spacing form walls. Patented ties consist of a rod that passes through the wall with holding devices at each end. The two basic patented tie designs are the continuous single-member types and internal disconnecting types. Their working loads range from 1000 lb to 50,000 lb.

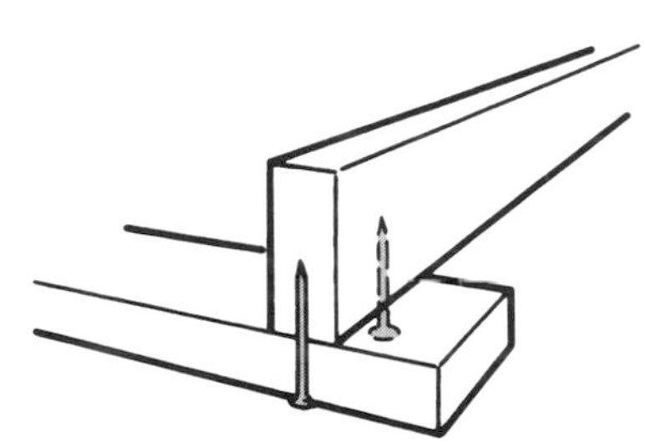

Withdrawal resistance of nail is greatest when nail is driven into edge grain.

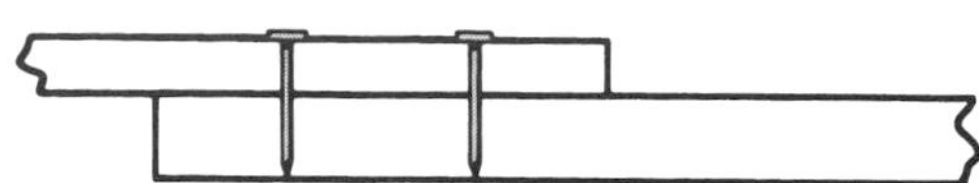

Always nail through thinner piece into thicker piece.

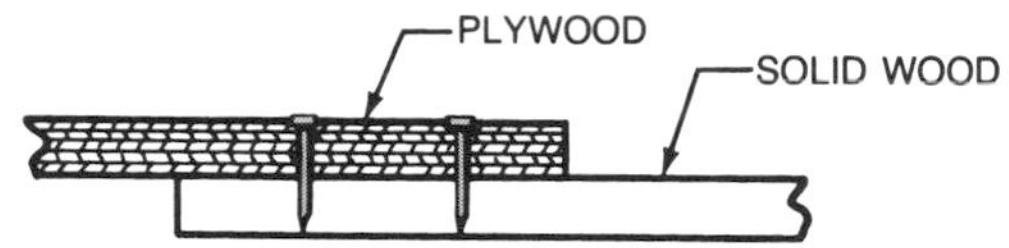

When nailing plywood to solid wood, drive the nail through the plywood into the solid wood.

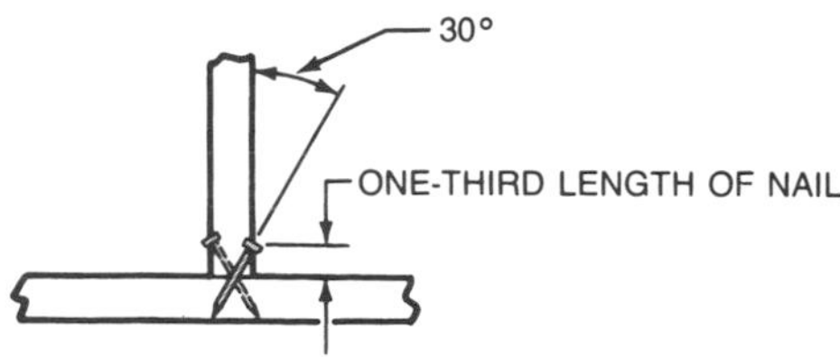

When toenailing, drive a nail at a 30° angle with one-third of nail length penetrating piece being fastened.

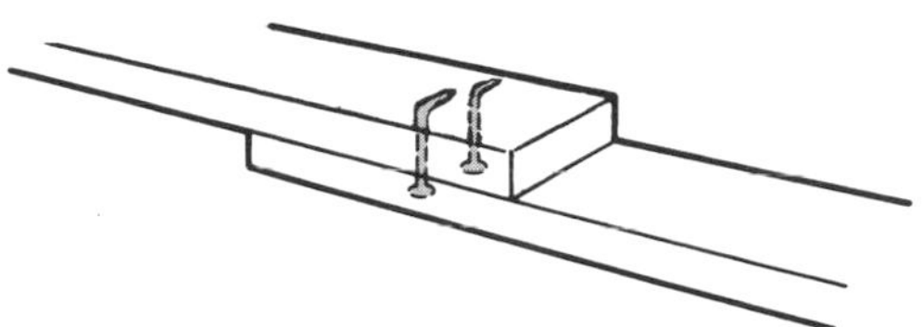

More withdrawal resistance is achieved by clinching across the grain.

Figure 2-23. Proper nailing methods are essential when building forms.

Continuous Single-Member Ties. A snap tie is the most common type of continuous single-member tie. A *snap tie* is a patented wall tie device with cones acting as form spreaders. The small plastic cones or metal washers placed in the section of the tie passing between the form walls act as spreaders holding the walls the correct distance apart. Snap ties are used with both double and single waler systems and are available for wall thicknesses ranging from 6″ to 26″. Various wedge and wedge-bracket devices are used to secure the snap ties. The ties are tightened by driving slotted metal wedges behind buttons at the ends of the ties. After the forms are stripped from the completed wall, the sections of the snap tie protruding from the wall are snapped off at the breakback. A *breakback* is a grooved section between the spreader cones.

A loop end tie is another commonly used continuous single-member tie. It is secured with a tapered steel wedge driven against a metal waler plate and through the loop at the end of the tie. Loop end ties are frequently used in prefabricated forms such as steel-framed plywood panels.

Adjustable flat ties are used with prefabricated and plank wall forms. **See Figure 2-24.** An *adjustable flat tie* is a patented wall tie device that consists of a flat piece of metal set on edge between the metal side rails. A series of uniformly spaced slots makes it possible to use the tie for different wall thicknesses. A wedge is driven through the appropriate slot to secure the tie in place.

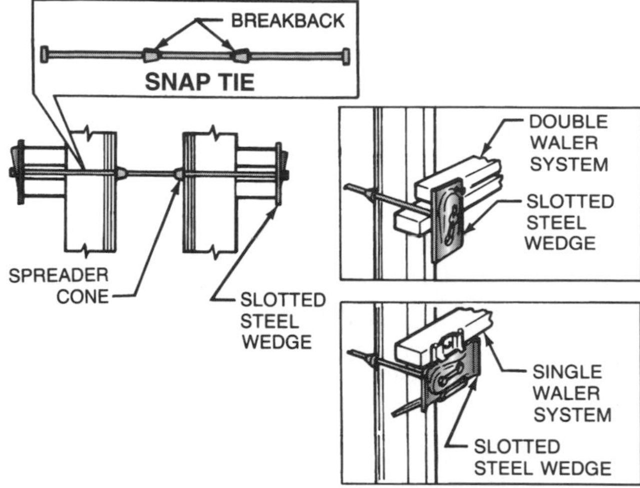

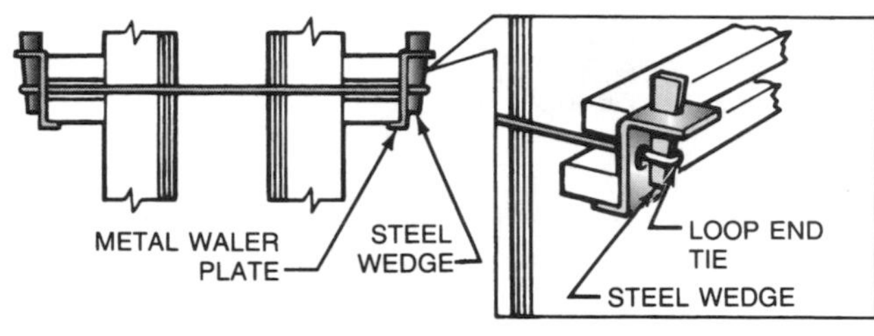

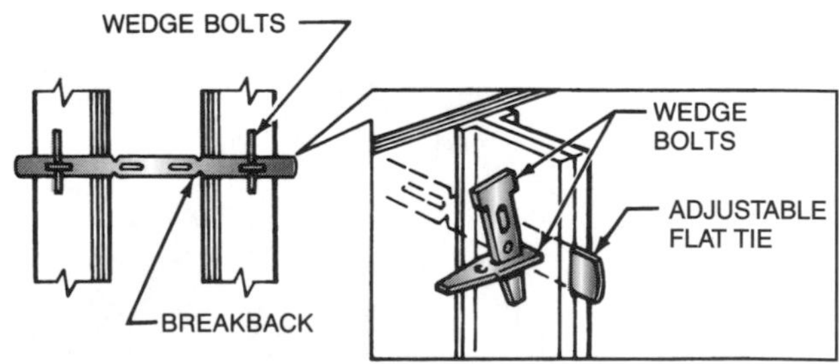

Figure 2-24. Continuous single-member ties are used with single and double waler systems.

Internal Disconnecting Ties. Internal disconnecting ties are used for heavier construction work where greater loads are anticipated. These form tie systems feature external sections that screw into an internal threaded section. The waler rod tie system and the coil tie system are two examples of internal disconnecting ties. **See Figure 2-25.** Metal spreaders hold the form walls the correct distance apart. Internal disconnecting ties are used for wall thicknesses ranging from 8″ to 36″, although they can also be assembled to accommodate much greater thicknesses.

The waler rod systems (she bolts) are composed of an inner rod threaded at each end that screws into two waler rods. The inner rod comes in various lengths for different wall thicknesses. The waler rods are fastened to the walers with large hex nut washers. The waler rod tapers, which facilitates removal. After the concrete sets, the waler rods are unscrewed and removed, and the inner rod remains in the concrete.

The coil tie system features external bolts that screw into an internal device consisting of metal struts with helical coils at each end. The coil assembly remains inside the concrete after the bolts are removed.

Placing Wall Ties. The type of tie to be used and the spacing between the ties are determined by several factors:

- Anticipated concrete pressure during placement
- Size and spacing of studs
- Size and spacing of walers, if necessary

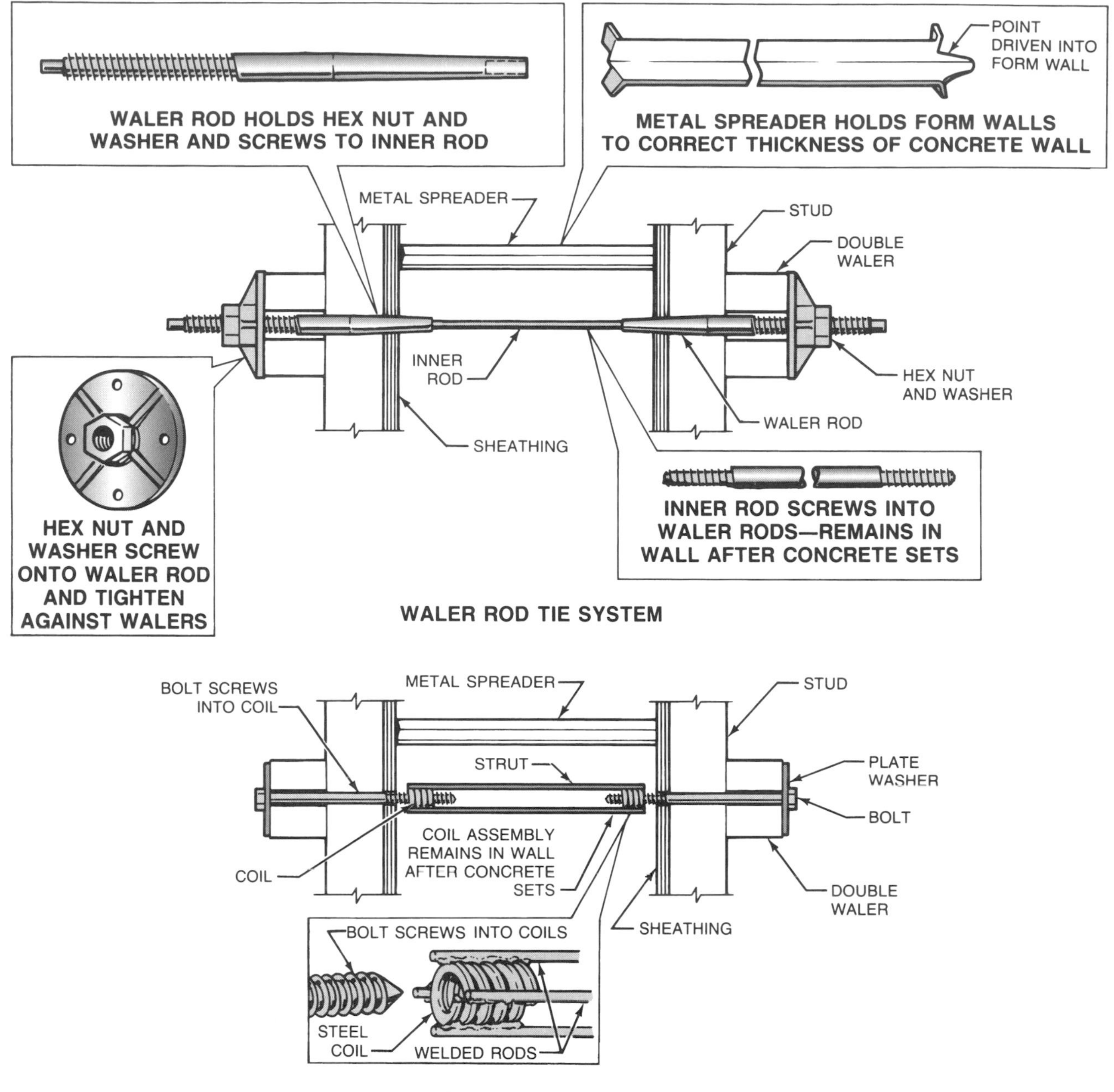

Figure 2-25. Internal disconnecting ties are used with heavier wall forms.

The load capacity of the tie must be considered in the spacing of the ties. The total load on a tie is determined by the contributing area of form around the tie. The contributing area equals one-half the vertical distance between ties (which is the distance between walers) times one-half the horizontal distance between the ties. Standard horizontal tie spacing is 24″ OC; however, greater pressure and other structural factors may require a shorter distance between the ties.

When tie holes are required, form panels are predrilled by drilling a number of panels at one time with an electric drill. The holes must be drilled square to the plywood surface. Holes for snap ties should be slightly larger than the end buttons, but smaller than the spreader cone. **See Figure 2-26.**

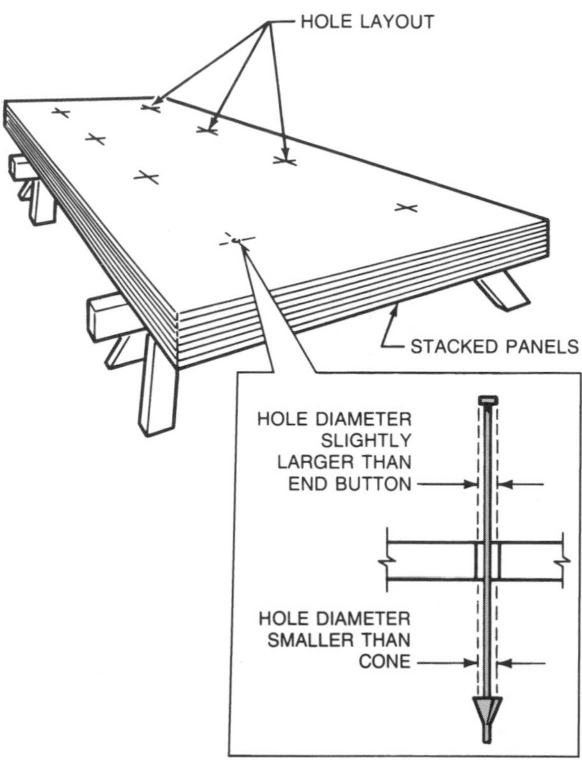

Figure 2-26. A number of panels may be predrilled at the same time to ensure proper placement of the ties.

Snap ties are usually placed after the outside wall forms have been constructed. Each tie is slipped through a predrilled hole, and one end of the tie is secured to the walers with brackets and/or wedges. The inside wall panels are set in place (doubled up) by tilting each panel and guiding the snap ties into the holes as the panel is straightened into its final position. The ties are then tightened against the inside walers with brackets and/or wedges. **See Figure 2-27.**

Figure 2-27. Snap ties are placed after the outside wall forms have been constructed.

In another method for placing wall ties, the outside wall form is constructed in the conventional manner with full panels, and the inside wall form is made up of smaller panels laid horizontally. The first row of tie holes in the outside wall form are laid out and drilled 12″ from the bottom, with the following rows 24″ OC. When constructing the inside wall form, a 12″ wide bottom panel is placed horizontally and 24″ wide panels are placed horizontally on top of it. As the inside panels are set in place, the snap ties are inserted into the tie holes that have been drilled in the outside wall form and hammered into the top edges of the inside form panels. **See Figure 2-28.**

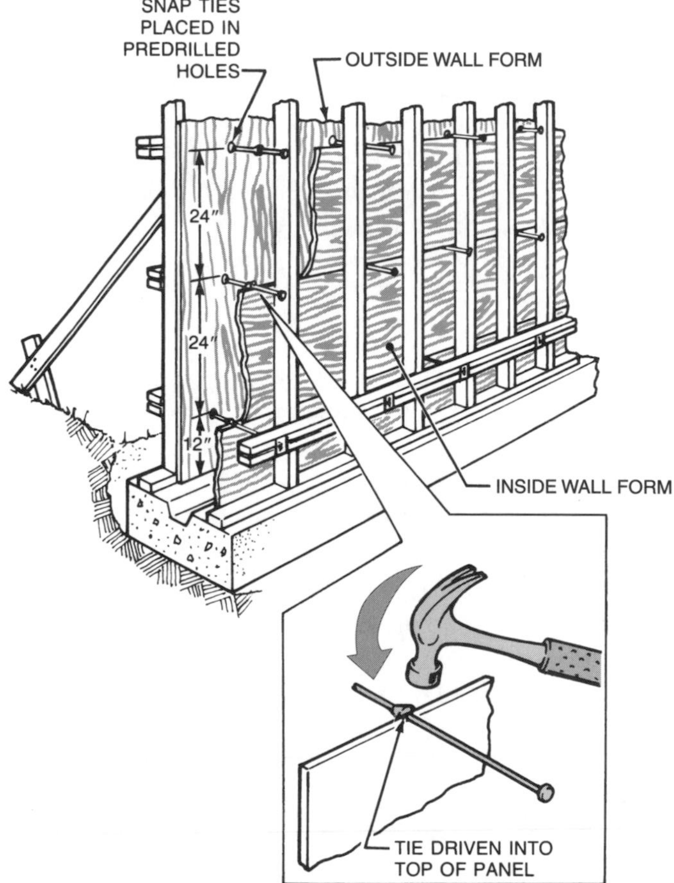

Figure 2-28. The inside form wall may be constructed by using 24″ wide panels laid horizontally. The ties are driven into the top edge of the panel as the form wall is being erected.

Form ties hold forms together to produce a smooth, straight wall with no wire ends protruding from the concrete. Wire ends can be removed from the wall with a hammer.

When waler rod or coil tie systems are used, the waler rods or bolts are slipped through predrilled holes after both the inside and outside wall forms have been erected. The rods or bolts must engage the threads of the mating section enough to provide maximum strength. A few rules regarding form tie placement are as follows:

- Drill tie holes so that the holes in the panels are directly across from each other when the inside and outside form walls are in place. Slanted ties lose considerable holding power.
- When using stud and waler stiffeners, place the ties close to studs to avoid panel deflection caused by the tightened tie.
- Maintain a uniform tightness for all wall ties. If one tie is tighter than others it carries more of the concrete pressure. This could cause the tie to break, which in turn could cause other ties to break because of the pressure increase. Although uneven tightness may not cause any ties to break, the form panels may deflect, producing bulges in the finished concrete walls.

Patching Tie Holes. When removing forms that have been reinforced with continuous single-member ties or internal disconnecting ties, a shallow hole in the surface of the wall will be present. The hole must be patched and sealed in order to give the concrete wall a finished appearance and prevent moisture from reaching the tie ends. Moisture penetrating the tie ends eventually causes rust stains to appear on the surface of the concrete.

Holes are patched with a nonshrink, moisture-resistant grout mixture or dry-pack mortar. Some manufacturers offer precast cement or rubber compound plugs shaped to fill the holes created by snap tie spreader cones. The plugs are secured in place with a fast tack waterproof neoprene adhesive. Another method is to inject a pressurized epoxy resin into the tie hole, then place a plastic cap insert in the hole. **See Figure 2-29.** Both the cement and plastic plugs can be placed either flush with or recessed from the surface of the concrete wall.

Preset Form Anchors. Preset form anchors are used to fasten formwork to previously placed concrete. A straight coil loop insert that receives a coil bolt, and a ferrule loop insert that receives a threaded machine bolt, are common form anchors. **See Figure 2-30.** The insert along with a temporary bolt and washer are set in place at the time of concrete placement. After the concrete sets, the insert remains embedded in the concrete. It will later be used to bolt down additional formwork to the existing concrete.

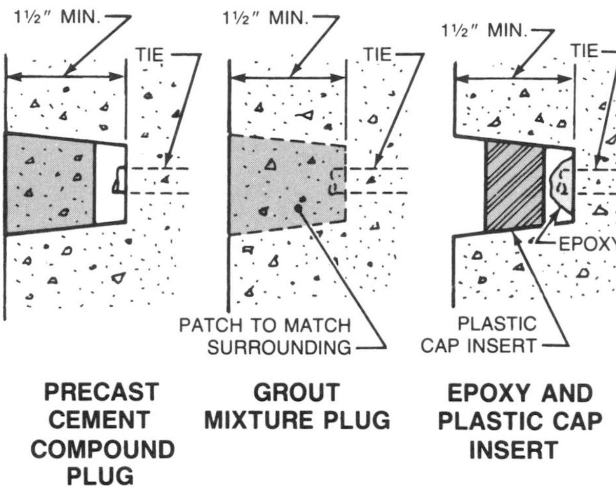

Figure 2-29. Holes created by the snap tie spreader cones must be filled or plugged after the snap ties have been broken off.

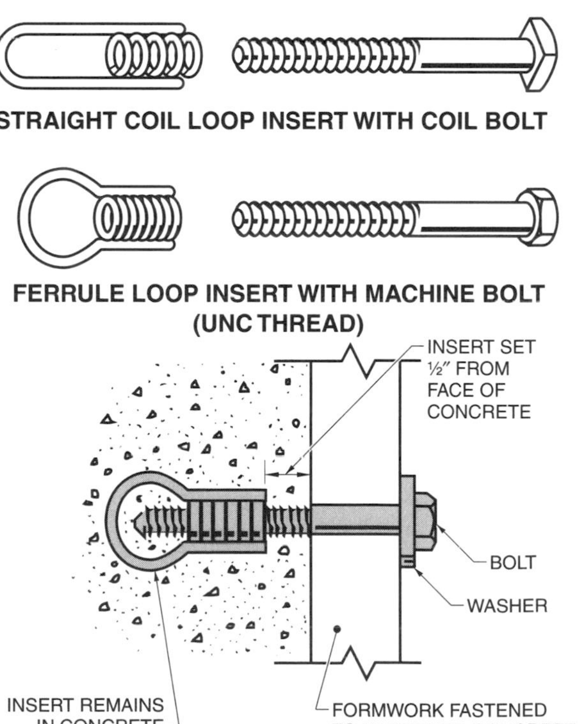

Figure 2-30. Preset form anchors are used to fasten formwork to concrete that has been placed. The form anchors are positioned when the concrete is placed.

Fiberglass Form Ties

High tensile strength fiberglass rods that are placed after the forms have been erected may also be used as form ties. **See Figure 2-31.** The procedure for installing and stripping fiberglass form ties is as follows:

1. The rods are obtained in bulk and cut to their proper length with an abrasive blade in a circular saw. The standard color of the rods is gray; however, the rods can be obtained in almost any color for special purposes. The fiberglass form tie procedure can be used with built-in-place forms as well as most prefabricated forming systems.
2. When the wall forms have been constructed and braced, the fiberglass rods are placed in predrilled holes.
3. A wedge is placed against the walers on each side of the form.
4. A self-gripping device is then placed on the highest point of the wedge and the form is held in place. Fiberglass spreaders can be clipped to a form tie where necessary.
5. To strip the forms after the concrete has hardened, the grippers on the outside of the form walls are bent until the rod breaks. The broken section inside the gripper can then be pulled out and the gripper stored for future use.
6. After the form walls have been pulled away, the protruding rods are cut flush with the surface with a grinding tool. The gray color of the rods blends well with the concrete; therefore, patching over the ends of the rods is not necessary. The fiberglass rods do not rust, which eliminates the concern about rust stains.

PREFABRICATED WALL FORMS

Prefabricated wall forms are constructed from prebuilt panel sections and other form components. Prefabricated wall forms are used when numerous reuses of the panels are anticipated. The use of prefabricated panels lowers labor and material costs. Prefabricated forming systems can be rented or purchased from various manufacturers or suppliers. Special-purpose custom-made forms, also produced at prefabricated plants, are available for special forming operations.

Panel Systems

The prefabricated wall forms most commonly used consist of modular panel sections, usually 2′ to 4′ in width and 2′ to 8′ in height. Smaller filler pieces of various sizes are also available. Panel manufacturers also provide the accessories (ties, walers, wedges, braces, etc.) for their particular systems. Although metal-framed plywood panels are the most common type of prefabricated panel sections used today, there is a growing use of all-metal panel sections. **See Figure 2-32.**

Metal-Framed Plywood Panels. A metal-framed plywood panel consists of ½″ or ⅝″ Plyform® set in an aluminum or steel frame. Horizontal metal stiffeners spaced approximately 1′ apart provide additional support. The frames are designed so the panels can be easily replaced when worn or damaged. The panel sections, often called hand-set forms, are light and easy to handle. Slots in the metal side rails are for wedge bolts that join the panels together. Walers are secured with metal waler ties. Wire, flat, or round ties space and hold the walls together. Braces are secured to the walls with wedge-shaped metal plates.

All-Metal Panels. Prefabricated all-metal panels are made of aluminum or steel. Aluminum hand-set forms consisting of an aluminum face stiffened with an aluminum frame are frequently used in the construction of residential foundations. Steel forms are used in heavy construction. Steel forms are combined in ganged panel forms and are also widely used in the precast industry. Accessories are provided by the manufacturer to assemble, align, and brace the steel and aluminum forms. With proper care and maintenance, steel and aluminum forms can be used indefinitely.

RJD Industries, Inc.

High tensile strength fiberglass rods can be used with job-built forms as well as with most prefabricated forming systems.

FIBERGLASS FORM TIES

1. Rods are cut to length.

2. Rods are placed in predrilled holes.

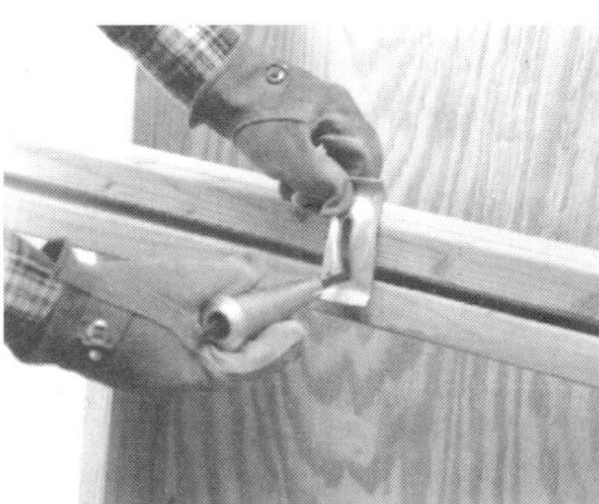

3. Wedges are placed against the walers.

4. Grippers are placed on the wedges.

INSTALLING

5. Grippers are bent, breaking the rods.

6. The rods are then ground flush.

STRIPPING

RJD Industries, Inc.

Figure 2-31. High tensile strength fiberglass rods can be used with job-built forms as well as most prefabricated forming systems.

METAL-FRAMED PANELS

ALL-METAL PREFABRICATED FORMS

Symons Corporation

Figure 2-32. Metal-framed plywood panels can be replaced when worn or damaged. All-metal prefabricated forms are commonly used on heavy construction projects.

Reinforcing Steel

A concrete wall is subjected to both compressive and lateral pressures. Concrete without reinforcement has a great deal of compressive resistance to vertical loads, but far less resistance to lateral loads. The lateral resistance of concrete walls is strengthened by placing reinforcing steel bars (rebar) in the walls. *Rebar* is steel reinforcing bar with deformations on the surface to allow the bar to interlock with concrete. The combination of concrete and steel reinforcement is generally referred to as reinforced concrete construction.

The uneven surface of rebar helps bond the concrete to the steel. Standard size rebar range from 3⁄8″ to 2¼″ in diameter and are identified by numbers from #3 to #18. **See Figure 2-33.** The diameter of the rebar is found by multiplying the number designation by 1⁄8″. For example, a #6 rebar is 6⁄8″, or ¾″, in diameter. The size of the rebar required for a wall, as well as their placement and spacing, is shown in section view drawings of prints.

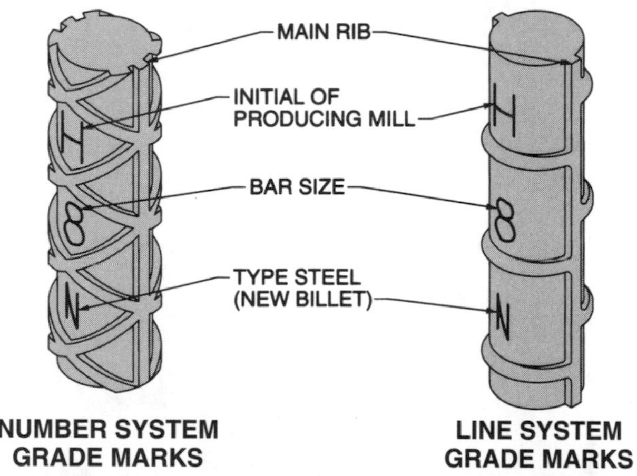

STANDARD REBAR SIZES						
Bar Size Designation	Weight per Foot		Diameter		Cross-Sectional Area Squared	
	lb	kg	in.	cm	in.	cm
#3	0.376	0.171	0.375	0.953	0.11	0.71
#4	0.668	0.303	0.500	1.270	0.20	1.29
#5	1.043	0.473	0.625	1.588	0.31	2.00
#6	1.502	0.681	0.750	1.905	0.44	2.84
#7	2.044	0.927	0.875	2.223	0.60	3.87
#8	2.670	1.211	1.000	2.540	0.79	5.10
#9	3.400	1.542	1.128	2.865	1.00	6.45
#10	4.303	1.952	1.270	3.226	1.27	8.19
#11	5.313	2.410	1.410	3.581	1.56	10.07
#14	7.650	3.470	1.693	4.300	2.25	14.52
#18	13.600	6.169	2.257	5.733	4.00	25.81

American Society for Testing and Materials

Figure 2-33. Steel reinforcing bars (rebar) are used to strengthen lateral resistance of walls.

Concrete buildings and other large concrete projects are heavily reinforced throughout the structure. Residential and other light construction foundations may not require rebar if the buildings are not located in a seismic (earthquake) risk zone. In areas where earthquakes occur, local building codes require steel reinforcement. The U.S. Department of Housing and Urban Development (HUD) *Minimum Property Standards for One and Two Family Dwellings* states the following regarding foundation walls:

1. Where earthquake design is required, and in seismic zone 2 or 3, reinforce concrete walls under the following conditions:
   - concrete walls when height exceeds 6 times thickness.
   - masonry walls when height exceed 4 times thickness.
2. Size and spacing of reinforcement shall be in accordance with accepted engineering practice.

Placing Rebar. Proper rebar placement is critical to ensure maximum resistance to lateral loads. A low foundation wall may need only a few rebar placed horizontally. High walls require horizontal and vertical rebar. The horizontal and vertical rebar are tied together to form a steel curtain. Larger and thicker walls may require two or more curtains of steel within the wall. Where only a few rows of rebar are required in low forms, the steel is placed by the worker constructing the forms. However, large quantities of rebar are commonly installed by reinforcing-metal workers. The placement of rebar requires careful coordination with other trades. The rebar are usually positioned after the outside wall forms have been set and the openings framed. Other formwork, such as positioning conduit, sleeves, anchors, straps, and inserts, must be completed prior to placing rebar. After the rebar have been placed, the inner wall forms are constructed. **See Figure 2-34.** In the case of walls that are part of a core, such as an elevator shaft or stairwell, the inner walls are constructed first.

Rebar should be free of loose rust, scale, paint, oil, grease, mortar, or other foreign matter that may weaken the ability of the concrete to grip the steel. The steel curtains must remain in their proper position within the form while the concrete is being placed. Rebar near the surface of the walls must be protected against corrosion and fire by an adequate layer of concrete. Rusting can occur if the steel is too close to the surface, thus producing cracks in the concrete. **See Figure 2-35.** Several methods are used to maintain proper spacing between the wall forms and the rebar. These include spacer blocks and plastic snap-on devices, as well as wood strips that are removed as the concrete is being placed.

Figure 2-34. Rebar is placed after the outside form walls have been constructed.

REQUIRED CONCRETE PROTECTION FOR REBAR		
Application		**Minimum Cover***
Against ground without forms		3
Exposed to weather or ground but placed in forms	Greater than $^5/_8$″ diameter rebar	2
	Less than $^5/_8$″ diameter rebar	$1\frac{1}{2}$
Slabs and walls (no exposure)		$^3/_4$
Beams, girders, columns (no exposure)		$1\frac{1}{2}$

* in in.

Figure 2-35. Rebar must be covered with an adequate layer of concrete to protect it from fire and corrosion.

Architectural Form Liners

Plastic form liners are glued, nailed, or screwed to prefabricated panel systems and are widely used to produce many different finishes on architectural concrete. They are available for one-time or multiple applications. Some of the wall patterns produced with plastic form liners include different types of brick, masonry, flagstone, as well as wood textures, rib designs, and others.

INSULATING CONCRETE FORMS

Insulating concrete forms (ICFs) are a more recently developed, specialized forming system. *Insulating concrete forms (ICFs)* are a specialized forming system that consists of a layer of concrete sandwiched between layers of insulating foam material on each side. ICFs are primarily used to construct concrete homes and other small to medium-size concrete buildings. The ICFs act as the inside and outside form walls into which concrete is placed. **See Figure 2-36.**

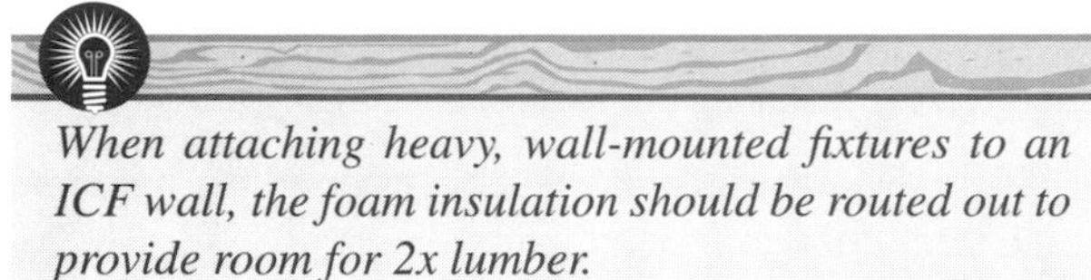

When attaching heavy, wall-mounted fixtures to an ICF wall, the foam insulation should be routed out to provide room for 2x lumber.

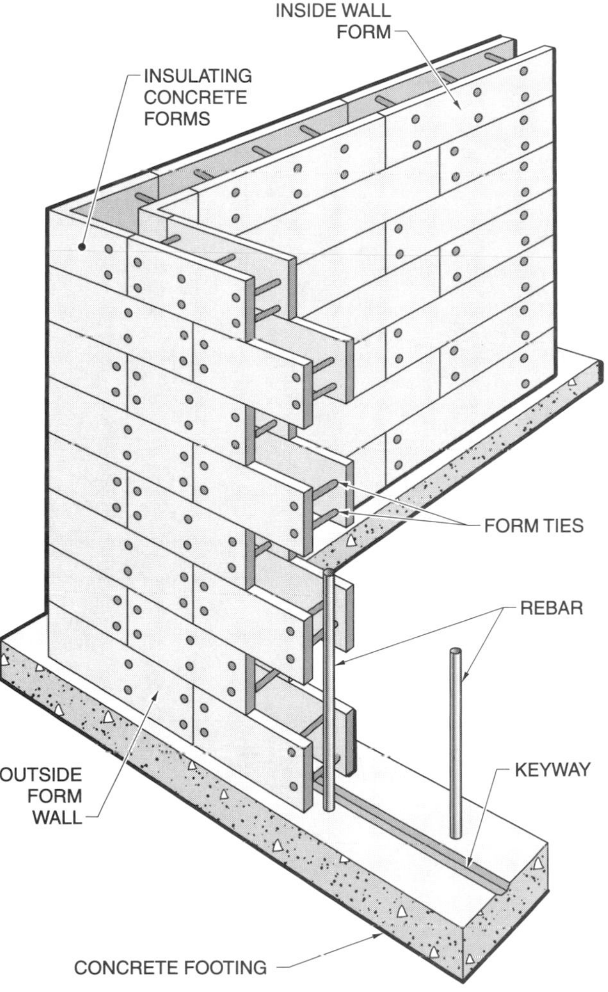

Figure 2-36. Insulating concrete forms (ICFs) act as the inside and outside form walls into which concrete is placed. ICFs are primarily used to construct concrete homes and other small to medium-size concrete buildings.

The forms remain permanently in place after the concrete has set, providing an insulating cover on both sides of the wall. Combined with the concrete, ICFs provide a continuous insulation system and an effective sound barrier. The inside forms will include wood, metal, or plastic backing material for attaching gypsum board or plaster lath. The outside forms provide attaching devices to fasten exterior finish such as siding, brick, and stucco. If desired, the exterior walls can be finished with wood, plastic, or metal materials to provide a similar appearance to traditional wood-framed structures.

ICF construction offers many advantages. Since the forms remain in place, the labor cost of stripping forms is eliminated. There is minimal air infiltration into the completed walls, reduced heating and cooling loads, and better fire resistance.

ICF Systems

ICF systems are made up of block, plank, or panel components. **See Figure 2-37.** The systems usually fit together with tongue-and-groove joints, but some methods use adhesives to bond the joints. ICF systems are most often made of expanded polystyrene insulation material, although other materials such as polyurethane, recycled wood, and cement mixtures are also used.

A number of building codes, particularly in the South, require that below-grade ICF wall sections be treated to resist infestation from termites and/or carpenter ants. **See Figure 2-38.** ICFs are usually constructed over cast-in-place concrete footings.

Block Forms. Block forms are similar to concrete masonry units (CMUs). Block forms are delivered to the job site ready for wall construction, with plastic or steel ties already in place.

Block forms used most often are 16″ high and 48″ long, with other dimensions available. The expanded polystyrene insulation on each side is 2″ to 2½″ thick. Plastic or steel form ties hold the two sides together. The length of the ties determines wall thickness. Wall thicknesses range from 4″ or more for above-grade applications to 12″ or more for foundations.

Plank and Panel Forms. Plank and panel forms are shipped flat to the job site. The form ties are installed as the walls are constructed. Common dimensions for plank forms are 1′ wide by 8′ long. Panel forms can be up to 4′ wide by 8′ long.

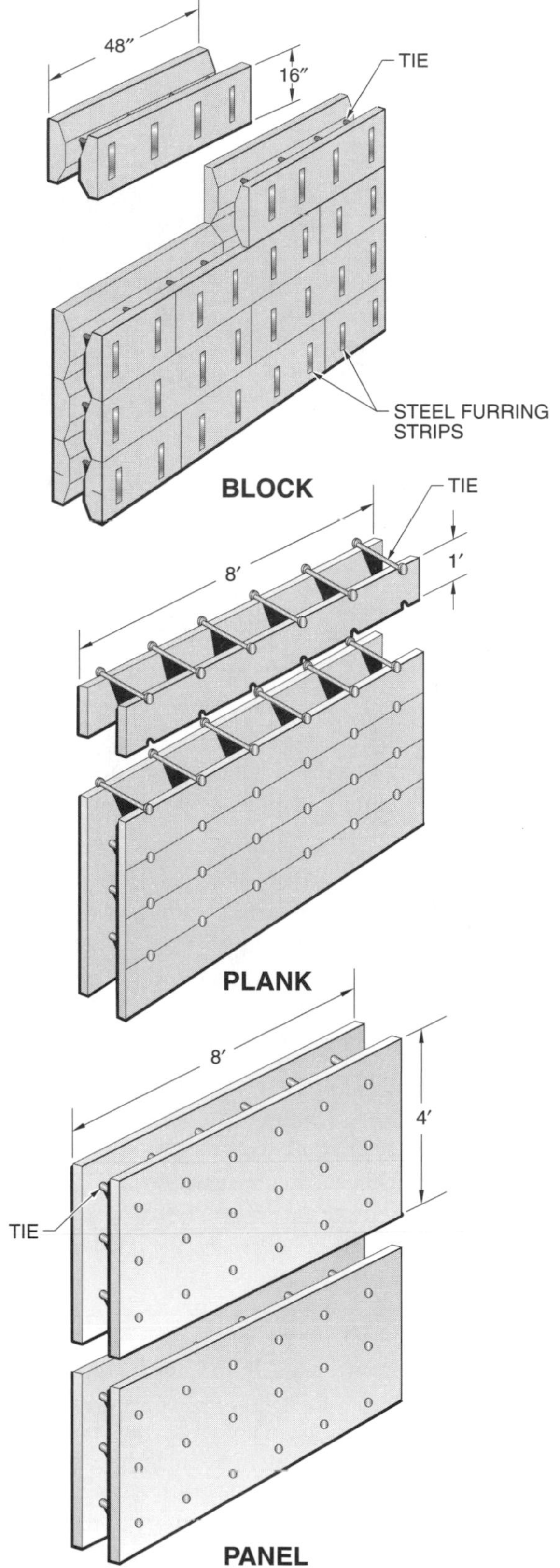

Figure 2-37. The three main types of insulating concrete form components are blocks, planks, and panels.

Figure 2-38. Many codes require that below-grade ICF wall sections be treated to resist infestation by pests.

ICF Wall Designs

Flat core, waffle grid, and screen grid are the three basic ICF wall designs. **See Figure 2-39.**

Flat Core Walls. Flat core walls have a layer of insulation attached to the concrete surfaces, and they resemble traditional cast-in-place concrete walls. Flat core walls are available in different thicknesses, commonly 4″, 6″, 8″, or 10″.

Waffle Grid Walls. A waffle grid wall uses less concrete than a flat core wall. The vertical core and horizontal core thicknesses are commonly 6″ or 8″. Web thickness between the cores should be no less than 2″. The maximum spacing between cores (on the vertical and on the horizontal) is 12″ OC.

Screen Grid Walls. Screen grid walls, also known as post-and-beam forms, feature columns spaced approximately 4′ OC and horizontal beams spaced 4′ or 8′ OC. Column and beam thicknesses are usually 6″ or 8″. Screen grid walls differ from waffle grid walls in that they do not have webs between the columns and the beams.

ICF Tools and Handling

ICFs weigh approximately 1 lb to 2 lb per square foot, making them much lighter and easier to handle than traditional concrete forms. Typical layout tools, such as a tape measure, framing square, chalk line, builder's level, or laser transit-level, are used to ensure that the walls are plumb and square, and are set to the proper measurements. A screwgun or a hammer can be used to fasten materials to the faces of the ICF. A foam glue gun is commonly used to fill gaps between the forms.

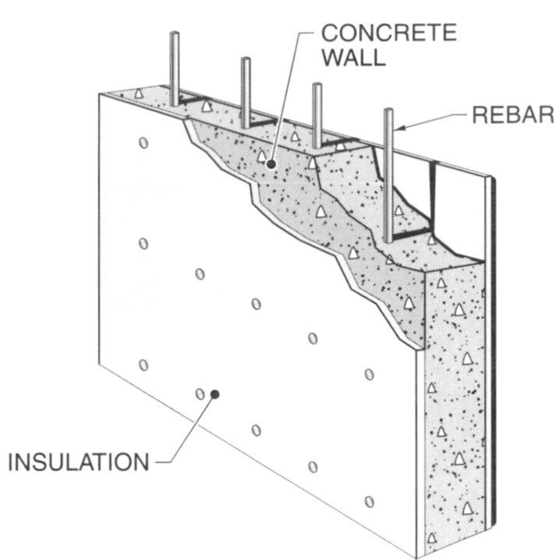

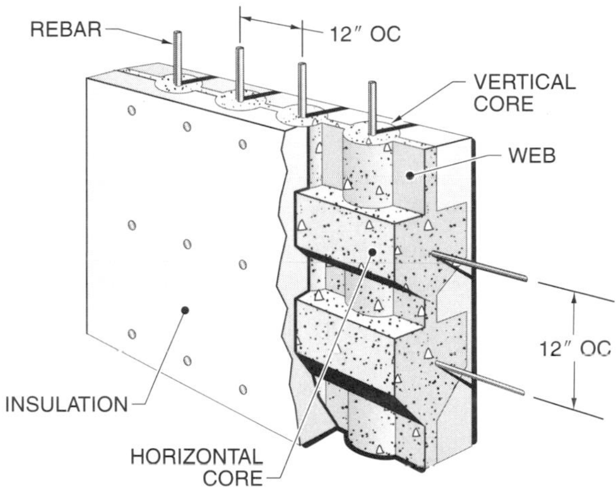

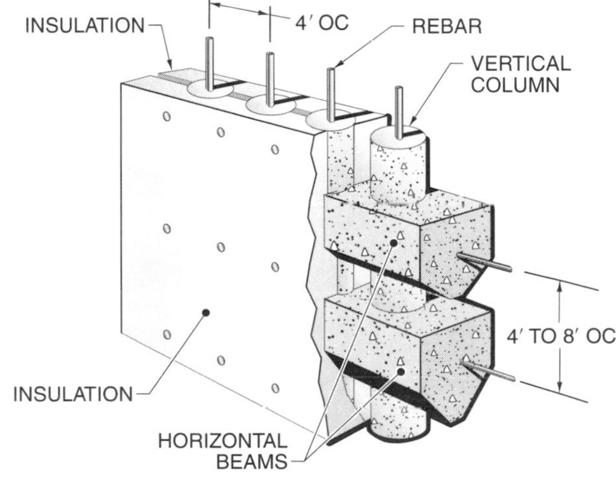

Figure 2-39. The three basic insulating concrete form designs are flat core, waffle grid, and screen grid.

ICFs can be cut with standard power tools. Long, straight cuts are made on a table saw with a fine-toothed blade. A circular saw or reciprocating saw can be used for cutouts when the walls are in place. Most ICF materials may also be cut with a curved pruning saw.

Aligning and Bracing. Minimal bracing is required to align low ICF wall forms. Corners are held in place with 2 × 6s secured with 2 × 4 diagonal braces. **See Figure 2-40.**

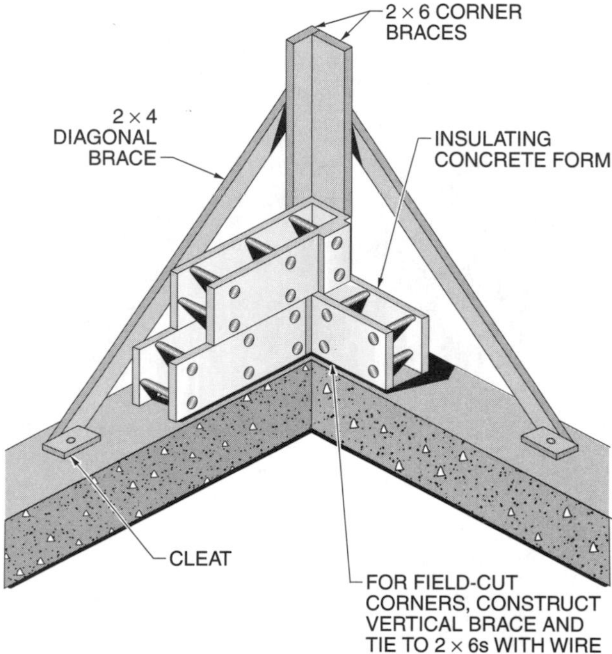

Figure 2-40. Corners are held in place with 2 × 6s secured with 2 x 4 diagonal braces.

One method used to align and brace forms for higher walls is to attach a scaffold to the inside form. Braces nailed to the uprights of the scaffold are secured at their lower ends to stakes in the ground. **See Figure 2-41.** Planks supported by brackets serve as a working platform.

Window and Door Bucks

Wood and metal window and door bucks are set in the ICFs to create openings for doors and windows. Ribbed flanges attached to the bucks secure the bucks to the concrete. Once in place, the bucks serve as the frames for doors and windows. Proper bracing and support is required to prevent buck movement as the concrete is being placed in the forms.

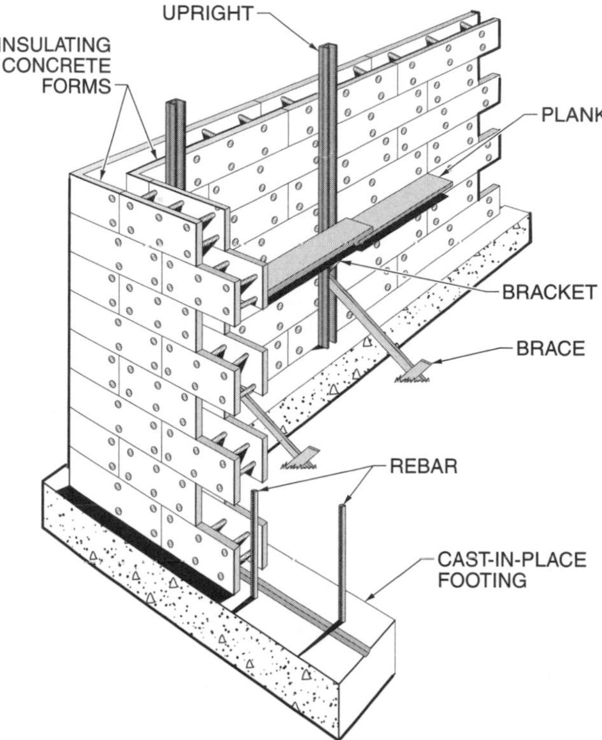

Figure 2-41. For higher walls, scaffolding can be designed to align and brace the insulating concrete forms.

Attaching Materials to ICFs

Finish materials are attached to the ICFs through a variety of devices and methods. The same exterior materials (wood siding, brick, stone, or stucco) used on wood-framed houses can be attached to the exterior of ICF surfaces. Interior wall finish materials such as gypsum board, solid wood board, and wall paneling can be fastened to the interior of ICF surfaces. Wood or metal floor joists, ledgers, and beams are supported by metal hangers and other devices.

Wall Finish. Adhesives are commonly used to apply interior finish materials to the inside surface of ICFs. Always consult the manufacturer for the proper type of adhesive. Another frequent method of applying interior finish is to embed plastic or metal furring strips into the form surface as a base for attaching finish materials. Self-tapping screws or ring-shank nails should be used.

Some applications may use sheets of 24 ga, 25 ga, or 26 ga galvanized sheet metal placed under the gypsum board to provide backing and support for heavy items. The sheet metal, using self-tapping screws or pop rivets, is fastened to previously placed metal furring strips. Heavier items such as cabinets are secured to the walls using self-tapping sheet metal screws with sufficient length to penetrate the sheet metal backing.

Metal Hangers and Connectors. Metal hangers and connectors can be mounted before or after concrete is placed in the forms. Hangers and connectors set before the concrete is placed have ribbed flanges that extend into the concrete. Another type of connector frequently installed is ledger board connectors that support wood or metal ledger boards to which joist hangers are fastened. Metal ledger boards are fastened directly to a connector. Wood ledger boards are installed with a ledger support fastened to a ledger board connector. **See Figure 2-42.**

When installing hangers and connectors after the concrete has set and hardened, expansion shields or self-tapping screws are used. The shields or screws must sufficiently penetrate the concrete. Also available is lighter hardware with flanges that are held in place by the foam insulation before the concrete is placed.

Placing Rebar and Concrete

The type of ICF form used determines the procedure for positioning vertical and horizontal rebar. Rebar installation in plank and panel forms resembles rebar placement in traditional forming systems. The vertical and horizontal rebar are joined by tying the bars together with tie wires. In block ICFs, rebar placement is similar to placement of rebar in concrete masonry units (CMUs). Some block forms are designed with cradles, making it more convenient to install the horizontal rebar.

Insulating concrete forms are not braced with the same system of uprights and walers used for traditional forms, which poses an increased risk of pillowing (bulging) in the walls. When using a concrete pump and hose to place concrete, the concrete should be placed in the forms very slowly. **See Figure 2-43.** A 2″ reducer can be attached to the hose to narrow the discharge. A mechanical vibrator is normally used to consolidate the concrete and fill voids.

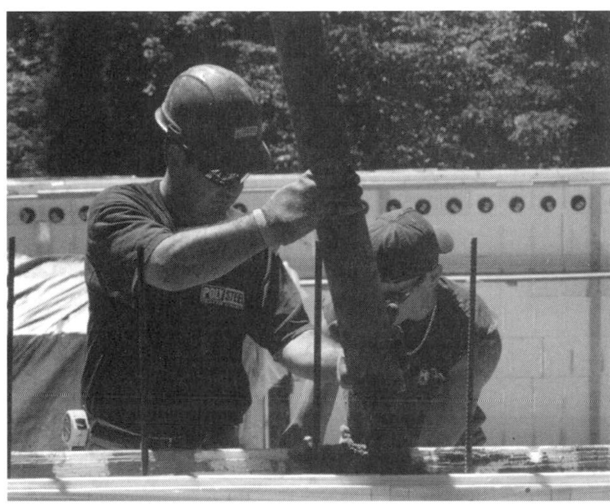

American Polysteel

Figure 2-43. Concrete must be pumped slowly and steadily into insulating concrete forms to prevent pillowing.

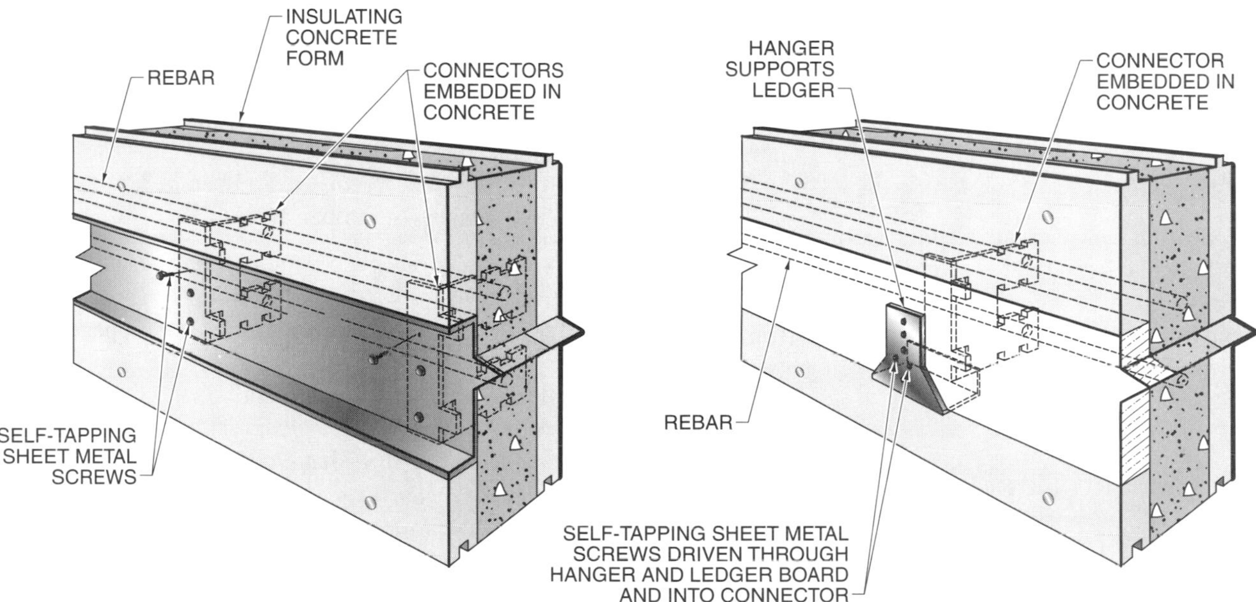

Figure 2-42. Metal hangers and connectors are used to attach wood or metal ledger boards to insulating concrete forms.

Name ______________________________ **Date** ______________

Completion

______ **1.** A(n) ___ panel system consists of a number of prefabricated panel forms tied together.

______ **2.** The construction of many similar panel sections can be facilitated by a(n) ___ table.

______ **3.** A(n) ___ is a rectangular column joined to a concrete wall.

______ **4.** A(n) ___ is one of an odd number of thin layers of wood that are glued together.

______ **5.** The most common form panel thicknesses are ___″, ___″, and ___″.

______ **6.** The standard size plywood panel most commonly used in form construction is ___′ × ___′.

______ **7.** Interior plywood is bonded with ___ glue.

______ **8.** The actual size of a 2 × 6 piece of lumber is ___″ × ___″.

______ **9.** The recommended spacing of braces for walls 10′ high is every ___′.

______ **10.** A 10d nail is ___″ long.

______ **11.** ___ plywood is used to produce special surface effects on concrete walls.

______ **12.** ___, or bending, of a wall form occurs when stiffeners are placed too far apart.

______ **13.** A liberal amount of ___ should be applied to plywood panels to reduce moisture penetration.

______ **14.** Plywood that is ⅝″ thick or more should be fastened with ___d nails.

______ **15.** ___ nails are made of case-hardened steel to resist bending.

Multiple Choice

______ **1.** A ___ nail is commonly used in form construction.
A. common
B. box
C. double-headed
D. all of the above

__________ **2.** ___ is a softwood species used for form lumber.
A. Douglas fir
B. Spruce
C. Pine
D. all of the above

__________ **3.** The type of continuous single-member tie used most often is a ___ tie.
A. snap
B. loop end
C. flat metal
D. spreader

__________ **4.** Waler rods and coil ties are ___.
A. types of snap ties
B. broken off after the concrete hardens
C. internal disconnecting ties
D. secured with wedges

__________ **5.** Snap ties are most often placed after constructing ___ wall(s).
A. the inside form
B. the outside form
C. both form
D. none of the above

__________ **6.** Waler rods and coil ties are secured to the walers with ___.
A. wedges
B. screws
C. double-headed nails
D. large hex nut washers

__________ **7.** The prefabricated panel sections most commonly used are constructed of ___.
A. plywood panels and wood studs
B. plywood panels and metal frames
C. metal panels and wood frames
D. all metal

__________ **8.** A ___ is a grooved section between the spreader cones of a snap tie.
A. helical coil
B. patented tie
C. breakback
D. none of the above

__________ **9.** The main purpose of rebar in concrete is to ___.
A. increase the compressive strength
B. resist vertical loads
C. increase the lateral strength
D. all of the above

__________ **10.** The diameter of a #4 rebar is ___″.

A. ¼
B. ⅜
C. ½
D. ⅝

Identification

__________ **1.** Temporary ribbon board

__________ **2.** Stud

__________ **3.** Footing

__________ **4.** Stake

__________ **5.** Keyway

__________ **6.** Brace

__________ **7.** Base plate

__________ **8.** Sheathing

__________ **9.** Waler

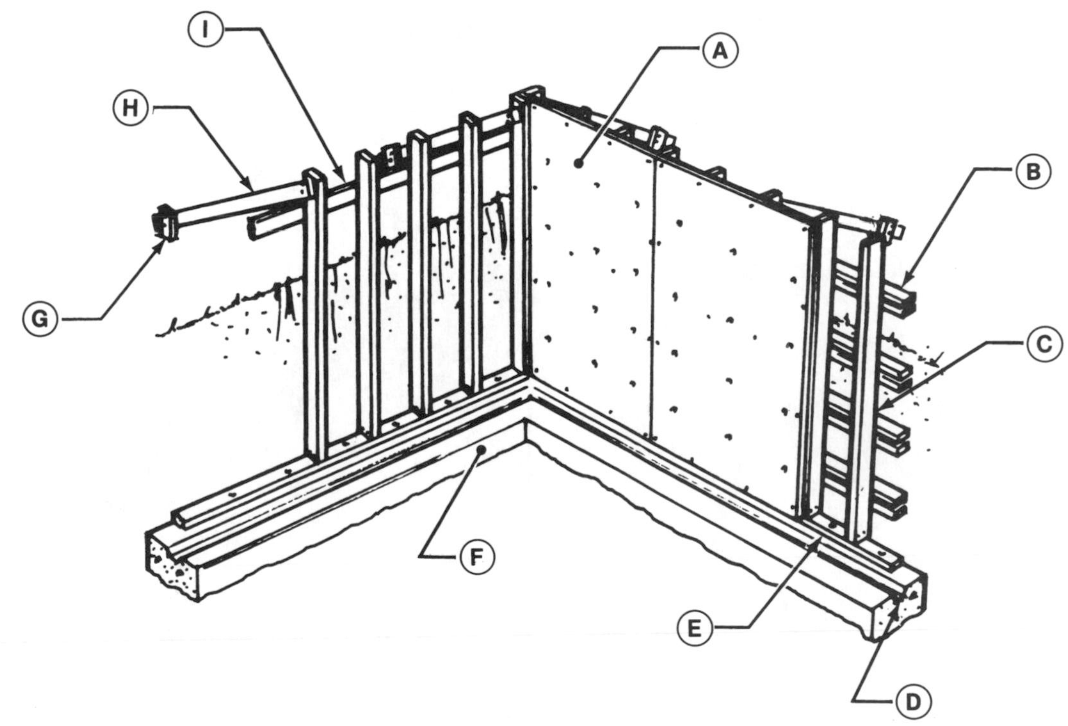

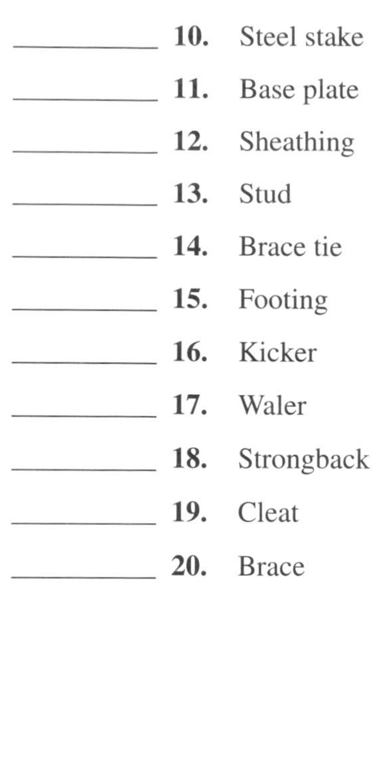

_________ **10.** Steel stake

_________ **11.** Base plate

_________ **12.** Sheathing

_________ **13.** Stud

_________ **14.** Brace tie

_________ **15.** Footing

_________ **16.** Kicker

_________ **17.** Waler

_________ **18.** Strongback

_________ **19.** Cleat

_________ **20.** Brace

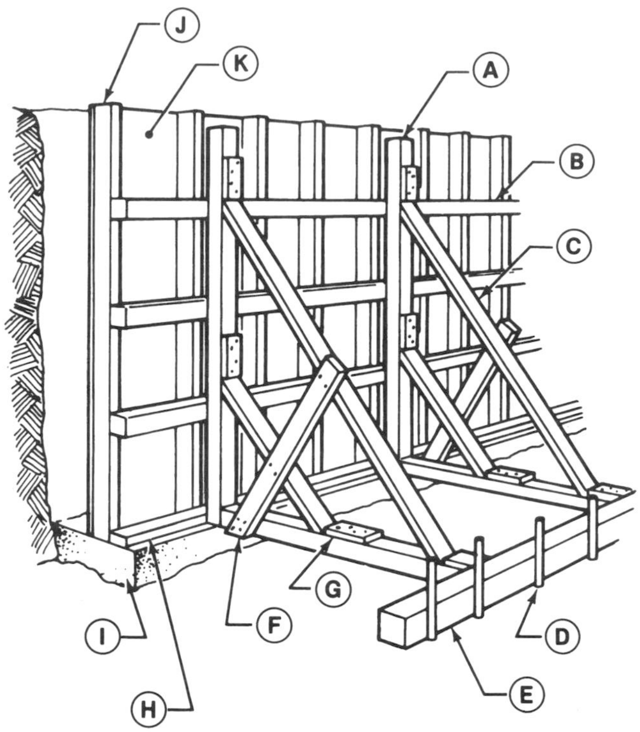

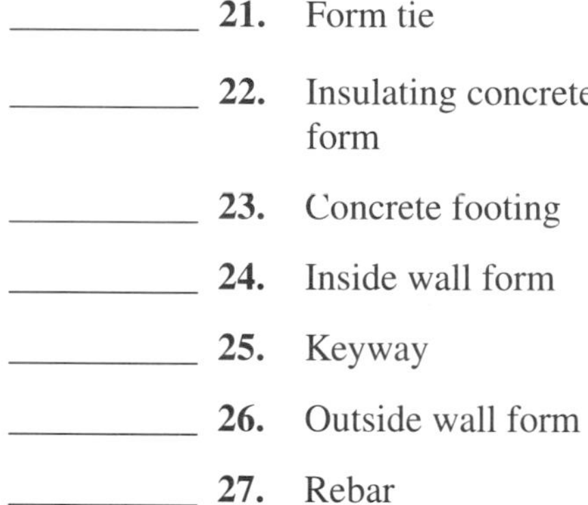

_________ **21.** Form tie

_________ **22.** Insulating concrete form

_________ **23.** Concrete footing

_________ **24.** Inside wall form

_________ **25.** Keyway

_________ **26.** Outside wall form

_________ **27.** Rebar

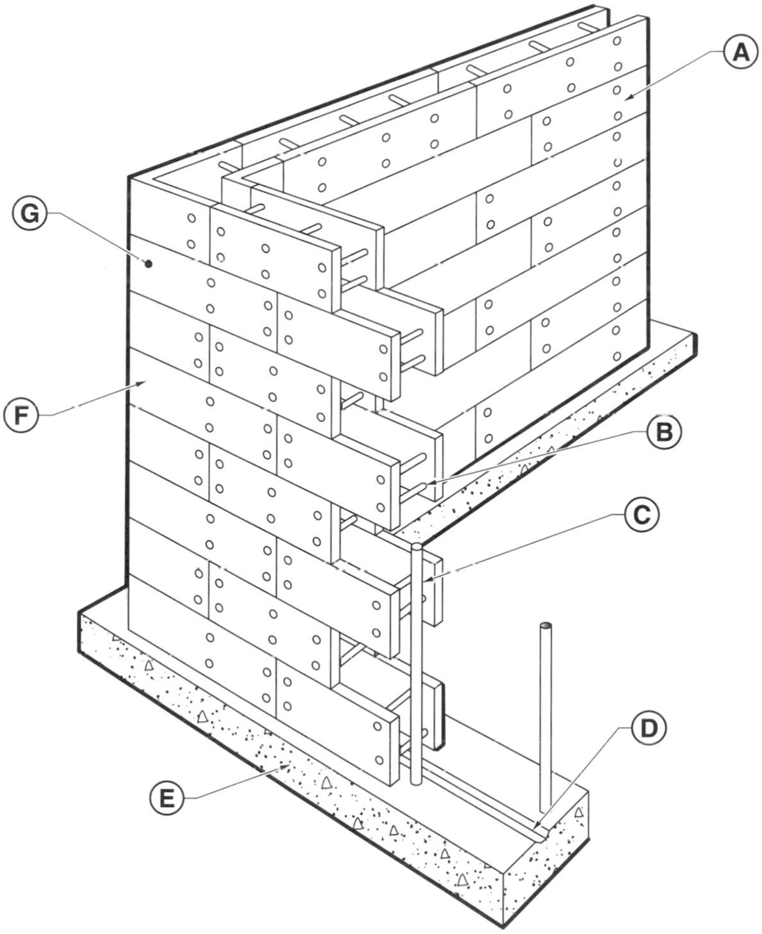

Chapter 3
Residential Foundations

Foundations support residential structures such as one-family and multifamily dwellings. Many types of foundations used for residential structures are also used for light commercial structures. Most foundations are constructed of cast-in-place concrete. The dimensions and shape of a foundation must conform to local building codes and accepted engineering practices.

Full basement, crawl space, and slab-on-grade foundations are commonly used for residential and light commercial structures. Full basement and crawl space foundation designs require foundation wall and footing forms to be constructed. Concrete for slab-on-grade foundations is placed monolithically, or the slab and the foundation are placed separately.

Various foundation shapes are used for residential and light commercial structures, including T-, L-, battered, and rectangular shapes. The design of a foundation is determined by the weight of the foundation, imposed load of the building's superstructure, and bearing capacity of the soil.

Under certain conditions, grade beam or stepped foundations support the superstructure. Grade beam foundations are used with ramped or stepped foundations erected on a sloped surface, or when the bearing capacity of the soil does not provide adequate support for conventional foundation footings. Stepped foundations are commonly constructed on steeply sloped building sites or with crawl space or full basement foundations.

RESIDENTIAL FOUNDATIONS

The major elements of a foundation system are the foundation footings, foundation walls, and pier footings. A *foundation footing* is the part of a foundation that rests directly on the soil, acts as a base for the foundation wall, and distributes the entire foundation load over a wide soil area. The *foundation wall* is a load-bearing wall that extends above and below the ground level. It rests on top of the footing and is secured to the footing. The foundation walls and footings support the entire superstructure of a building.

A *pier footing* is the part of the foundation system that supports wood or metal posts bearing girders. Pier footings, posts, and girders are normally used to provide intermediate support for joist floor systems.

The thickness and width of footings, and thickness and height of foundation walls are determined by soil conditions below the footings and vertical and lateral pressure against the foundation walls. **See Figure 3-1.**

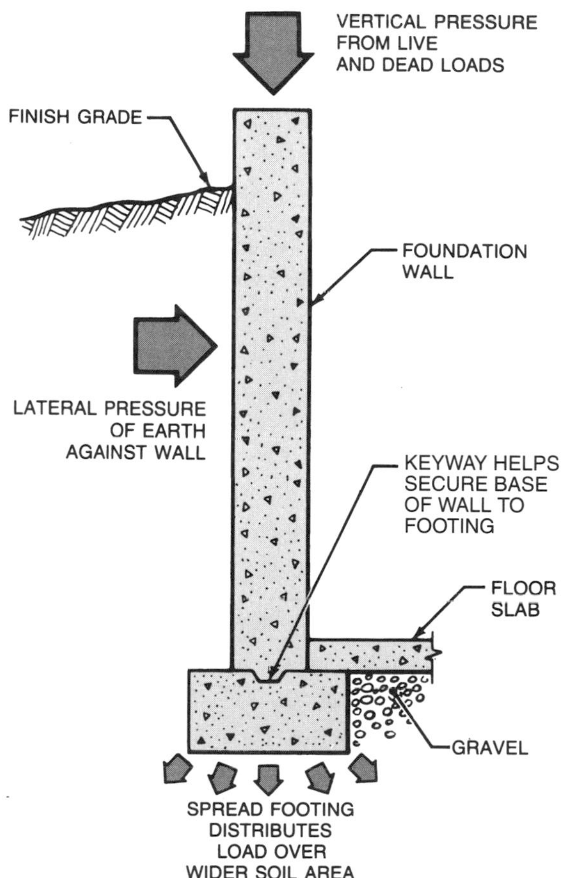

Figure 3-1. A foundation is designed to withstand lateral and vertical pressure. The spread footing distributes the load over a wider soil area.

Vertical pressure on foundation walls is created by dead and live loads that bear down on the walls. A *dead load* is the constant weight of the entire superstructure. Dead loads are calculated by adding the total weight of materials in the building. A *live load* is a varying load supported by a structure. Live loads are loads such as the weight of people, furniture, and snow.

Lateral pressure against a foundation wall is created by the force of the earth against the wall. A high foundation wall must be thicker than a low foundation wall to resist the greater lateral pressure exerted on it.

Bearing capacity and compressibility of the soil are also considered in foundation design. Foundations are expected to settle a small amount over a period of time. However, too much uneven settlement causes structural damage to the building. Therefore, the foundation must be designed to minimize the amount of settlement.

FOUNDATION CONSTRUCTION

The location of a building on the lot must be determined before constructing the foundation. Dimensions for locating the building are determined from the site or plot plan. After the building location is determined and the building lines have been set, the groundwork is started. The groundwork may involve deep excavation work for a full basement foundation or minor grading and shallow trenching for a crawl space foundation.

For full basement foundations, the footing forms are first constructed and the concrete is placed in them. When the concrete in the footings has set sufficiently, the wall forms are constructed and the concrete is placed in them. Low forms used for crawl space foundations may be constructed monolithically. In monolithic construction, an inverted T-shaped form is constructed and the concrete for the footings and walls is placed at the same time.

Foundation Shapes

The T-foundation (inverted-T), L-foundation, rectangular foundation, and battered foundation are commonly used foundation shapes. **See Figure 3-2.** A *T-foundation* is a foundation consisting of a wall placed above a spread footing that extends on both sides of the wall. The T-foundation is the foundation design used most in residential and light commercial construction. In the T-foundation, a wall is placed on a spread footing that rests directly on the soil. A *spread footing* is a base for the wall above that distributes the load of the building over a wider area. Footings for T-foundations should be placed below the frost line, extend below any fill material, and rest on firm, undisturbed soil.

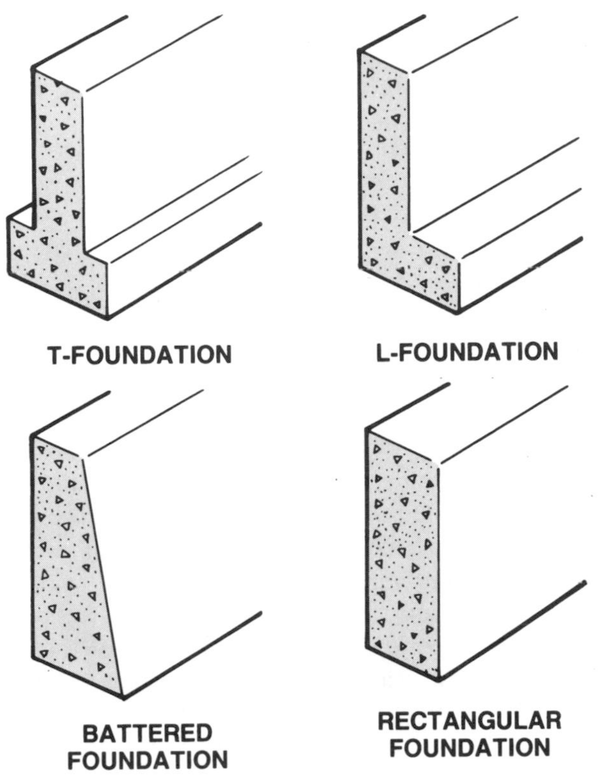

Figure 3-2. Several shapes are commonly used for residential foundations. The T-foundation is used most often.

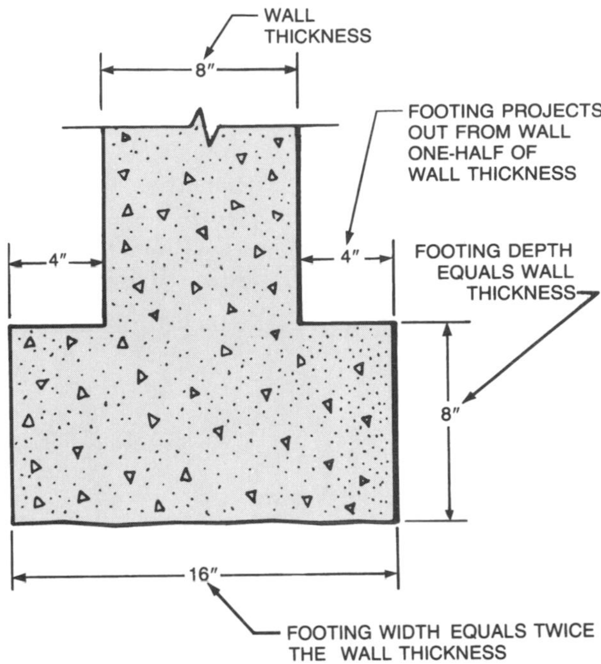

Figure 3-3. A formula is used for calculating the dimensions of a footing resting on soil of average bearing capacity.

The footings for T-foundations commonly support a cast-in-place concrete wall or a wall constructed of load-bearing solid or hollow concrete masonry units. Thickness and width of footings must conform to local code requirements established for soil conditions in the area.

A formula used to determine dimensions for residential and light commercial foundation footings that rest on soil of average bearing capacity states the following: the depth of footings equals the wall thickness, and the width of footings equals twice the wall thickness. **See Figure 3-3.** Another general rule is that footings should never be less than 6″ thick and not less than 1½ times the footing projection.

The walls of a T-foundation rest on the center of the spread footing. Lateral movement at the base of the foundation wall is prevented by a keyway along the top surface of the footing. In most cases, the foundation wall is secured by rebar extending vertically from the footing and tied to rebar placed in the walls.

A keyway along the top surface of a footing prevents lateral movement at the base of a foundation wall.

A foundation wall must be thick enough to support the vertical load of the building and the lateral pressure from the surrounding earth against the wall. A tall foundation wall is exposed to greater lateral pressures than a short foundation wall. The tall foundation wall must be thicker to provide adequate support for the greater pressures exerted upon it. **See Figure 3-4.** Consult the International Residential Code (IRC) or local building codes for specific wall thickness requirements for residential buildings.

When placing concrete, prevent fresh concrete from mounding on rebar, particularly where rebar intersects in beams and girders and whenever a low-slump mixture is used in walls. To prevent concrete from mounding on the rebar, vibrate the concrete around rebar (without touching rebar) to allow concrete to slip into the forms.

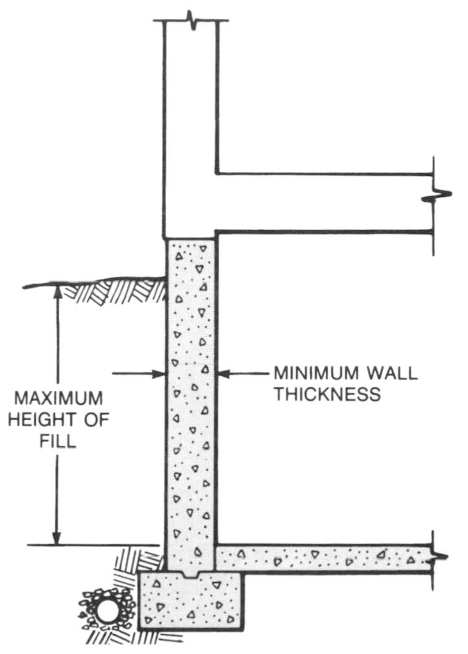

FOUNDATION WALL DIMENSIONS			
Foundation Wall Construction	**Maximum Height of Unbalanced Fill***	**Minimum Wall Thickness**	
		Frame Construction†	**Masonry or Veneer†**
Hollow masonry	3	8	8
	5	8	8
	7	12	10
Solid masonry	3	6	8
	5	8	8
	7	10	8
Plain concrete	3	6	8
	5	6	8
	7	8	8

* in ft
† in in.

Figure 3-4. Concrete and masonry wall dimensions increase as the wall height increases.

Design considerations for the T-foundation also apply to the L-foundation. An *L-foundation* is a foundation that has a footing on only one side of the foundation wall. The main difference between the L-foundation and the T-foundation is that, in the L-foundation design, the wall rests on one edge of the footing rather than at the center. An L-foundation is used where an existing building foundation next to the new building site does not allow room for a T-foundation. The L-foundation may also be used for concrete retaining walls that hold back earth banks.

Rectangular and battered foundations do not have a spread footing as the base. A *rectangular foundation* is a monolithically placed structural support consisting of two vertical faces with no dimensional changes. Small rectangular foundations may be adequate to support light structural loads. They may also be incorporated with slab-on-grade foundation and floor systems. Large rectangular foundations are used as grade beams.

A *battered foundation* is a monolithic structural support consisting of a wall with a vertical exterior face and a sloping interior face. The wide base provides sufficient bearing for the entire wall. Battered foundations may also be used when a new foundation is close to an existing foundation.

Foundation Layout

Foundation layout is based on measurements provided on the plot plan of the prints. A plot plan shows the exact location of the building on the job site. A plot plan commonly indicates the front setback of the building. The *front setback* is the distance from the building to the front property line. The distance from the sides of the building is measured to the side property lines. **See Figure 3-5.**

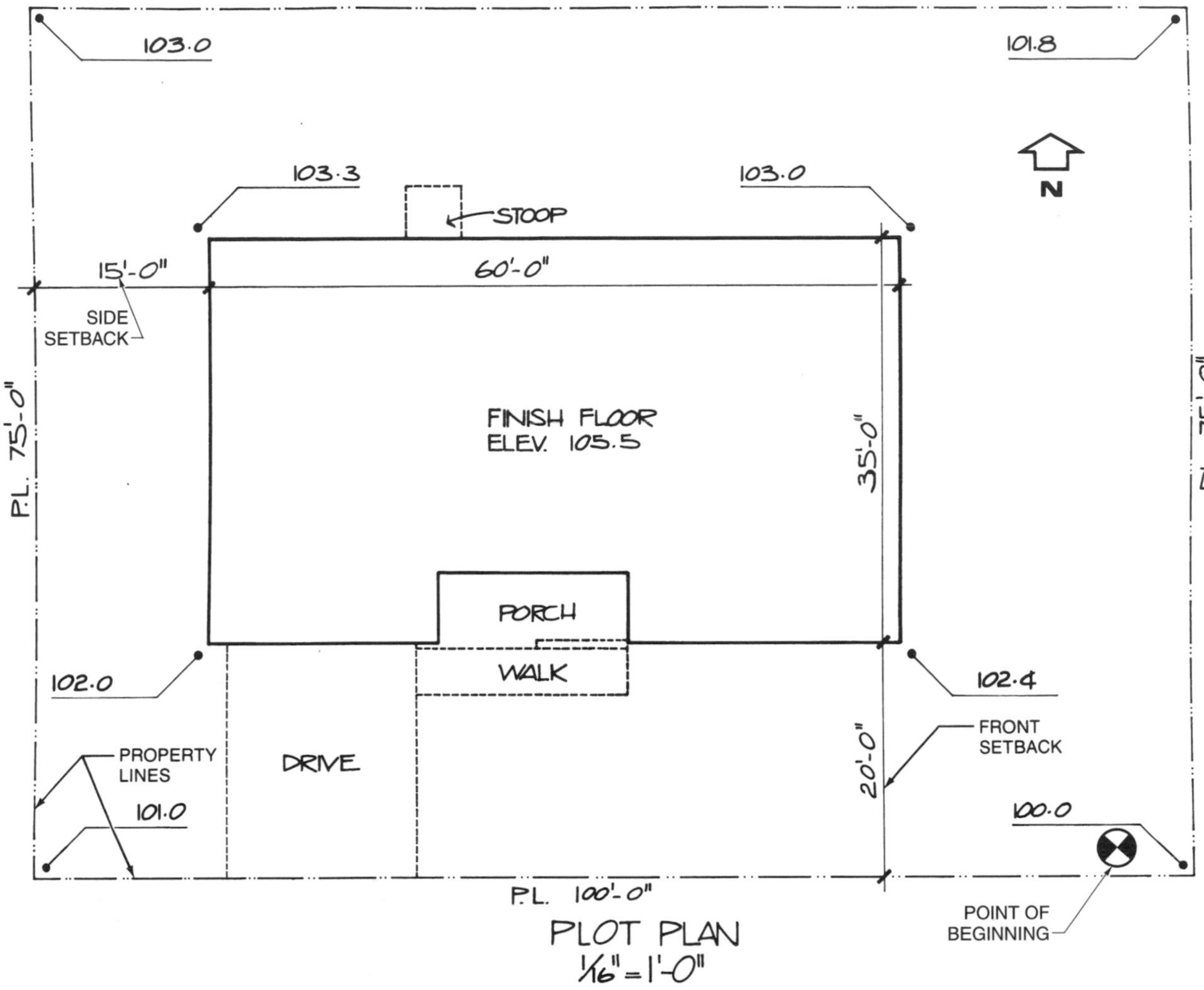

Figure 3-5. The plot plan includes the front setback of the building and the distance from the building to the side property line.

Building lines are set up on the job site, based on information on the plot plan, to establish exact boundaries of the foundation walls. Building lines show the area where the ground must be excavated before foundation construction begins. The building lines also indicate where the foundation forms are to be constructed.

Building corners are located using a leveling instrument before establishing building lines. A transit-level is commonly used to establish building corners. (See Appendix E, Leveling Instruments, for information regarding transit-levels.) Building corners must be located accurately because many measurements are taken from these points. When the building corners have been located, wood stakes are driven at each corner of the building and a nail marking the exact position of the corner is driven into the top of each stake. **See Figure 3-6.**

When a transit-level is not available, building corners may also be established using the 3-4-5 method. The *3-4-5 method* is a layout method used to establish right angles for the building corners. Larger multiples of 3-4-5 (for example, 2 × 3-4-5 = 6-8-10) may also be used when squaring lines over a greater distance. **See Figure 3-7.**

Transit-levels and laser levels must be properly tested and calibrated to ensure accurate layout of the job site. Vibration from passing machinery or other shocks to the instrument may cause it to be knocked out of level. The level should be checked against the job site benchmark(s) throughout the day.

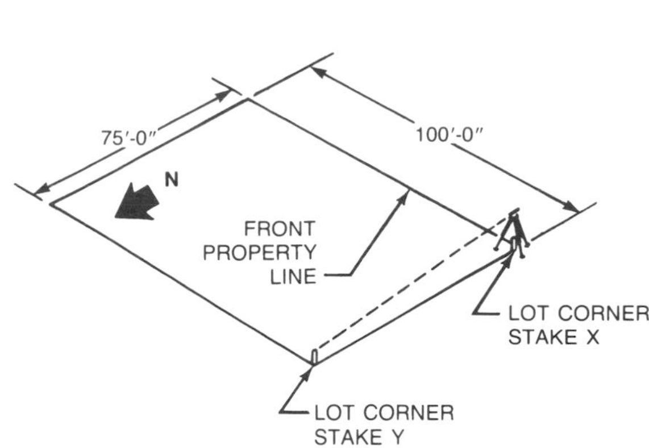

1. Level and plumb transit-level over lot corner stake X. Sight down to lot corner stake Y.

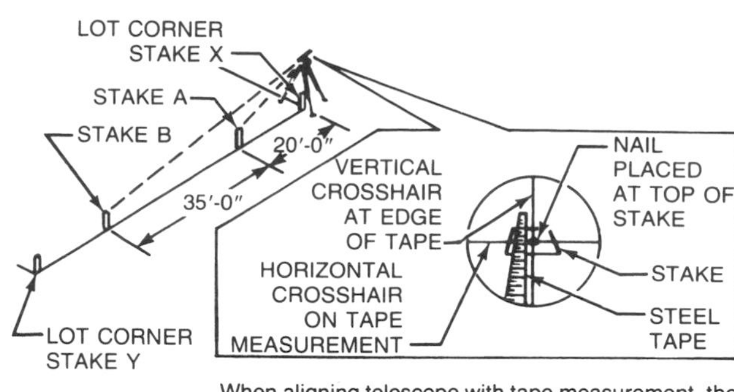

When aligning telescope with tape measurement, the vertical crosshair is at one edge of the tape and horizontal crosshair is at the measurement.

2. Measure front setback (20'-0") using steel tape. Lower telescope until vertical and horizontal crosshairs align with 20'-0" mark. Drive stake A and place a nail. Measure width of building (35'-0"). Raise telescope until crosshairs align with 35'-0" mark. Drive stake B and place a nail.

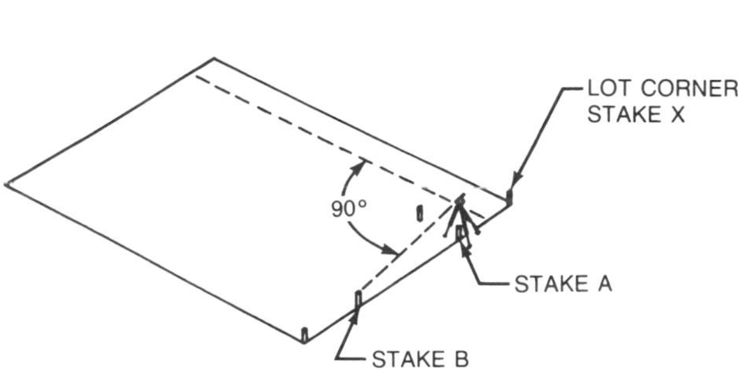

3. Level and plumb transit-level over stake A. Sight down to stake B. Swing telescope 90° to the right.

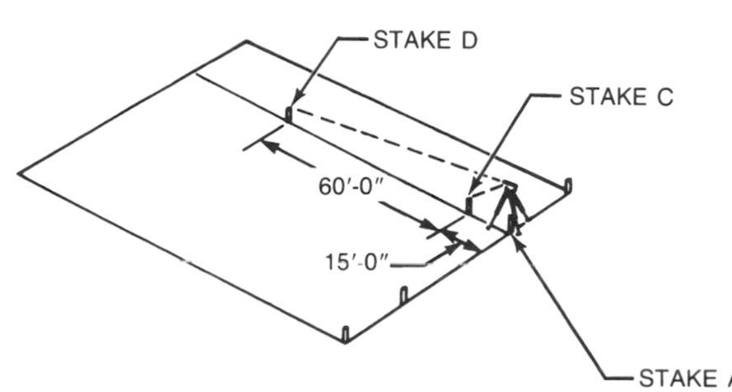

4. Measure distance from side property line to building (15'-0"). Lower telescope until crosshairs align with 15'-0" mark. Drive stake C and place a nail establishing first building corner. Measure length of building (60'-0") from stake C. Raise telescope until crosshairs align with 60'-0" mark. Drive stake D and place a nail establishing second building corner.

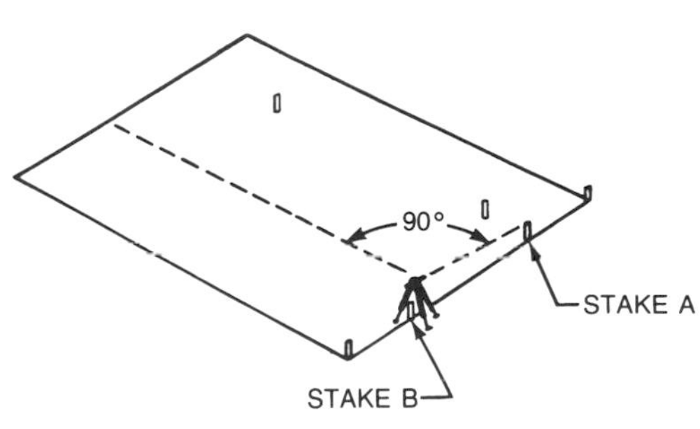

5. Level and plumb transit-level over stake B. Sight back to stake A. Swing telescope 90° to the left.

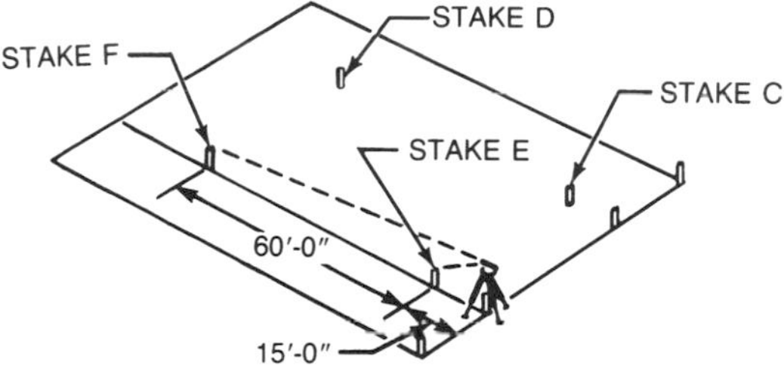

6. Measure distance from side property line to building (15'-0"). Lower telescope until crosshairs align with 15'-0" mark. Drive stake E and place a nail establishing third building corner. Measure length of building (60'-0") from stake E. Raise telescope until crosshairs align with 60'-0" mark. Drive stake F and place a nail establishing fourth building corner.

Figure 3-6. Building corners may be laid out using a transit-level. The dimensions used are shown on the plot plan in Figure 3-5.

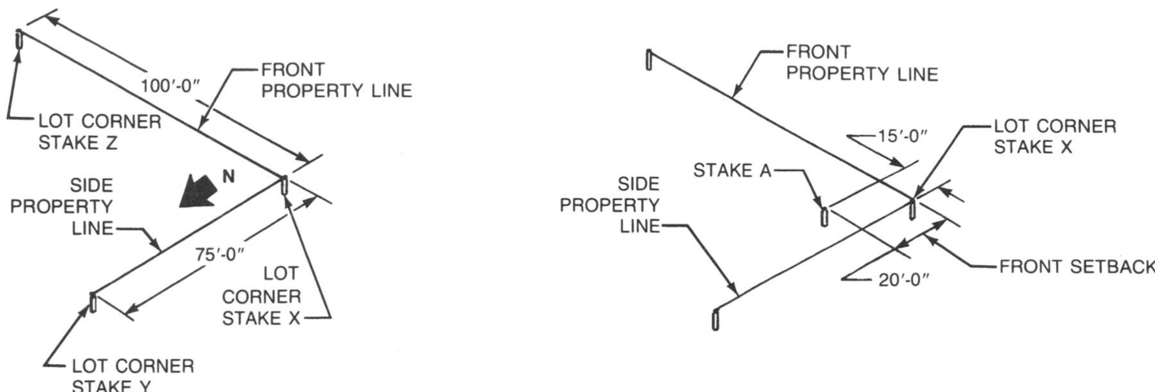

1. Stretch lines from lot corners X, Y, and Z.

2. Measure the front setback (20'-0") from the front property line and the distance from the side property line to the building (15'-0") at the same time. Drive stake A and place a nail establishing first building corner.

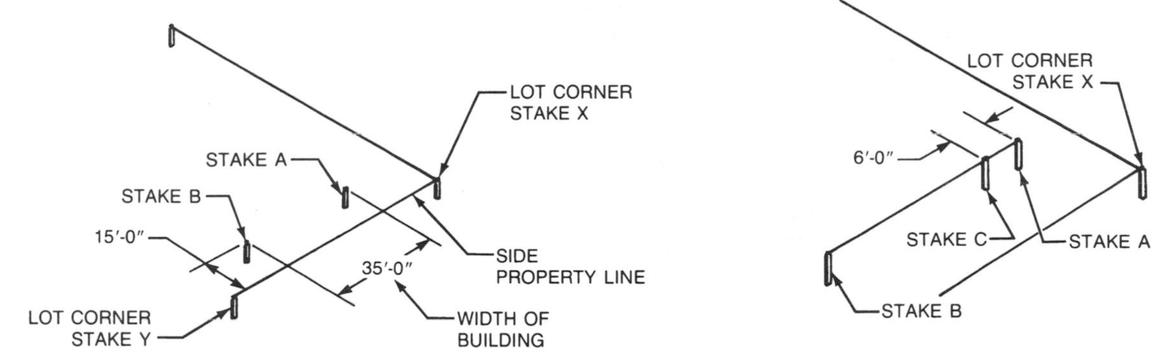

3. Measure the distance from the side property line to the building (15'-0"). Measure the width of the building (35'-0") from stake A. Drive stake B and place a nail establishing second building corner.

4. Stretch a line between stakes A and B. Drive stake C 6'-0" from stake A and align with stakes A and B. Drive a nail exactly 6'-0" from the nail on stake A.

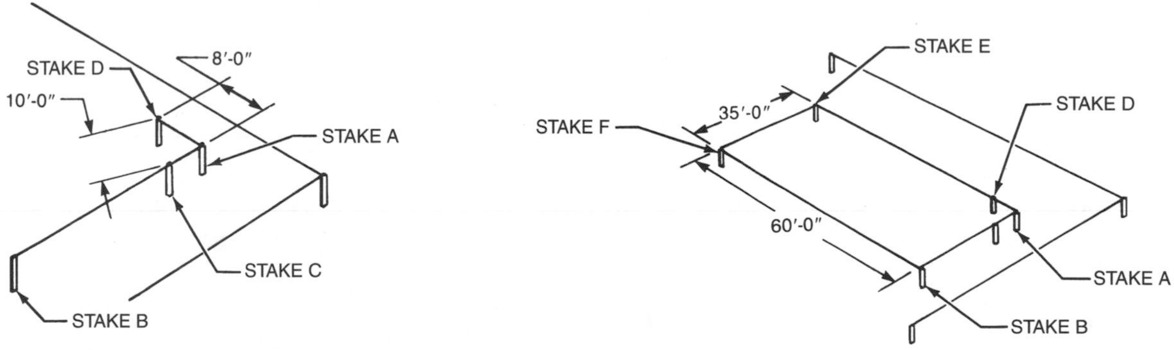

5. Measure 8'-0" from stake A and 10'-0" from stake C. Drive stake D and place a nail exactly where the measurements intersect. Angle DAC is a 90° angle.

6. Stretch line from stake A and over stake D. Measure length of building (60'-0") from stake A. Drive stake E and place a nail establishing third building corner. Measure length of building from stake B (60'-0"). Measure width of building from stake E and place a nail establishing fourth building corner.

Figure 3-7. The 3-4-5 method can be used to lay out building corners. The dimensions used are shown on the plot plan in Figure 3-5.

Batterboards. After building corners have been established, building lines (nylon string or light wire) are stretched to show the exact boundaries of the building. Since it is not practical to stretch building lines by nailing them directly to the tops of the corner stakes, batterboards are used to hold the building lines in place. A *batterboard* is a 1″ or 2″ level piece of dimensional lumber formed to hold the building lines in position and to show the exact boundaries of a building. Batterboards are usually erected 4′ to 6′ behind each building corner to provide room for excavation or form construction. Batterboards are nailed to 2 × 3 or 2 × 4 wood or metal stakes that have been driven around the corner stakes. **See Figure 3-8.**

Excavation of a building site occurs after the building lines have been established.

On level lots, the batterboards should be set level with each other at all four corners. The heights of the batterboards are established on the stakes by sighting through a builder's level or transit-level. Stakes are driven solidly into the ground and braced in all directions. Shifting of a batterboard after the building lines have been set up may result in inaccurate foundation layout.

Setting up Building Lines. Building lines are stretched from the four building corners and secured to the tops of the batterboards. A plumb bob or straightedge and level are used to ensure the building lines intersect directly over the corner stakes of the building. **See Figure 3-9.** The measurements between the lines should be verified to ensure that they are accurate and conform to the dimensions on the foundation plan. Even though the building line dimensions may be accurate, the lines may still be out of square.

Building codes require that foundation walls extend above the outside grade level to prevent water from entering the basement. The International Residential Code (IRC) states that foundation walls should extend at least 8″ above the finish grade. Extended foundation walls must be fully waterproof. Foundation walls also hold the sill plate high enough to prevent damage from moisture and insects.

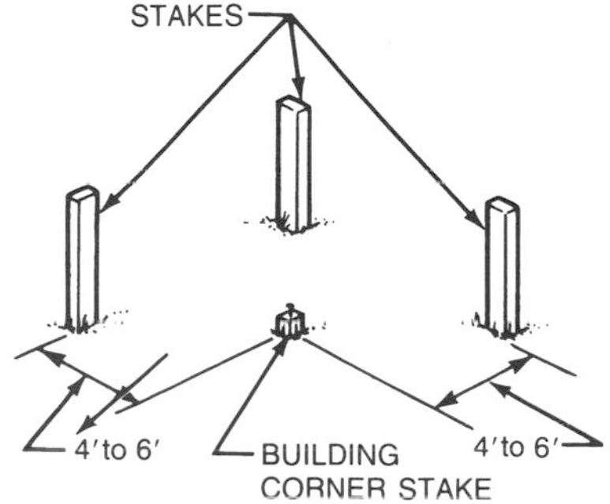

1. Drive three stakes 4′ to 6′ behind the building corner stakes.

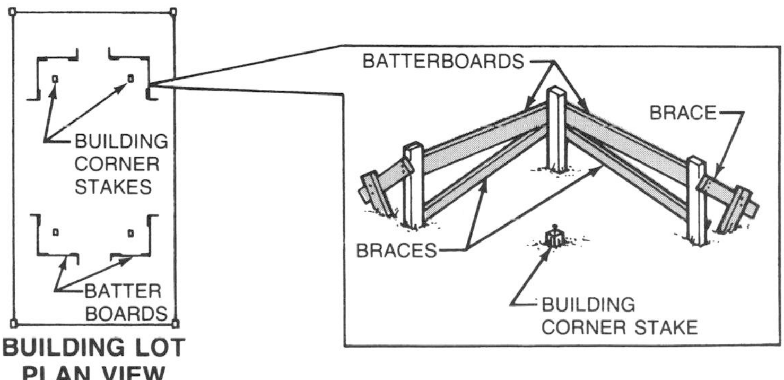

2. Level and nail batterboards to the stakes. Nail braces between the stakes and to the outside stakes.

Figure 3-8. Batterboards hold the building lines that establish the exact position of the foundation.

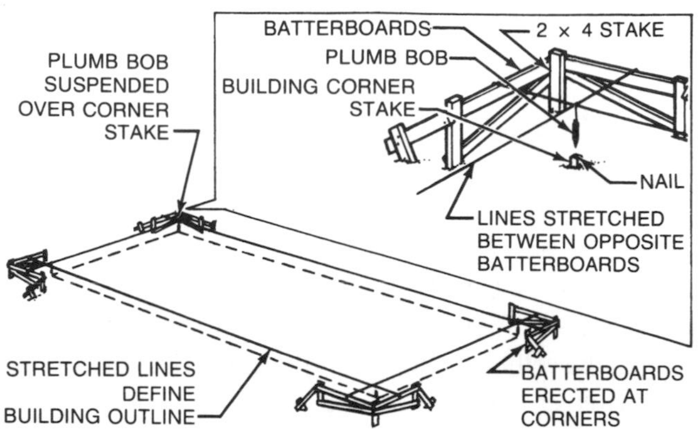

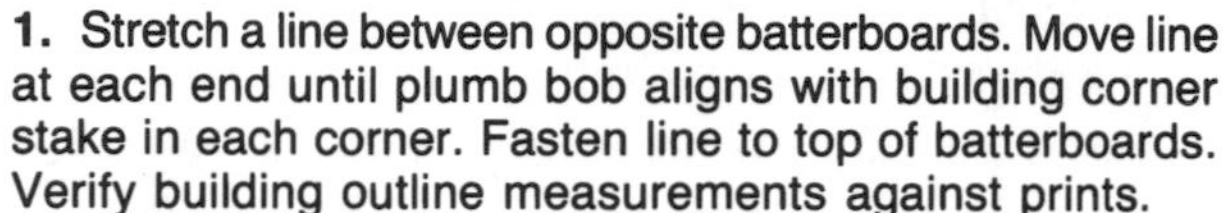
1. Stretch a line between opposite batterboards. Move line at each end until plumb bob aligns with building corner stake in each corner. Fasten line to top of batterboards. Verify building outline measurements against prints.

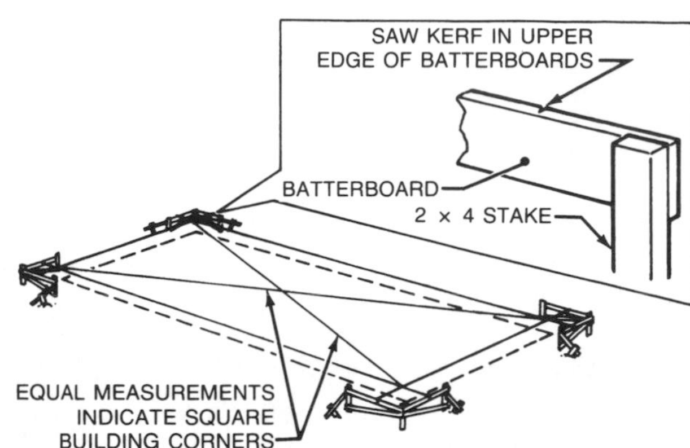

2. Measure diagonal corners of building lines. Equal diagonal measurements indicate that building lines are square. Cut saw kerfs in upper edge of batterboards to hold lines in place.

Figure 3-9. Building lines must be set up carefully to ensure accurate foundation layout.

To verify the squareness of the building lines, measure diagonally across the building lines. If the diagonal measurements are equal, the building lines are square to each other. If the diagonal measurements are different, the building lines are out of square and adjustments must be made. The actual amount that the building lines are out of square is one-half the difference between the diagonal measurements. The correction can often be made by shifting the lines at two corners of the building layout a small amount. For example, if the difference in the diagonal measurements is ¾″, the line must be shifted ⅜″ (¾″ ÷ 2 = ⅜″). By moving the building lines a small amount, one of the diagonal measurements is shortened and the other is lengthened. **See Figure 3-10.**

A ⅛″ kerf (saw cut) is made in the upper edge of the batterboards to secure the building lines. A *kerf* is a cut or groove made by a saw blade. This guarantees that the lines will stay in their proper position and makes it convenient to reset the lines during foundation construction.

FULL BASEMENT FOUNDATIONS

A full basement foundation provides an area below the superstructure of the building for living space or storage. The basement area is commonly below the ground surface. **See Figure 3-11.** Concrete pier footings act as bases for posts or columns that support girders. Girders provide central support for the floor directly above. Basement floors have concrete slabs that are a minimum of 4″ thick. Information regarding walls, piers, stairways, and window and door openings is provided on the prints. **See Figure 3-12.**

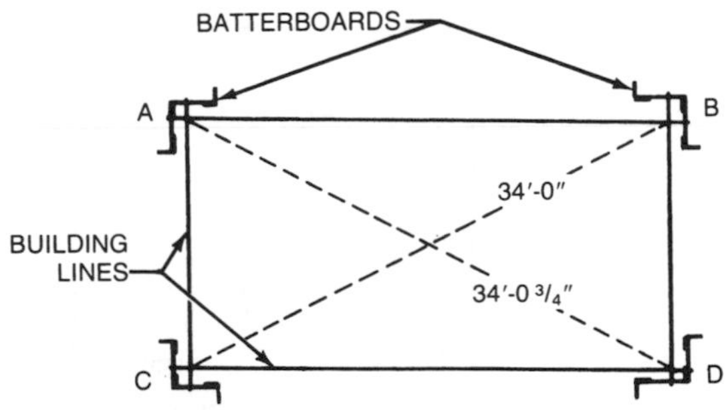

1. Measure distance between diagonal corners. In this example, C-B measures 34′-0″ and A-D measures 34′-0 3/4″. The difference, 3/4″, is twice the amount of adjustment to be made (34′-0 3/4″ − 34′-0″ = 3/4″).

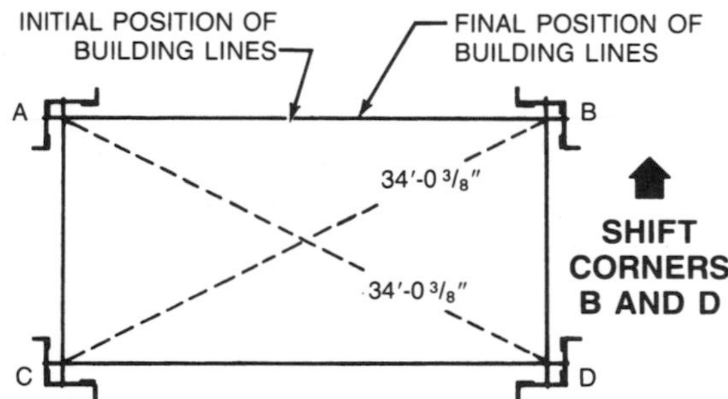

2. Shift lines at B and D 3/8″ (3/4″ ÷ 2 = 3/8″). This adjustment shortens diagonal A-D and lengthens diagonal C-B to obtain equal measurements (34′-0 3/8″).

Figure 3-10. Diagonal measurements are taken to determine if the building lines are square. The corner stakes are adjusted if an error is encountered.

Figure 3-11. The basement area of a full basement foundation extends below the ground level and, after fill is placed around the exterior, the wall should extend at least 8″ above finish grade.

Building codes require that foundation walls extend above the outside grade level to prevent water from entering the basement. A common requirement is that foundation walls extend at least 8″ above the finish grade. The extended foundation wall also holds the sill plate high enough to minimize damage from moisture and insect attack. A *sill plate* is a wood member placed on top of the wall. The foundation wall should measure a minimum of 7′-0″ from the floor slab to the bottom of the ceiling joists to allow adequate headroom in the basement. Eight-foot high walls above the footing are commonly used to provide a clearance of 7′-8″ above a 4″ floor slab. The basement walls should be well insulated to resist water and vapor penetration. A bituminous material such as asphalt is often applied to the outside surface of the wall for waterproofing. **See Figure 3-13.**

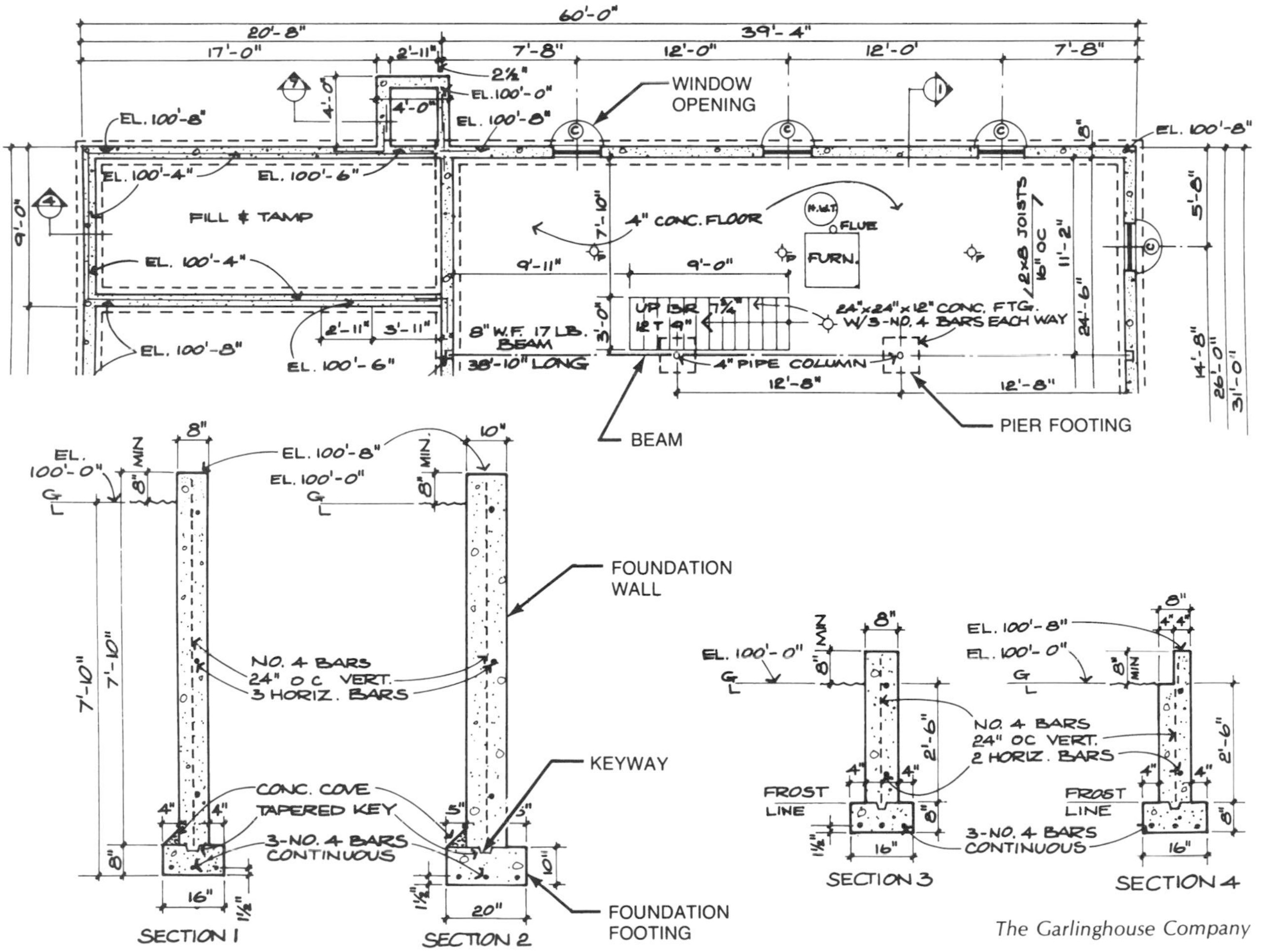

The Garlinghouse Company

Figure 3-12. A set of prints provides information for a full basement residential foundation. The details and foundation section drawings are referenced by numbers on the foundation plan.

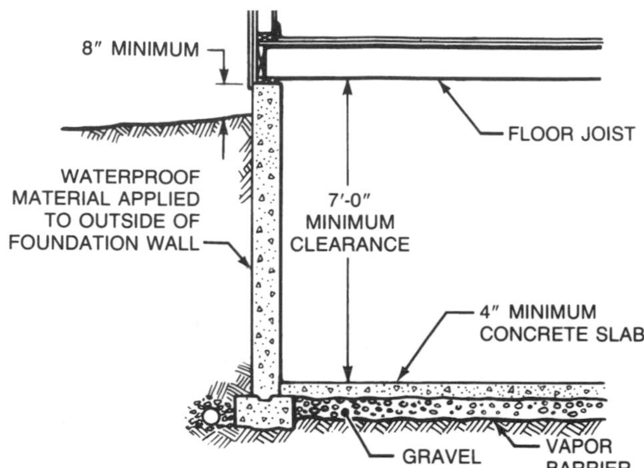

Figure 3-13. The foundation wall should measure a minimum of 7′-0″ from the floor slab to the bottom of the ceiling joists to allow adequate headroom in the basement. The basement walls should be waterproofed to resist water and vapor penetration.

Constructing Footing Forms

Footing forms for a full basement foundation are constructed after the excavation work has been completed. Excavations for below-grade basements should extend at least 2′ outside the building lines to provide ample working space for the formwork. More than 2′ may be necessary in loose or porous soil. An excavation should extend to the bottom of the floor slab and is determined by the height of the foundation wall and how much the wall extends above the finish grade. The bottom of the excavation must be level. Therefore, the highest elevation point around the perimeter of the excavation should be used as the reference point for determining the depth of an excavation. **See Figure 3-14.**

Footing forms are positioned by measuring the required distance from the building lines. A transit-level, builder's level, or laser transit-level, is used to establish the elevations of the footings. (See Appendix E, Leveling Instruments, for information on reading grades and elevations.) Footings are formed using an earth-formed or constructed footing method. Earth-formed footings can be used in firm and stable soil. An *earth-formed footing* is a footing formed by digging a trench to the dimensions of the footing and filling it with concrete. Precautions must be taken to avoid collapse of the sides of the trench.

Constructed wood forms are required in loose and unstable soil conditions. After the stakes are driven into the ground, 1″ or 2″ boards are nailed to the stakes and held the correct distance apart with form ties. **See Figure 3-15.** When using boards, stakes are placed farther apart and less bracing is required. Although wood is commonly used for stakes and bracing, many metal devices are available.

Keyways and Reinforcement. Keyways and rebar tie the footings to the foundation walls. A *keyway* is a tapered groove formed in concrete at the top surface of a spread footing. Keyways are formed by pressing key strips (chamfered 2 × 4s) into the concrete immediately after the concrete has been placed. **See Figure 3-16.** Keyways may also be formed by securing a key strip to the top of the footing form with a crosspiece before placing the concrete. Keyways help to secure the bottom of the foundation wall to be placed on top of the footing. Keyways are generally used for foundation walls and are optional for crawl space walls.

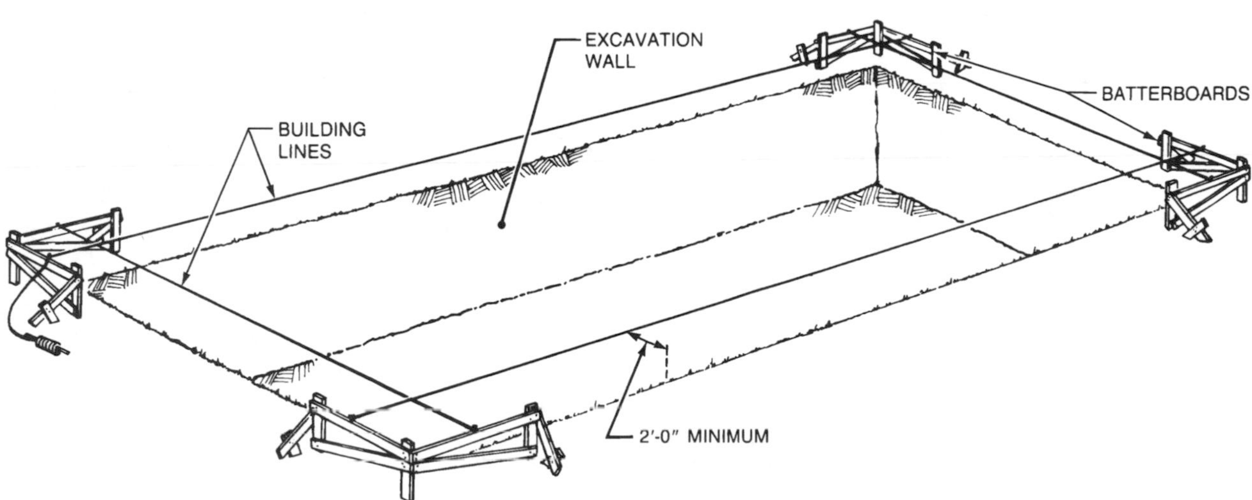

Figure 3-14. Excavation for below-grade basements should extend at least 2′-0″ outside the building lines to allow sufficient room for form construction.

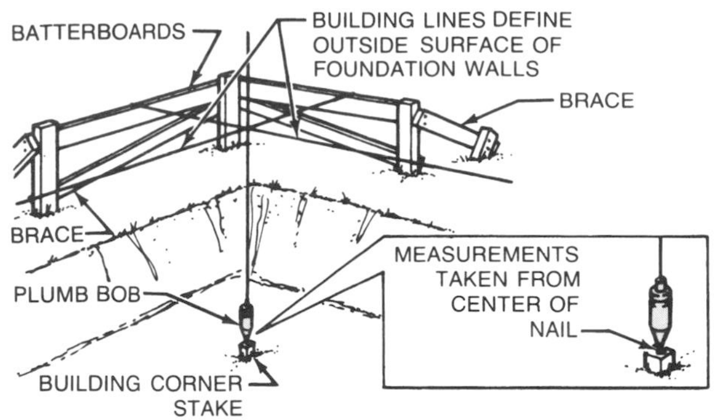

1. Stretch building lines on batterboards. Suspend a plumb bob from each building corner intersection. Drive stakes and place nails to establish building corners.

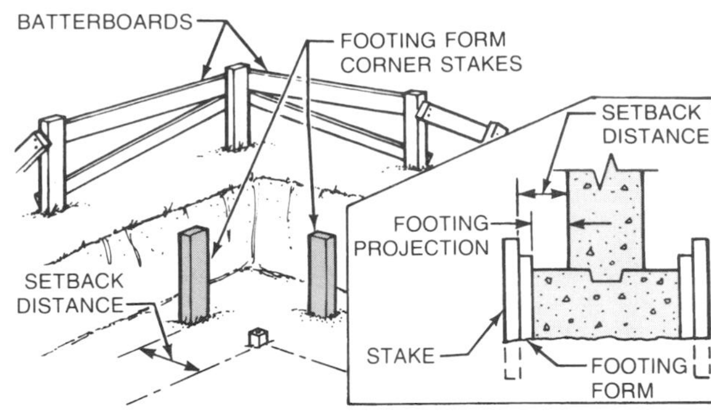

2. Measure the distance that the footing projects beyond the foundation wall plus the thickness of one form board. Drive two footing corner stakes and remove building corner stake.

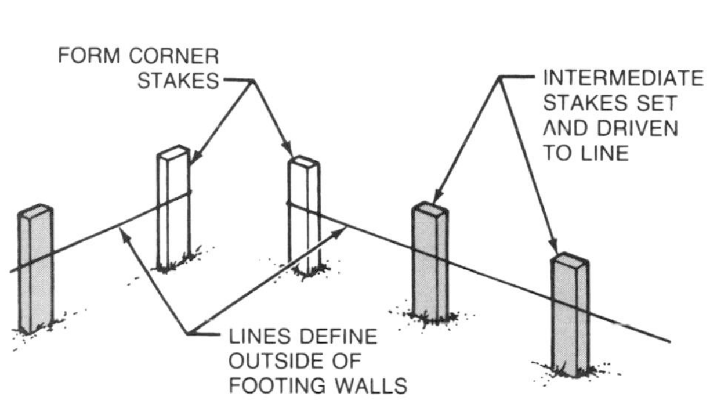

3. Stretch lines between footing corner stakes. Set intermediate stakes to lines and drive into position.

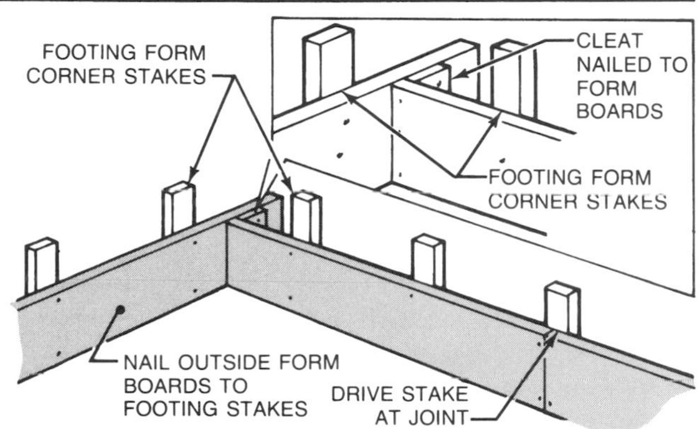

4. Mark the footing elevation on all stakes using a transit-level or builder's level. Nail form boards to stakes, running one side past the adjoining side. Nail a cleat to the corner for strength.

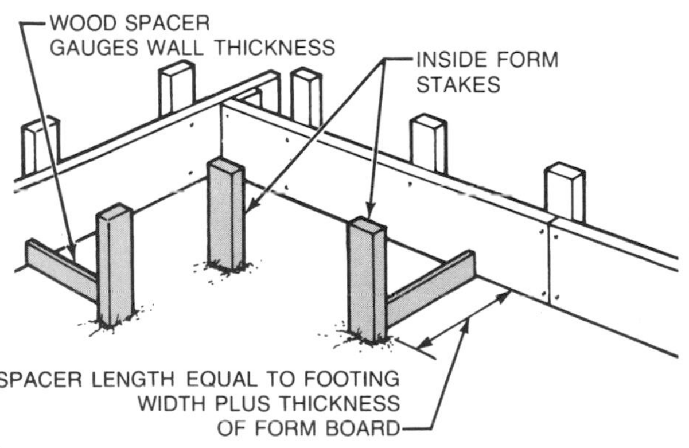

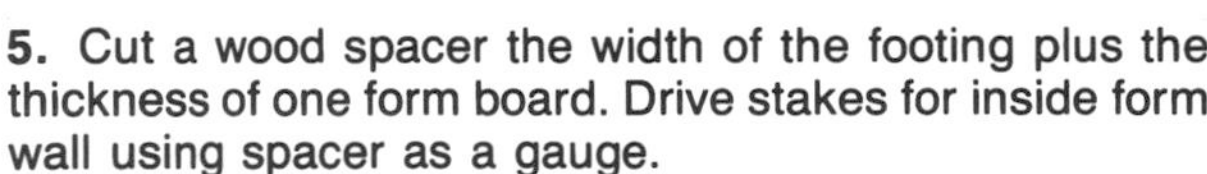

5. Cut a wood spacer the width of the footing plus the thickness of one form board. Drive stakes for inside form wall using spacer as a gauge.

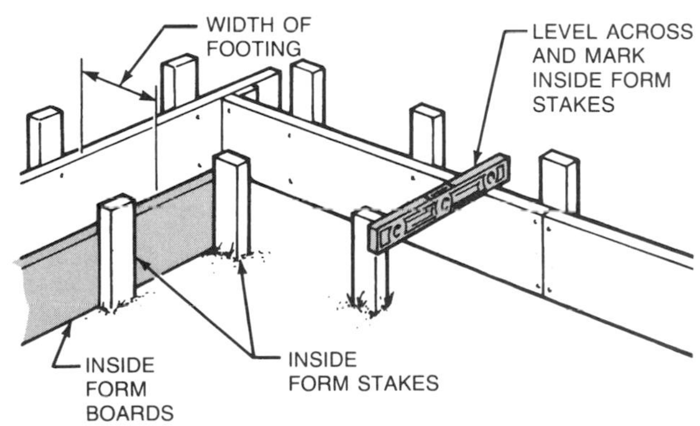

6. Level across from the outside form boards with a hand level and mark the inside form stakes. Nail inside form boards to stakes.

Figure 3-15. Wood-formed footings are constructed for loose and unstable soil conditions. (continued)

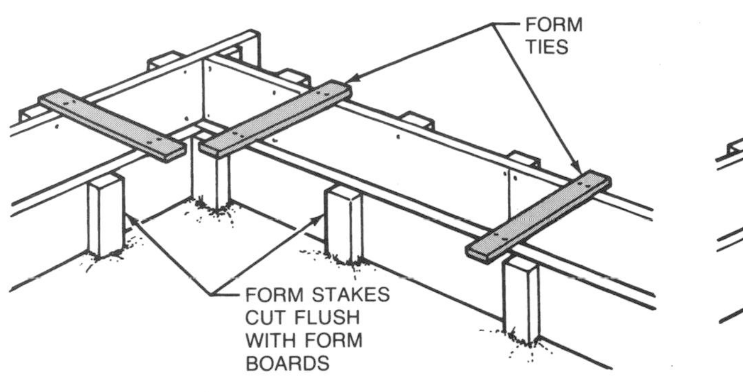

7. Nail form ties across tops of form boards. Cut form stakes flush with top of form boards.

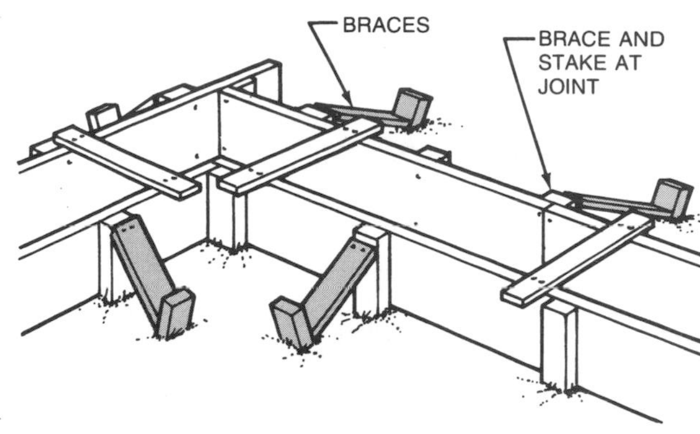

8. Drive stakes and brace all form walls.

Figure 3-15 (continued). Wood spacers are used to establish consistent footing form width. Form ties secure the form boards in position.

Figure 3-16. The keyway is formed by pressing a key strip into the soft concrete.

Vertical rebar extending above the footing, and a keyway are used to secure the foundation wall to the footing. The vertical rebar later tie into rebar placed in the concrete or masonry walls. **See Figure 3-17.** The size and number of rebar required is provided on section views of the foundation plan. The local building code should also be consulted to verify rebar requirements for the area.

Constructing Wall Forms

Wall forms are constructed after the concrete has set in the footings. Outside form walls are usually constructed first to facilitate placement of reinforcement.

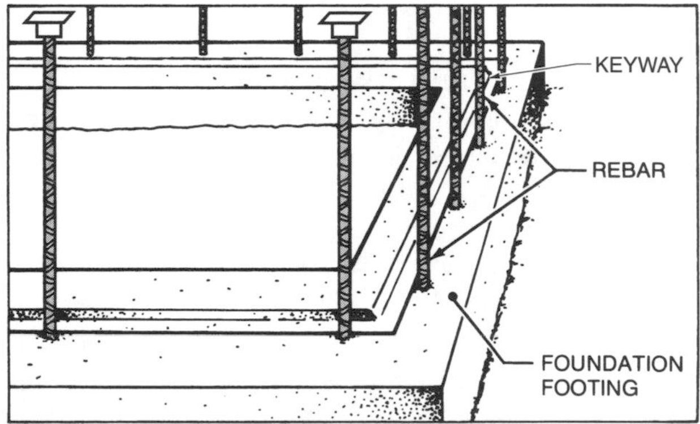

Figure 3-17. Vertical rebar extending from the footing will tie into rebar placed in the wall constructed over the footing.

Base plates are used as a base for outside form walls of built-in-place or panel forming systems. After the base plates are secured, outside form walls are erected and carefully plumbed and aligned. **See Figure 3-18.** Rebar and electrical conduit and other utilities are then positioned.

Wall forms are erected after the concrete for the footings has properly set.

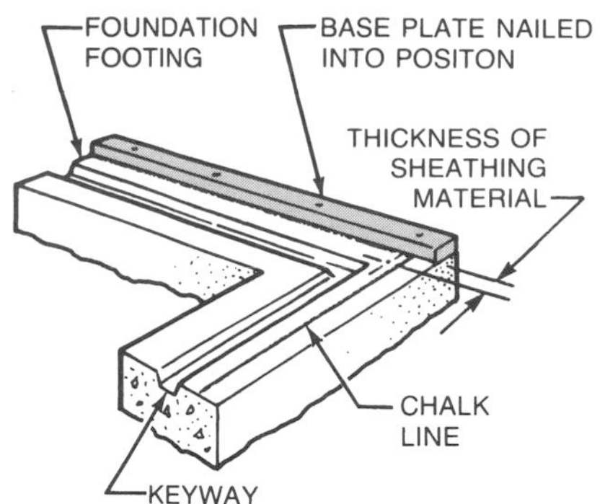

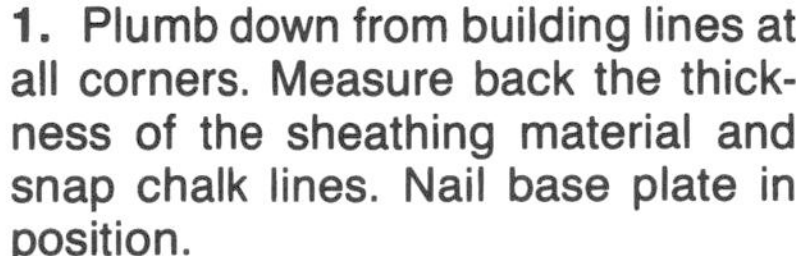

1. Plumb down from building lines at all corners. Measure back the thickness of the sheathing material and snap chalk lines. Nail base plate in position.

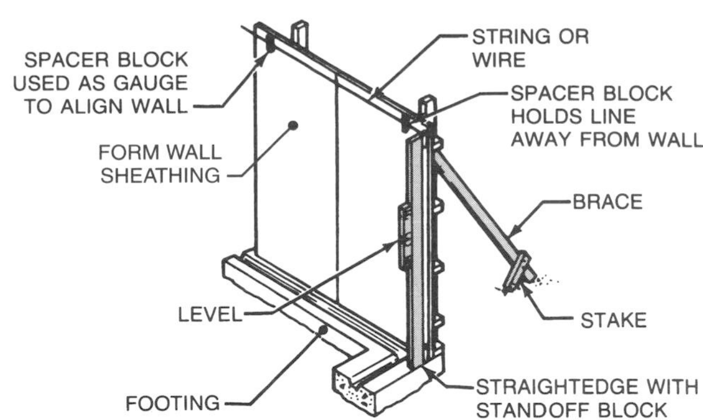

2. Construct form wall. Plumb walls with level and straightedge with 3/4″ × 3/4″ × 4″ standoff block nailed to edge. Brace ends of form walls. Stretch a line from one end of wall forms to the other and hold away with 3/4″ spacer blocks. Align form walls by using a third spacer block as a gauge. Brace where necessary.

Figure 3-18. Form walls must be plumbed and aligned properly.

If window or door bucks are required to provide an opening for windows and doors, they are positioned and fastened to the outside form walls. If beam pockets are required, they are also formed at this time. A *beam pocket* is a space in the foundation wall that receives a beam. The top surface of the beam is flush with the top of the foundation wall. When wood beams are used, the pockets should provide 4″ minimum bearing surface and ½″ clearance around the sides and end of the beam.

After the outside form walls are positioned, the inside form walls are doubled up and tied together. *Doubling up* is the placing of the second or opposite form wall. **See Figure 3-19.**

CRAWL SPACE FOUNDATIONS

A crawl space foundation is often used as an alternate foundation design in a building that does not require a full basement. In crawl space foundations, or basementless foundations, short walls are constructed over spread footings. A crawl space foundation facilitates the placement of plumbing, electrical, and other utilities. A T-foundation is commonly used and is placed monolithically. The perimeter of the superstructure is supported by the T-foundation, and pier footings provide intermediate support. Information required for the construction of a crawl space foundation is found in the foundation plan. **See Figure 3-20.**

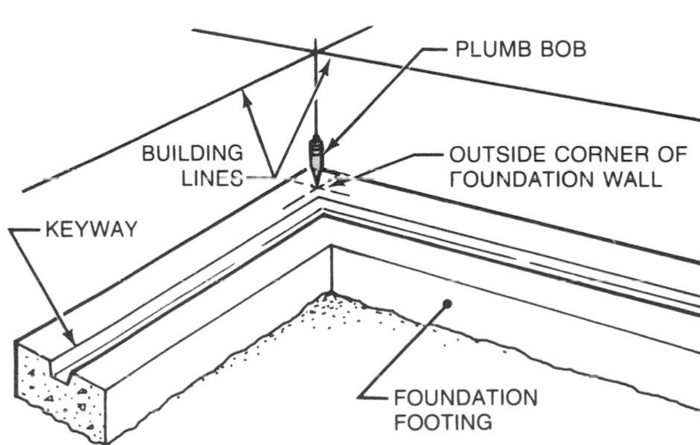

1. Plumb down from the building lines and mark the outside corners of the foundation wall on the footing.

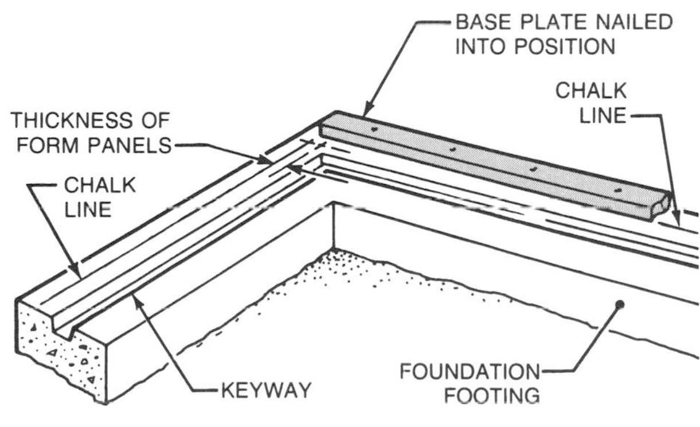

2. Measure the thickness of the form panels to the outside of the corner and snap lines. Nail base plates next to the chalk line with concrete nails.

Figure 3-19. The position of full basement wall forms is laid out on top of the foundation footing. (continued)

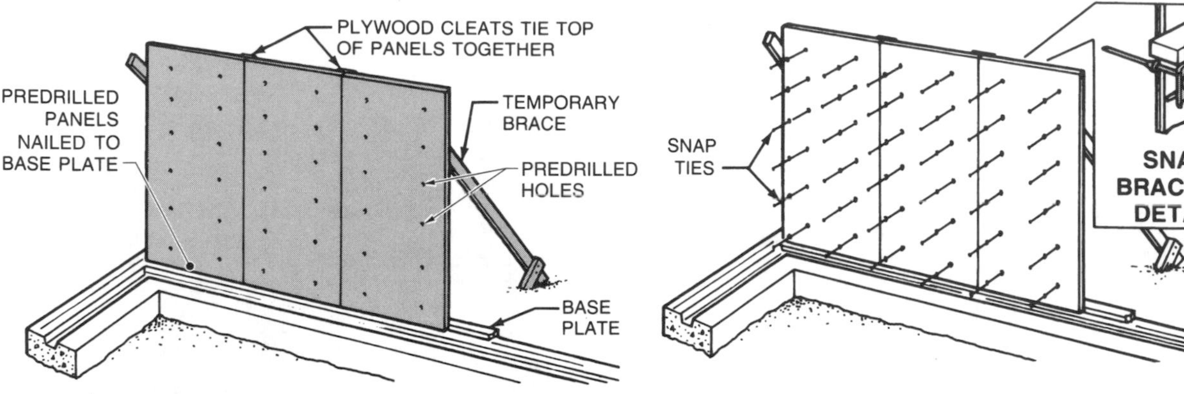

3. Nail a predrilled panel into place and support with a temporary brace. Nail additional panels into place and tie tops together with plywood cleats. Brace every third or fourth panel.

4. Place snap ties into predrilled holes and secure with snap brackets and wedges.

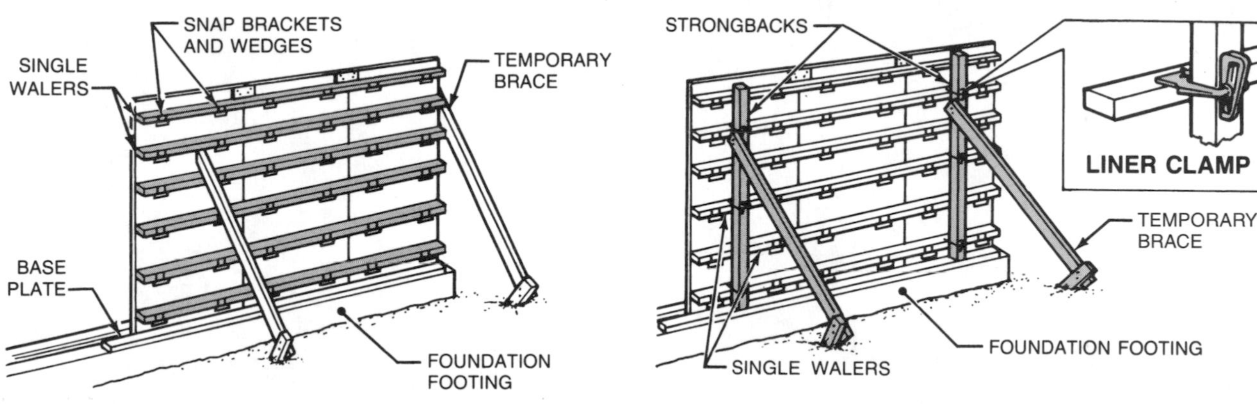

5. Place a row of walers and hand tighten wedges. Place remaining walers and tighten with a hammer.

6. Attach strongbacks to walers. Align top of wall to a line and nail braces to strongbacks.

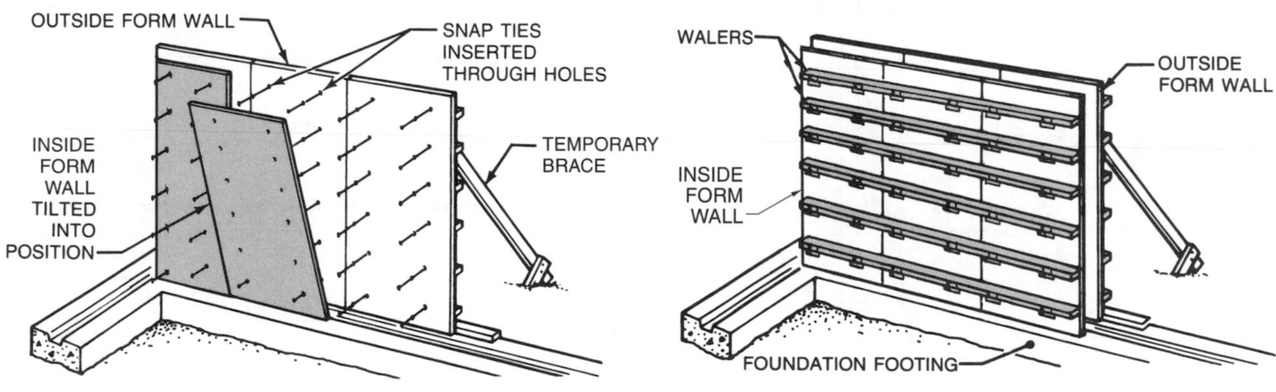

7. Tilt inside form wall into position while inserting ties through the panels.

8. Place snap brackets, wedges, and walers for inside form wall.

Figure 3-19 (continued). Full basement wall forms are reinforced with strongbacks and walers.

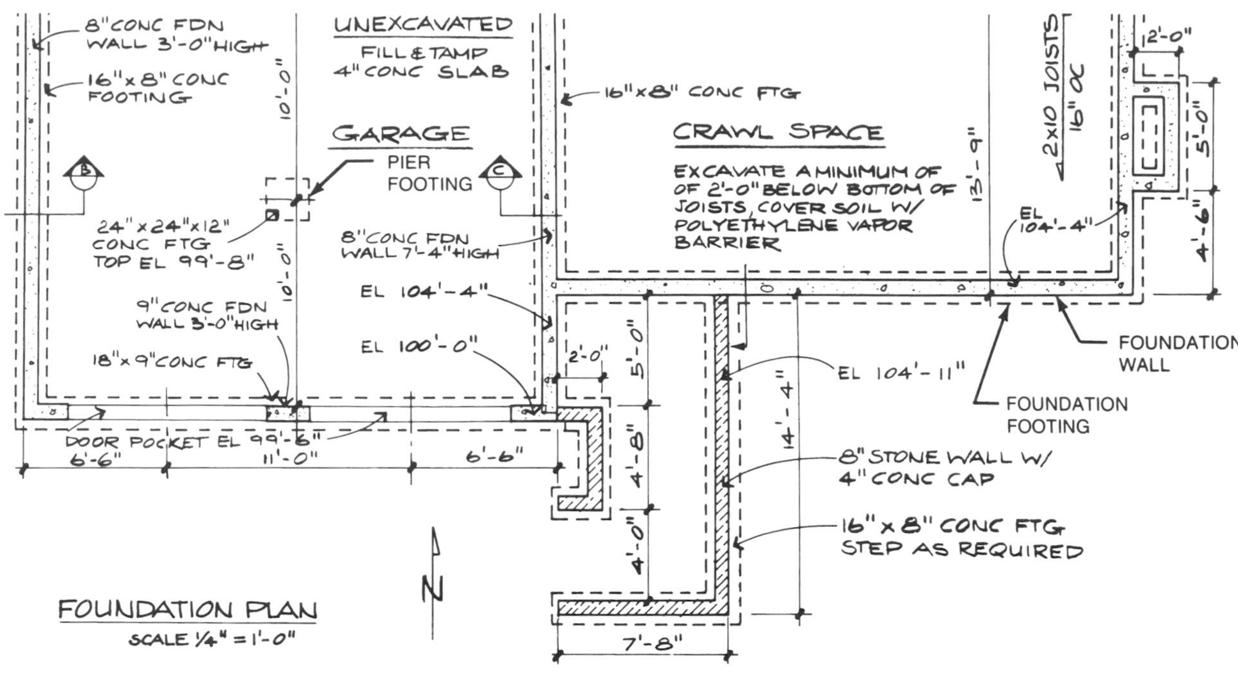

8" CONC FDN WALL 3'-0" HIGH
16" x 8" CONC FOOTING
UNEXCAVATED
FILL & TAMP
4" CONC SLAB
GARAGE
PIER FOOTING
24" x 24" x 12" CONC FTG TOP EL 99'-8"
8" CONC FDN WALL 7'-4" HIGH
9" CONC FDN WALL 3'-0" HIGH
EL 104'-4"
EL 100'-0"
18" x 9" CONC FTG
DOOR POCKET EL 99'-6"
16" x 8" CONC FTG
CRAWL SPACE
EXCAVATE A MINIMUM OF OF 2'-0" BELOW BOTTOM OF JOISTS, COVER SOIL W/ POLYETHYLENE VAPOR BARRIER
2x10 JOISTS 16" OC
EL 104'-4"
FOUNDATION WALL
FOUNDATION FOOTING
EL 104'-11"
8" STONE WALL W/ 4" CONC CAP
16" x 8" CONC FTG STEP AS REQUIRED
FOUNDATION PLAN
SCALE 1/4" = 1'-0"

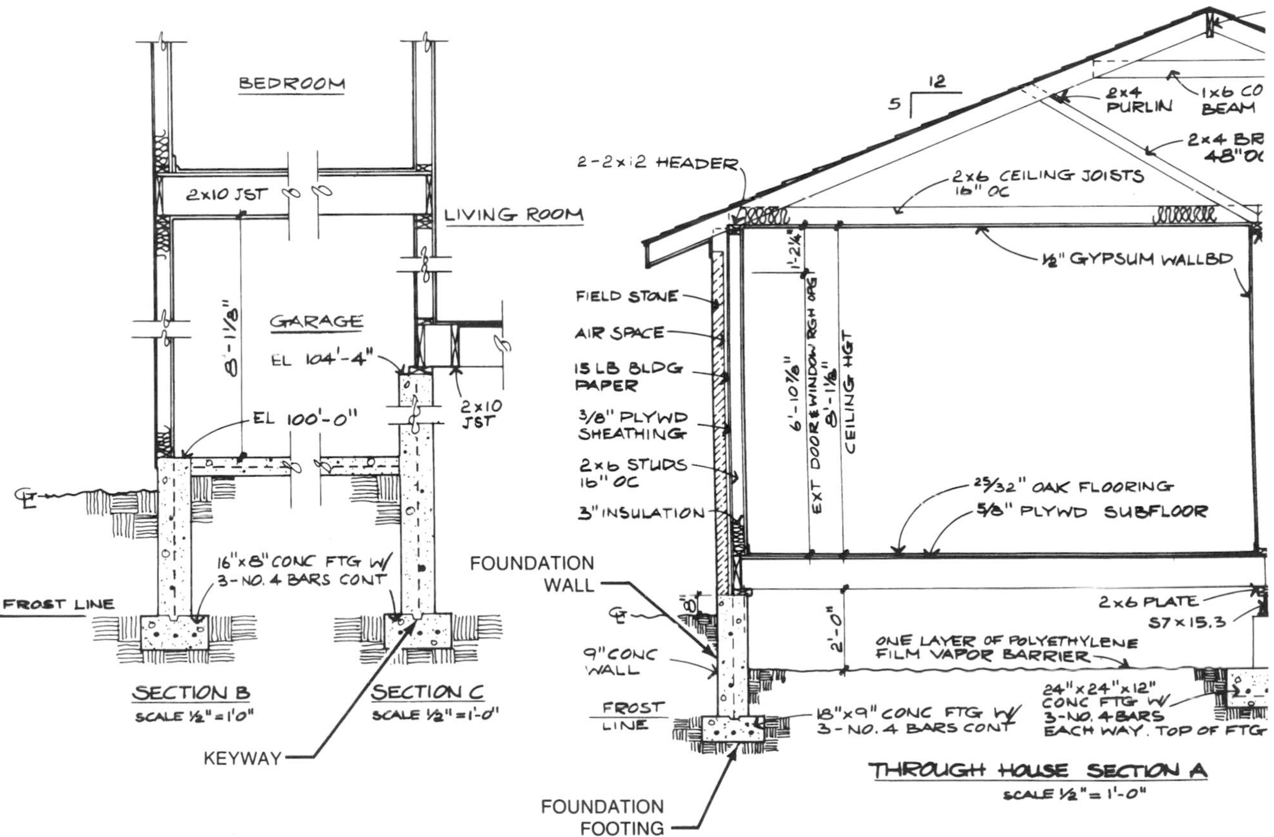

BEDROOM
2x10 JST
LIVING ROOM
GARAGE
EL 104'-4"
EL 100'-0"
2x10 JST
16" x 8" CONC FTG W/ 3-NO. 4 BARS CONT
FROST LINE
SECTION B
SCALE 1/2" = 1'-0"
SECTION C
SCALE 1/2" = 1'-0"
KEYWAY
FOUNDATION WALL
2-2x12 HEADER
2x4 PURLIN
1x6 CO BEAM
2x4 BR 48" OC
2x6 CEILING JOISTS 16" OC
1/2" GYPSUM WALLBD
FIELD STONE
AIR SPACE
15 LB BLDG PAPER
3/8" PLYWD SHEATHING
2x6 STUDS 16" OC
3" INSULATION
EXT DOOR & WINDOW RGH OPG
CEILING HGT
25/32" OAK FLOORING
5/8" PLYWD SUBFLOOR
2x6 PLATE
S7 x 15.3
ONE LAYER OF POLYETHYLENE FILM VAPOR BARRIER
9" CONC WALL
FROST LINE
18" x 9" CONC FTG W/ 3-NO. 4 BARS CONT
24" x 24" x 12" CONC FTG W/ 3-NO. 4 BARS EACH WAY. TOP OF FTG
THROUGH HOUSE SECTION A
SCALE 1/2" = 1'-0"
FOUNDATION FOOTING

Figure 3-20. A set of prints for a crawl space foundation provides information regarding foundation walls and footings.

A *crawl space* is the space between the bottom of the floor joists resting on top of the foundation walls and the ground below. Information regarding the minimum crawl space distance for a building in a given area is usually provided by the local building code. Crawl spaces range from 18″ to 24″ in height. A 24″ clearance allows easier access for plumbing, electrical, or other utility repairs. Local building codes should also be referred to for the recommended distance the foundation wall must extend above the finish grade. **See Figure 3-21.**

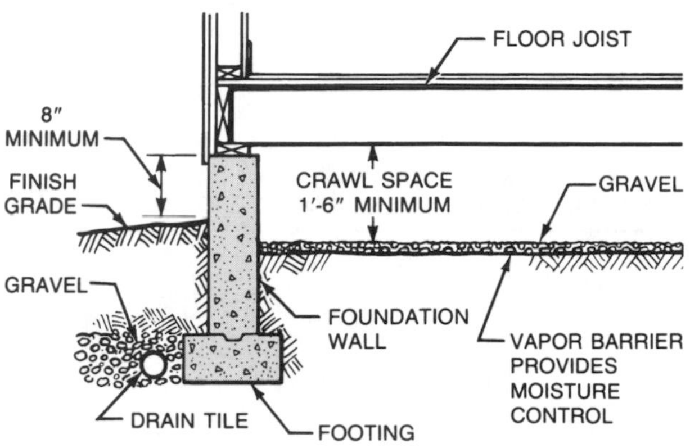

Figure 3-21. The foundation wall must extend a minimum of 8″ above the finish grade. The floor joists must be a minimum of 1′-6″ above the ground. A vapor barrier covered by 2″ of gravel provides moisture control.

Moisture in the soil often results in moisture accumulating in a crawl space. A polyethylene film vapor barrier covered with a layer of gravel is placed over the earth in the crawl space area to reduce moisture accumulation in the crawl space. Vent openings placed in the crawl space walls can also be used to reduce moisture accumulation.

Constructing Crawl Space Foundations

Crawl space foundations are often constructed using monolithic forms. Monolithic forms allow the concrete for the footing and walls to be placed at the same time. Using a monolithically formed foundation is a fast and efficient method of construction and prevents the formation of cold joints. A *cold joint* is a joint formed when concrete for a wall is placed over a concrete footing that has already set. A monolithically formed wall also prevents outside moisture from seeping into the crawl space.

The layout for a crawl space foundation is the same as the layout for a full basement foundation. The batterboards are built to the actual height of the walls, and building lines are set in place.

Although major ground excavation is not required for a crawl space foundation, trenches are dug for the footings. Building lines serve as a guide for the width of the trench. The depth of the trench is measured vertically from the building lines. **See Figure 3-22.** When earth-formed footings cannot be used, the trenches must be dug wide enough for constructed footing forms.

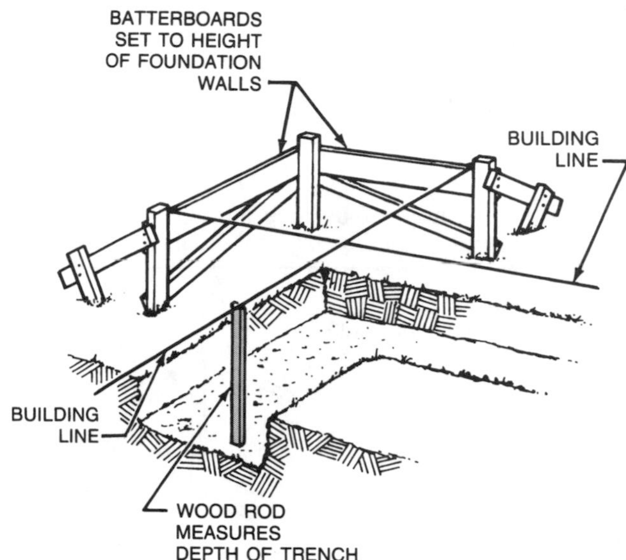

Figure 3-22. Crawl space foundation trenches are laid out according to the building lines. The depth of the trenches is measured from the building lines.

Crawl space foundation wall forms can be constructed over earth-formed footings by securing plywood panels with flat metal stakes. A single waler is placed toward the top of the form and is braced with stakes. Temporary wood spacers are placed between the form walls, and wood form ties are nailed across the walers to hold the walls together. The temporary wood spreaders and metal stakes must be removed during the initial setting period. The panels can be reused after the forms are stripped. **See Figure 3-23.**

On many building projects, formwork must be removed (stripped) as soon as possible after placement in order to reuse the formwork materials. Minimum curing time information may be specified in local building codes.

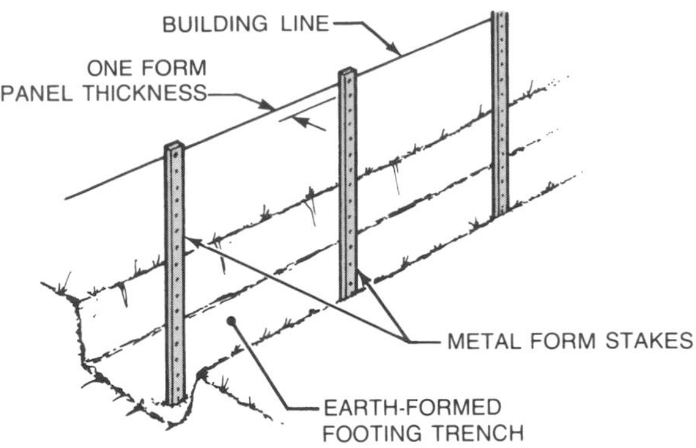

1. Plumb down from the building lines. Drive stakes one panel thickness to the outside of the building lines.

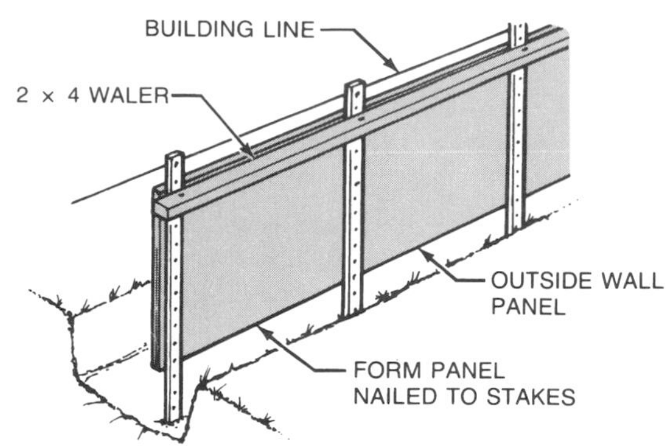

2. Mark the top elevation of the wall on the stakes. Nail the panels to the stakes and place a top waler.

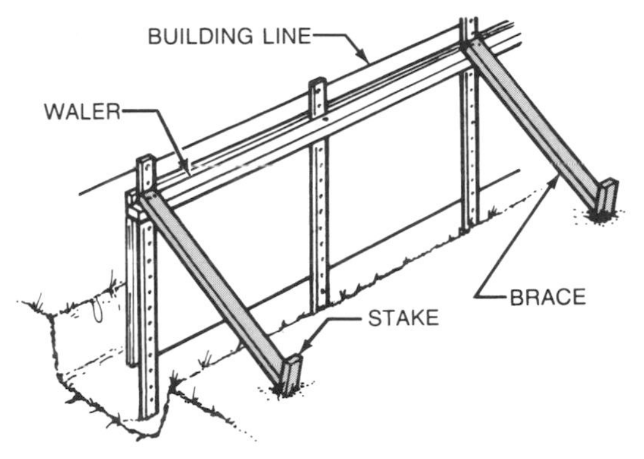

3. Plumb and brace all wall corners. Align the wall to a line and brace where necessary.

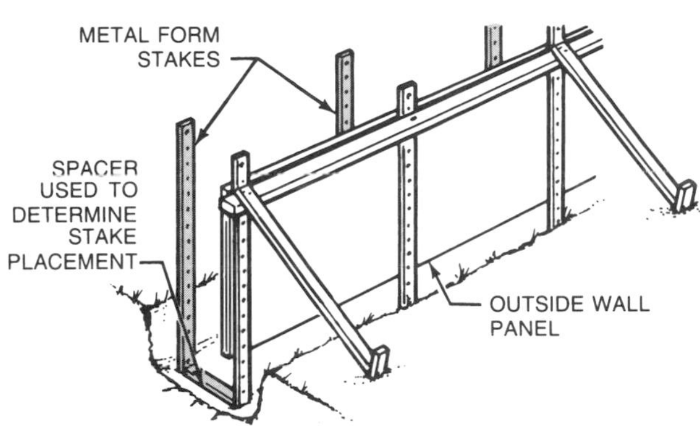

4. Cut a spacer block equal to the wall thickness plus the thickness of two panels. Drive the inside form stakes using the spacer block as a gauge.

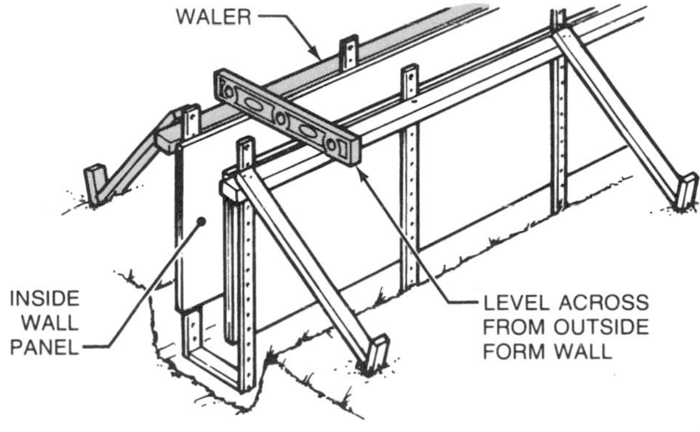

5. Level across from the outside form wall and nail the inside wall panels to the stakes. Place a top waler.

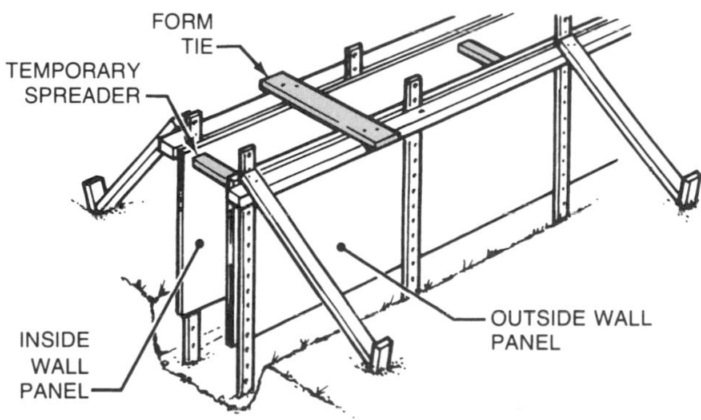

6. Place temporary spreaders (equal to the wall thickness) between the wall forms. Nail form ties across the tops of the walers. Brace inside wall form.

Figure 3-23. A low panel wall form can be constructed over earth-formed footings. Flat metal stakes are recommended to hold the form wall in position.

Crawl space foundation wall forms can also be constructed by nailing 2″ thick planks to short form studs. The form walls are suspended over the trench, which eliminates removing stakes from the concrete. Horizontal 2 × 4s are staked to the ground and nailed to the bottom of the form studs. Braces run from the top of the studs to the horizontal 2 × 4s. Plywood cleats are used to strengthen the tie between the horizontal 2 × 4 and the brace. A plywood template is used for spacing the forms while they are being secured in place. After the forms are stripped, the members can be reused. **See Figure 3-24.**

PLYWOOD TEMPLATE SPACES FORMS WHILE NAILING INTO PLACE
2″ THICK PLANKS
FORM STUDS
PLYWOOD CLEAT
STAKE
BRACE
STAKE
HORIZONTAL 2 × 4
TRENCH FOR EARTH-FORMED FOOTING

Figure 3-24. Two-inch thick planks are suspended over an earth-formed footing. This method eliminates the need to remove the stakes from the concrete.

Occupational irritant contact dermatitis is an inflammation of the skin caused by irritants found on the job site that come in contact with the skin. Symptoms include redness of the skin, blisters, scales, or crusting of the skin. To prevent occupational irritant contact dermatitis, workers should wash hands before putting on gloves, avoid wearing jewelry at work, and launder work clothes separately from street clothes.

In loose or porous soil conditions, monolithic forms can be constructed by using prefabricated plywood sections framed with plates and short studs. The footing forms are constructed with footing boards nailed to stakes that extend above the footing boards. The prefabricated plywood sections are then placed on top of the footing boards and nailed to the stakes. The width of the low wall is determined by the width of the framing material. For example, 2 × 4 framing material produces a wider wall than 2 × 6 framing material over the same footing. **See Figure 3-25.** This forming method is useful for tract home construction over level terrain where a series of similar foundations are being built. The framed plywood sections can be reused many times.

Portland Cement Association

Slab-on-grade foundations are used for light commercial construction or where conditions do not allow a crawl space or basement foundation to be dug.

Monolithic forms can also be constructed using 2″ thick planks tied together with wood cleats. The footing form is constructed of planks that are staked to the ground. The planks are reinforced with cleats to resist lateral pressure. Horizontal supports are leveled, placed on top of the footing form, and nailed to the stakes and wall cleats. The supports suspend the wall form over the footing form. The thickness of the wall is determined by the amount that the horizontal supports project past the footing form. **See Figure 3-26.**

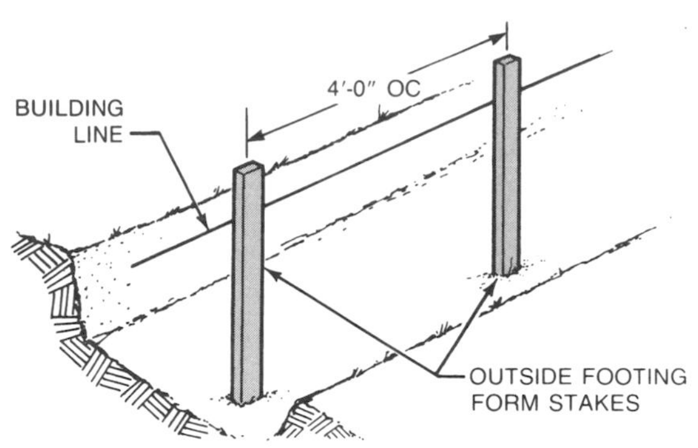

1. Stretch a line for the outside footing form stakes. Plumb and align stakes and drive into ground at 4'-0" OC.

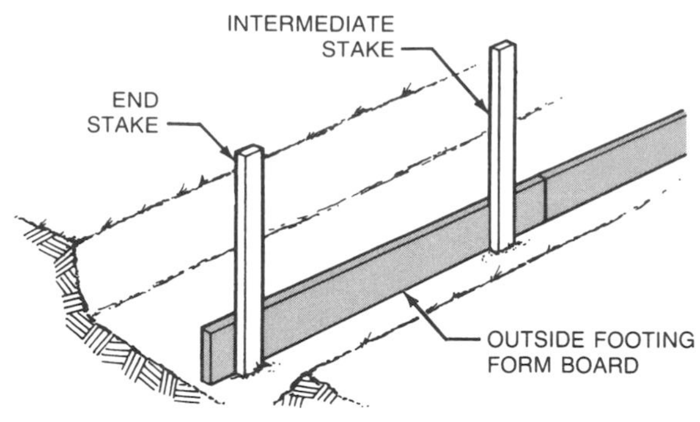

2. Establish the height of the footing on two end stakes. Snap a chalk line on intermediate stakes. Nail the outside footing form boards to the stakes.

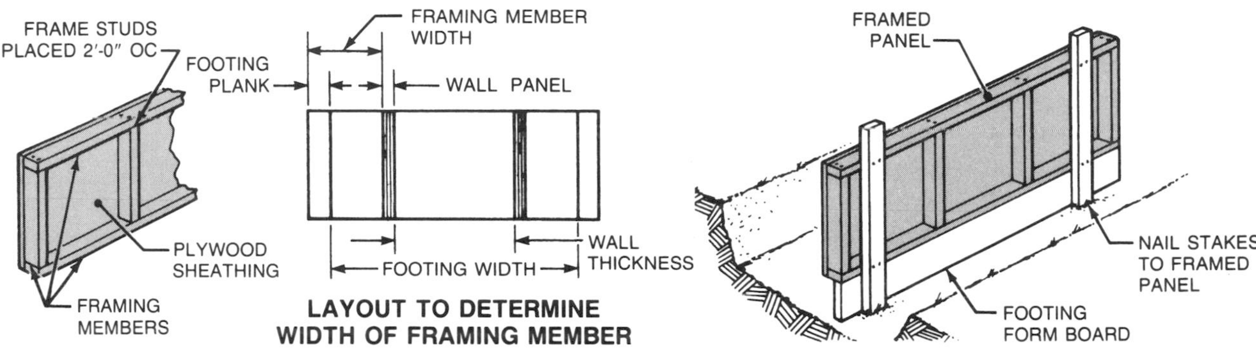

3. Construct panel sections with plywood sheathing nailed to frames. Frame width determines the wall thickness.

4. Place the framed panel on top of the footing form boards. Nail the stakes to the panel frame.

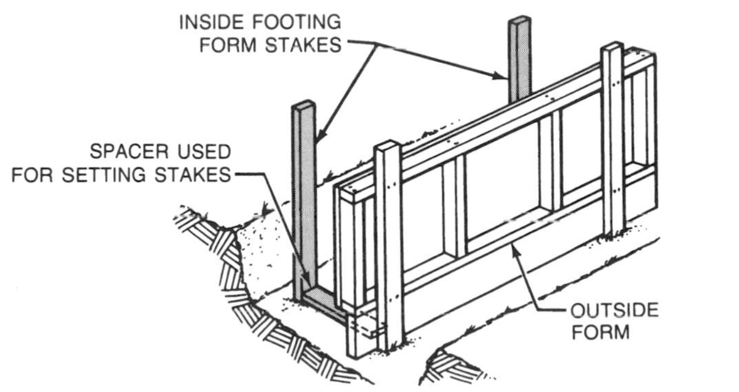

5. Cut spacers equal to the footing width plus the form board thickness and use as a gauge for setting inside form stakes. Drive stakes and nail the inside footing form boards to them.

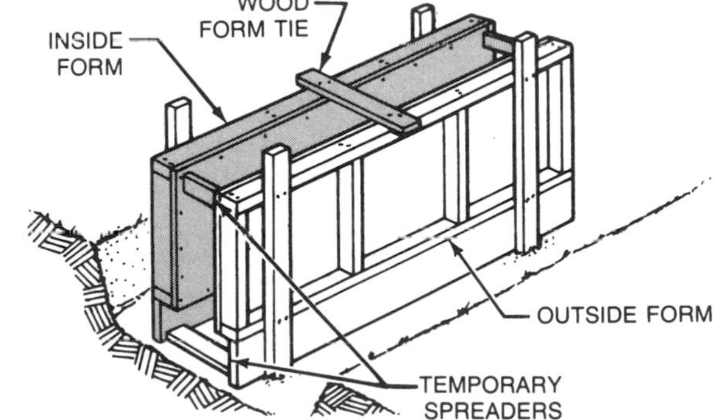

6. Place the framed panel on top of the footing form. Use temporary spreaders to position the panel. Nail the stakes to the panel frame. Nail wood form ties across the top of the form.

Figure 3-25. A framed panel wall form is used to construct a foundation. The size of the frame material will determine the width of the low wall.

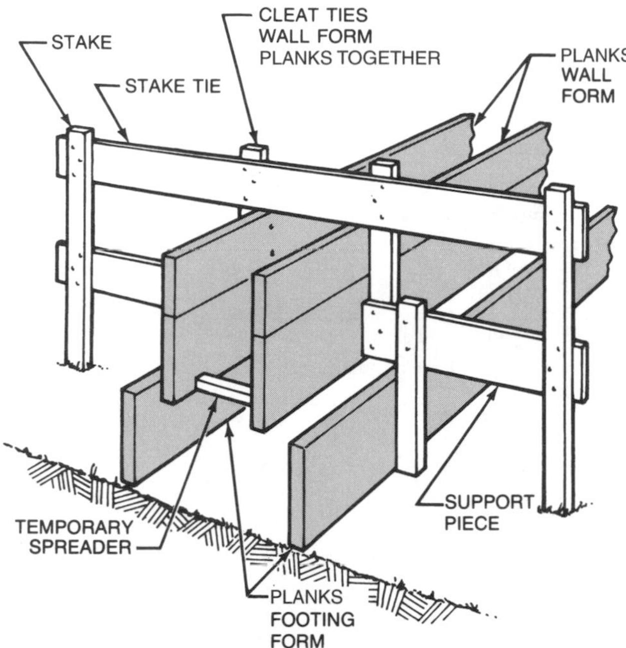

Figure 3-26. A low wall may be formed by using 2″ planks tied together with cleats.

SLAB-ON-GRADE FOUNDATIONS

Slab-on-grade foundations combine foundation walls with a concrete floor slab. The top surface of the floor slab is at the same elevation as the top of the foundation wall. The slab receives its main support from the ground directly below and is reinforced with welded wire reinforcement or rebar. **See Figure 3-27.**

Figure 3-27. A slab-on-grade foundation consists of the foundation walls and a floor slab. The top surface of the floor slab is at the same elevation as the top of the foundation wall.

Slab-on-grade foundations are commonly used in residential and other light construction. A slab-on-grade foundation is a cost-effective method that eliminates deep excavations, high foundation walls, and costly wood floor systems. Slab-on-grade foundations are not practical over steeply sloped lots, or where the water table is close to the ground surface. Slab-on-grade foundation systems are shown on the prints. **See Figure 3-28.**

The water and electrical utility hookups are positioned before the concrete floor slab is placed.

Per OSHA 29 CFR 1926.102, Eye and Face Protection, *the employer is responsible for ensuring that each employee uses appropriate eye or face protection when exposed to potential eye or face hazards, such as when placing concrete.*

A commonly used slab-on-grade foundation design is a floor slab constructed independently of a T-, rectangular, or battered foundation. The slab butts against the foundation wall or rests on a shoulder at the top of the wall. Because the floor slab is placed independently of the foundation, cracks around the perimeter of the slab are prevented. The independent wall and slab method is recommended in cold climates so that insulation may be placed around the perimeter of the floor. **See Figure 3-29.**

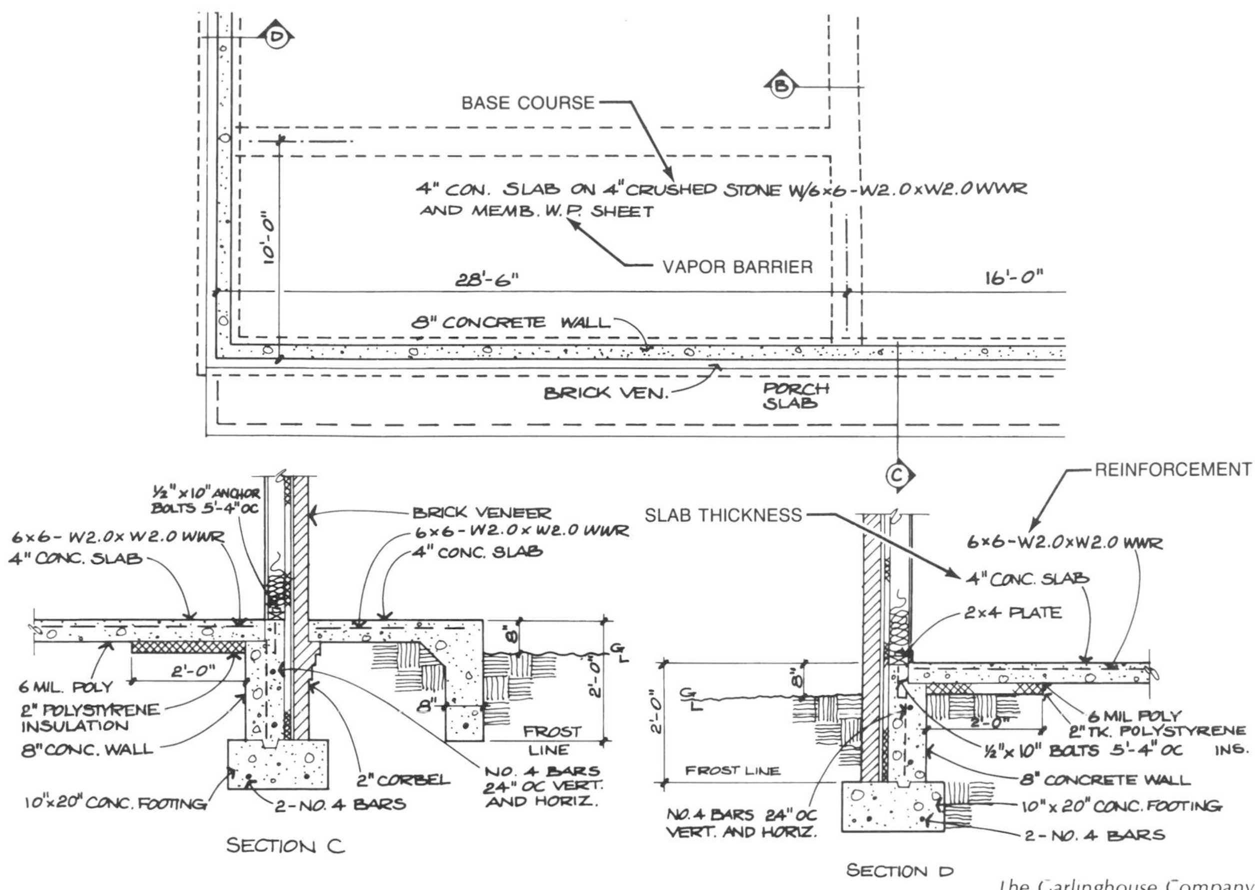

The Garlinghouse Company

Figure 3-28. The print for a slab-on-grade foundation includes the slab thickness and type and size of reinforcement.

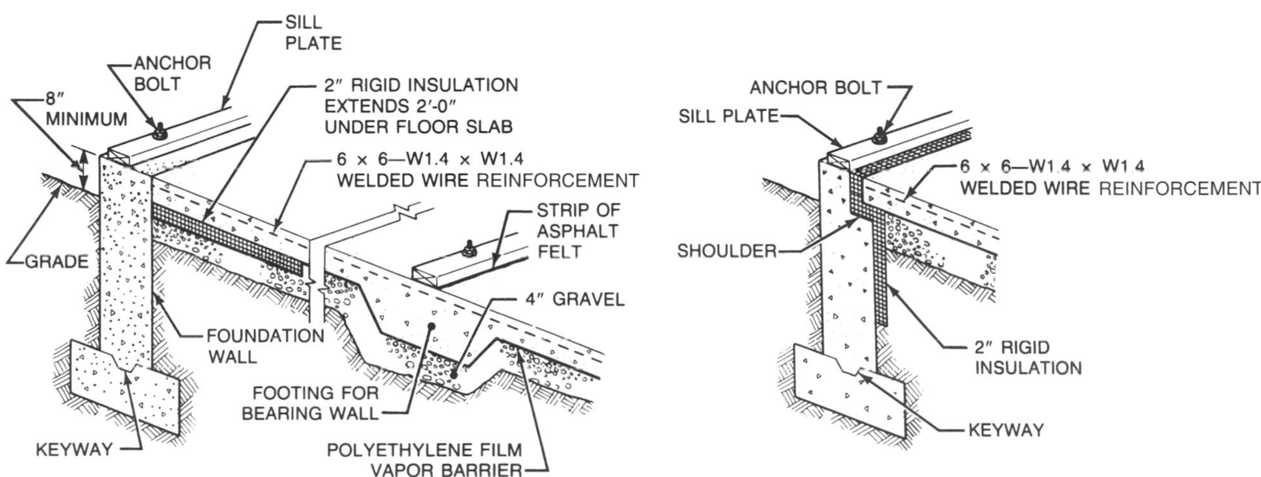

Figure 3-29. A slab-on-grade foundation is formed with the floor slab butting against the wall or resting on a shoulder. Rigid insulation is placed around the perimeter of the slab.

Another commonly used design for a slab-on-grade foundation is a monolithically placed floor slab and foundation. This design, in which the walls and floor slab are placed at the same time, is common in warm climates. Rebar or welded wire reinforcement extending from the floor slab into the foundation walls makes the floor slab an integral part of the foundation and helps distribute the building load to the soil.

Constructing Slab-on-Grade Foundations

When the floor slab and foundation walls are placed independently, forms for conventional T-, rectangular, or battered foundation walls are constructed. The floor slab may either butt against the inside of the foundation wall or rest on a shoulder at the top of the wall. A shoulder is formed by nailing wood members or Styrofoam® equal to the depth and width of the shoulder to the top of the inside wall form.

A monolithic slab-on-grade foundation is formed by constructing an edge form along the outside edge of a trench. The edge form is constructed of plywood members held in place with stakes and braces. A footing is formed by the concrete placed between the edge form and the opposite wall of the trench. **See Figure 3-30.**

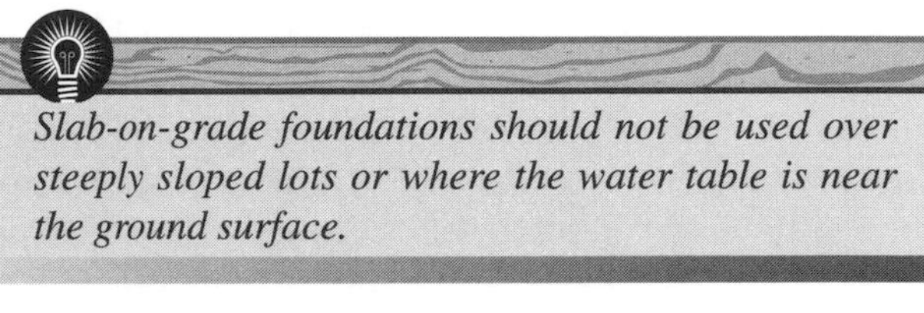

Slab-on-grade foundations should not be used over steeply sloped lots or where the water table is near the ground surface.

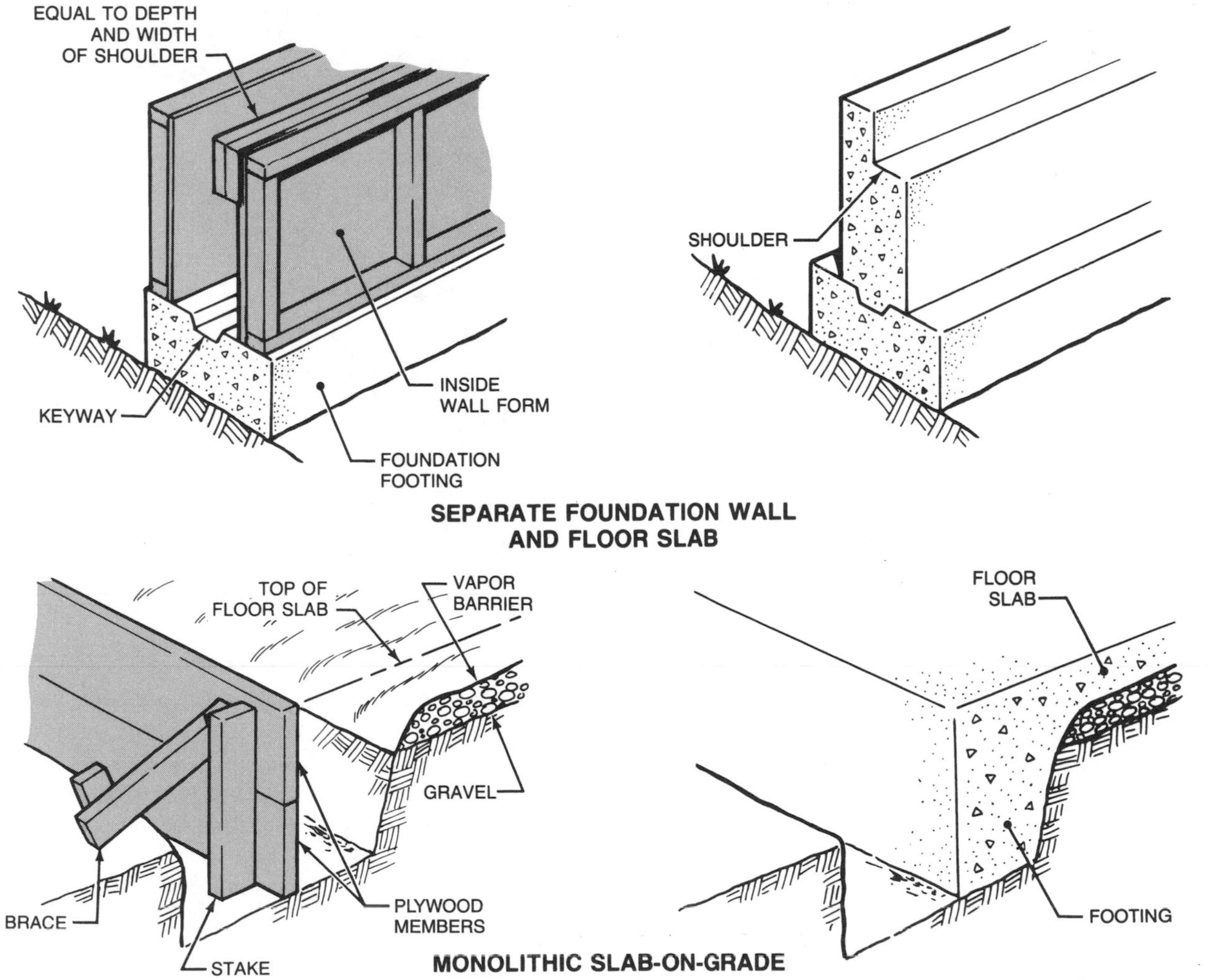

Figure 3-30. Two basic designs may be used when constructing slab-on-grade foundations. The foundation system may be formed separately or monolithically.

GRADE BEAM FOUNDATIONS

In grade beam foundations, walls are supported by reinforced concrete piers that extend deep into the ground. A grade beam foundation is commonly used with stepped or ramped foundations erected on hillside lots where soil conditions do not provide adequate support for conventional footings. Grade beams should extend a minimum of 8″ above the finish grade when supporting wood-frame construction over average soil conditions. The design of grade beams and piers should conform to accepted structural engineering practices and local code requirements.

The bottoms of grade beams should extend below the frost line. The soil directly under the grade beams should be removed and replaced with coarse rock or gravel. Coarse rock or gravel reduces the chance of the ground freezing, which could cause beam movement. **See Figure 3-31.** Other materials that drain water away from the bottom of the beam may also be used. The grade beam should be at least 6″ thick and 14″ deep. However, in a crawl space foundation, the beam must be deep enough to provide 18″ clearance between the ground and the bottom of the floor joists. Grade beams should be reinforced with four #4 horizontal rebar.

The piers beneath the grade beams should have a minimum diameter of 10″ and may be flared at their base to cover a wider soil area. They should be spaced no more than 8′-0″ OC and extend below the frost line into firm soil. A #5 rebar should run the full length of the pier and extend into the grade beam. In seismic risk areas, additional vertical rebar extend from the piers and tie into horizontal rebar placed in the grade beam.

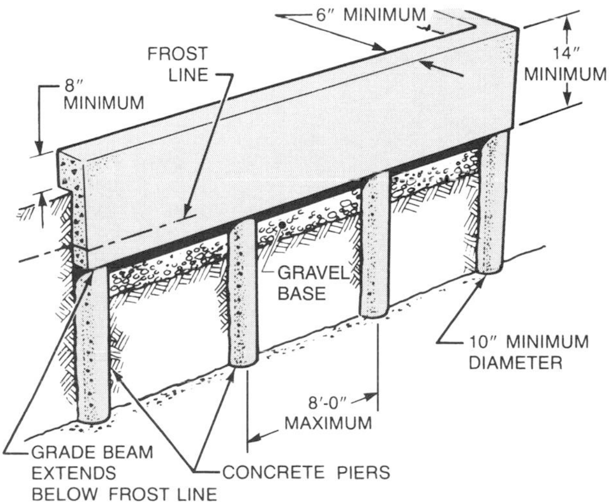

Figure 3-31. Grade beams are supported by concrete piers. A gravel base drains water away from the beam.

Constructing Grade Beam Foundations

The design and construction of a grade beam foundation form is similar to a rectangular wall form built over a foundation footing. However, holes for supporting piers are dug before the formwork begins. In firm soil, the concrete is placed directly into the holes. In soft, unstable soil, tubular fiber forms are used to form the piers. After the concrete has set, the tubular fiber forms are stripped from the piers. The wall forms are then constructed directly over the concrete piers. **See Figure 3-32.**

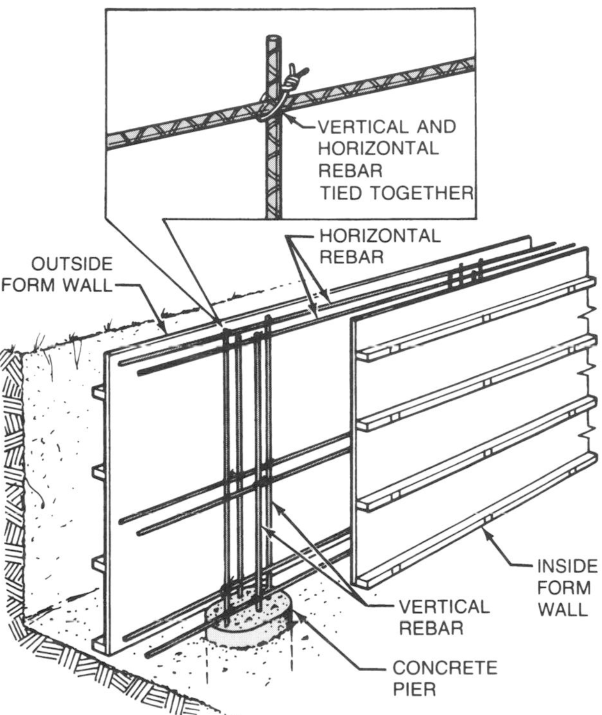

Figure 3-32. Grade beam forms are constructed over completed piers. Horizontal and vertical rebar reinforce the foundation.

STEPPED FOUNDATIONS

Stepped foundations are commonly constructed on steeply sloped lots and used with crawl space or full basement foundations. A stepped foundation is shaped like a series of long steps. A stepped foundation requires less labor, material, and excavation than a level foundation on a sloped lot.

The footings of a stepped foundation must be level. Many building codes require a minimum distance of 2′-0″ between horizontal steps. The thickness of the vertical portion of the footings must be at least 6″ and no higher than three-fourths the distance between horizontal steps. **See Figure 3-33.**

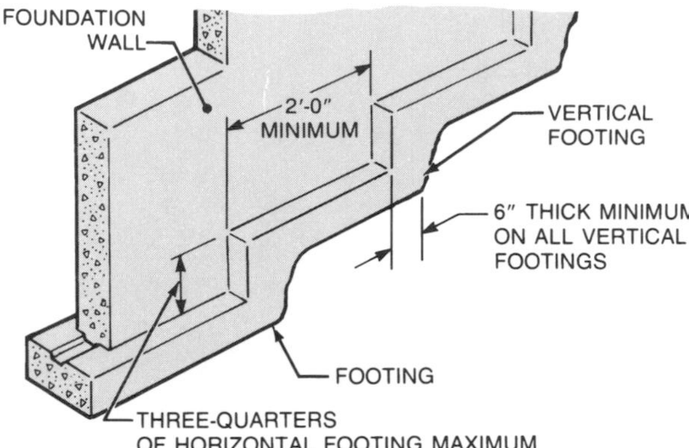

Figure 3-33. Stepped footings are constructed on sloped lots. Consult the local building code for stepped footing requirements.

Wall forming methods used for stepped foundations are similar to those used for constructing rectangular wall forms. Forms for high walls are constructed on previously placed footings. Plywood or planks form the walls and footings. A bulkhead or shutoff is placed at the end of each step to hold the concrete. A *bulkhead* is a wood member installed inside or at the end of a concrete form to prevent concrete from flowing into a section or out of the end of the form. Shutoffs are secured in place with cleats and must be able to withstand a great amount of pressure when the concrete is being placed. The cleats are nailed to the shutoff and the whole unit is positioned to a line established for the end of the step. The cleats are then nailed from the outside of the form with duplex nails. **See Figure 3-34.**

STAIRWAYS AND ENTRANCE PLATFORMS

A low entrance platform or stairway may be required to gain access to residential buildings or other light commercial structures. Concrete is commonly used to construct entrance platforms and stairways because of its durability and ability to withstand damp and wet conditions.

The risers in a stairway are all the same height, and the treads are all the same depth. Recommended riser heights range from 7″ to 7½″, and tread depths from 10″ to 12″. Risers on an exterior concrete stairway should slope in from the top between ¾″ and 1″, and treads should slope between ⅛″ and ¼″ from back to front. (Riser and tread calculations are covered in chapter 5, Heavy Construction.)

Low entrance platforms generally require a few steps. Therefore, a monolithic form is constructed for low entrance platforms and stairways. Low entrance platforms are usually formed against the foundation wall. An isolation strip placed between the platform and foundation wall prevents cracking caused by movement of the platform. The movement results from expansion and contraction of the concrete or soil settlement beneath the platform. An *isolation strip* is a piece of ½″ thick premolded asphalt-impregnated material placed before the concrete is placed.

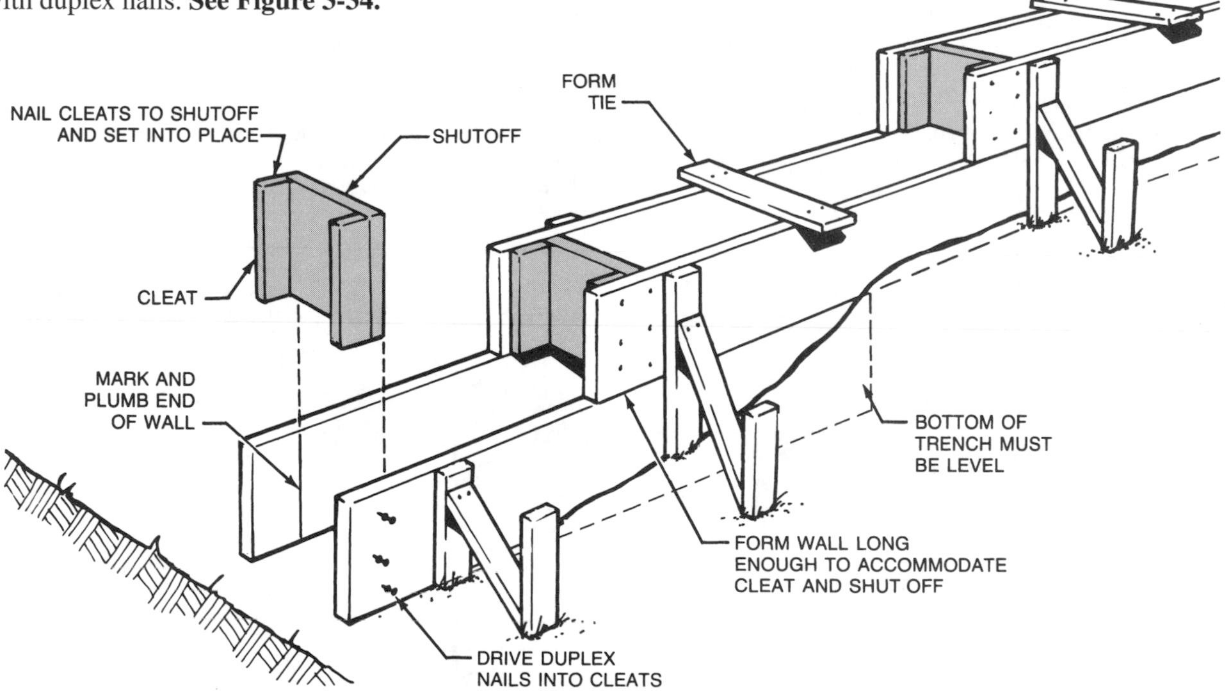

Figure 3-34. Shutoffs must be firmly secured to withstand the pressure of the concrete during placement. Cleats are nailed to the shutoff and duplex nails secure the assembly in place.

The outside form walls for entrance platforms and stairs are constructed of plywood panels stiffened with braced studs and/or walers. After the treads and risers are laid out on the panels, cleats supporting riser form boards are nailed to the panels. Riser form boards should be beveled at the bottom to facilitate troweling of the steps after placing the concrete. **See Figure 3-35.**

Pier footings are a commonly used structure in highway systems as well as in commercial and residential foundations. Pier footings support the loads of structures and must be properly protected from freezing to prevent upheaval and displacement of the footing.

PIER FOOTINGS

A pier footing is a foundation footing for a pier or column. Pier footings are a base for wood posts, steel columns, and masonry or concrete piers supporting wood beams or steel girders. The beams and girders provide intermediate support for the framed floor above. **See Figure 3-36.** Pier footings are used with all types of foundation construction, including crawl space, stepped, and grade beam foundations.

Most pier footings are independent structures. However, pier footings may be joined to foundation footings that support chimneys, fireplaces, or pilasters. In some sections of the U.S., pier foundations may be used to support the entire superstructure. A *pier foundation* is a foundation in which the exterior and interior walls of the building are supported by beams, posts, and pier footings.

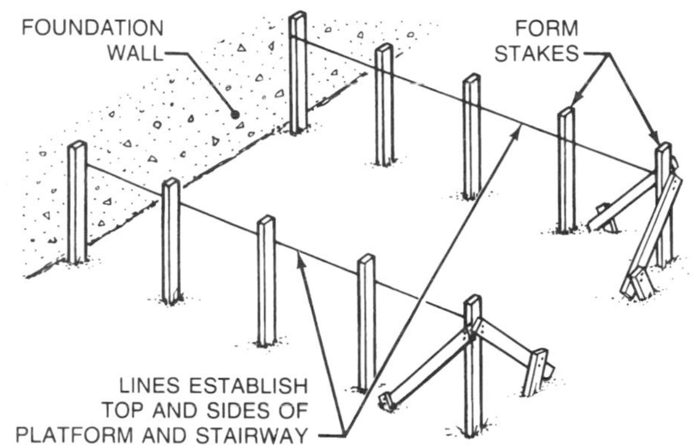

1. Stretch lines establishing the top and sides of the platform and stairway. Position form stakes one panel thickness to the outside of the lines. Plumb and drive stakes.

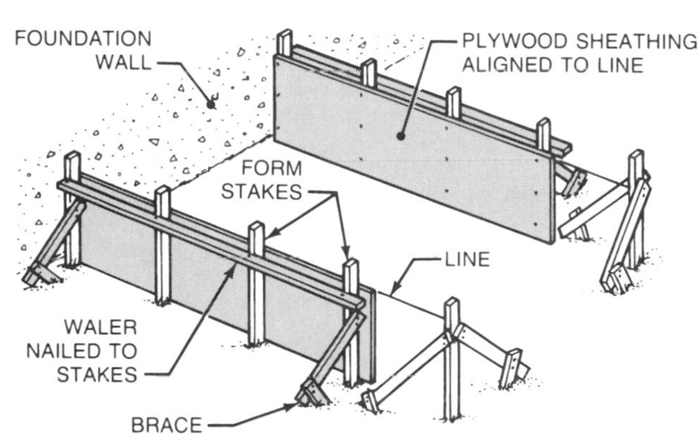

2. Align top of plywood sheathing to line and nail to the form stakes. Nail a waler 3″ to 4″ below the top of the sheathing. Brace the form.

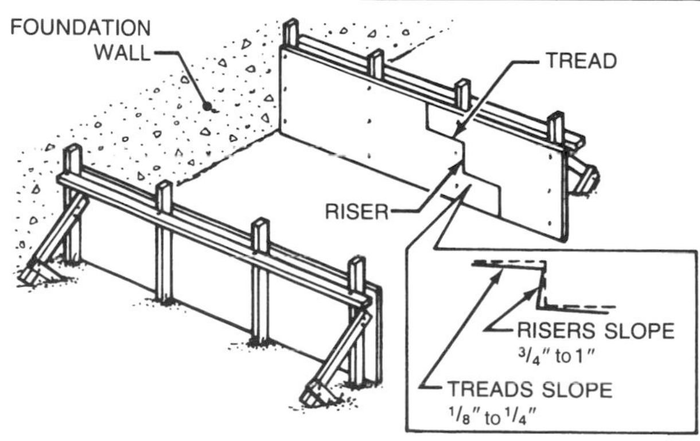

3. Lay out risers and treads on plywood sheathing. Slope the risers 3/4″ to 1″ from top to bottom. Slope the treads 1/8″ to 1/4″ from front to back.

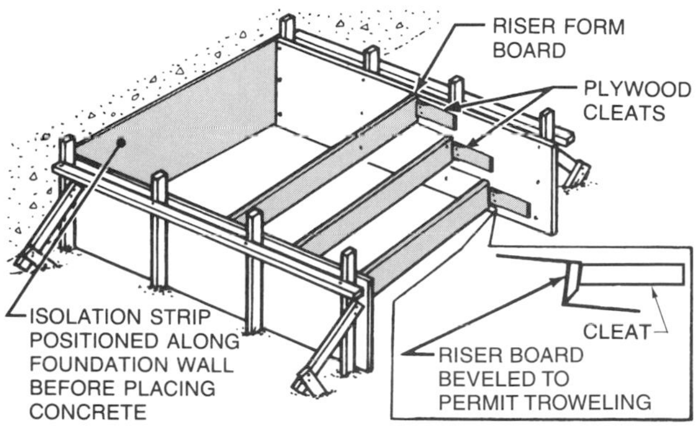

4. Position plywood cleats back from the riser mark one riser form board thickness. Nail cleats into place. Nail riser form boards to the cleats.

Figure 3-35. A low entrance platform and stairway are formed monolithically.

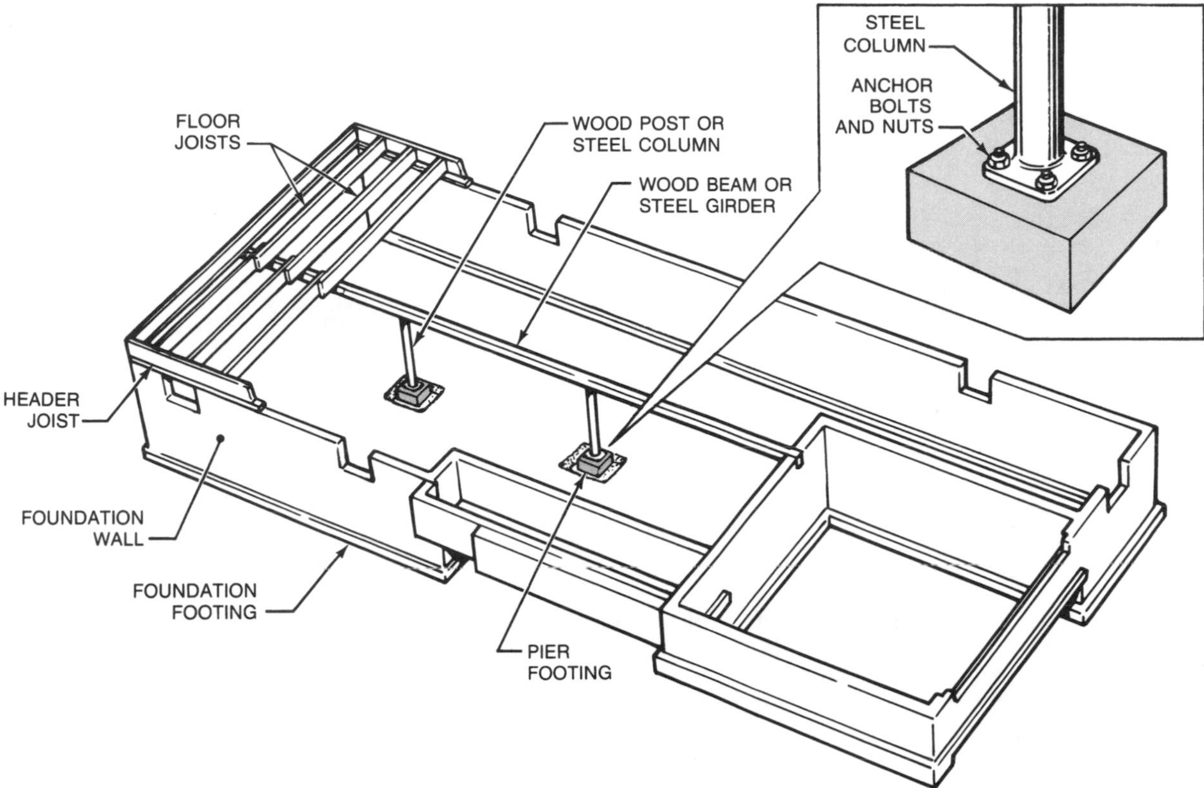

Figure 3-36. Concrete pier footings are a supporting base for posts and columns.

Pier footings distribute loads over a larger soil area than other footings. The size of a pier footing is determined by the weight of the live and dead loads and the bearing capacity of the soil. Various pier footing designs are used, including rectangular, stepped, tapered, and circular piers. **See Figure 3-37.** Information regarding the size, shape, and reinforcement of pier footings is included in the section view drawings of the print. A local building code or structural engineer should be consulted if this information is not included in the prints.

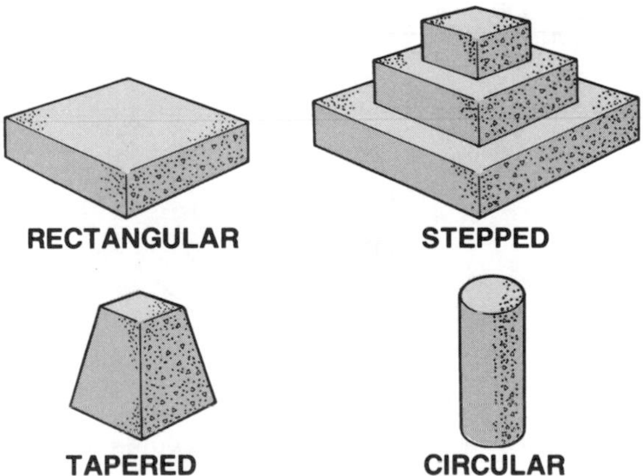

Figure 3-37. The imposed load of the building and the bearing capacity of the soil must be considered when determining pier design.

Constructing Pier Footing Forms

A *pier box* is a form for pier footings. Pier footings are fabricated and set to lines that establish the exact positions of the piers. Forms for square, battered, and stepped pier footings are made of planks or plywood. Forms for circular pier footings are often constructed of fibrous material, such as treated waterproof cardboard, or circular metal forms. The bottoms of all pier footings rest on firm soil and extend below the frost line. The forms are held in place by stakes to prevent uplift or movement. Post or column anchors may be positioned before placing the concrete, or embedded in the concrete during the initial set.

Rectangular and Square Pier Footings. Rectangular and square pier footings are commonly placed under steel columns, chimneys, and fireplaces. Rectangular and square pier forms are usually built with 2″ thick members. After the pieces have been cleated and nailed together, they are placed and held in position with stakes. **See Figure 3-38.**

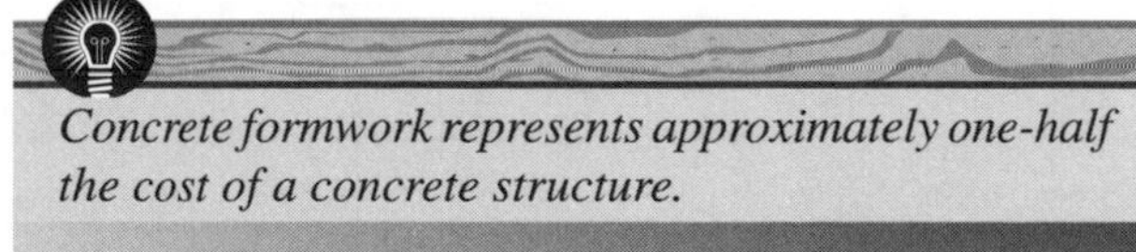

Concrete formwork represents approximately one-half the cost of a concrete structure.

Stepped Pier Footings. A stepped pier footing is designed for conditions where the imposed structural load per square foot is greater than the bearing capacity of the soil. It is used as a base for wood posts or steel columns. The construction of each level of a stepped pier footing is the same as the construction of an individual rectangular or square form. However, two sides of the upper level forms must be long enough to rest on the form below. The upper forms are held in place with cleats. Rebar are positioned in the forms to tie the steps together. **See Figure 3-39.**

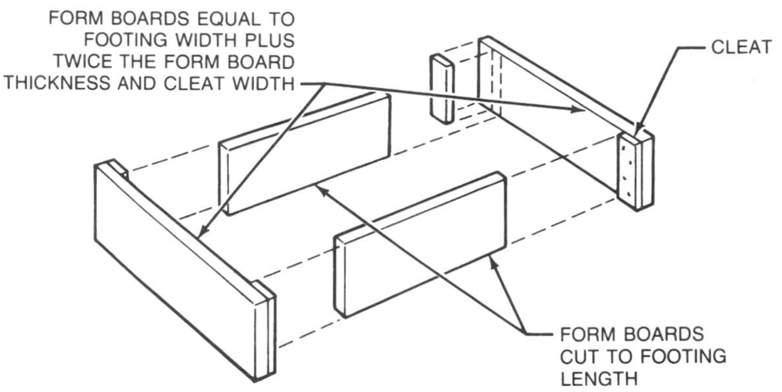

1. Lay out and cut two form boards equal to the footing length. Lay out and cut two form boards equal to the footing width plus twice the form board thickness and cleat width. Nail cleats to the longer pieces.

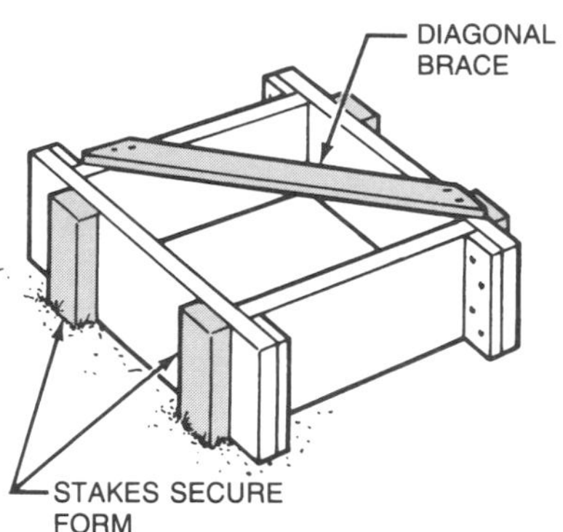

2. Nail the sides together. Square the form and nail a diagonal brace across the top. Position the form and drive stakes to secure it in place.

Figure 3-38. Rectangular or square pier footings are commonly used to support Lally columns or fireplaces.

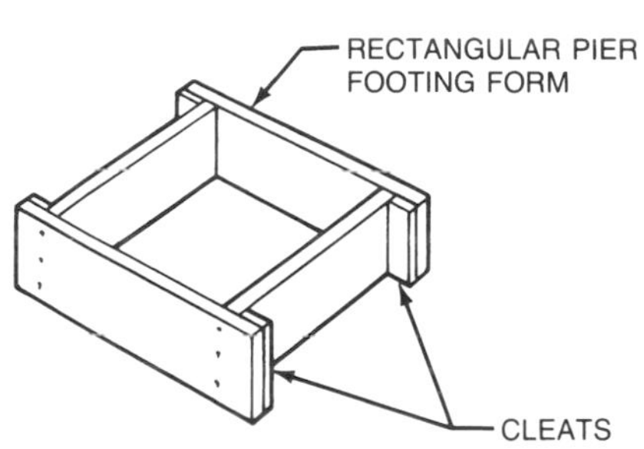

1. Construct a rectangular pier footing form for the base of the stepped pier.

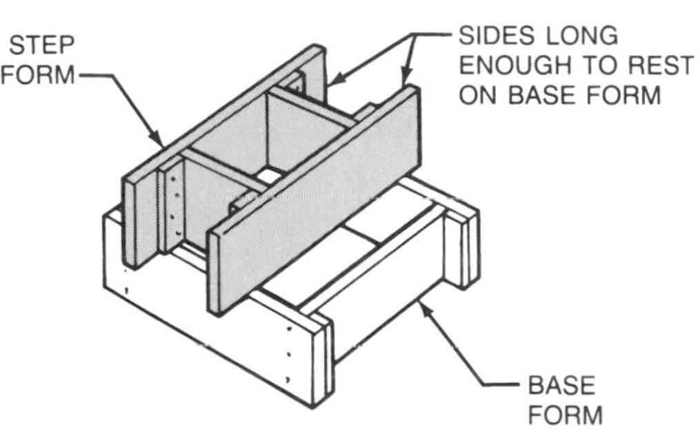

2. Construct a smaller rectangular pier footing form for the step with two sides long enough to rest on the base form.

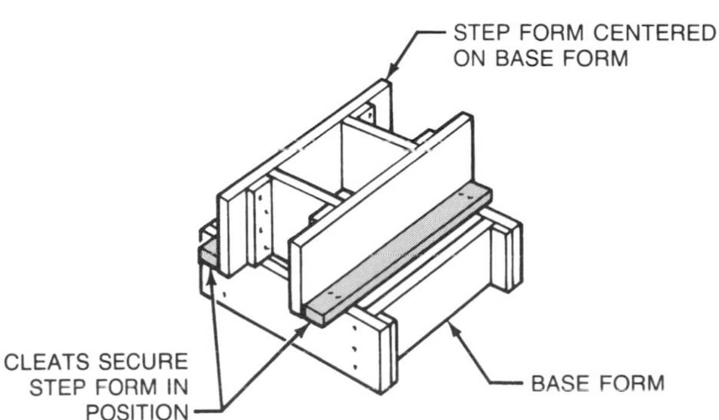

3. Center the step form on the base form. Nail cleats along the sides to secure the step form.

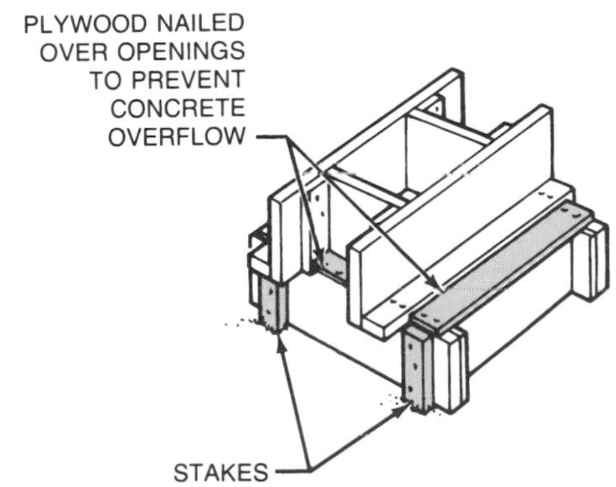

4. Nail plywood over open sections of the base form. Drive stakes to secure the stepped footing form in position.

Figure 3-39. Stepped pier footings are designed to support a structural load that is greater than the bearing capacity of the soil.

Tapered Pier Footings. Tapered pier footings have a wide base that distributes the load over a large area of soil. Tapered pier footings require less concrete than rectangular piers with the same size base. A 60° taper angle from the horizontal should be maintained to provide a safety margin based on a 45° shear stress angle.

Tapered pier forms are commonly constructed of plywood. Two sides of the form are cut to the exact width and height of the pier footing and the other two sides are cut wider to accommodate the plywood thickness and cleats. After the form has been assembled and set in place, it is staked securely to the ground to prevent uplift. Tapered pier forms are subjected to greater uplift when placing concrete than rectangular or square pier forms. **See Figure 3-40.**

Portland Cement Association

The tops of pier footings may be flush with the surface of the floor slab.

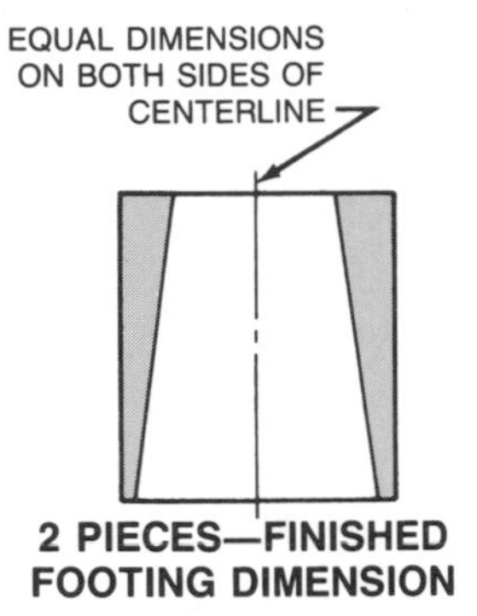

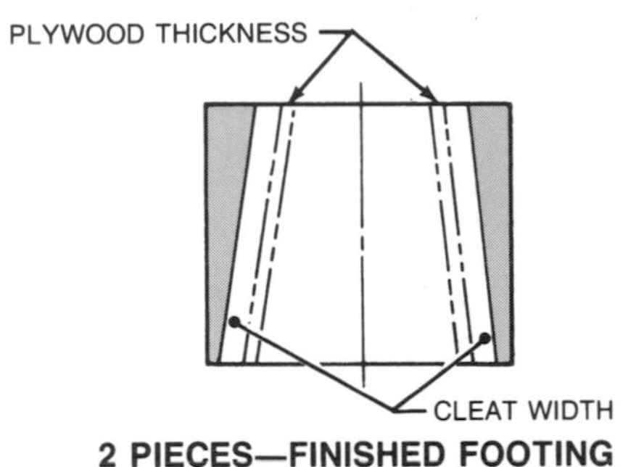

1. Lay out and cut two pieces of plywood to the finished footing dimensions. Lay out and cut two pieces of plywood to the finished footing dimensions plus twice the plywood thickness and cleat width.

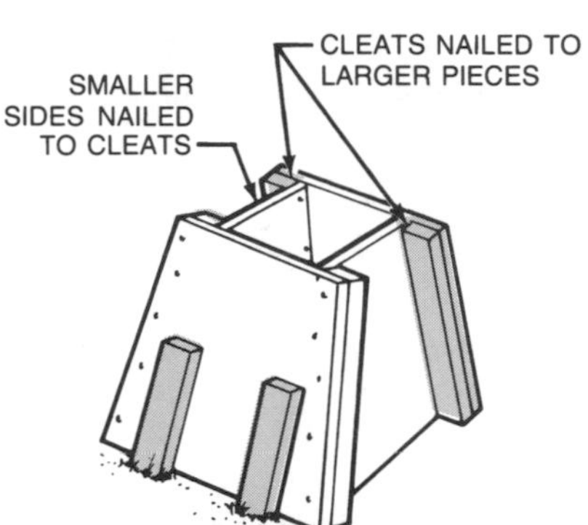

2. Nail cleats to the larger pieces. Nail the smaller sides to the cleats. Secure the form with stakes.

Figure 3-40. Tapered pier footings provide a wide base to withstand heavy structural loads.

Circular Pier Footings. Deep circular pier footings are commonly used to support residential and light grade beam foundations. Shallow circular pier footings are placed beneath wood posts supporting floor beams. Circular pier footings are also used as part of the supporting structure beneath porches, decks, and stair landings.

Circular pier footings may be formed with fiber or metal forms. Fiber forms are cut from tubes made of spirally constructed fiber plies. Metal forms are often made of one-piece spring steel that clamps together at the ends. After the forms are placed in the area excavated for the footing, they are plumbed and staked in position. Post anchors or post bases are positioned before the concrete is placed. **See Figure 3-41.**

Figure 3-41. Concrete is placed in a round fiber form when constructing a circular pier footing. A metal post base is embedded in the concrete.

Pier Footing Layout. The layout of pier footings is shown on the plan view of the foundation prints. The location of pier footings is indicated with dimensions from the foundation wall to the center of the closest pier. Other pier footings are located with center-to-center dimensions. When laying out the positions of the pier forms, lines are stretched to establish the centerlines of the pier forms and are fastened to stakes, batterboards, or wall forms. The pier forms are set to the lines and leveled. The forms are then held in place with stakes and/or soil thrown against the forms. **See Figure 3-42.**

ANCHORING DEVICES

Anchoring devices are embedded in the top surfaces of foundation walls and pier footings. Anchoring devices fasten sill plates to the tops of foundation walls. They also are used to anchor the bottoms of wood posts or steel columns to pier footings.

Sill Plates

Sill plates, or mudsills, are fastened to the tops of foundation walls to provide a nailing surface for floor joists or wall studs. In residential or other light construction, sill plates are usually 2 × 4s or 2 × 6s. **See Figure 3-43.** Larger and heavier buildings may require 4 × 6s. In previous years, foundation-grade redwood was frequently used for sill plates because of its superior resistance to decay and insect attack. Today, other pressure-treated softwoods provide the same advantages and are more commonly used. *Pressure treating* is a process in which chemical preservatives are forced into the wood under intense pressure. Common preservatives include ammoniacal copper arsenate (ACA), chromated copper arsenate (CCA), and pentachlorophenol. Only ACA-treated lumber can be used for residential applications. After a preservative is applied, the surface of the wood is generally clear, odorless, and easy to paint. However, pentachlorophenol may affect the surface color of the treated material.

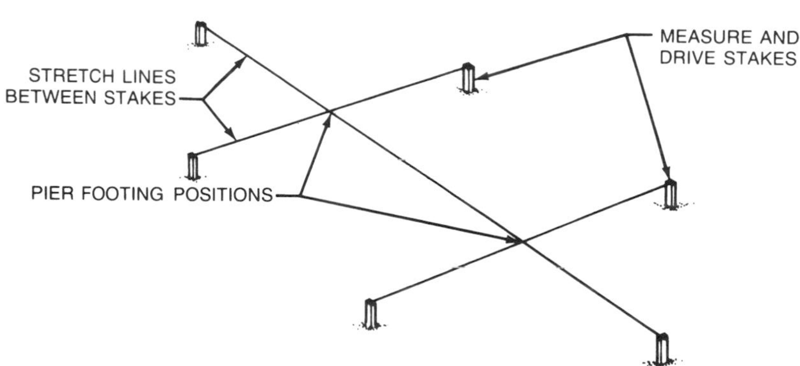

1. Drive stakes and stretch lines indicating the centerlines of the pier forms.

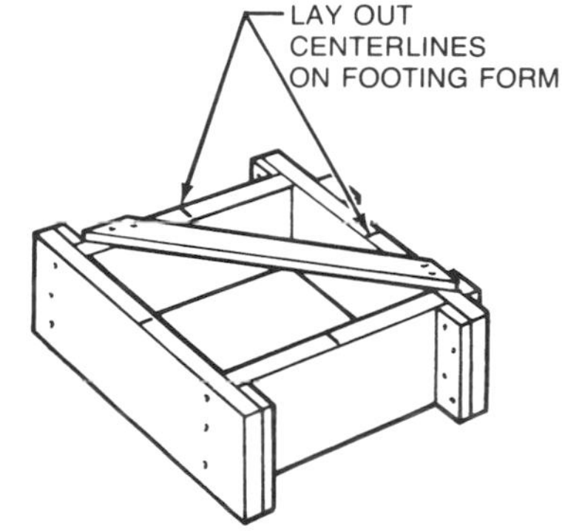

2. Lay out centerlines on all four sides of the pier form.

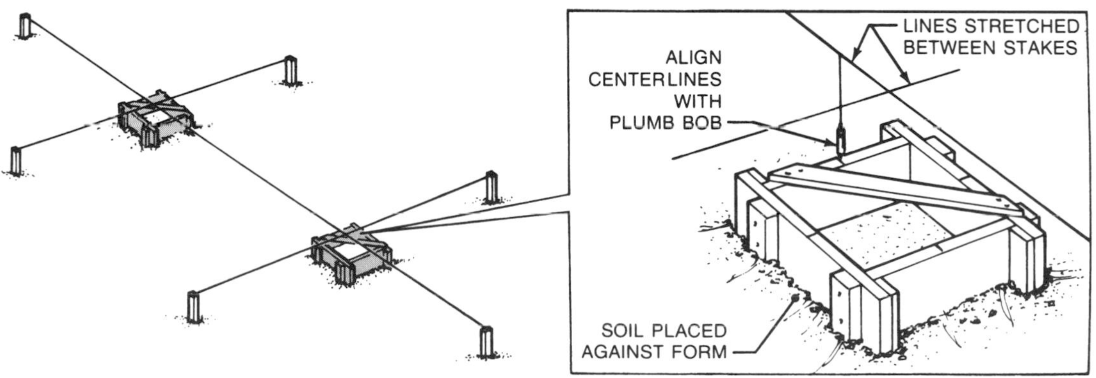

3. Align the centerlines of the pier form with the stretched lines using a plumb bob or hand level. Drive stakes and place soil against the pier form.

Figure 3-42. Lines are stretched between layout stakes and are used to locate the position of pier footing forms.

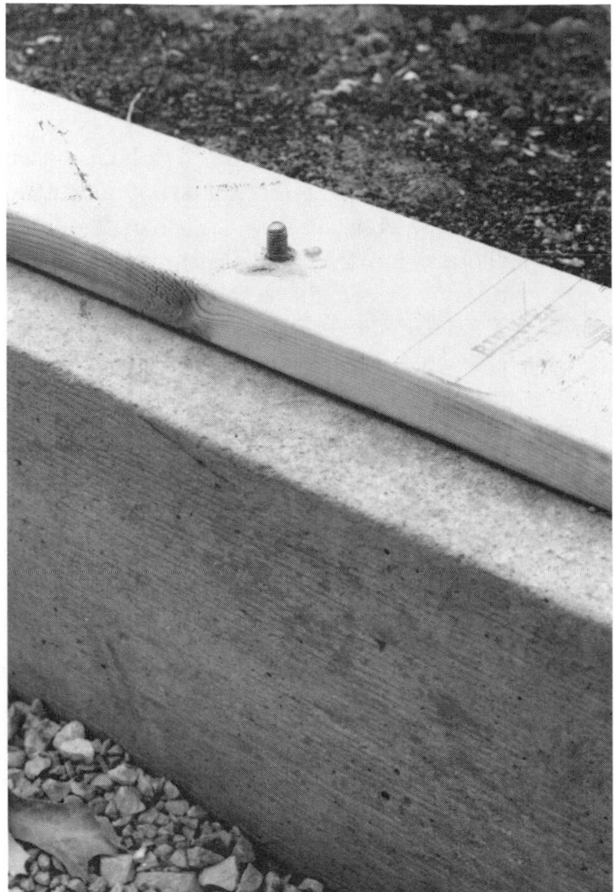

Figure 3-43. Sill plates are positioned over anchor bolts. They are later tightened down with washers and nuts.

Inhalation of sawdust from ACA- and CCA-treated wood should be avoided, as should direct contact with pentachlorophenol-treated wood. When sawing or machining treated wood, a particulates mask (dust mask) and goggles should be worn. Also, wear long-sleeved shirts, long pants, and gloves impervious to chemicals. Pentachlorophenol is highly toxic to fungi and insects.

Sawing and machining operations should be performed outdoors to prevent accumulation of sawdust within a building. Wood treated with waterborne preservatives may be used inside residential structures as long as all sawdust and scrap are cleaned up and properly disposed of after construction. ACA- and CCA-treated and pentachlorophenol-treated wood should not be burned; rather, they should be disposed of in the trash or buried. After working with treated wood, exposed body areas should be washed thoroughly.

The edge of the sill plates may be flush with the outside of the foundation wall or held back the thickness of the wall sheathing. Anchor bolts or anchor clips are used to fasten sill plates to foundation walls. An *anchor bolt* is a bolt used to secure sill plates, columns, and beams to concrete or masonry. An *anchor clip* is a strap-like device embedded in the top of a foundation wall and used to secure sill plates. **See Figure 3-44.** In some cases, sill plates are fastened with powder-actuated fasteners. A *powder-actuated fastener* is a special concrete nail driven with a powder-actuated fastening tool.

Fastening Sill Plates. Anchor bolts are the most effective anchoring device used to fasten sill plates to concrete. One end of the anchor bolt is threaded and extends above the sill plate to receive a nut and washer. The end embedded in the concrete is usually bent into an L or J shape for greater holding ability.

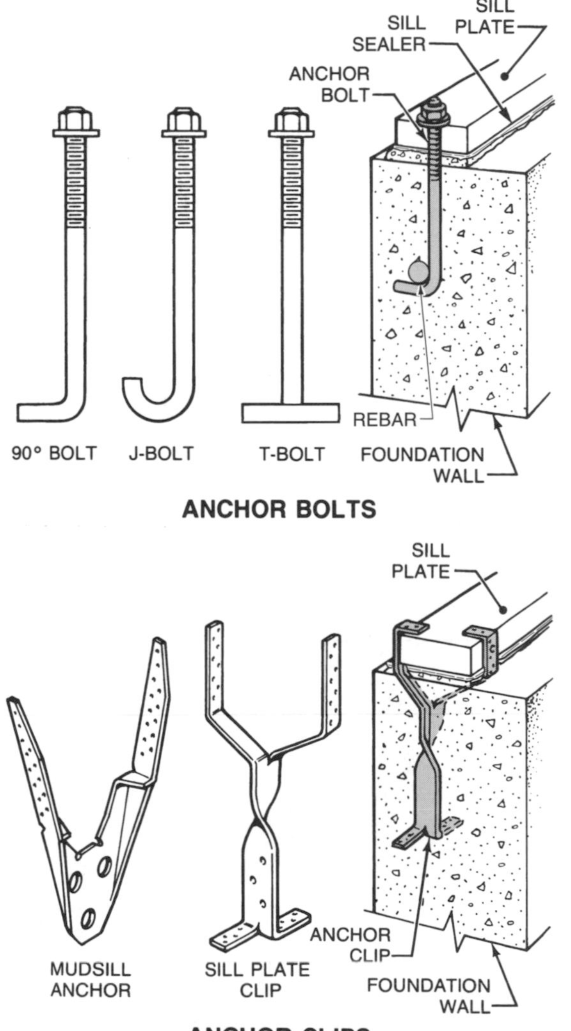

Figure 3-44. Anchoring devices secure the sill plate to the foundation wall.

The depth and spacing of anchor bolts in foundation walls is shown on foundation section view drawings of the prints. If the information is not available on the prints, consult the local building code. A typical building code recommendation regarding anchor bolts might be as follows:

- Steel anchor bolts should be at least ½″ diameter.
- The bent end of the anchor bolt should be embedded a minimum of 7″ into the concrete of reinforced masonry, or 15″ into unreinforced grouted masonry.
- Anchor bolts must be spaced not more than 6′ apart. There should be a minimum of two bolts per sill plate, with one bolt located within 12″ of each end.

Setting Anchor Bolts and Sill Plates. Anchor bolts may be set before or after the concrete is placed. When setting anchor bolts before concrete is placed, the anchor bolts are suspended in the wall form by securing them to wood templates. **See Figure 3-45.** The location of the anchor bolt is marked so that the bolt will be at the center of the width of the sill plate. The hole is then drilled. The anchor bolt fits through the hole, and a nut and washer are placed on the threaded end of the anchor bolt. A nail driven into the template is bent over the nut and bolt to secure them to the template. The template is positioned and nailed to the top of the forms. After the template is secured, a horizontal rebar should be placed over the lower ends of the anchor bolts and tied. Using a template provides greater accuracy in the placement of anchor bolts. Always consult local building codes for specific requirements.

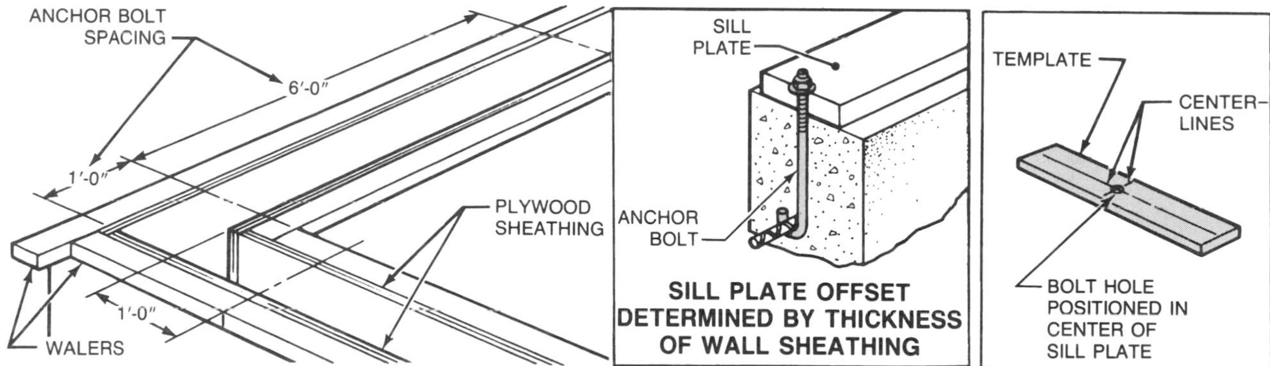

1. Mark the spacing of the anchor bolts on top of the wall form. Cut wood templates and determine the amount of offset required. Lay out the centerlines and drill a snug hole for the anchor bolts.

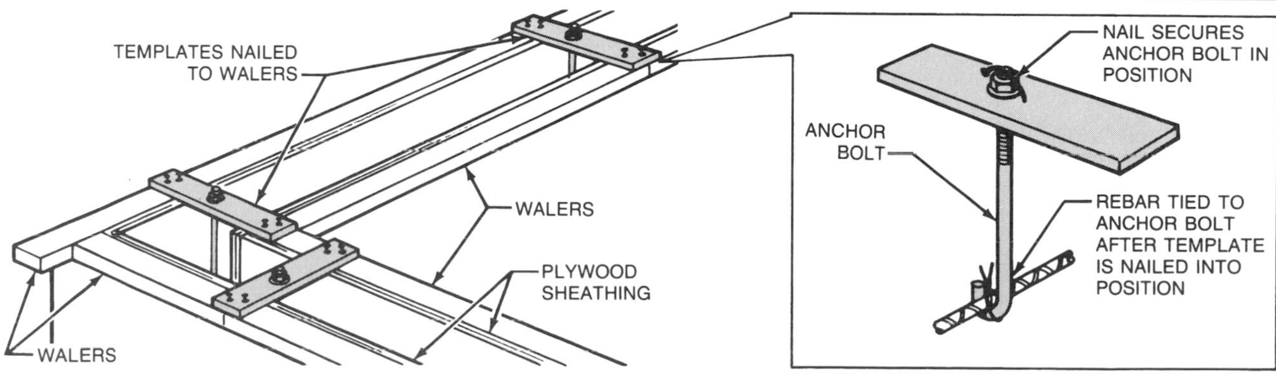

2. Insert the anchor bolt in the hole and place a nut and washer on the threaded end of the bolt. Drive a nail and bend it over to secure the bolt in position. Align the templates to the centerlines on top of the form and nail into position.

Figure 3-45. Anchor bolts are set before the concrete is placed. The anchor bolts are centered in the width of the sill plates and are offset in the foundation wall.

Once the concrete has set, the nuts are unscrewed and the templates removed. The sill plates are cut to length, drilled to accept anchor bolts, and fitted over the bolts. **See Figure 3-46.** A layer of grout (a mixture of sand, cement, and water) is often applied to the foundation wall to provide an even base for the sill plates. Some codes may require a strip of water-resistant material or metal termite shield beneath the sill plates.

When setting anchor bolts and sill plates after concrete placement, the anchor bolts and sill plates are prepared by cutting the sill plates to length and drilling holes for the anchor bolts. The anchor bolts are placed through the holes. Nuts and washers are screwed onto the threaded ends of the anchor bolts. When the concrete reaches the top of the wall form, the sill plates and anchor bolts are pressed into the concrete. The nuts are tightened after the concrete has set. **See Figure 3-47.**

Anchor bolts are also used to fasten sill plates for interior walls to concrete slabs. Lines are stretched immediately after the concrete is placed, and the anchor bolts are set to the lines. Sill plates for interior walls are commonly pinned down with powder-actuated studs.

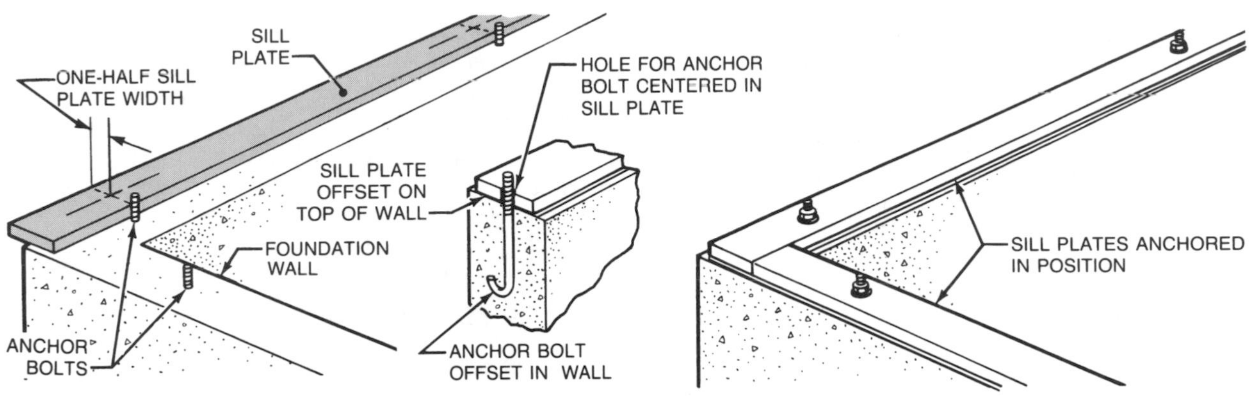

1. Cut the sill plates to length and place against anchor bolts. Square lines across the sill plate. Lay out one-half the sill plate width on the lines.

2. Drill holes in the sill plate. Position the sill sealer and sill plate and tighten down with washers and nuts.

Figure 3-46. Sill plates are placed over the anchor bolts and tightened down.

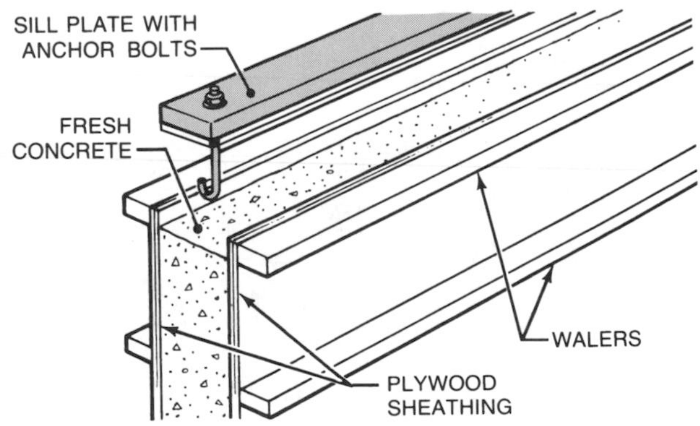

1. Press the sill plates with sill sealer and anchor bolts into concrete when it reaches the top of the form. Offset sill plate from outside of wal to compensate for wall sheathing. Place a wood piece over the bolts when tapping them into the concrete.

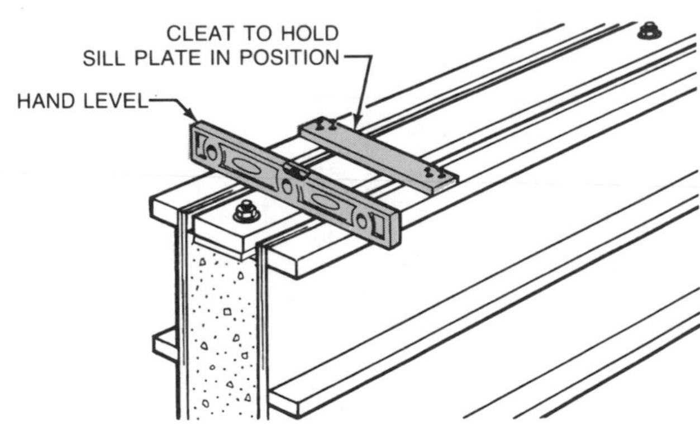

2. Level the sill plate with a hand level. Nail cleats across the form to hold the sill plate in position while the concrete sets.

Figure 3-47. Sill plates may be set into fresh concrete.

Post Anchors

Common post anchor methods use a metal post base, an adjustable metal post base, or anchor bolts. Post anchors are positioned in the piers while concrete is placed in the pier forms.

Metal Post Base. A metal post base is used to fasten wood posts to pier footings. The prongs of the metal post base are set in the concrete during concrete placement. The upper section of the metal post base is nailed to the post when the post is positioned. The metal post base must be at the correct elevation and position to ensure the proper position of the wood post.

Adjustable Metal Post Base. An adjustable metal post base is secured to a pier footing with a ½″ J-bolt that had been set during the placement of the concrete. After the concrete has set, the adjustable metal post base is fastened to the anchor bolt. A standoff plate provides a flat bearing surface for the post and keeps the post off the surface of the concrete, protecting it from wood rot and termite damage. **See Figure 3-48.**

Anchor Bolts. Anchor bolts are used to fasten the bearing plates of steel (Lally) columns to concrete piers. The bolts must be set precisely to align with the bearing plate at the base of the steel column. Holes are laid out and drilled in the template that holds the anchor bolts. The template is centered and nailed to the top of the pier form. **See Figure 3-49.** Anchor bolts are used for other devices, such as holddowns and tiedowns, and/or structural members that must be securely fastened to the surface of concrete. Holddowns and tiedowns are used to secure shear walls to foundations. A *shear wall* is a panel-sheathed wall used to withstand severe seismic activity and heavy wind loads. **See Figure 3-50.**

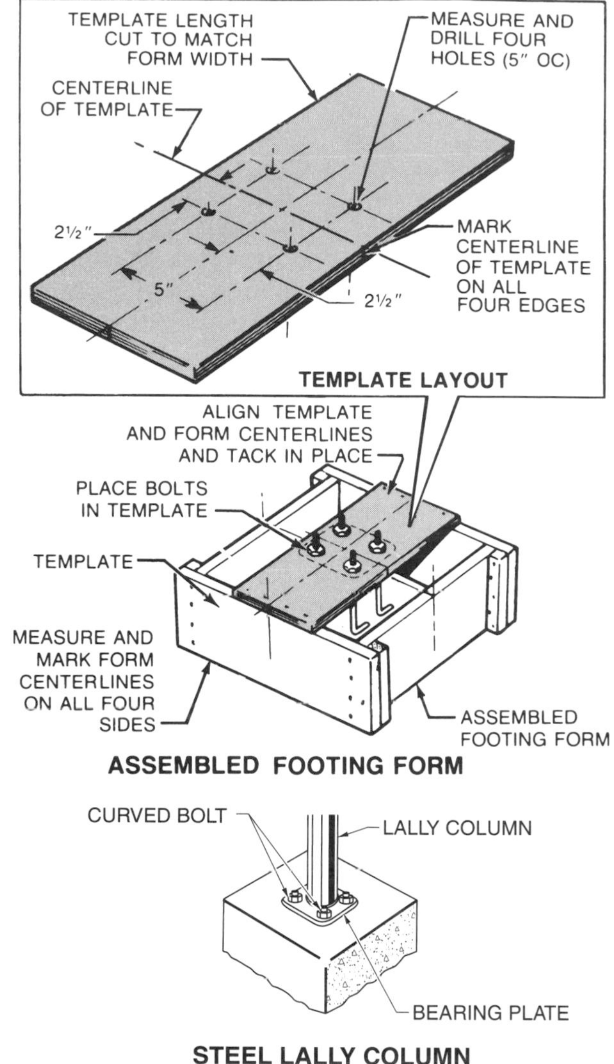

Figure 3-49. A template is used to lay out steel column anchor bolts. The bolts must be positioned accurately to align with the bearing plate of the column.

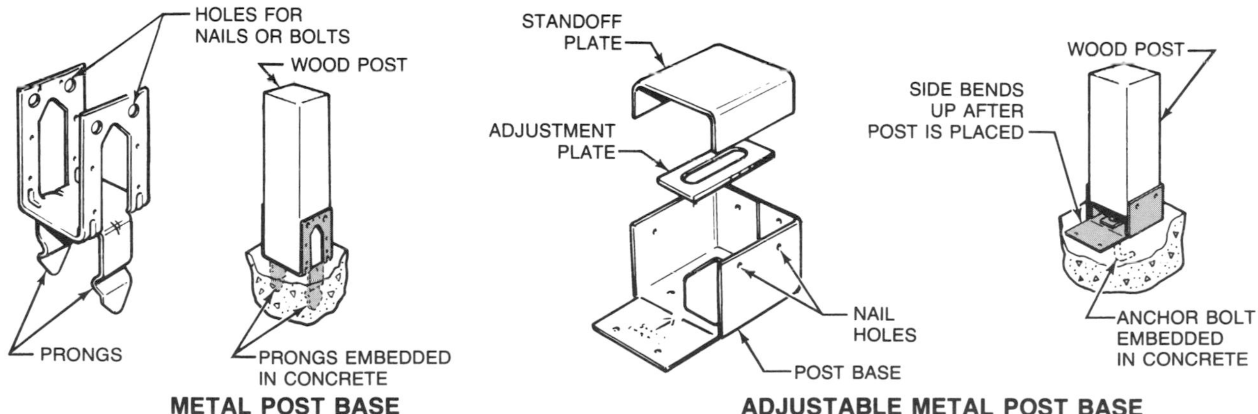

Figure 3-48. Metal post bases are used to anchor wood posts to concrete pier footings.

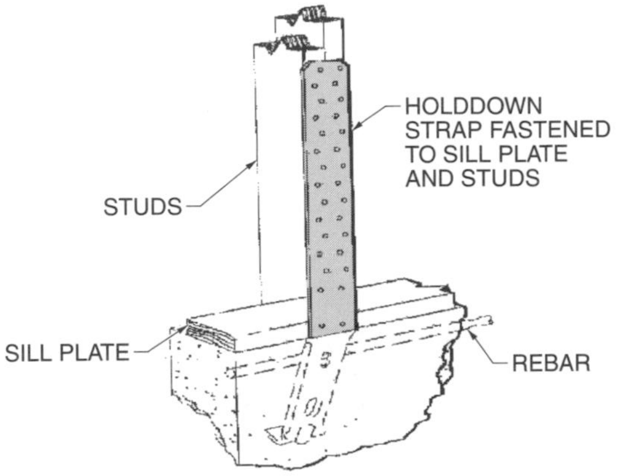

STUDS
HOLDDOWN STRAP FASTENED TO SILL PLATE AND STUDS
SILL PLATE
REBAR

STRAP EMBEDDED INTO CONCRETE

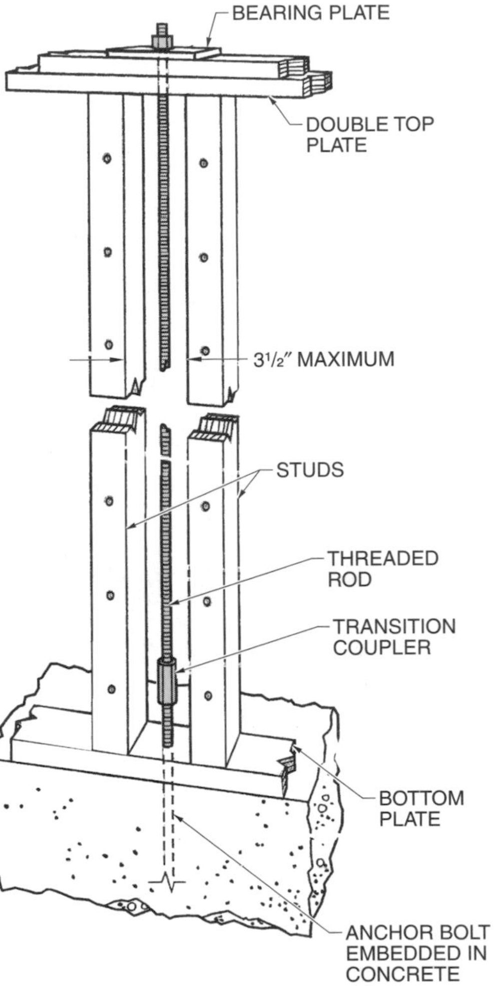

CORNER POST
ANCHOR BOLT
HOLDDOWN
SILL PLATE
FOUNDATION

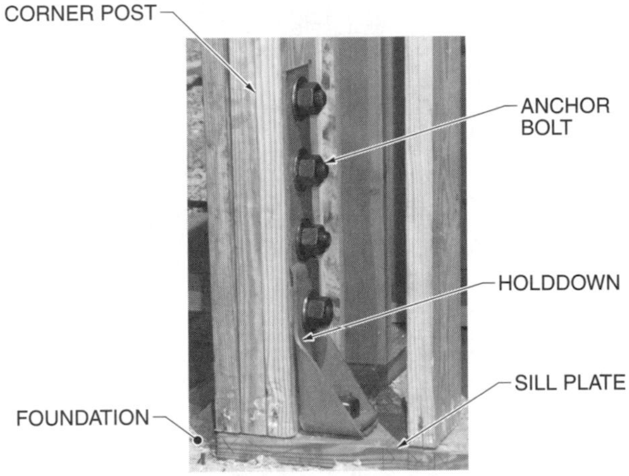

ANCHOR BOLT SECURES THE HOLDDOWN TO SILL PLATE AND FOUNDATION

BEARING PLATE
DOUBLE TOP PLATE
3½″ MAXIMUM
STUDS
THREADED ROD
TRANSITION COUPLER
BOTTOM PLATE
ANCHOR BOLT EMBEDDED IN CONCRETE

TIEDOWN IS SECURED TO ANCHOR BOLT

Figure 3-50. Posts or columns are secured to piers using steel dowels, pier blocks, or anchor bolts.

The "pull out" strength of any embedded anchor is determined by the embedded depth of the anchor, the bonding strength of the concrete to steel, and the tensile strength of the steel.

In recent years, additional types of anchoring devices have been developed. Two examples are pigtail bolts and coil inserts. A *pigtail bolt* is an anchoring device that has curved and angular shapes that increase holding power. **See Figure 3-51.** A *coil insert* is an anchoring device that allows bolts to be inserted after concrete has set. They are available in different designs to provide the desired holding power. **See Figure 3-52.** A type of template similar to that used with traditional anchor bolts is used to set the pigtail and coil anchors.

Holddowns are used to secure shear walls to a foundation.

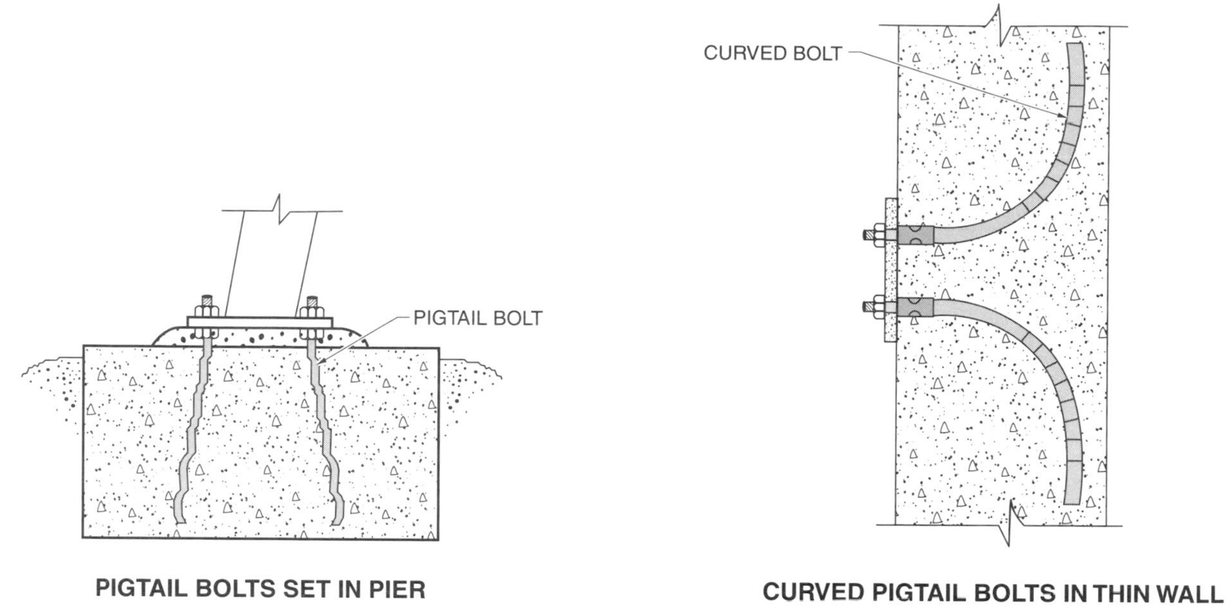

Figure 3-51. A pigtail bolt is an anchoring device that has curved and angular shapes to increase holding power.

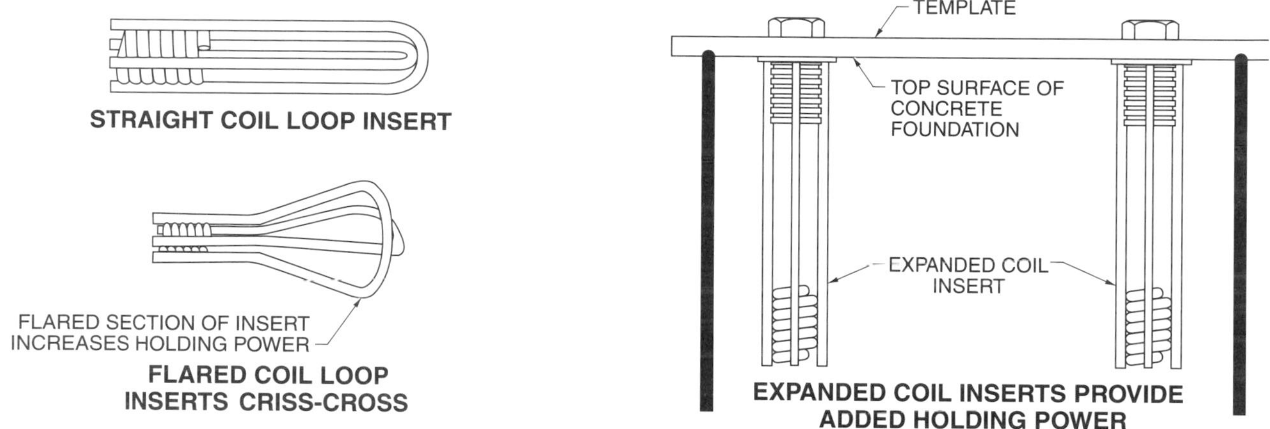

Figure 3-52. Coil inserts allow bolts to be inserted after concrete has set.

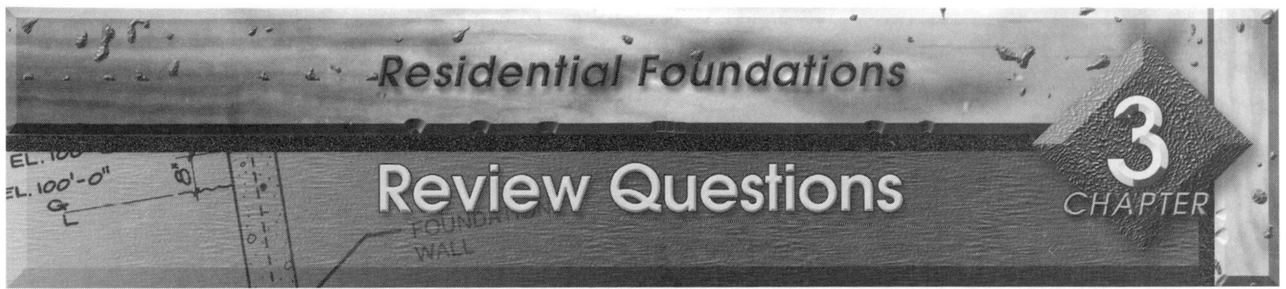

Name ______________________ Date ____________

Completion

__________ **1.** A(n) ___ rests directly on the soil and acts as the base for a foundation wall.

__________ **2.** A T-foundation is placed on a(n) ___ footing that rests on bearing soil.

__________ **3.** The width of a footing must be ___ the thickness of the wall.

__________ **4.** A(n) ___ plate is a wood member placed on top of the wall.

__________ **5.** The front setback of a building is measured from the front ___ line.

__________ **6.** The ___ method is used for squaring building lines without a transit-level.

__________ **7.** Batterboards hold the ___ lines in position.

__________ **8.** If the diagonal measurements of a square or rectangular shape are the same, all sides are ___ to each other.

__________ **9.** A(n) ___ is a saw cut made in the batterboards to secure the building lines.

__________ **10.** ___-formed footings are used in firm and stable soil.

__________ **11.** A(n) ___ is a tapered groove at the top surface of a footing.

__________ **12.** ___ views provide information regarding the size and spacing of rebar.

__________ **13.** Beam pockets should provide ___″ clearance around the sides and ends of wood beams.

__________ **14.** A(n) ___ foundation may be used to facilitate the placement of plumbing, electrical, and other utilities.

__________ **15.** A(n) ___ pier footing is constructed when the imposed structural load per square foot is greater than the bearing capacity of the soil.

__________ **16.** A(n) ___° stress angle should be used for the slope of a tapered pier.

__________ **17.** Crawl space foundations are also called ___ foundations.

__________ **18.** A(n) ___-foundation design is used most often for crawl space foundations.

__________ **19.** The minimum distance required from the bottom of the floor joists resting on crawl space foundations to the ground is ___.

__________ **20.** ___ pressure against a foundation wall is caused by the force of the earth against the wall.

__________ **21.** A(n) ___ load is the constant weight of the entire superstructure.

__________ **22.** Excavations for below-grade basements should extend at least ___′ outside the building lines.

__________ **23.** The size of a pier footing is determined by the weight of the live and dead loads and the ___ of the soil.

__________ **24.** ___ are chamfered 2 × 4s used to form keyways.

__________ **25.** A(n) ___ joint is formed when fresh concrete is placed over a concrete surface that has set.

Multiple Choice

__________ **1.** The advantage of a monolithic form for a crawl space foundation is that ___.
A. concrete for the walls and footing is placed at the same time
B. a cold joint will not be formed
C. water cannot seep through from outside
D. all of the above

__________ **2.** When placing concrete for an entrance platform against an existing foundation wall, ___.
A. place welded wire reinforcement first
B. place an isolation strip between the platform and wall
C. place a vapor barrier between the platform and wall
D. embed steel dowels in the foundation wall

__________ **3.** A grade beam is ___.
A. placed under a row of square pier footings
B. a foundation level with the finish grade
C. a foundation wall supported by concrete piers
D. a low foundation

__________ **4.** In seismic risk areas, piers should be tied to grade beams with ___.
A. angle iron
B. rebar
C. welded wire reinforcement
D. all of the above

__________ **5.** Slab-on-grade foundations ___.
A. can be an integral wall and slab foundation system
B. do not require foundation walls
C. are more costly to construct
D. do not require reinforcement

__________ **6.** A pier footing is a pedestal for ___.

A. wood posts
B. steel columns
C. concrete piers
D. all of the above

__________ **7.** Steel columns are usually anchored with ___.

A. anchor bolts
B. angle iron
C. wood posts
D. all of the above

Identification

__________ **1.** String

__________ **2.** Footing

__________ **3.** Stake

__________ **4.** Spacer block

__________ **5.** Brace

__________ **6.** Level

__________ **7.** Straightedge

__________ **8.** Form wall sheathing

__________ **9.** Battered foundation

__________ **10.** L-foundation

__________ **11.** Rectangular foundation

__________ **12.** T-foundation

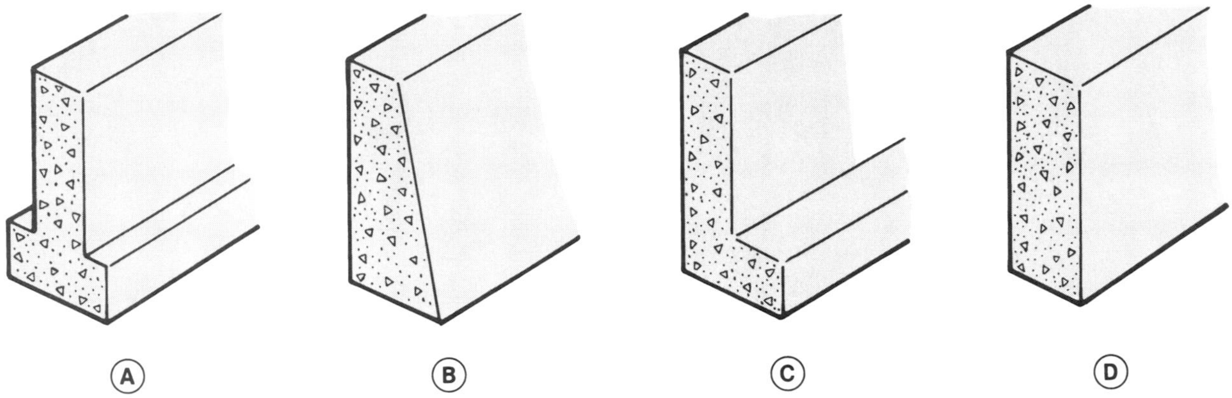

A

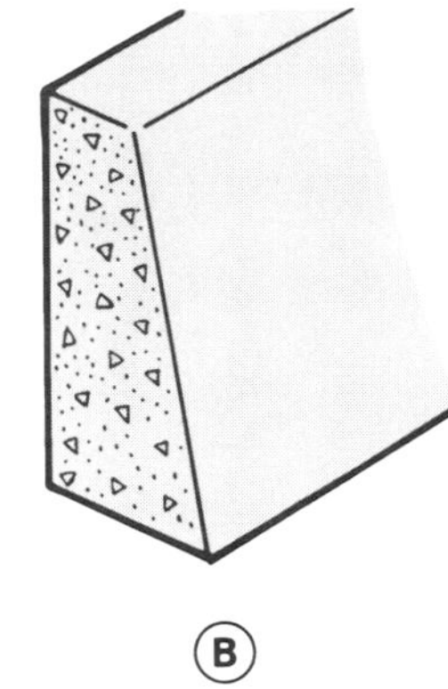

B

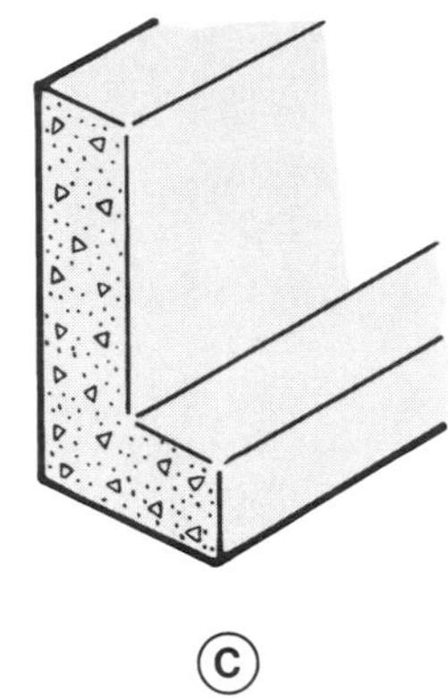

C

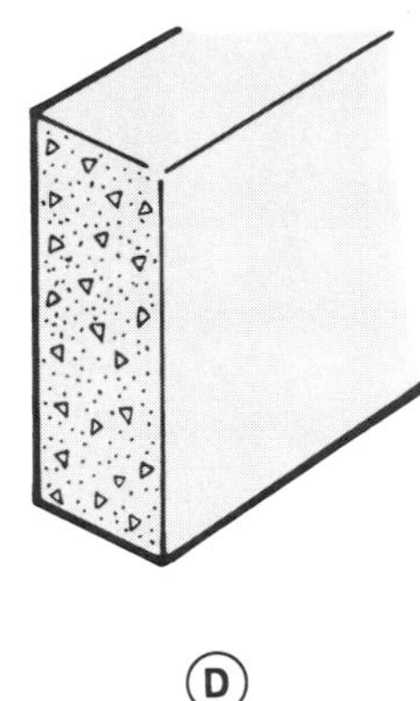

D

__________ **13.** Circular pier footing

__________ **14.** Stepped pier footing

__________ **15.** Tapered pier footing

__________ **16.** Rectangular pier footing

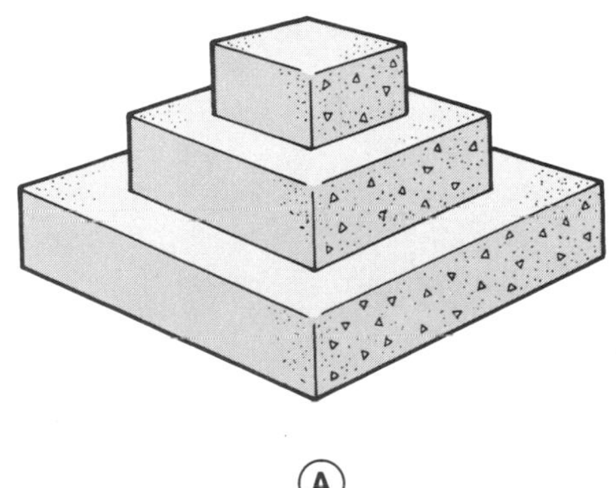

A

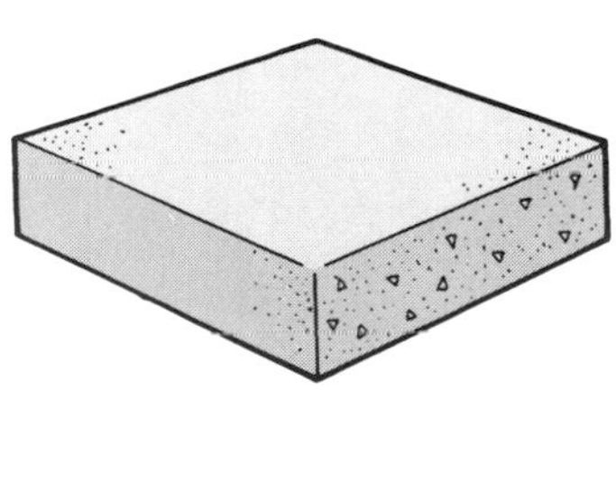

B

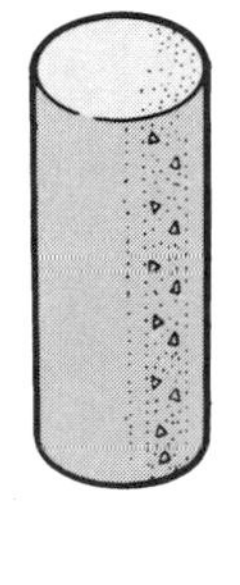

C

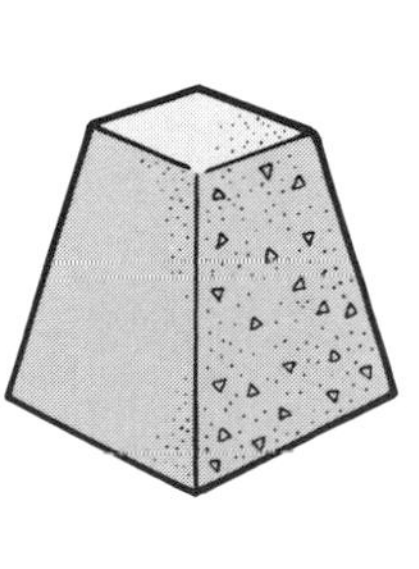

D

CHAPTER 4

Flatwork

Flatwork is the construction of indoor and outdoor concrete slabs, patios, walkways, and other flat, concrete, horizontal surfaces. The site preparation and concrete placement is done by laborers. The edge forms for the slabs are constructed by carpenters, and cement masons work and finish the concrete as it is placed.

Flatwork preparation includes accurate layout of the finished slab elevation with a builder's level, transit level, or laser level. When the elevations are established, a screed system and/or edge forms are constructed. Edge forms are used in flatwork such as driveways or walkways to retain the concrete in a specific area. A screed system is used to maintain proper floor elevations in the interior areas of the flatwork.

Preparations must also be made for construction joints, isolation joints, and expansion joints. Construction joints are used where fresh concrete butts against the edge of a section of concrete that is set. Isolation joints are used to prevent cracking between two adjacent sections of concrete. An isolation joint is usually filled with caulking compound or asphalt-impregnated material. Expansion joints are used in large sections of flatwork where expansion and contraction forces are anticipated.

Driveways and walkways are constructed for vehicle and pedestrian use. Driveways are 4″ to 6″ thick and are reinforced with welded wire reinforcement or rebar. Public sidewalks, front walks, and service walks are commonly located around residential structures. Walkways are usually 4″ thick, although thicker slabs reinforced with rebar may be required where heavy vehicles pass over.

GROUND-SUPPORTED SLABS

Ground-supported slabs are commonly placed for ground-level floors of residential and commercial buildings that do not have basements. For these types of structures, ground-supported slabs are usually less costly to build than wood-framed floors. Ground-supported slabs are also used in the construction of garage floors and below-grade basement floors.

The design and construction of ground-supported slabs are based on soil properties at the job site. In addition, moisture and thermal conditions and the shape and slope of the lot are considered in the design of ground-supported slabs.

Properly drained, dense soil mixtures such as gravel, sand, and silt generally provide a good base for ground-supported slabs. The suitability of soil conditions on a job site is often determined by past practice in the area. However, if there is any question about the soil composition at a particular job site, a qualified soil engineer should conduct a soil investigation.

Moisture conditions are also considered in the construction of ground-supported slabs. The amount of predictable surface water from precipitation (rain and snow) and the amount of groundwater can result in a volume change and/or reduction of the bearing capacity of the soil. Problems may also result from the combination of moisture and temperature conditions. Low temperatures cause groundwater to freeze and pose the danger of frost heave below the slab.

Ground-supported slabs are placed on level ground. Steeply sloped lots are not practical for ground-supported slabs because of high excavation costs and potential water drainage problems.

Slab-on-Grade Floors

Slab-on-grade (also called slab-on-ground) floors are usually integrated with a slab-on-grade foundation system. Slab-on-grade floors are placed after the foundation has been constructed, or are placed monolithically with the foundation walls. The floor slab is at the same elevation as the top of the foundation wall. In most slab-on-grade floors, the top surface of the slab is at least 8″ above the finish grade level at the perimeter of the foundation walls. Floor slabs for residential buildings are a minimum of 4″ thick. Thicker floor slabs may be required in commercial structures that support heavy loads.

When a slab-on-grade floor is placed after the foundation has been constructed, rigid insulation is recommended around the perimeter of the slab to reduce heat loss. The insulation should be at least 1″ thick and extend 24″ vertically below grade level or 24″ horizontally under the concrete slab. **See Figure 4-1.**

Site Preparation. Site preparation, including all groundwork, must be completed before the concrete is placed for a slab-on-grade floor. Site preparation may only require removing the topsoil to reach undisturbed soil, or it may require excavating deep enough to place a layer of compacted fill and a gravel base course. **See Figure 4-2.**

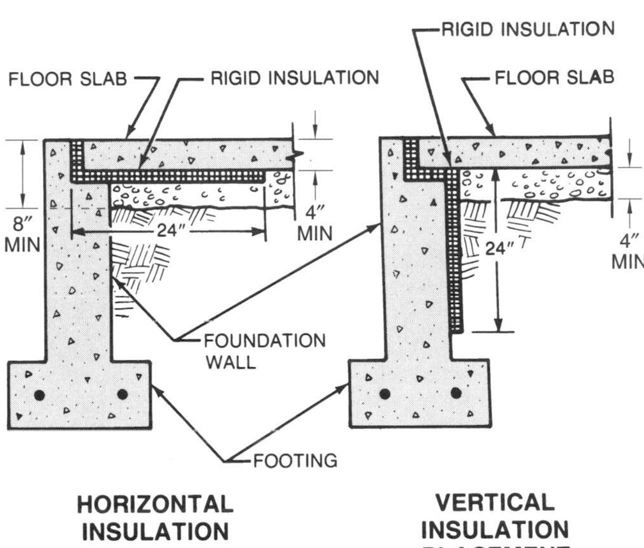

Figure 4-1. Rigid insulation extends 24″ horizontally or vertically from the perimeter of a slab-on-grade foundation. The top surface of the slab is a minimum of 8″ above the outside grade level.

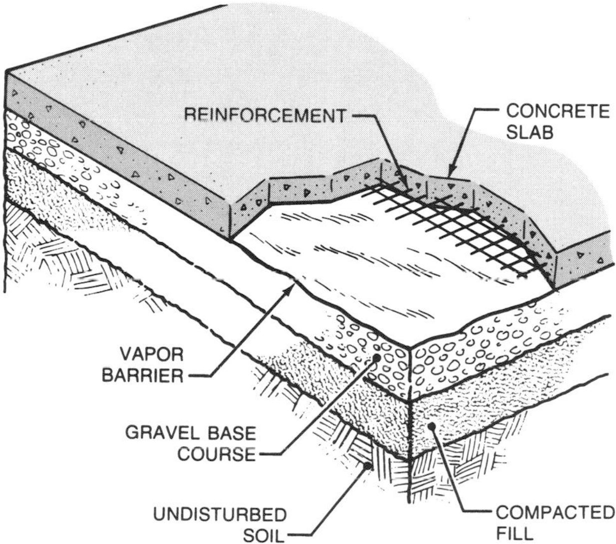

Figure 4-2. The building site for a slab-on-grade floor must be excavated to undisturbed soil. Compacted fill, a gravel base course, and a vapor barrier are placed in the excavation.

Groundwork provides support for the slab and controls ground moisture. Vapor barriers are used to contain ground moisture beneath the slab. All pipes and ducts to be embedded in the concrete must be set in place before the concrete slab is placed. Site preparation for a slab-on-grade floor requiring fill and a gravel base course is as follows:

1. Remove all topsoil and excavate to firm, undisturbed soil. Excavation must be deep enough to hold the layers of fill and gravel.
2. Place and compact fill material. Fill is required when the ground surface is uneven or where a gradual slope must be leveled. The fill should be compacted in 4″ to 12″ courses by using hand or power equipment. Fill material should be free of vegetation and other foreign material that might cause uneven settlement.
3. Install pipes, drains, ducts, and other utility lines.
4. Erect formwork.
5. Place a gravel base course at least 4″ thick to control the capillary rise of water through the slab bed. The base course also provides uniform structural support for the concrete slab and reduces the amount of heat lost to the ground.
6. Place a moisture-resistant vapor barrier over the base course, directly beneath the floor slab, to prevent moisture from seeping through the slab. Six-mil polyethylene film is commonly used as a vapor barrier. All joints are lapped at least 6″ and should fit snugly around all projecting pipes and other utility openings. Precautions must be taken so that the vapor barrier is not punctured during construction work.
7. Place perimeter insulation along the foundation walls if required.
8. Place reinforcement in the slab according to the foundation section views. The reinforcement is usually at the center of the slab or approximately 1″ to 1½″ from the top surface.

Forming Slab-on-Grade Floors. When placing a floor slab independently of the foundation walls, forming the edges of the slab is not required because the perimeter of the slab butts up against the foundation walls. When the concrete slab is placed monolithically with the foundation walls, the outside foundation form boards also form the edges of the slab. This forming method often consists of 2″ thick planks held in place with stakes and braces.

Construction Joints. A *construction joint* is a joint used where two successive placements of concrete meet, across which a bond is maintained between the placements. A construction joint is formed where a fresh concrete section butts up against the edge of a concrete section that has already set. A construction joint is formed by staking down a 2″ thick bulkhead at the outer edges of the concrete placement area. The top of the bulkhead is positioned at the height of the floor surface and a beveled key strip is fastened to the bulkhead. Metal, wood, and premolded key strips are commonly used to form a keyway for the floor. The keyway secures the edge of the next section in position. **See Figure 4-3.**

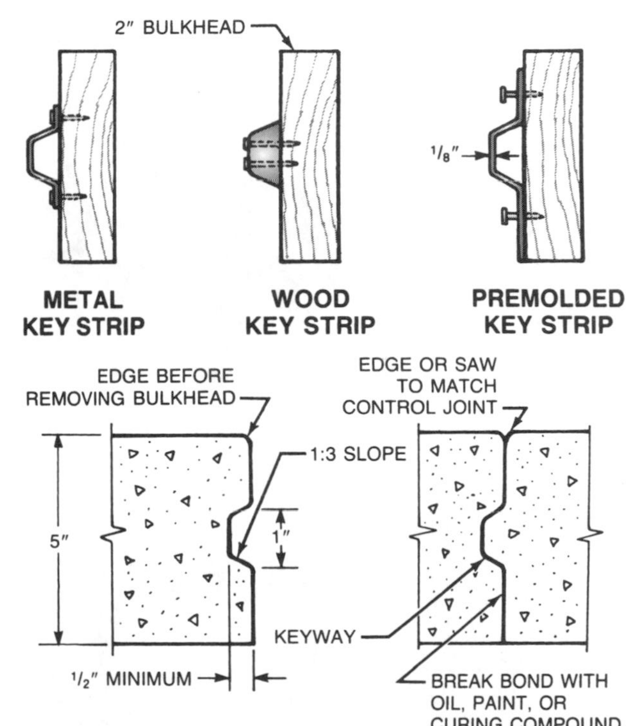

Figure 4-3. Bulkheads are used to form construction joints when a concrete slab is placed in sections. Premolded key strips are permanently embedded in the slab.

Gomaco Corporation

Long expanses of concrete, such as on a highway, require a bulkhead to be used to contain the concrete at the end of each day's pour.

Concrete for large floor areas in commercial buildings such as warehouses, factories, and stores is placed in sections. Therefore, provision must be made for construction joints when placing large floor slabs. **See Figure 4-4.**

The Euclid Chemical Company

Figure 4-4. Concrete is placed in sections for large industrial or commercial concrete slabs.

On large concrete slabs, placement, screeding, and troweling operations often occur at the same time.

Screeding. Placing concrete for large floor sections requires the use of a screed system to maintain proper floor elevations in the interior areas of the slab. The screed is positioned with its bottom edge at the finish elevation of the floor surface. Wood stakes or screed supports hold the screed off the ground and allow rebar to be positioned. Lines are stretched from the top of the outside walls or form boards to adjust the screeds to their proper height. A strike board acting as a straightedge to level the concrete is placed between the screed boards. A *strike board* is a wood or metal straightedge used for screeding concrete. Strike boards are held at the same level as the screeds by cleats nailed at opposite ends. **See Figure 4-5.** As the concrete is being placed and consolidated, cement masons strike off the concrete by moving the strike board along the screeds with a saw-like motion. The screeds and their supports are then removed from the concrete.

Screeds can also be placed with the top edge flush with the finish surface of the concrete. Two screeding methods may be used with this system. In one method, a section of the floor slab is placed and struck off to the screeds. The screeds are then removed and the concrete is placed for the next floor section. In the second method, the screeds remain in place until the entire slab has been placed. The screeds and their supports are removed and the cavities are filled with concrete. Metal pipe screeds supported by wooden stakes or adjustable chairs are often used with this method. **See Figure 4-6.** Mechanical equipment is also available for screeding operations and is often used when placing larger slabs.

When setting steel reinforcement for a slab, reinforcement must be set to the proper depth. Setting steel reinforcement as little as ½″ above its design position in a 6″ slab can reduce the live load–bearing capacity of the concrete by as much as 20%.

Control Joints. Control joints (also called relief or contraction joints) confine and control cracking in concrete slabs caused by expansion and contraction. A *control joint* is a groove made in a horizontal or vertical concrete surface to create a weakened plane and control the location of cracking. A control joint is one-fourth the slab thickness. Cracks occurring in the future will be confined to the area beneath the control joints. Control joints may be formed with a special grooving tool when the concrete is being finished. They may also be cut into the slab after the concrete has set using a power saw equipped with an abrasive blade. Recommended spacing for control joints is 15′ to 20′. **See Figure 4-7.**

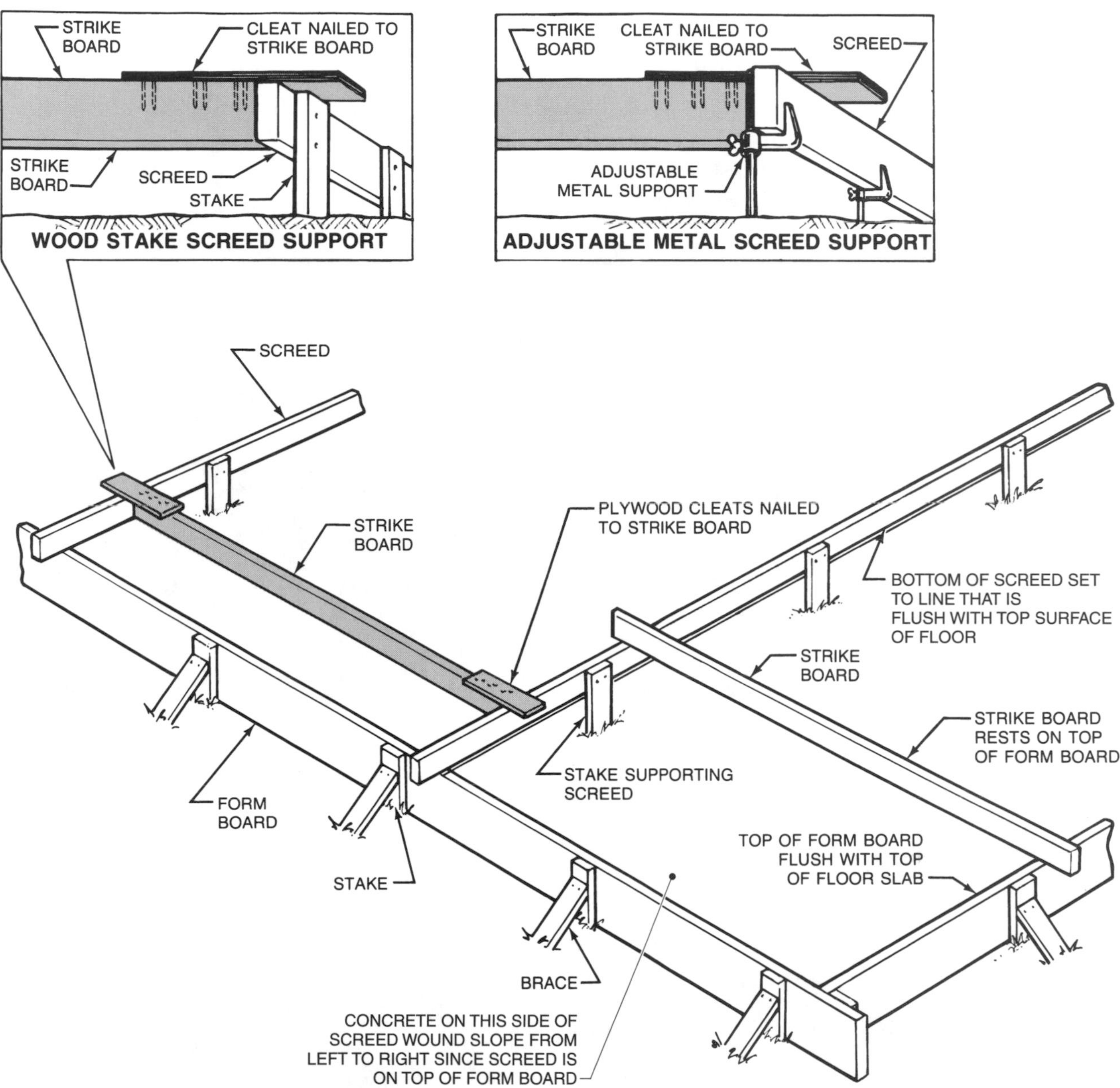

Figure 4-5. A screed system is used to strike off concrete placed for a concrete floor slab. The screed is supported by wood stakes or adjustable metal supports.

Concrete structures and slabs are subjected to many stresses as a result of shrinkage and movement. Some stresses in concrete can be controlled using construction joints. Construction joints are formed when concrete placement must be interrupted. Plan the location of construction joints so that they align where a control joint would have been cut had concrete placement continued.

Expansion Joints. An *expansion joint* (isolation joint) is a joint that separates adjoining sections of concrete to allow for movement caused by expansion and contraction of the slabs. Expansion joints are used in slabs that cover large areas of commercial buildings, and where a great amount of expansion and contraction is anticipated. Expansion joints run through the complete thickness of the slab. One common method is to place a piece of preformed asphalt-impregnated material in the joint. The fiber material is tacked to the form board before the concrete is placed and remains in place when the form board is removed. **See Figure 4-8.**

An expansion joint is also used to prevent cracking between a slab and foundation walls. Without an expansion joint, there is a weak bond between the floor slab and foundation wall since the slab is not placed monolithically (at the same time) with the wall. Ground heaving may also cause cracks in and around the slab, allowing ground moisture to seep into the basement. Expansion joints usually consist of caulking or asphalt-impregnated material placed around the perimeter of the slab to separate the floor slab from the foundation walls.

When using caulking for an expansion joint, an oiled, wedge-shaped strip is placed against the foundation wall before the slab is placed. The strip should be ½″ thick with the width equal to the thickness of the floor slab. After the slab has been placed and the concrete has set, the oiled strip is removed. The area between the slab and the wall is then filled with a caulking material.

When using an asphalt-impregnated strip, a ½″ thick premolded strip is placed against the foundation wall. After the slab is placed, the asphalt-impregnated strip remains in the concrete. **See Figure 4-9.**

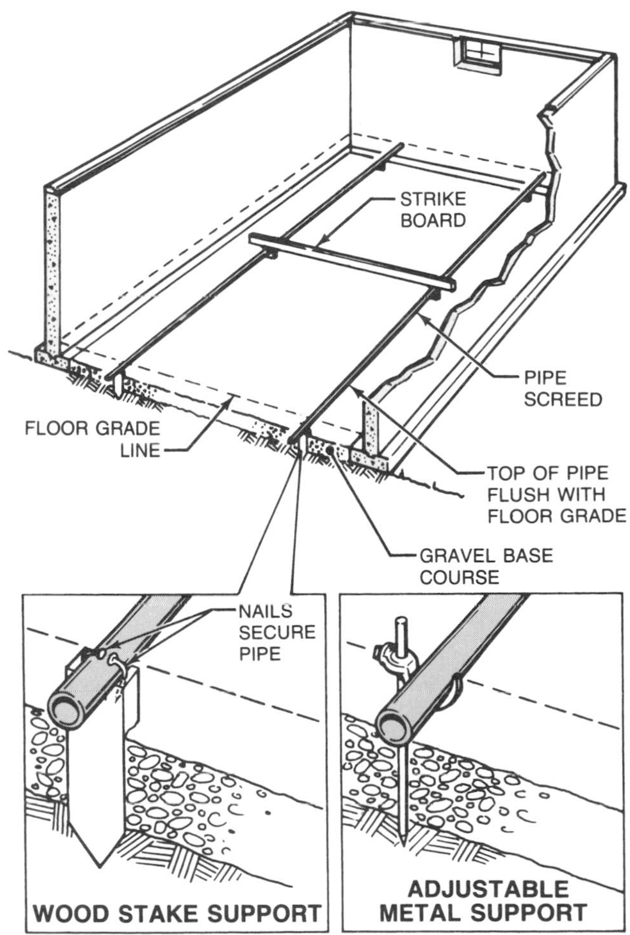

Figure 4-6. Metal pipe screeds are used to support a strike board.

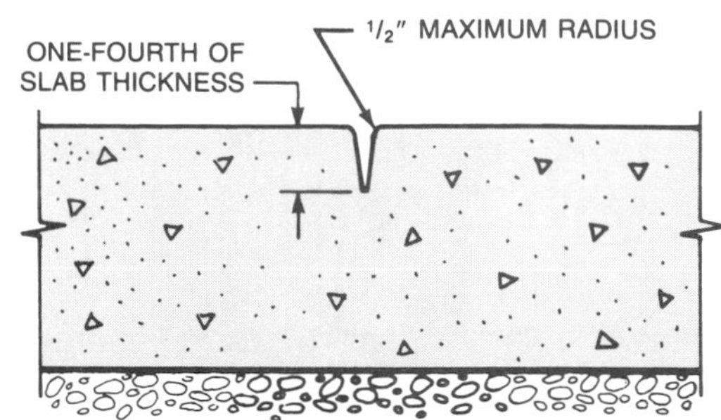

HAND-TOOLED CONTROL JOINT

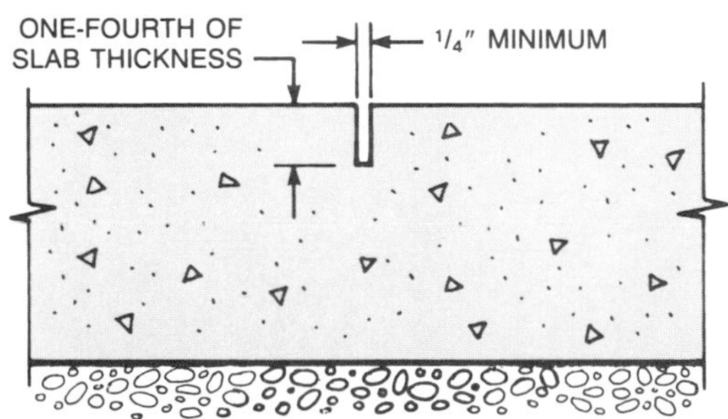

SAWED CONTROL JOINT

Figure 4-7. Control joints in a concrete slab confine cracking resulting from expansion and contraction of the slab. Control joints are hand tooled or cut into the slab.

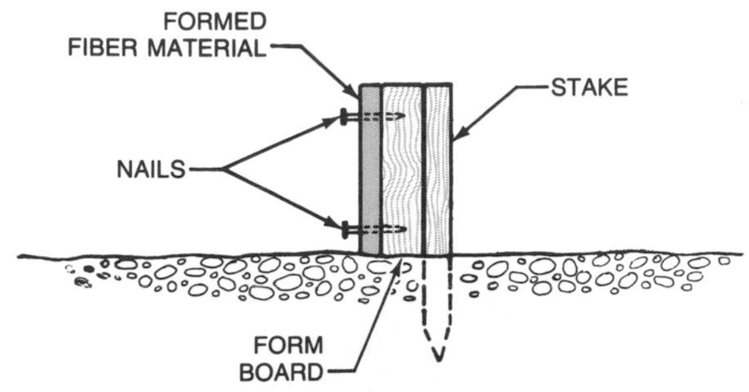

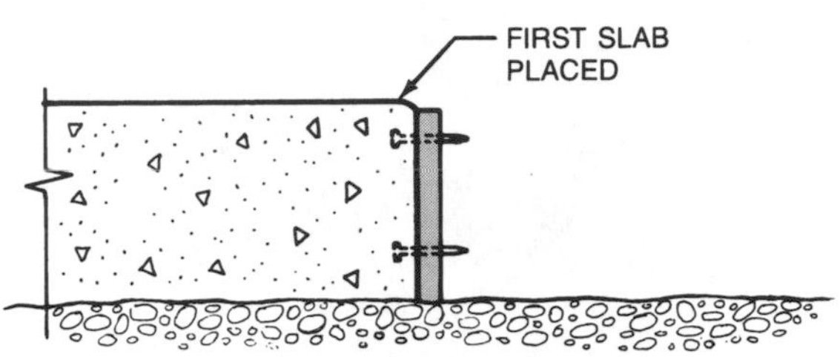

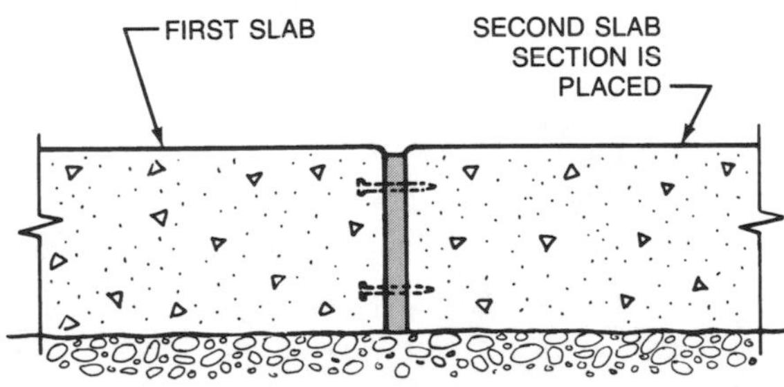

Figure 4-8. An expansion joint is used when a great amount of expansion and contraction is anticipated.

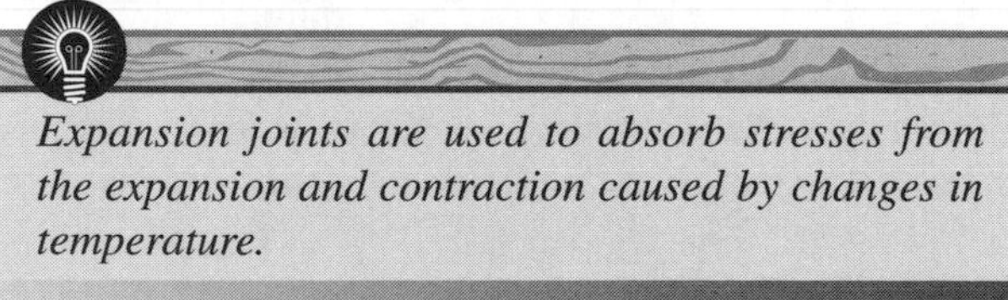

Expansion joints are used to absorb stresses from the expansion and contraction caused by changes in temperature.

Concrete Reinforcement

A wall made of concrete has a great deal of compressive strength, which is its ability to hold up under vertical pressure. Vertical pressure is exerted on a foundation by live and dead loads, including the weight of the structure, furniture, and appliances. However, concrete has far less resistance to lateral forces, which push against the wall sides. Lateral pressure is exerted on a foundation by the soil.

Reinforcement such as rebar, welded wire reinforcement, or plastic or steel fibers is used to improve resistance of concrete to lateral pressure. Unreinforced concrete has little tensile strength, allowing it to crack easily when bending stress is applied.

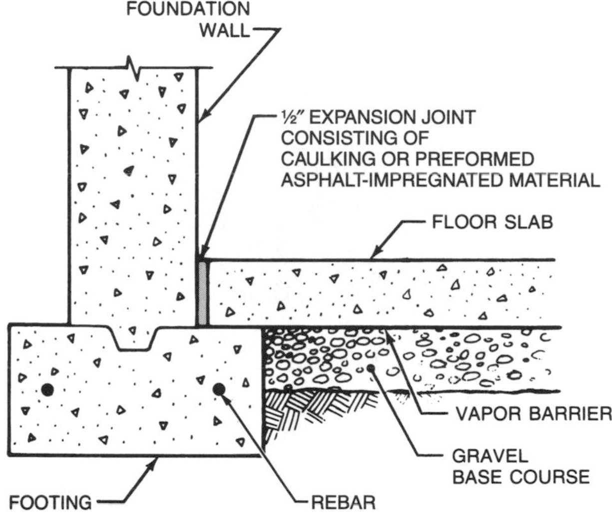

Figure 4-9. An expansion joint contains caulking or a preformed asphalt-impregnated strip.

Welded Wire Reinforcement

Welded wire reinforcement (WWR), or wire mesh, is heavy-gauge wire joined in a grid and used to reinforce and increase the tensile strength of concrete. WWR helps prevent cracks in the concrete from occurring later due to settlement. Welded wire reinforcement is available in a variety of sizes with either smooth or deformed wire. Wire size is denoted by numbers and letters. The first two numbers specify wire spacing; the second two numbers specify wire size. A letter in front of the wire size number specifies whether a smooth (W) or deformed (D) wire is used. For example, 9 gauge wire has a diameter of .1483 (1/8″).

The WWR is laid in position before the concrete is placed. Information on the size and type of welded wire reinforcement to be placed in the slab is shown on the foundation section view of the prints. Welded wire reinforcement is represented by a long dashed line in a section drawing. Spacing of the wire, type of wire, and size of wire required are identified. For example, a drawing calling for 6 × 6—W2.0 × W2.0 indicates that the spacing between wires is 6″ longitudinally and 6″ transversally. The W indicates a smooth wire, and the 2.0 × 2.0 indicates wire size, or cross-sectional area of the wire. **See Figure 4-10.**

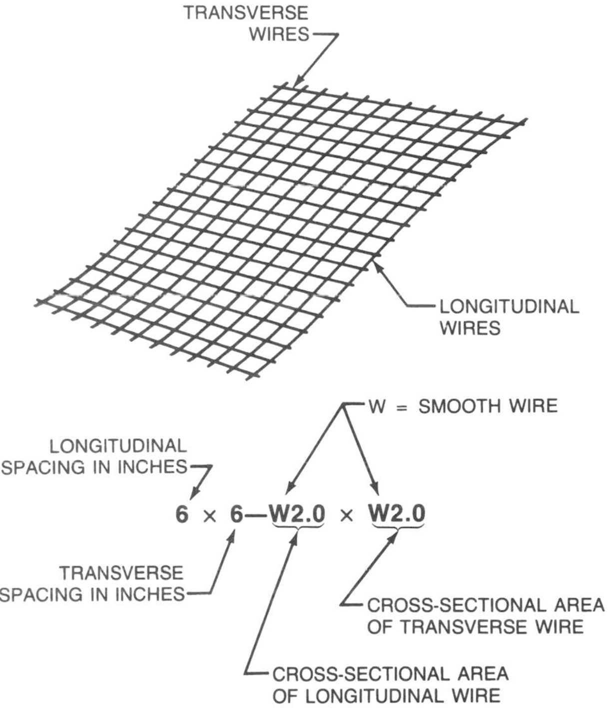

COMMON STOCK SIZES OF WELDED WIRE REINFORCEMENT				
Style Designation		Steel Area sq in. per ft		Weight Approx. lb per 100 sq ft
New Designation (by W-Number)	Old Designation (by Steel Wire Gauge)	Long.	Trans.	
ROLLS				
6 × 6—W1.4 × W1.4	6 × 6—10 × 10	.028	.028	21
6 × 6—W2.0 × W2.0	6 × 6—8 × 8*	.040	.040	29
6 × 6—W2.9 × W2.9	6 × 6—6 × 6	.058	.058	42
SHEETS				
6 × 6—W2.9 × W2.9	6 × 6—6 × 6	.058	.058	42
6 × 6—W4.0 × W4.0	6 × 6—4 × 4	.080	.080	58
6 × 6—W5.5 × W5.5	6 × 6—2 × 2†	.110	.110	80

* Exact W-number size for 8 gauge is W2.1
† Exact W-number size for 2 gauge is W5.4

Wire Reinforcement Institute

Figure 4-10. Welded wire reinforcement is used to reinforce concrete floor slabs.

Setting Welded Wire Reinforcement. Welded wire reinforcement should be set no more than 2″ from the top surface of concrete. In a 4″ slab, welded wire reinforcement is set in the middle of the slab. Setting welded wire reinforcement near the bottom of the slab adds no additional bending strength to concrete and allows surface cracks to open. During placement, welded wire reinforcement must be protected from being pushed toward the bottom of the slab. **See Figure 4-11.**

Figure 4-11. To ensure maximum strength of concrete welded wire reinforcement must be set no more than 2″ from the top surface of the concrete, or in the middle of the slab.

Welded wire reinforcement is held in place with chairs. A *chair* is a support structure made from metal, plastic, or precast concrete used to provide an accurate, consistent spacing between welded wire reinforcement or rebar and subgrade.

Chairs are classified according to the rust protection they provide. Class 1 chairs are either all plastic or have plastic-covered feet or legs. Class 1 chairs are used in extreme exposure conditions. Class 2 chairs are made entirely from stainless steel or have stainless steel feet. Class 2 chairs are used in concrete that is to be sandblasted or in areas where moderate weather conditions prevail. Class 3 chairs are made of plain carbon steel and provide no protection against rust. Class 3 chairs are used in concrete that is to have other structures erected on top of it or where a blemished surface is acceptable. Chairs are set perpendicular to the formwork, and rebar or welded wire reinforcement is placed on and tied to the chairs. Pieces of wood, soda cans, rocks, etc. should never be used as a replacement for chairs. These materials will decompose or react negatively with the concrete.

When welded wire reinforcement is used for flat floors, every other longitudinal wire should be cut along the control joints, allowing the joints to open, which relieves tensile stresses and prevents random cracking.

Rebar

Rebar is a steel bar containing lugs (protrusions) that allow the bars to interlock with concrete. Rebar increases the tensile strength of concrete and is available in various sizes, strengths, types of steel, and protective coatings.

Yield strength is the maximum load that a material will bend or stretch to accommodate a load and still return to its original size or shape. A number 50 or 60 located on the bar designates a tensile strength of 50,000 psi or 60,000 psi, respectively. Rebar that is 60,000 psi may have an extra rib running the length of the bar instead of a number. A yield strength of 40,000 psi should be assumed if no number or extra rib is present.

Rebar is identified by markings located at one end of the bar. Markings include a manufacturer letter or symbol and numbers representing diameter, type of steel, and grade. Identification marks should be checked before or during rebar placement to ensure that proper rebar is used. Protective coatings are used on rebar to decrease corrosion.

Setting Rebar. Rebar and other reinforcing members are set as specified in the prints. The load-carrying capacity of concrete is diminished if reinforcement is not set carefully at the proper elevation. Rebar set toward the center of a slab can reduce load carrying capacity by 20%. A slab with a load-carrying capacity of 50,000 lb is reduced to 40,000 lb when rebar is moved even a few inches.

Rebar is held securely in place during concrete placement using continuous chairs, which hold rebar off the ground and away from forms. Rebar is set on the chairs perpendicular to the direction the chairs are set. Rebar is tied together using wire ties at alternating intersections. **See Figure 4-12.** The ties hold rebar together and in place during concrete placement. Rebar is spliced when the area of concrete to be placed is larger than the length of the rebar.

Portland Cement Association

Figure 4-12. Rebar is held in place using chairs and is tied together using wire ties. Protective coatings are used to decrease corrosion.

Fiber-Reinforced Concrete

Fiber-reinforced concrete (FRC) is a concrete mixture that uses glass, metal, or plastic fibers mixed with concrete to provide extra strength. **See Figure 4-13.** Fibers can be used as the sole reinforcement or they can be used in combination with rebar and WWR.

The characteristics of the FRC are affected by the type of fiber used, the volume percent of fiber, the aspect ratio of the fibers, and the orientation of the fibers in the mix. The *aspect ratio* is the length of the fiber divided by its diameter. The materials used as fibers must be resistant to the acids and alkalis to which they are exposed, including the chemical reaction that occurs during hydration.

Figure 4-13. Fibers in concrete reduce cracking caused by drying shrinkage and thermal expansion.

Portland Cement Association

Chairs and tie wire are used to secure rebar in position to ensure that rebar does not move when concrete is placed.

Fibers in concrete reduce cracking caused by drying shrinkage and thermal expansion. As concrete dries, it hardens and shrinks, developing microscopic cracks. When the microscopic cracks intersect the fibers mixed into the concrete, the crack is stopped and prevented from increasing in size. Fibers in concrete also prevent cracks from becoming long and continuous.

Fibers also increase impact capacity, add abrasion resistance, provide tensile strength, and reinforce concrete against shattering. Fibers should be evenly distributed in concrete to provide reinforcement for crack control.

Plastic and steel fibers are available in various lengths and diameters, ranging in length from 1/4″ to 2 1/2″. Short fibers are commonly used for thin-wall applications. Longer fibers are used for thick slab-on-grade floors.

Basement and Garage Floors

Basement and garage floors may be placed after the foundations have been constructed and are often placed after the building has been framed. This protects the slab from weather damage and other possible damage from construction work.

In basement and garage floors, the perimeter of the floor butts against the foundation walls. The ground area below the basement and garage slabs is excavated to firm undisturbed soil. It is filled and compacted if necessary, and a base course of gravel is laid down. A vapor barrier is placed over the gravel and should extend over the foundation footings. Rebar and welded wire reinforcement are often used to reinforce basement and garage slabs.

Basement Floors. Most residential basement slabs are 4″ thick and are often reinforced with welded wire reinforcement. However, basement slabs that support heavy loads have thicker slabs reinforced with rebar. If floor drains are installed, the areas around the drains are slightly pitched toward the drain to facilitate water removal.

Basement floor information is included in section view drawings of the foundation plans. These drawings show slab thickness, type of reinforcement used, and the distance between the surface of the slab and the ceiling joists above. **See Figure 4-14.**

Before placing the concrete slab, floor elevations must be established by measuring down from the bottom of the ceiling joists. Screeds are set up close to the foundation walls and at intervals throughout the slab area. **See Figure 4-15.**

Garage Floors. Specifications for a garage floor elevation are included in a section view of the garage. In a residential structure with an attached garage, the section of the garage floor adjacent to the building is lower than the finish floor elevation of the building. The garage floor is sloped ⅛″ to ¼″ per foot toward the front of the garage for proper drainage.

Once points for the floor elevation at the rear of the garage are established, the slope from back to front is calculated by multiplying the length of the floor by the slope per foot. This amount is subtracted from the elevation at the rear of the garage and marked at the front wall. For example, a garage floor measuring 16′ from front to back with a ⅛″ per foot slope has a total floor slope of 2″ (16′ × ⅛″ = 2″). After garage floor elevations are established, chalk lines are snapped along the back and side walls. A line is then stretched across the garage opening and a form is constructed to this line. No other forms are required because the slab is formed on three sides by the foundation walls. **See Figure 4-16.**

Although the orientation of fibers in concrete is generally random, steel fibers in precast columns or beams can be aligned using a magnetic field.

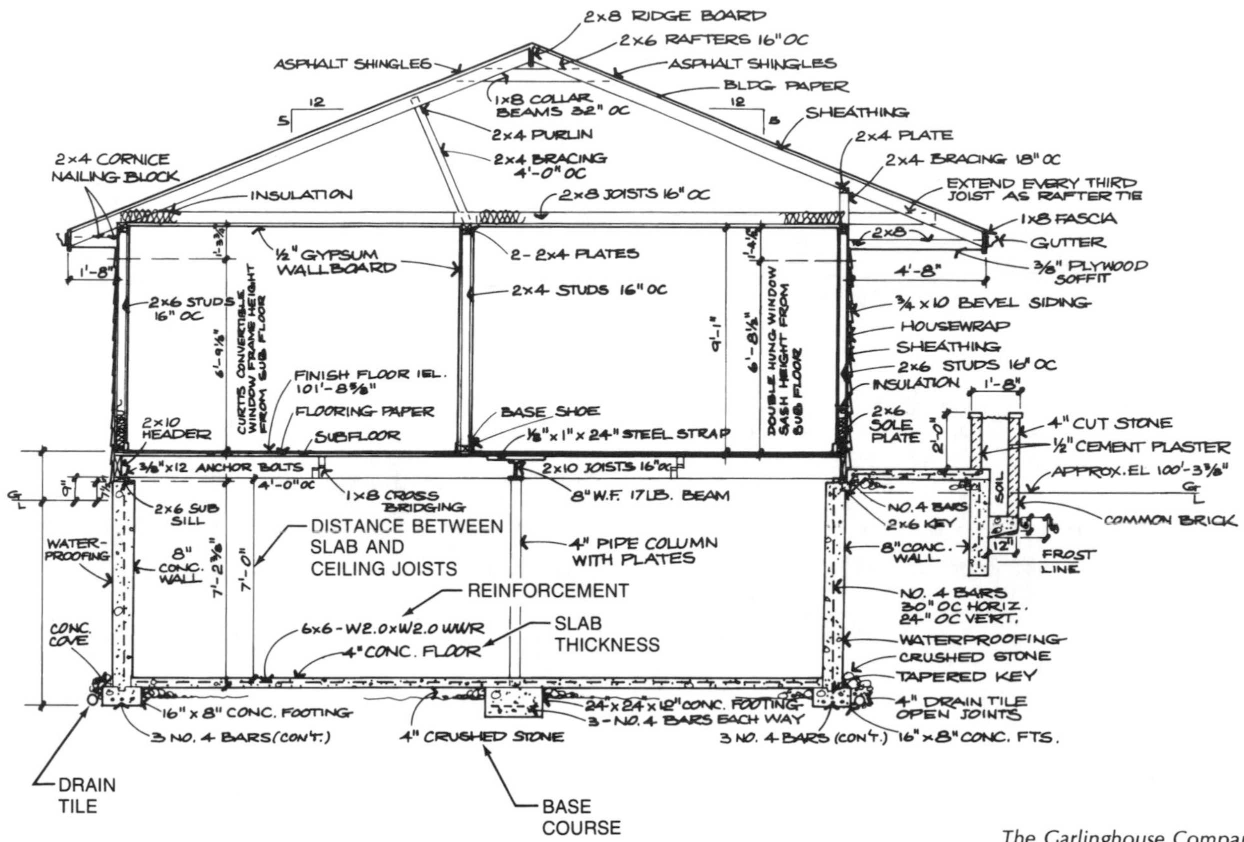

The Garlinghouse Company

Figure 4-14. The section view of a print provides information regarding the location, thickness, and reinforcement of the basement floor slab.

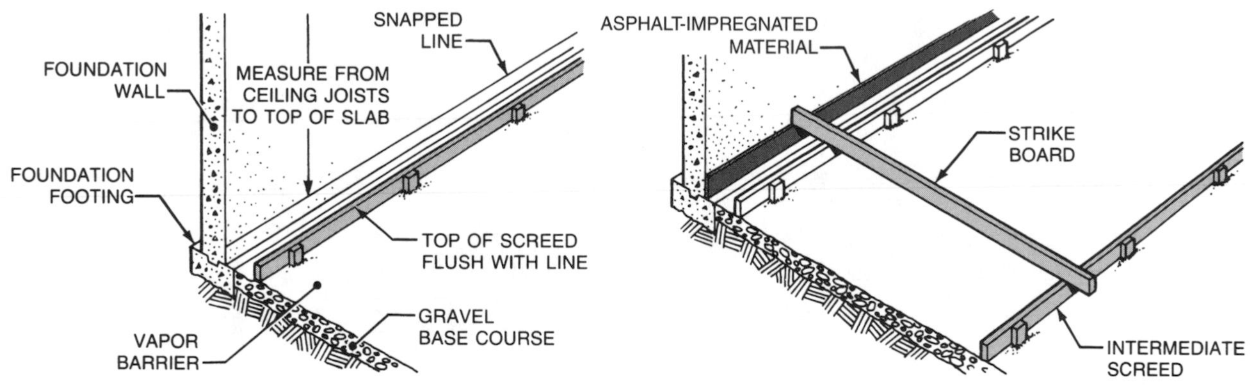

1. Measure the distance from the ceiling joists to the top of the slab and snap a line. Set a screed close to the wall and stake in position, with the top of the screed flush with the line.

2. Set and stake intermediate screeds. Cut the strike board to length and place across the screeds.

Figure 4-15. Screeds are set up close to the foundation walls when concrete is placed for a basement floor slab.

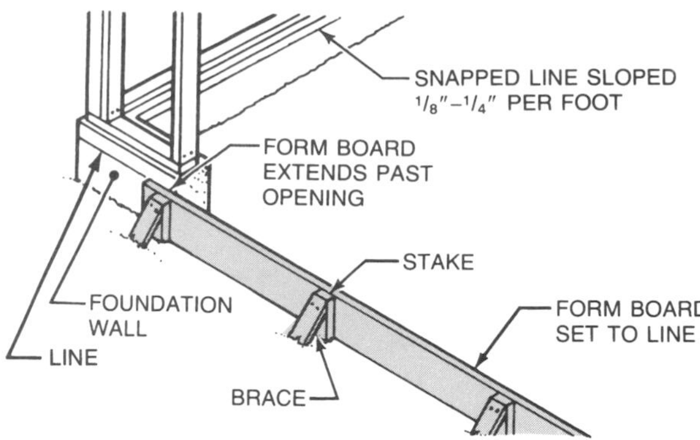

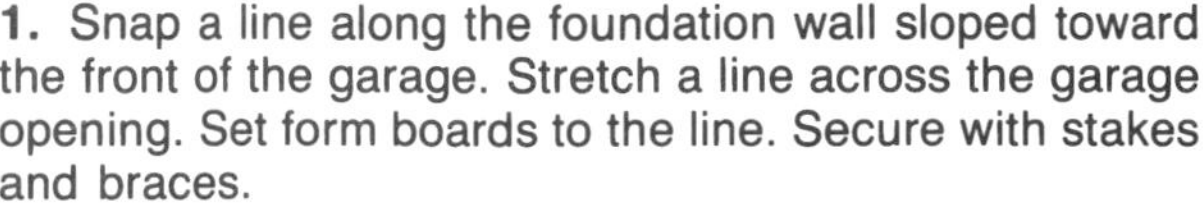

1. Snap a line along the foundation wall sloped toward the front of the garage. Stretch a line across the garage opening. Set form boards to the line. Secure with stakes and braces.

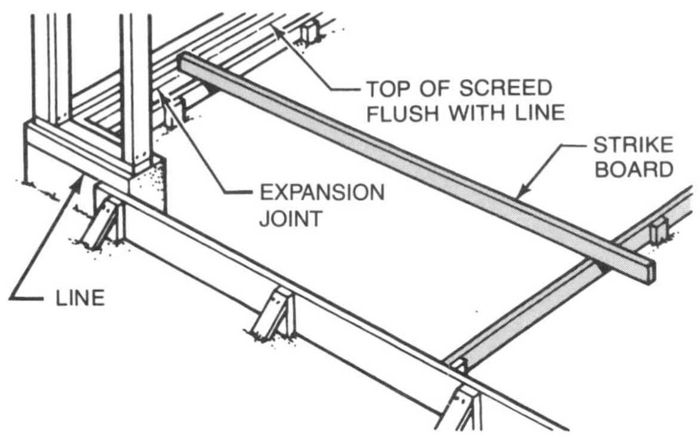

2. Set and stake screeds with the tops flush with the lines. Cut the strike board to length and place across screeds. Construction expansion joints along the walls.

Figure 4-16. Garage floors are sloped ⅛" to ¼" per foot toward the front of the garage. The garage floor is formed on three sides by the foundation walls and on one side by a form board.

EXTERIOR FLATWORK

Exterior flatwork includes driveways, walks, patios, concrete curbs and gutters, and other exterior concrete slabs placed around a building. Forms for driveway, walk, and patio slabs are laid out and placed so that the top edges of the form are set to the finish surface of the slab. Excavation may be required for fill and a gravel base course beneath the slab. The required depth for ground excavation below the slab is measured down from the top edge of the form.

Rebar or welded wire reinforcement is used where the slabs are subjected to great pressures such as driveways or other areas supporting moving vehicles. Proper sloping of exterior flatwork is very important for water drainage. Expansion and control joints are necessary to control cracking. **See Figure 4-17.** Exterior flatwork is usually the last concrete work performed on a construction project.

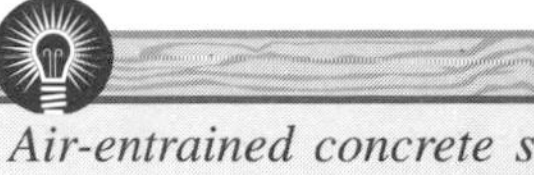

Air-entrained concrete should be used in locations prone to freezing and thawing cycles. The use of air-entrained concrete can reduce cracking, flaking, and spalling caused by temperature changes.

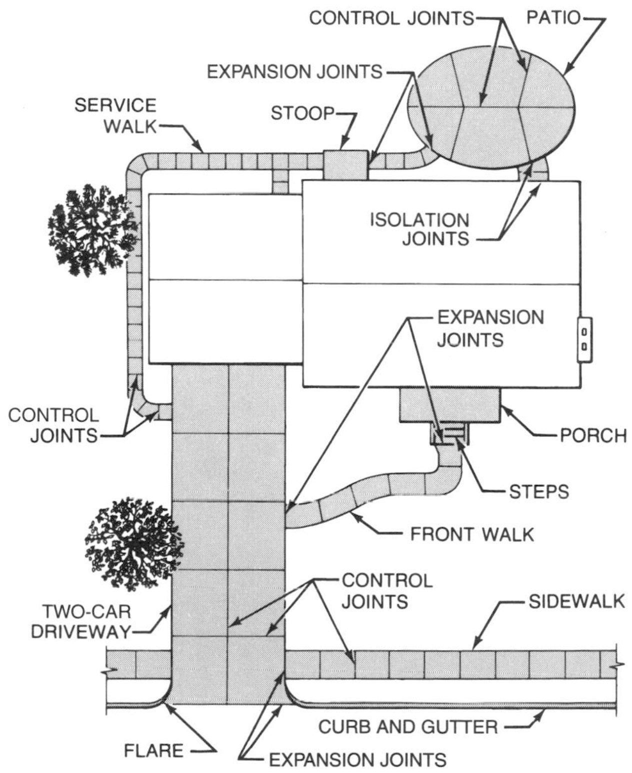

Figure 4-17. Exterior flatwork includes driveways, walkways, and patios. Expansion and control joints are used to control cracking.

Driveways

Driveway slabs for passenger cars are usually 4″ thick. Driveways that support truck movement, such as in commercial and industrial structures, should be 6″ thick. Reinforcement consisting of welded wire reinforcement or rebar is normally required for both passenger cars and truck traffic. The finished surface of the garage (or carport) end of the driveway should be ½″ below the surface of the garage floor for proper water drainage. **See Figure 4-18.**

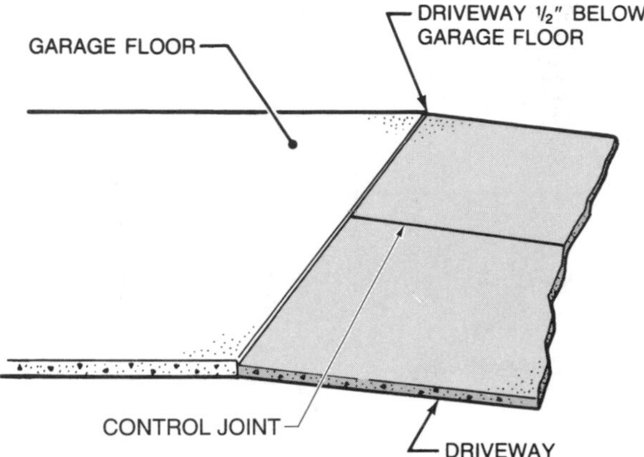

Figure 4-18. The finished surface of the garage or carport end of a driveway should be ½″ below the surface of the garage floor. The driveway is sloped away from the garage to facilitate drainage.

Manufacturers have developed many options to produce a decorative driveway, including stamping concrete, engraving concrete, coloring concrete, and creating exposed aggregate finishes.

Most one-car driveways are 8′ to 12′ wide, and two-car driveways are 15′ to 20′ wide. Consult local building codes for the minimum widths of driveways. Driveways should be flared and widened at the curbs because the back wheels of a vehicle turn in a smaller radius than the front wheels. The width of the driveway is sloped ⅛″ to ¼″ per foot from side to side. Wide driveways are pitched from the center in both directions. This requires placing a screed at the center of the driveway area before the concrete is placed.

A gradual slope from the garage area to the street curb is recommended. However, a garage considerably above or below the street level may have a steep slope. In conditions where a steep slope exists, an abrupt grade change should be avoided to prevent scraping of bumpers and undersides of cars on the driveway or sidewalk.

Driveways that slope down to the garage area should have a drain directly in front of the garage entrance. A recommended drain system involves installing a removable grate over a concrete trough that runs the full width of the garage. **See Figure 4-19.** Water collected in the trough is discharged by plastic pipe or drain tile into the surrounding soil. The water can also be discharged through a culvert or storm sewer, if allowed by local code.

Control joints are placed in the driveway to control cracking. Control joints that run the width of the driveway should be spaced no more than 10′ apart. A driveway 10′ or more in width should also have a control joint running down the center of the drive.

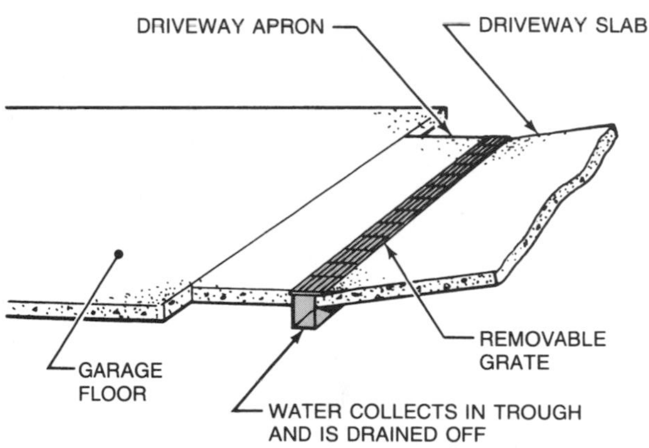

Figure 4-19. A concrete trough with a removable grate diverts water away from a garage that has a driveway sloping toward the garage.

Constructing Driveway Forms. Driveway forms usually consist of staked 2 × 4s or 2 × 6s placed on edge running the length of the driveway. Form boards are not required at the garage end or at the opposite end if the garage and front sidewalk have been placed. However, provision should be made for expansion joints where the driveway butts up against the garage floor or sidewalk. **See Figure 4-20.**

Walkways

The types of walkways generally found around buildings are public sidewalks, front walks, and service walks. Public sidewalks run along the street that borders the building lot. Front walks extend from a driveway or public sidewalk to the front entrance of a building. Service walks extend from a driveway or sidewalk to a rear entrance.

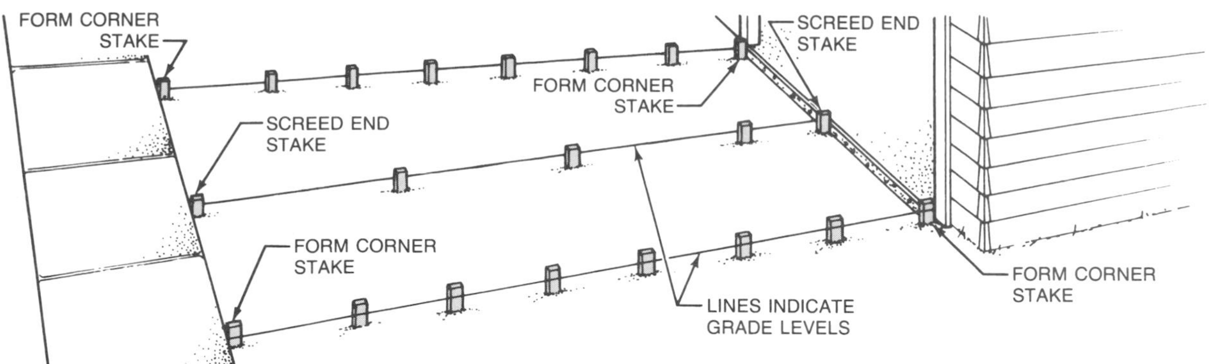

1. Drive form corner stakes. Drive screed end stakes. Lay out grade mark on the stakes and stretch lines. Drive intermediate form and screed stakes at 4′-0″ OC.

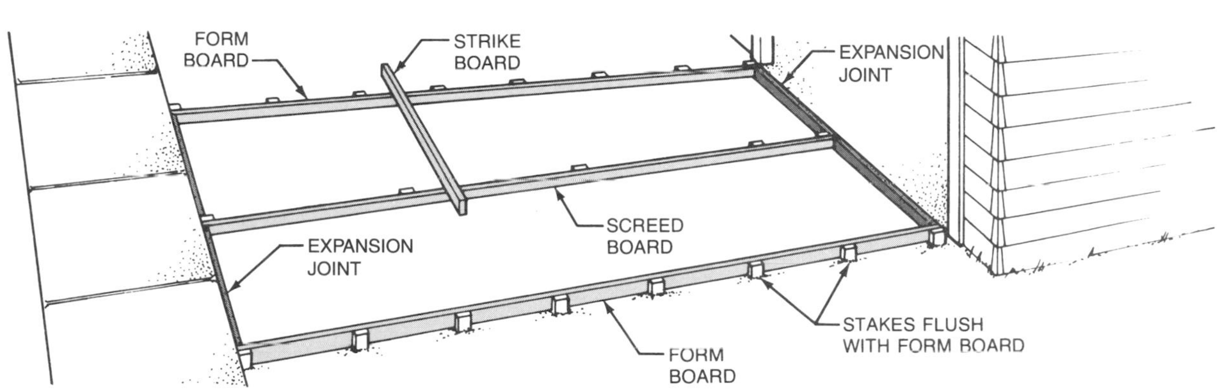

2. Nail form and screed boards to stakes. Cut the form stakes flush with the form boards. Cut a strike board to length and place on the screeds. Construct expansion joints along adjoining concrete surfaces.

Figure 4-20. Two edge forms are required to form a driveway. The garage floor and sidewalk are used to form the concrete on both ends.

Public sidewalks are usually 4′ to 5′ wide. They are placed next to the street curbs or separated from the curbs by a planter strip. Front walks are usually 3′ wide. Service walks are usually 2′-6″ wide and run along the side of a building. They are usually 2′ away from the foundation.

A walkway that butts against an entrance should be 5″ to 6″ below the surface of the stoop or door sill. If the walkway butts against a stairway, the distance from the surface of the walk to the first tread should be the same as the individual riser height.

Expansion joints should be constructed where walkways butt up against driveways, stoops, or steps. Control joints should be spaced not less than 40″ apart in walkways 2′ wide, and every 5′ in walkways 3′ or wider. Construction joints should be planned so that they occur where a control joint would have been cut had concrete placement continued.

A common thickness for walkways is 4″ although thicker slabs and rebar may be required where heavy vehicles, such as trucks, cross over the walks. Walkways should be sloped from side to side ⅛″ to ¼″ per foot to allow for water drainage. Specific dimensions for walkways are found in local building codes.

Constructing Walkway Forms. Walkway forms are usually constructed of 2 × 4s or 2 × 6s held in place by wood or metal stakes. **See Figure 4-21.** Reusable metal forms are also available. Control joints are tooled or cut into the surface of the walk to control cracking of the concrete.

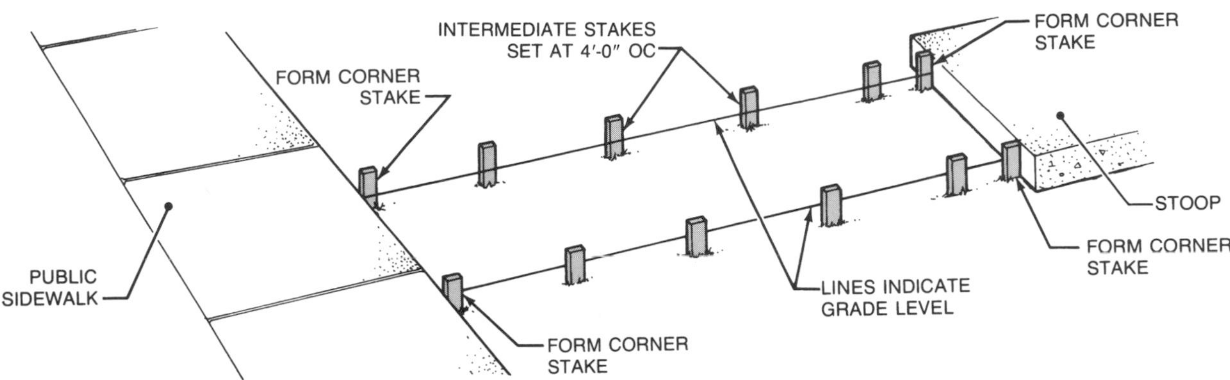

1. Drive stakes at the four corners of the walk. Mark the grade levels and stretch lines. Drive intermediate stakes at 4′-0″ OC.

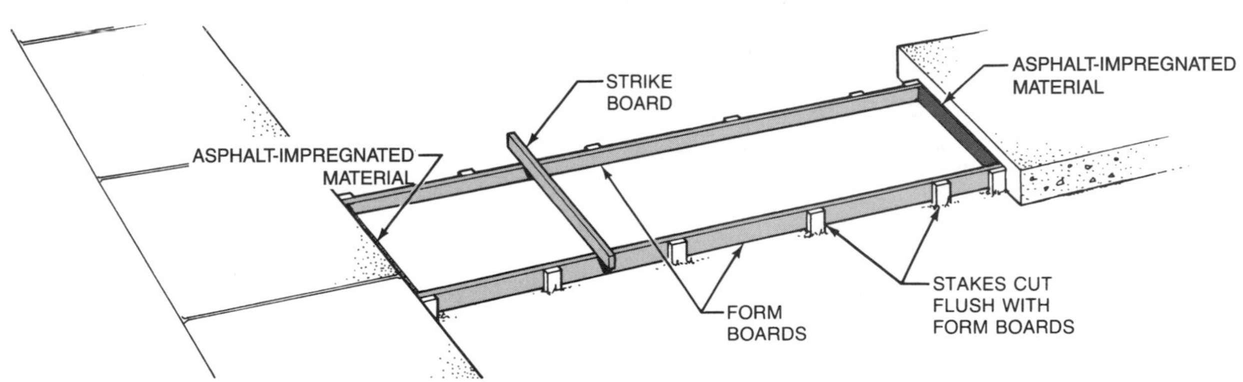

2. Snap grade levels on the stakes and nail form boards to the stakes. Cut the stakes flush with the form boards. Cut a strike board to length and place across form boards. Construct expansion joints at adjoining concrete surfaces.

Figure 4-21. Two-inch thick form boards are used to form the sides of walks. A public sidewalk and building stoop are used to form the concrete on both ends.

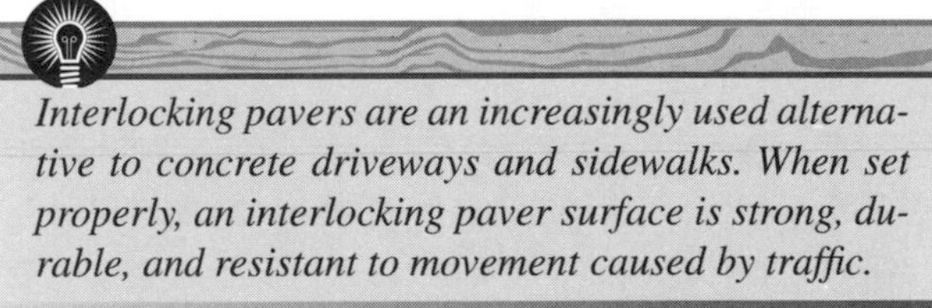

Interlocking pavers are an increasingly used alternative to concrete driveways and sidewalks. When set properly, an interlocking paver surface is strong, durable, and resistant to movement caused by traffic.

Laying Out and Forming Curves

A curve is laid out using a line that is equal in length to the curve's radius. A round stake is driven at the center point of the curve, and one end of the line is tied loosely around the stake. The other end of the line is tied to an individual flat stake that is driven along the edge of the curve. The line is then released from the flat stake and the procedure is repeated for the remaining stakes in the curve. **See Figure 4-22.**

The radius of the curve determines the material used to form the curve. A ¾″ piece of plywood can bend sufficiently for long-radius curves. Hardboard or ¼″ plywood is used for short-radius curves. Saw-kerfing ¾″ material is another method used to form short-radius curves. **See Figure 4-23.**

Patios

A concrete patio is an exterior slab constructed next to a building. It is primarily used for recreational purposes and can be designed in a variety of shapes. **See Figure 4-24.** Ground preparation is basically the same as for other slab work.

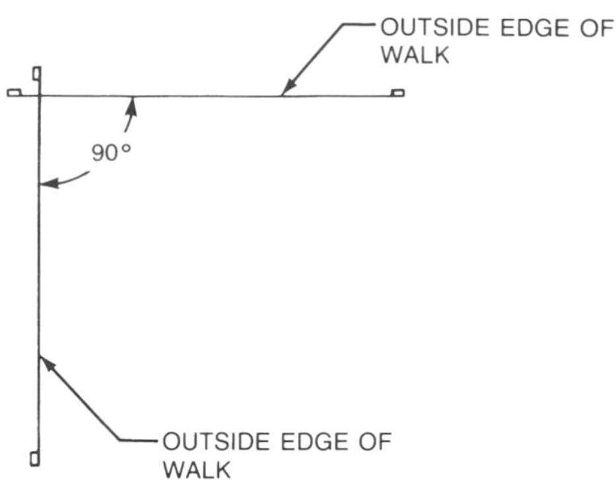

1. Drive stakes and stretch lines at a 90° angle to each other, representing the stake line for the outer edge of the walk.

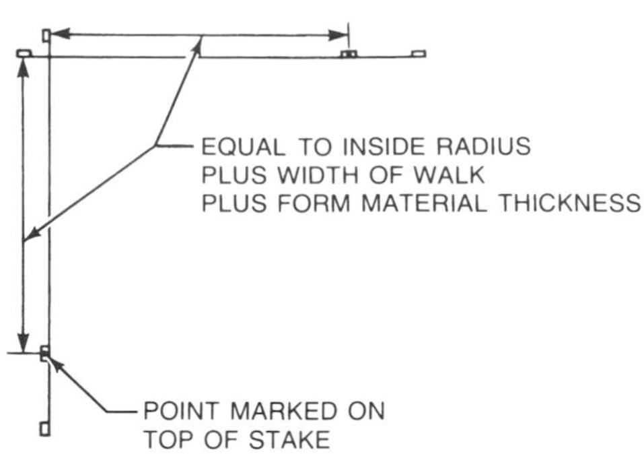

2. Measure the outside radius of the curve. This equals the curve radius (15′-0″) plus the width of the walk or driveway (5′-0″) plus the form material thickness. Drive stakes and mark the outside radius (20′-0$^1/_4$″) on top of the stakes.

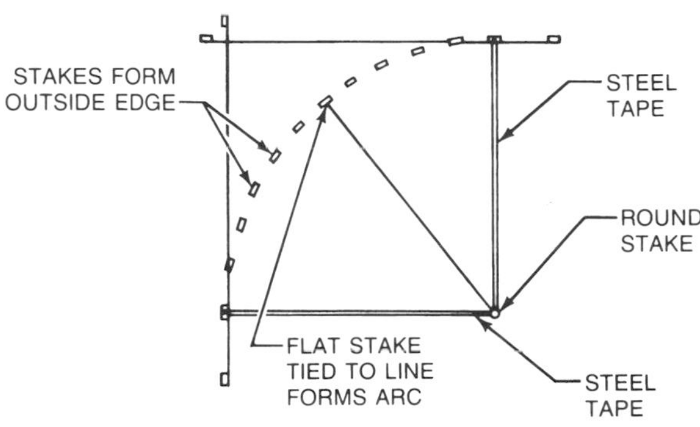

3. Measure the outside radius from both corner stakes using two steel tapes. Drive a round stake at the intersection. Tie a loose knot in one end of a line and secure around the round stake. Swing an arc (20′-0$^1/_4$″) using a flat stake tied to the opposite end of the line. Drive stakes along the arc.

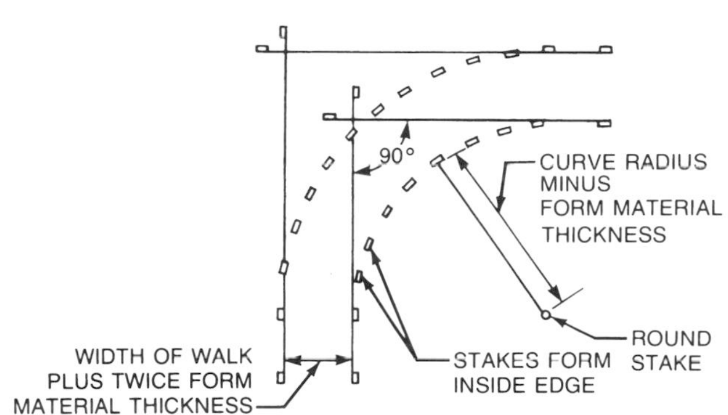

4. Measure in from the outside edge the width of the walk or driveway plus twice the form board thickness (5′-0$^1/_2$″). Drive stakes and set lines to form a 90° angle. Swing an arc measuring the curve radius minus the form material thickness (15′-0″ – $^1/_4$″ = 14′-11$^3/_4$″). Drive stakes for the inside edge of the walk or driveway.

Figure 4-22. Curved forms may be required to form curves for sidewalks or driveways. The example is a 5′-0″ wide sidewalk with a 15′-0″ radius.

Patio slabs should be at least 4″ thick and pitched ⅛″ to ¼″ per foot in one direction. Patio slabs are often reinforced with welded wire reinforcement. Expansion joints are constructed where any part of the patio meets adjacent concrete walls or walks. Control joints should be provided at a maximum of 10′ intervals in both directions.

Pressure-treated divider strips may be used in place of control joints in the patio. The divider strips are set to lines extending from the edge forms of the patio and are held in place by stakes driven below the surface of the dividers.

Patio surfaces can also be finished to resemble brick, flagstone, cobblestone, and other patterns. This is done by stamping the surface of the concrete before it has hardened. Colors can also be applied to the finished surface.

Concrete detailing tools, such as a concrete stamp and a finishing broom, are used to place a final texture or design on a concrete surface. Concrete must reach a certain consistency before detailing tools are used.

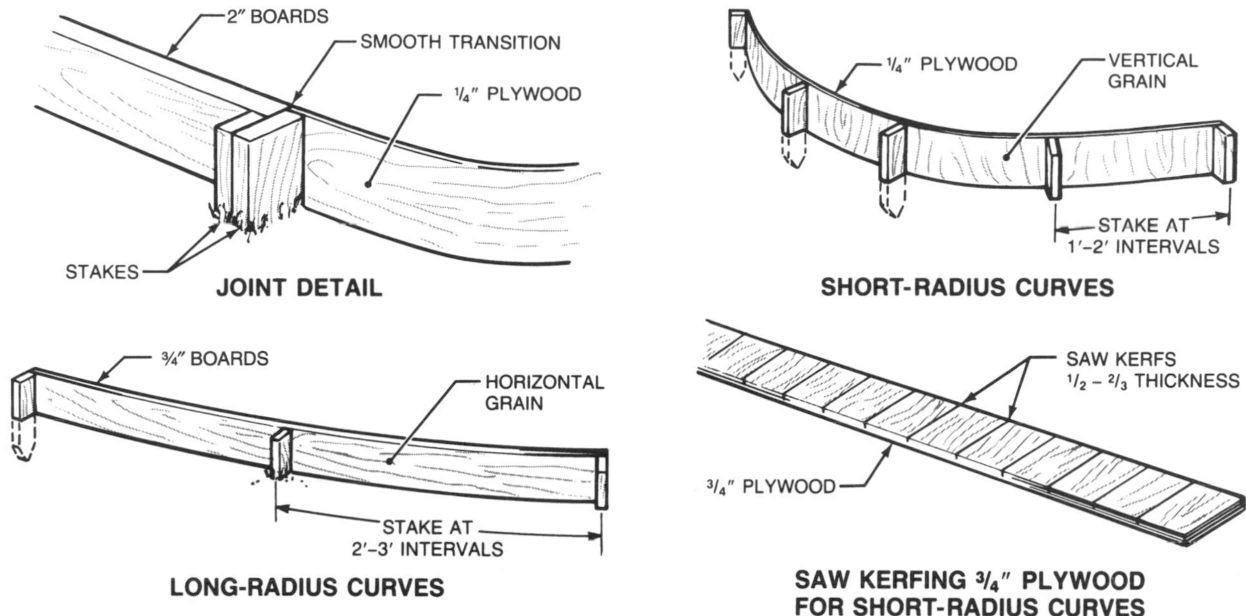

Figure 4-23. Plywood is used to form curves. Saw kerfs are made in ¾″ plywood to provide greater flexibility.

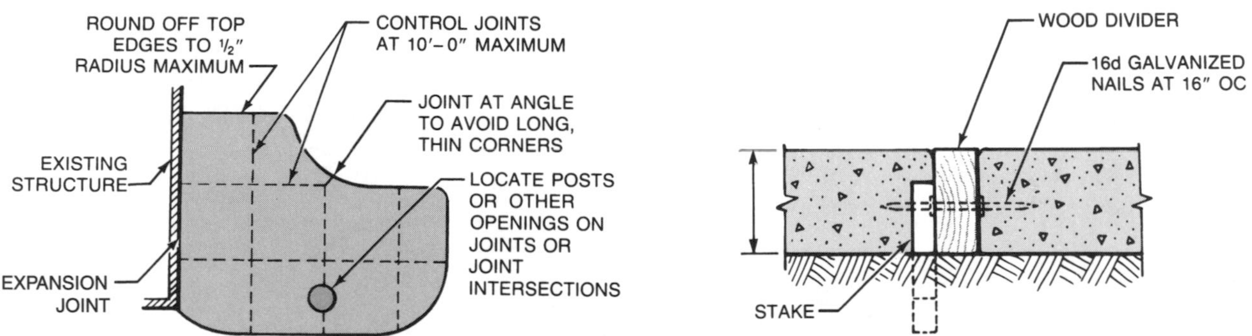

Figure 4-24. Patios serve as outdoor recreational areas in residential construction. Wood dividers add to the attractiveness of the patio and serve as screeds to strike off the patio.

Curbs and Gutters

Curbs and gutters bordering the street pavement are formed in various designs. Two basic designs are curb only and curb and gutter combination. The top of a curb is usually flush with the sidewalk and runs along the street edge except where driveways are located. Expansion joints are provided in the curbs to control cracking and are spaced according to local code requirements.

Wood forms are usually built with 2″ thick members. Stakes are driven to established lines. After the grade marks have been marked on the stakes, the form planks are nailed into place.

Prefabricated metal curb and gutter forms are commonly used to form curbs and gutters for large construction projects. **See Figure 4-25.** The edge forms are staked to the ground with metal stakes. The division plates are positioned and the curb face form is suspended from the division plates. After the concrete has initially set, the curb face form and division plates are removed and the curb and gutter are finished.

Post-Tensioned Slabs

Post-tensioned slabs are an improved development of residential and commercial slab-on-grade construction. High-strength cables (tendons) running in both directions are placed at 2′ to 5′ intervals in the slab area. They are covered with a corrosion-inhibiting grease and a plastic sheath that permits the tendons to move when they are stressed. Where unstable soil conditions exist beneath the slab area, post-tensioned ground beams are recommended 10′ to 20′ apart.

The concrete is placed after the tendons have been positioned. When the concrete has hardened to the required strength, the tendons are stressed with hydraulic jacks to an effective force of 25,000 lb or more. Afterward, anchoring devices placed at the ends of the tendons transfer the stressing force to the concrete slab. **See Figure 4-26.**

There are a number of advantages to post-tensioned slab designs. The construction process is faster because fewer reinforcing elements (rebar or WWR) are needed. Less excavation is required and the slab thickness can be reduced. The compressive force on the concrete improves the slab's resistance to cracking and deflection caused by weight pressing down on the slab.

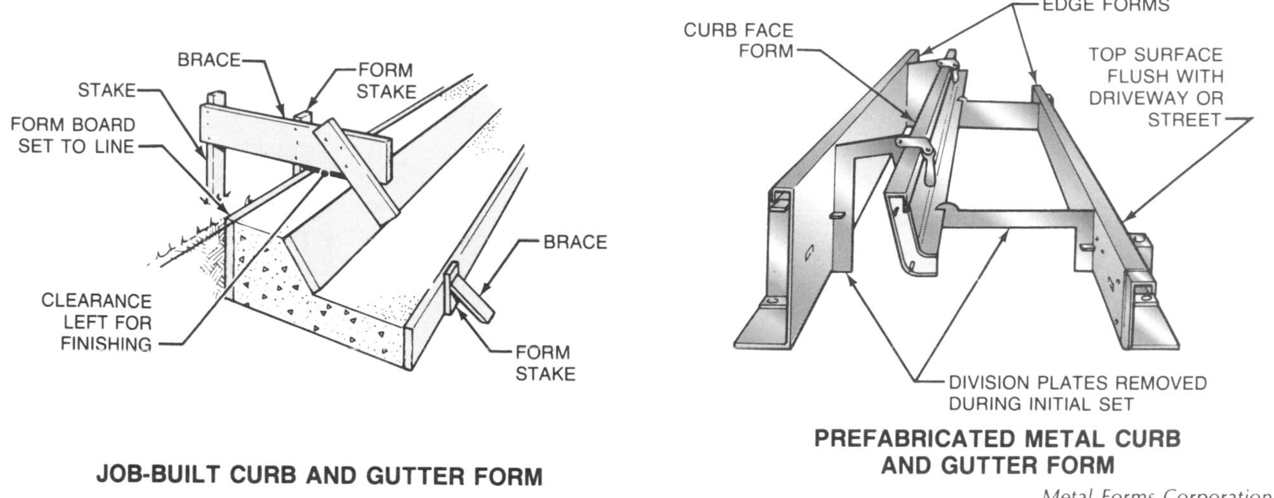

Figure 4-25. Job-built or prefabricated curb and gutter forms are used to form curbs and gutters. The top of the curb is usually flush with the sidewalk surface and the top of the gutter is flush with surface of the roadway.

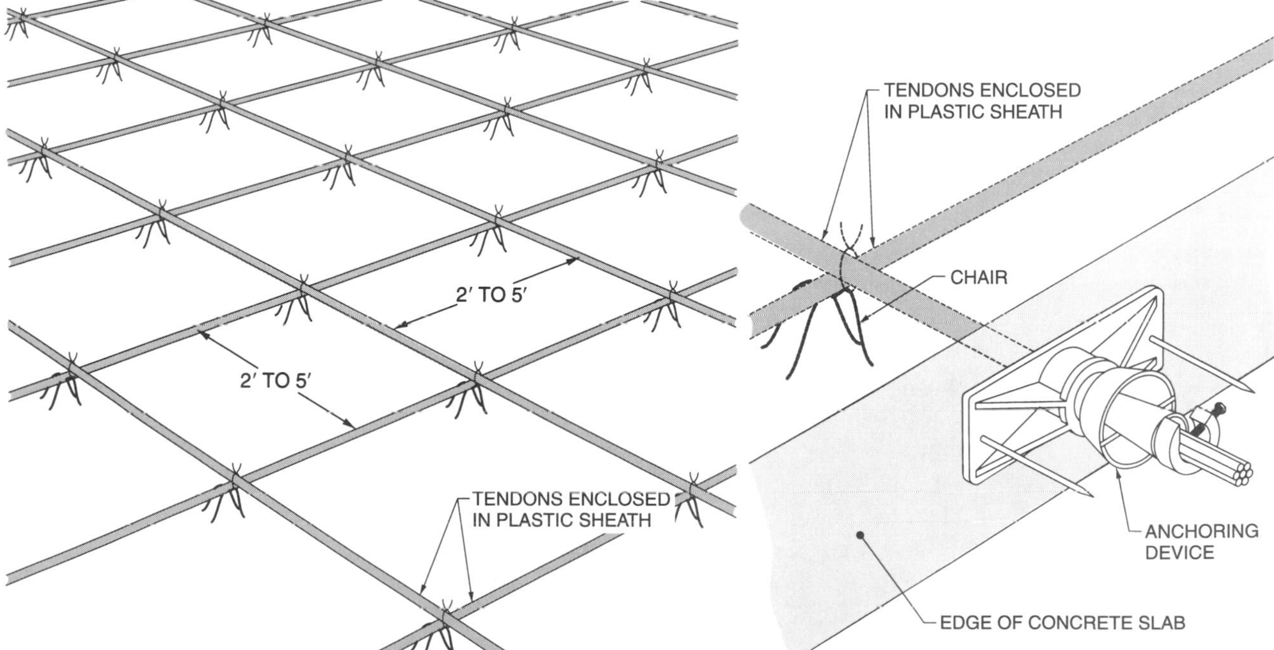

Figure 4-26. Tendons are placed in the slab area before the concrete is placed. After the concrete has hardened, the tendons are stressed with hydraulic jacks and held in place with anchoring devices.

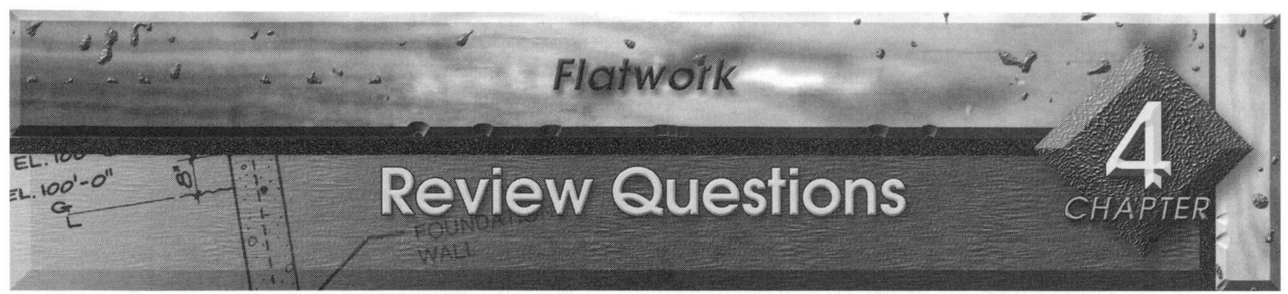

Name ______________________________ Date ______________

Completion

__________ **1.** The floor of a slab-on-grade system is at the same elevation as the top of the ___.

__________ **2.** The recommended minimum thickness of residential slab-on-grade floors is ___″.

__________ **3.** The types of reinforcement used in ground-supported floor slabs are ___ or ___.

__________ **4.** Insulation along the perimeter of floor slabs should be at least ___″ thick and extend ___″ vertically or horizontally.

__________ **5.** A(n) ___ joint is used between adjacent sections of concrete in large floor slab sections.

__________ **6.** A(n) ___ system must be set up to maintain proper floor levels when placing concrete for a concrete slab.

__________ **7.** A(n) ___ board is a straightedge placed between screeds to strike off the concrete.

__________ **8.** ___ joints help confine and control the cracking of concrete slabs.

__________ **9.** The spacing recommended for control joints in concrete slabs is ___′ to ___′.

__________ **10.** ___ drawings of a foundation plan provide information regarding basement floors.

__________ **11.** A(n) ___ joint is recommended between the slab perimeter and the foundation wall.

__________ **12.** Asphalt-impregnated isolation strips are commonly ___″ thick.

__________ **13.** Garage floors should slope ___″ to ___″ per foot toward the front of the garage.

__________ **14.** The total slope of a garage floor that is 20′ long with a ⅛″ per foot slope is ___″.

__________ **15.** The recommended thickness of driveways supporting truck movement is ___″.

__________ **16.** The surface of a driveway butting against a garage floor should be ___″ below the garage floor.

__________ **17.** One-car driveways range in width from ___′ to ___′.

__________ **18.** Two-car driveways range in width from ___′ to ___′.

__________ **19.** When forming construction joints, a 2″ thick ___ is staked at the edges of the concrete placement area.

__________ **20.** The letter D in 6 × 6—D2.9 × D2.9 welded wire reinforcement indicates that the wire is ___.

Multiple Choice

__________ **1.** Driveways should be flared and widened at the curbs to ___.
A. give it better appearance
B. conform with code requirements
C. allow for the back wheels of a vehicle to track a smaller radius than the front wheels
D. result in better water drainage

__________ **2.** Two-car driveways should be sloped ___.
A. ¼″ per foot from one side to the other
B. from the outside edges toward the center
C. from the front to the rear of the driveway
D. from the center to the outside edges in both directions

__________ **3.** The spacing of control joints across the width of a driveway should not exceed ___′.
A. 6
B. 8
C. 10
D. 12

__________ **4.** Driveways that are 10′ or more in width should have ___.
A. control joints spaced close together
B. a control joint running down the center
C. greater pitch
D. none of the above

__________ **5.** Public sidewalks are commonly ___″ thick.
A. 3
B. 4
C. 5
D. 8

__________ **6.** Walks should be sloped ___″ to ___″ per foot.
A. ⅛; ¼
B. ½; ¾
C. ⅝; ¾
D. ¾; 1

__________ **7.** The minimum width of front walks is commonly ___′.
A. 2
B. 3
C. 4
D. 5

__________ **8.** A front walk should be placed ___″ below an entrance.
A. 4
B. 5 to 6
C. 6 to 8
D. 9½

__________ **9.** ___ is recommended to form short-radius curves.

A. Three-quarter inch plywood

B. Hardboard

C. Kerfed ⅛″ plywood

D. A 1½″ form board

__________ **10.** A ___ controls the capillary rise of water and provides uniform structural support for a slab-on-grade floor.

A. vapor barrier

B. layer of fill

C. gravel base course

D. none of the above

__________ **11.** A keyway is formed with a ___ strip for a horizontal construction joint.

A. recess

B. key

C. chamfer

D. all of the above

__________ **12.** The W in welded wire reinforcement designated as 6 × 6—W4.0 × W4.0 indicates that the wire is ___.

A. rigid

B. deformed

C. smooth

D. thick

__________ **13.** When positioning a polyethylene film vapor barrier, overlap the joints ___″.

A. 2

B. 4

C. 5

D. 6

__________ **14.** The depth of a control joint is approximately ___ the slab thickness.

A. ⅛

B. ¼

C. ½

D. ¾

__________ **15.** The top surface of a slab-on-grade floor should be ___″ above the finish grade level.

A. 2

B. 4

C. 6

D. 8

Identification

_________ 1. Longitudinal spacing in inches

_________ 2. Transverse spacing in inches

_________ 3. Smooth wire

_________ 4. Cross-sectional area of longitudinal wire

_________ 5. Cross-sectional area of transverse wire

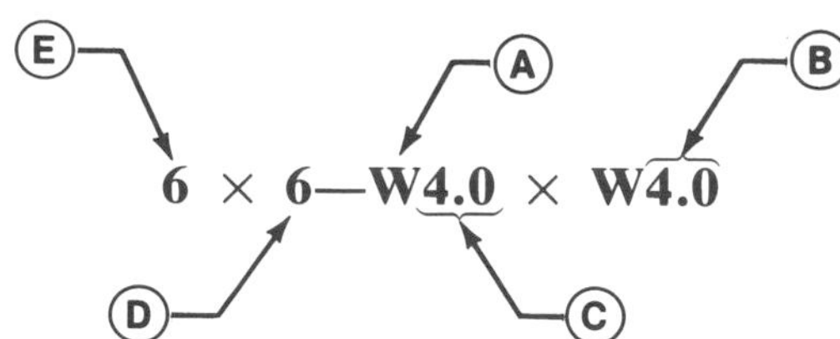

_________ 6. Reinforcement

_________ 7. Vapor barrier

_________ 8. Gravel base course

_________ 9. Compacted fill

_________ 10. Concrete slab

_________ 11. Undisturbed soil

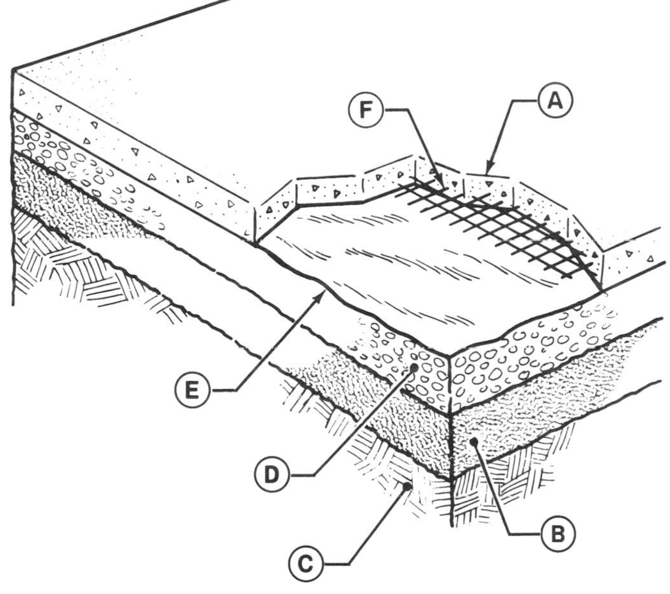

Chapter 5 Heavy Construction

Heavy construction techniques are used to construct large concrete structures such as office and apartment buildings, hospitals, and highways. Heavy construction equipment is often required for deep excavations for large concrete structures.

Large concrete structures are erected using cast-in-place concrete and/or precast concrete members. Structural members of large concrete buildings include foundation footings and walls, floor slabs, columns, beams, and girders. Forms for cast-in-place concrete are designed to frame into each other to produce monolithic structural members. The forms also resist movement during concrete placement and are easily stripped without causing damage to the structural member.

Highway construction includes the construction and maintenance of highways, bridges, overpasses, and ramps. Although most of the paving and curbing of road surfaces is performed by mechanical slipforming equipment, form construction is required for bridges, ramps, and overpasses. Formwork procedures and materials for highway construction are similar to those used in the construction of other heavy concrete structures.

Safe and established construction procedures must be followed on heavy construction projects. Form construction may occur at great heights or near heavy construction equipment. Consult the American Concrete Institute (ACI) or the Occupational Safety and Health Administration (OSHA) for information regarding safe construction procedures.

FOUNDATION FORMS

A foundation supports and transmits heavy loads to the soil below the structure. Building load and soil conditions are factors used to determine the type of foundation system. Foundation forms are laid out and built according to information supplied in the foundation plans and related section view drawings of a set of prints. **See Figure 5-1.**

Most heavy concrete structures rest either on T-foundations or on grade beam and pile systems. The main difference between foundations for heavy concrete structures and residential foundations is the size of the forms used when forming large structural members. Stiffeners, ties, and bracing members for heavy concrete structures are heavier and spaced closer than residential structures. Grade beams beneath heavy concrete structures rest on piles or caissons that are larger in diameter and extend deeper than piers used beneath residential foundations.

Mat and raft foundations (floating foundations) are also used in heavy construction projects constructed over low-bearing-capacity soils or in seismic risk areas. A *mat foundation* is a thickened reinforced slab that transmits the load of the structure as one unit over the surface of the soil. A *raft foundation* is a thickened reinforced slab placed monolithically with the walls. **See Figure 5-2.**

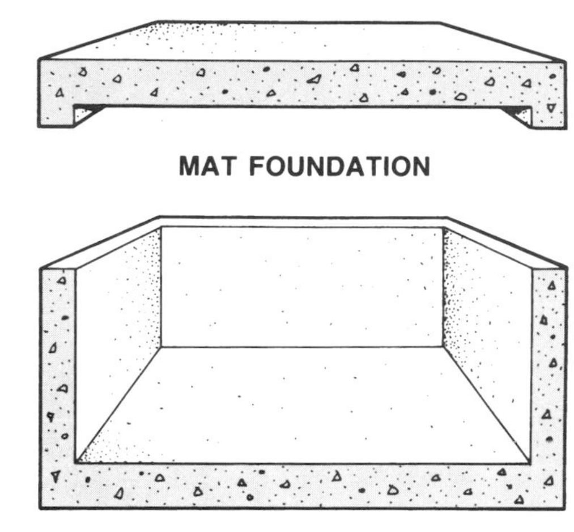

Figure 5-2. Mat and raft foundations support heavy concrete structures over low-bearing-capacity soils.

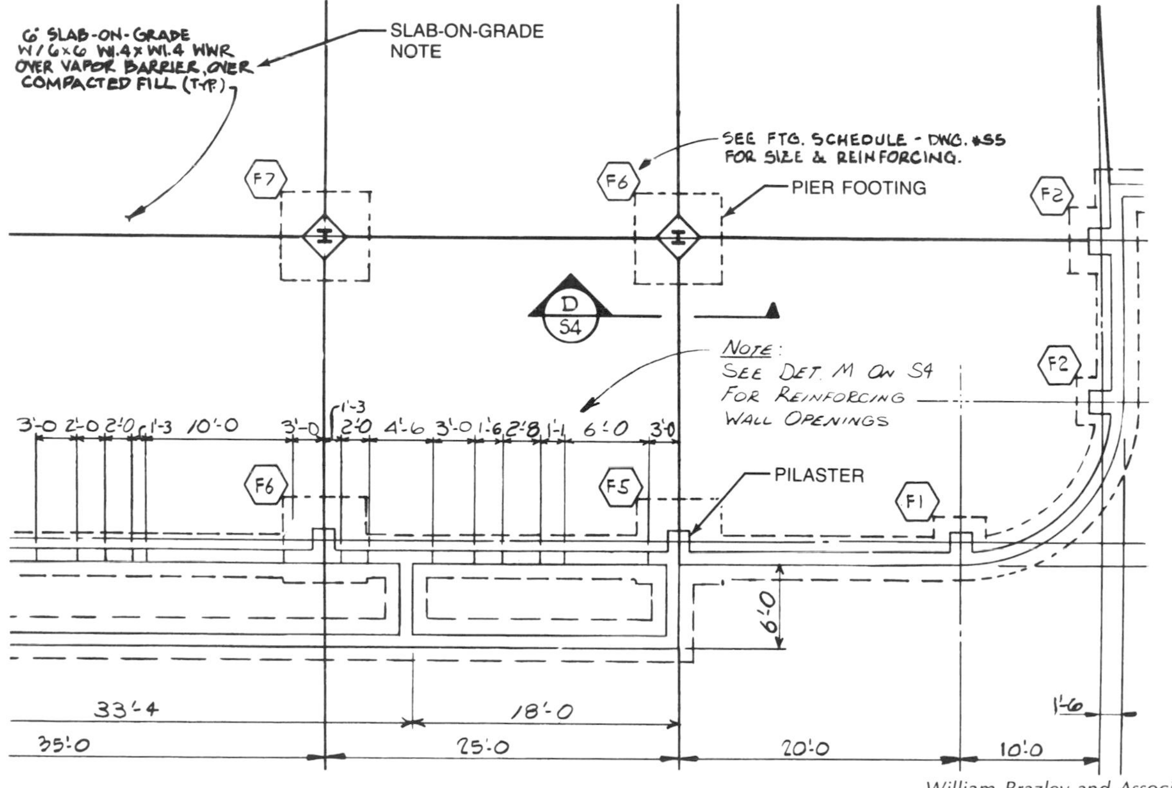

Figure 5-1. Foundation designs for heavy concrete structures may include pier footings, pilasters, and slab-on-grade floors.

Heavy construction foundations often use ground beams that tie wall or column footings together. A *ground beam* is a reinforced beam running along the surface of the ground that does not rest on supporting piers or piles.

Large concrete buildings may require deep excavation to reach load-bearing soil or to provide space for a below-grade basement. Typical heavy equipment used for excavation includes bulldozers, loaders, and excavators. **See Figure 5-3.** Various mobile or stationary cranes are also used, including freestanding and climbing tower cranes.

Freestanding tower cranes are secured to a concrete pad next to the building. Climbing tower cranes are set in position during the foundation work and move up as the height of the building increases. Climbing tower cranes are raised to each new position with hydraulic jacks and are supported by steel collars resting on the floor slabs. They are often used in the construction of high-rises. **See Figure 5-4.**

When working with cranes, a signalperson must be present to ensure accurate communication with the operator. The signalperson must understand and use proper crane handsignals.

Piles

A *pile* is a long structural member that penetrates deep into the soil. Factors such as soil conditions and adjacent buildings are considered when determining whether a pile-supported foundation should be used. Soil conditions may be too unstable to allow a conventional deep excavation, or the proximity of adjacent buildings may limit the depth of an excavation.

Pile-driving rigs equipped with either a drop, mechanical, or vibratory hammer drive solid piles or casings into the ground. Pressurized hydraulic fluid or diesel pistons are used as a power source for pile-driving rigs. Mobile cranes equipped with specialized machinery and attachments may also be used to drive piles. **See Figure 5-5.**

Bearing piles and friction piles are used to support foundations. Bearing piles are more frequently used and are driven completely through the unstable soil layers to rest on firm load-bearing soil. Friction piles do not have to penetrate to load-bearing soil. They must only be driven to a point where there is adequate soil resistance and pressure against the pile to support the imposed load.

David White Instruments

BULLDOZERS

John Deere Construction & Forestry Company

LOADERS

John Deere Construction & Forestry Company

EXCAVATORS

Figure 5-3. Heavy equipment is used for excavating heavy construction sites.

FREESTANDING TOWER CRANE

CLIMBING TOWER CRANE

Portland Cement Association

Figure 5-4. Tower cranes are used in the erection of heavy concrete structures.

© Case Foundation

Figure 5-5. Pile-driving rigs are used to drive wood, steel, or precast concrete piles.

Piles are placed beneath grade beams supporting bearing walls. Grouped piles may also be placed beneath concrete caps that act as a base for load-bearing columns. A *grouped pile* is numerous piles driven in close arrangement. Grouped piles are used when the main structural support of a building is provided by columns and the column load exceeds the load-bearing capacity of an individual pile.

Piles are constructed of wood, steel, or concrete. Wood piles are the oldest type of pile used and support wharves, docks, and other structures built over water. Steel piles placed beneath buildings are most often H-shaped or tubular. **See Figure 5-6.** Both types are driven into the ground with a pile-driving rig.

H-shaped piles are used as a foundation support and also in the construction of shoring around deep excavations. Although H-shaped piles are relatively expensive on a per foot basis, they have a number of advantages over other types of pilings. H-shaped piles are available in various sizes of roll shapes or structural steel plates welded together. They are made from high-strength and/or corrosion-resistant steels and can penetrate to bedrock, where other piles would be destroyed during pile driving. Tubular piles (pipe piles) may be filled with concrete after they have been driven. Tubular piles are a series of drill bits added to the pile-driving rig as the depth increases. When the specified depth is reached, concrete is pumped into the bottom of the pipe pile. Once the pile is filled with concrete, the pile pipes are raised and removed.

© Case Foundation

Figure 5-6. H-shaped piles may be driven into the ground using a pile-driving rig. H-shaped piles are used as foundation support.

Concrete piles are commonly used beneath foundations of heavy concrete structures. They may be precast or cast-in-place. Precast piles are commonly fabricated in a plant and heavily reinforced with rebar or prestressed cables. Precast piles are delivered to the job site by truck and are driven into place with a pile-driving rig.

The *pile head* is the upper surface of a precast pile in its final position. **See Figure 5-7.** The *butt* is the large upper portion of the pile. The *foot* is the lower section of the pile. The *tip* is the small lower end of the pile. The *driving head* is a metal device placed on top of the pile head to receive the pile-driving rig's blows and protect it from damage. The *pile cutoff* is the portion of the pile head that is removed after the pile is in its desired position. A *pile shoe* is a metal cone placed over the tip of the pile to protect it from damage while the pile is being driven. The pile shoe also allows the pile to penetrate through very hard materials in the ground.

Grade beam forms are constructed over piles after they have been driven. The grade beams are secured to the piles by tying rebar placed in the grade beams to steel dowels projecting vertically from the piles. Holes for the steel dowels are drilled into the butt after the pile has been driven and cut off. The steel dowels are then inserted into the holes and grouted. The dowels often penetrate 4′ or more into the pile and extend the height of the grade beam above the pile.

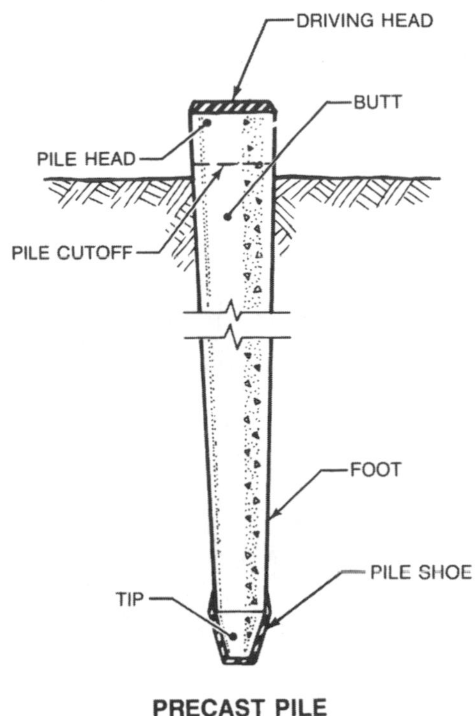

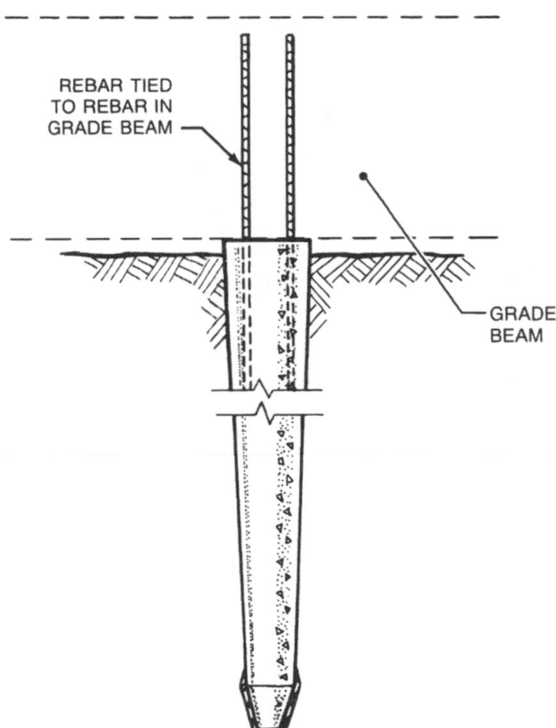

Figure 5-7. Precast piles are tied to grade beams with rebar after the piles are cut off.

Rebar cages provide the necessary reinforcement for caissons.

Cast-in-place piles are placed using a shell or shell-less method. In the shell method, a metal casing is driven into the ground and remains in position as the concrete is placed into the casing. In the shell-less method, the metal casing is removed after being driven and concrete is placed into the bored hole.

Caissons

Foundation designs may require caissons instead of piles. A *caisson* is a cast-in-place pile formed by drilling a hole, inserting reinforcement, and filling the hole with concrete. Caissons are larger in diameter and extend to greater depths than piles. Caissons are used where the building design and/or soil conditions make pile driving difficult or inadequate. The larger diameter of the caisson allows for greater load-bearing capacity, which means that caissons can be spaced further apart than piles, yet are able to carry more load. Columns, rather than walls, carry the main vertical loads transmitted to the foundation and soil below. The depth of soil penetration needed to support the column loads may not be possible using a grouped pile system. Caissons may be used because these large surface area supports the loads.

Caissons are drilled with a crane and drill rig attachment that drills a hole of a specified diameter into the soil. A *casing* is a metal cylindrical shell that is driven into the ground to restrain uncompacted soil near the surface. After the casing is inserted into the soil, the remainder of the caisson is drilled through compacted soil to the specified depth of the caisson. Once the caisson is drilled to the proper depth, a rebar cage is set in the hole and concrete is placed to the proper elevation. **See Figure 5-8.** Casings for deeper holes are constructed of sections that are added as the drilling proceeds.

A belled caisson features a greater bearing area at the base of the caisson. After the caisson hole has been bored to the desired depth, a belling tool is attached to the drilling head. The bottom of the hole is then dug out to the bell shape. **See Figure 5-9.**

WALL FORMS

Foundation wall forms are built over footings or grade beams. Basic form designs and methods used in large concrete buildings are similar to foundation wall designs for residential and light commercial construction. However, heavy construction requires much larger forming units, such as large panel forms or ganged panel forms, and mechanical equipment to position the forms. Materials commonly used for wall forms are plywood panels with wood stiffeners, metal-framed plywood panels, and all-metal panels and frames.

Foundation wall forms and wall forms for floor levels above the foundation differ in the way the form bottoms are secured. The bottom of an outside foundation wall form is commonly fastened in position by securing a plate to the top of the footing. The bottom of an outside wall form placed at floor level above the foundation is fastened with bolts or ties provided in the previous lift. A *lift* is a layer of concrete placed in a wall and separated by horizontal construction joints.

Caissons require steel reinforcement to provide additional load-bearing capacity. Reinforcement is usually provided by rebar cages. Rebar can rust when left outdoors. Rust does not affect the strength of the reinforcement as long as the rust coating is light. If the rust is scaling or flaking, rebar should be checked thoroughly before use to ensure rust has not eaten into metal.

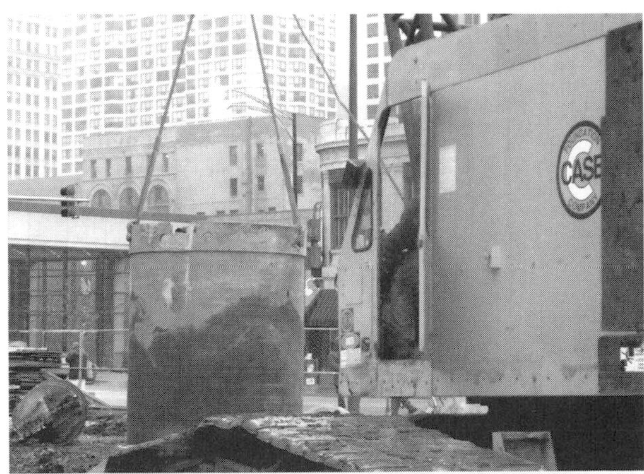

1. Insert casing.

2. Drill caisson hole to specified diameter and depth.

3. Use belling tool to form bell at caisson bottom.

4. Insert rebar cage.

5. Fill caisson with concrete and allow to harden.

6. Loosen and remove casing.

Figure 5-8. Caissons are larger in diameter and extend to greater depths than piles. Holes for caissons are bored with special drilling rigs.

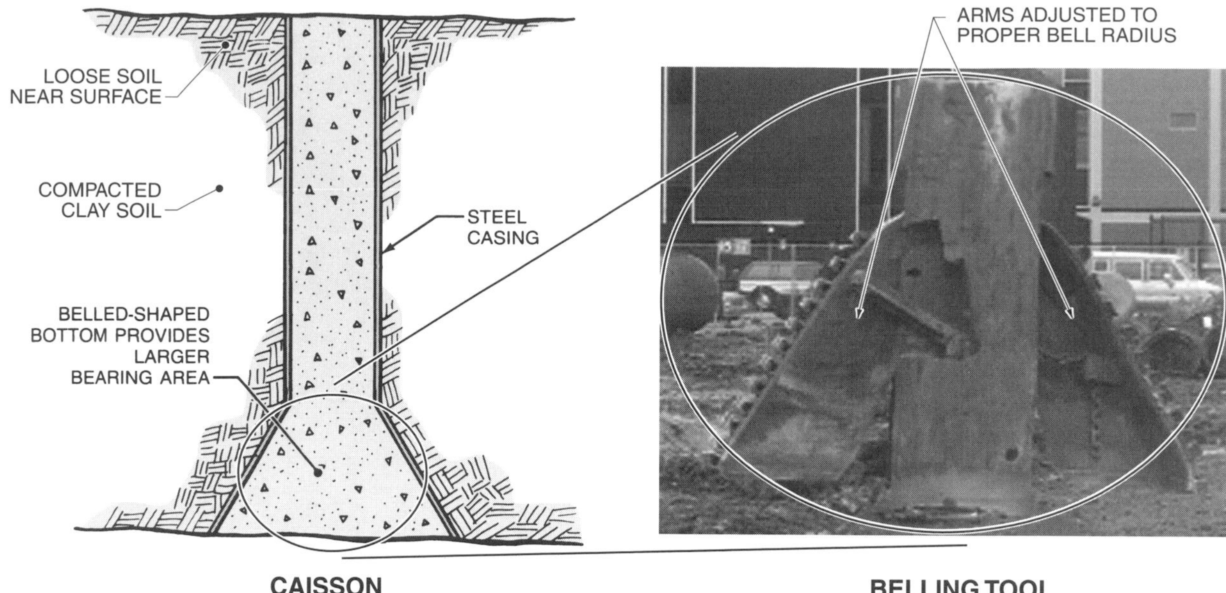

Figure 5-9. A belled caisson provides a larger bearing area at the base.

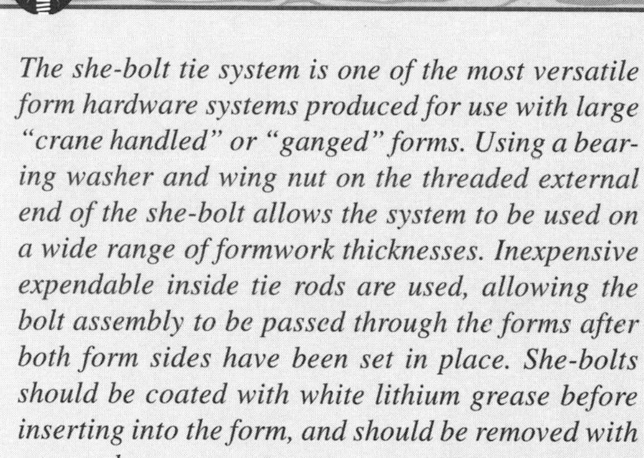

The she-bolt tie system is one of the most versatile form hardware systems produced for use with large "crane handled" or "ganged" forms. Using a bearing washer and wing nut on the threaded external end of the she-bolt allows the system to be used on a wide range of formwork thicknesses. Inexpensive expendable inside tie rods are used, allowing the bolt assembly to be passed through the forms after both form sides have been set in place. She-bolts should be coated with white lithium grease before inserting into the form, and should be removed with a wrench.

Information related to the layout and design of the walls is contained on the floor plans and working drawings of the prints for a building. On large construction jobs, formwork detail drawings and schedules are developed from the working drawings to aid in form construction. Detail drawings of prefabricated forms are commonly furnished by the form manufacturer. **See Figure 5-10.**

Large Panel and Ganged Panel Forms

Large panel forms and ganged panel forms, also referred to as climbing forms, are most efficient in constructing high walls covering large areas. *Large panel forms* are wall forms constructed in large prefabricated units. *Ganged panel forms* are wall forms constructed of many small panels bolted together. The forms are usually assembled in a shop and delivered by truck to the job site. Cranes and other lifting equipment raise and position the forms. After the concrete has been placed and has gained sufficient strength, the forms are released and raised for the next lift.

Steel-framed plywood panels or all-steel panels are used to construct the ganged panel forms, which may range in size up to 30′ × 50′. The inside and outside form walls are held together with internal disconnecting ties that are unscrewed when the ganged panel form is ready to be released from the wall. Ganged panel forms can be taken apart at the end of a job and reassembled for other shapes and sizes of walls.

A layout drawing is used to identify the components of a ganged panel form. The individual panels are placed on sleepers laid in a flat area and bolted together according to the layout drawing. Walers are then attached to the panels, and strongbacks are fastened to the walers. Lifting brackets are bolted to the ganged panel form for the crane attachment. **See Figure 5-11.**

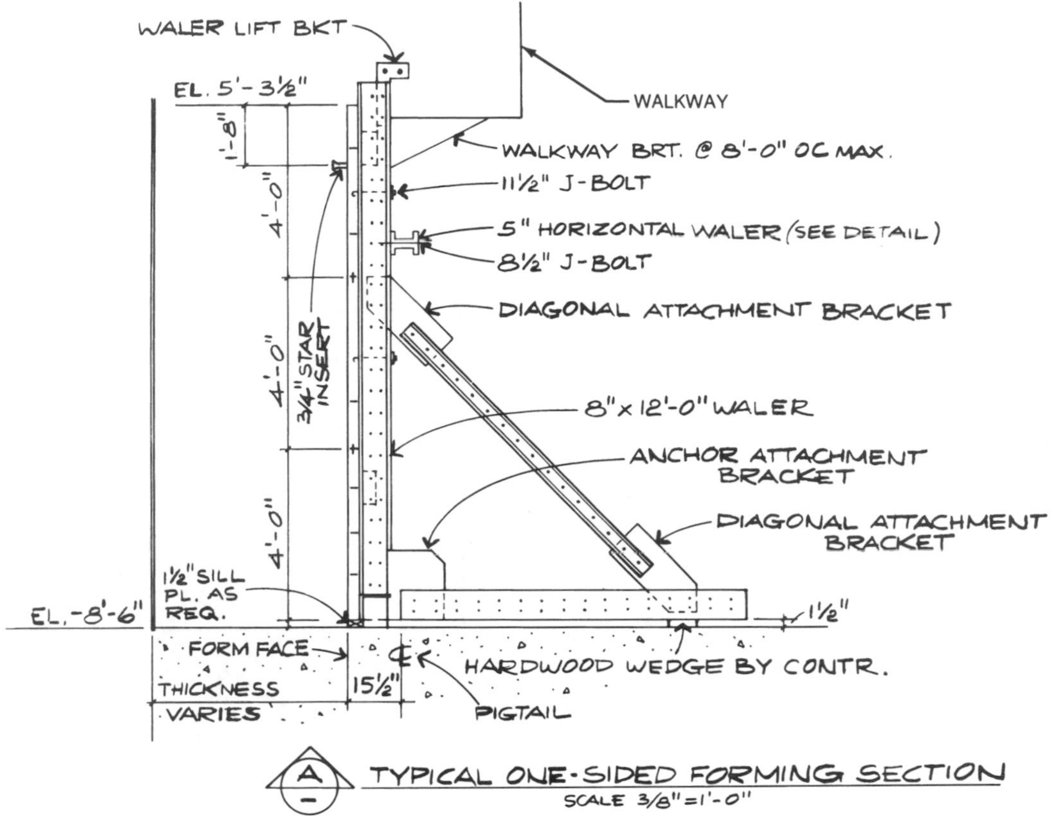

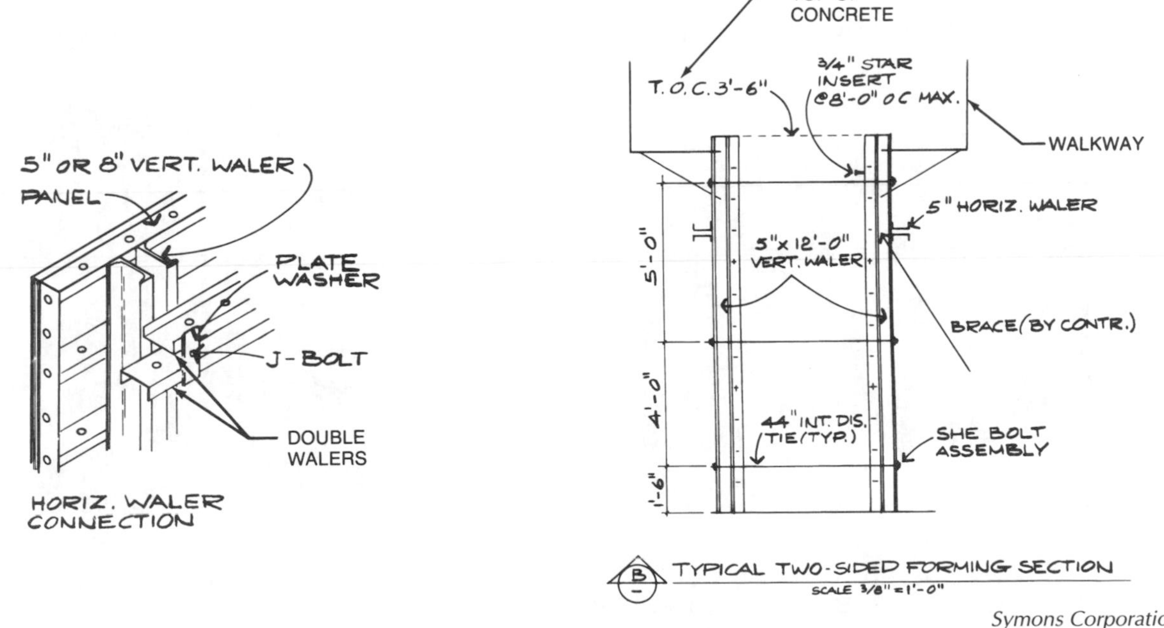

Symons Corporation

Figure 5-10. Formwork detail drawings provide information to the tradesworker for erection of forms and braces.

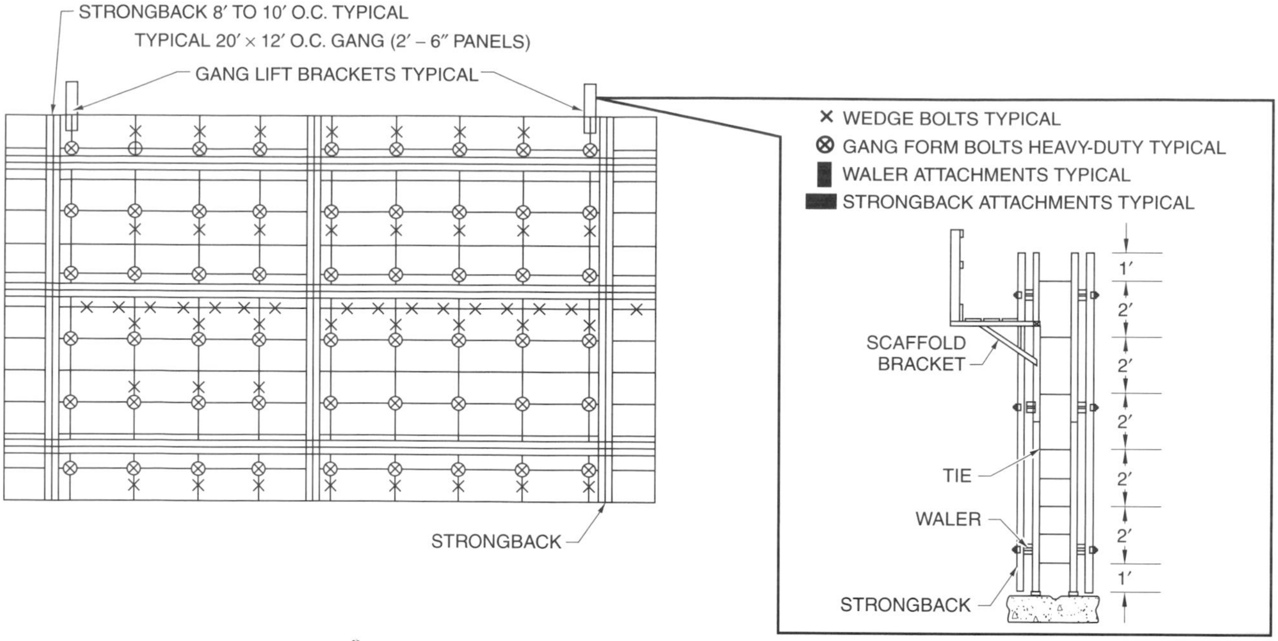

STRONGBACK 8′ TO 10′ O.C. TYPICAL
TYPICAL 20′ × 12′ O.C. GANG (2′ – 6″ PANELS)
GANG LIFT BRACKETS TYPICAL
STRONGBACK

× WEDGE BOLTS TYPICAL
⊗ GANG FORM BOLTS HEAVY-DUTY TYPICAL
WALER ATTACHMENTS TYPICAL
STRONGBACK ATTACHMENTS TYPICAL
SCAFFOLD BRACKET
TIE
WALER
STRONGBACK
1′ 2′ 2′ 2′ 2′ 2′ 1′

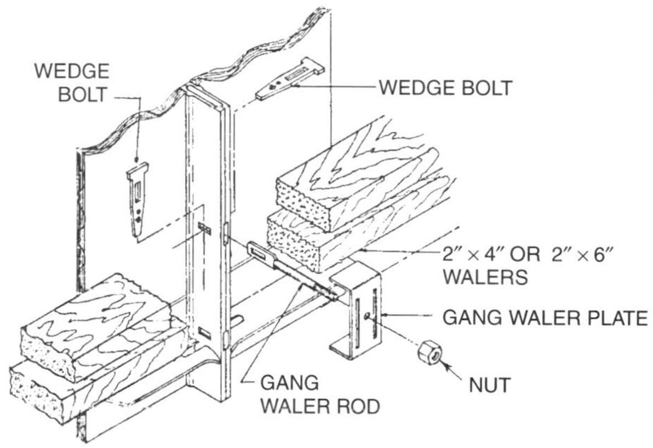

WEDGE BOLT
WEDGE BOLT
2″ × 4″ OR 2″ × 6″ WALERS
GANG WALER PLATE
GANG WALER ROD
NUT

WALER ATTACHMENT METHOD

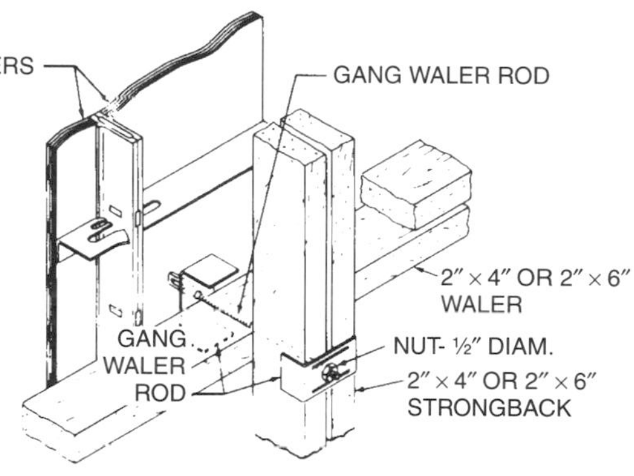

PANELS OR FILLERS
GANG WALER ROD
2″ × 4″ OR 2″ × 6″ WALER
NUT- ½″ DIAM.
2″ × 4″ OR 2″ × 6″ STRONGBACK
GANG WALER ROD

STRONGBACK ATTACHMENT METHOD

Symons Corporation

WALERS BOLTED TO PANEL SIDE RAILS AND STRONGBACKS BOLTED TO WALERS

Gates & Son, Inc.

GANGED PANEL FORM IN POSITION

Figure 5-11. A ganged panel form is assembled from smaller panels and lifted into position by crane.

Curved Wall Forms

Curved wall forms may be required for corners and wall sections of buildings that have an arc or other circular elements. Curved forms for completely circular walls may be required for projects such as storage tanks or silos. Curved wall forms are built in place or constructed with sections that are fabricated on the job. The forms for complex circular structures are often custom built in fabricating plants using all-wood, all-metal, or metal-framed plywood sections. Large prefabricated curved wall forms may be constructed of ganged panels. Curved walls may also be formed with all-steel panels consisting of a ⅛″ flexible skin and supported by 4″ wide vertical stiffeners. The panels are adjustable to form curved walls with a minimum radius of 5′-0″.

Plywood sheathing is commonly used to construct built-in-place curved wall forms. **See Figure 5-12.** The thickness of plywood used for curved wall form sheathing depends on its bending capacity in relation to the radius of the curve. Short-radius curve forms can be sheathed using two or more layers of ¼″ plywood panels. Thicker panels may be kerfed partially through to increase their bending capacity. The kerfs should be close, evenly spaced cuts. Although a kerfed thicker panel may be used, thinner panels are preferable because the kerfs weaken the thicker panel. Built-in-place curved wall forms can also be formed by using horizontal and vertical boards as form walls.

APA — The Engineered Wood Association

Figure 5-12. Plywood sheathing is used to construct curved wall forms.

Symons Corporation

Flexible forms can be ordered from the manufacturer preassembled to the required radius.

Laying out and establishing a line for a curved wall is similar to laying out a curved sidewalk. **See Figure 5-13.** Marks are made on top of the floor slab or footing and the wall lines are snapped. A common method for constructing curved wall forms uses 6″ wide top and bottom plates cut from ¾″ plywood. Studs spaced 8″ OC are nailed to the top and bottom plates. Walers are fabricated with 1″ pieces of laminated plywood and placed flat against the studs. Ties are placed midway between the studs and should not be overtightened. Overtightening may cause distortion. Curved pieces ripped out of 2″ thick stock can also be used as walers. In this case, the ties should be placed next to the studs to produce a smooth curve. **See Figure 5-14.** When constructing curved wall forms, the inside wall form is constructed first to facilitate rebar placement.

Construction Joints and Control Joints

Construction joints are constructed when the concrete of a wall section is placed on top of or adjacent to a previously placed section of wall. Construction joints extend through the entire thickness of a wall. Horizontal construction joints are constructed for walls placed in two or more lifts. Control joints are shallow grooves placed in the wall to control cracking that results from expansion and contraction in the set concrete.

Vertical Construction Joints. A vertical construction joint requires that a bulkhead be constructed inside the form at the end of the section of concrete being placed. Bulkheads are constructed between form walls with short boards nailed horizontally against vertical cleats. The horizontal boards are notched around rebar that extends past the bulkhead. **See Figure 5-15.**

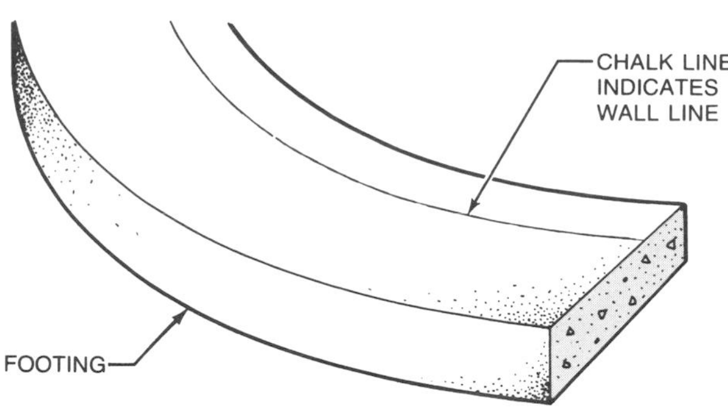

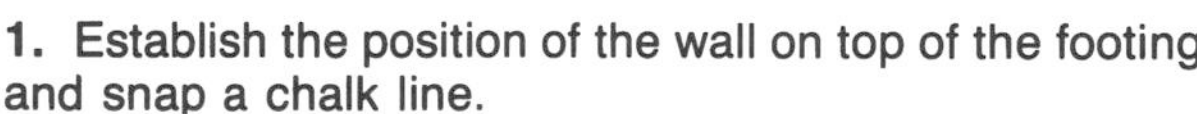

1. Establish the position of the wall on top of the footing and snap a chalk line.

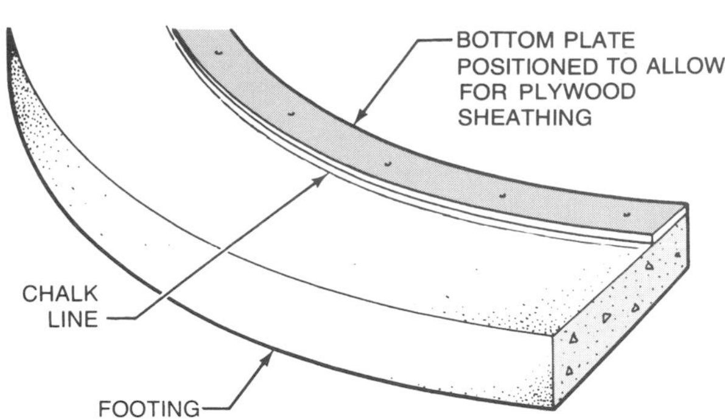

2. Position the bottom plate back from the line the thickness of the plywood sheathing. Nail the plate into position.

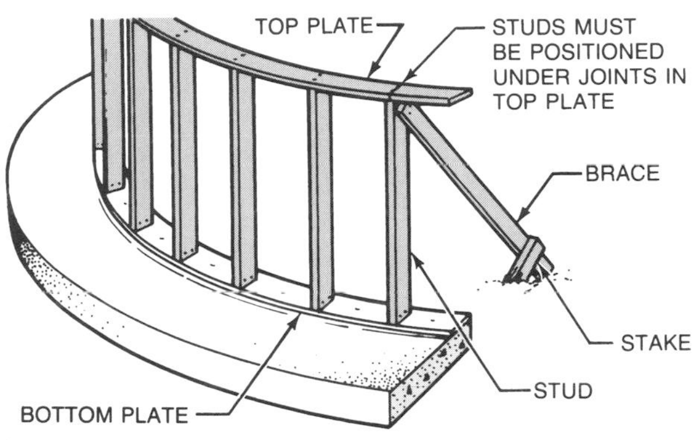

3. Toenail the studs to the bottom plate. Endnail the studs to the top plate. Plumb and brace the studs.

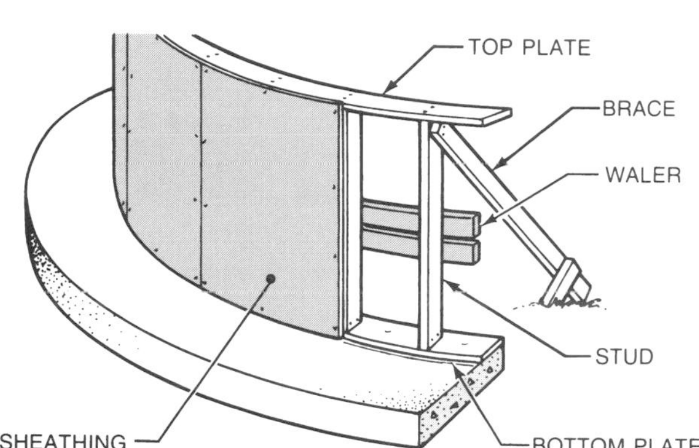

4. Nail walers and sheathing to the studs.

Figure 5-13. Curved wall forms are constructed using plywood sheathing reinforced with studs and walers.

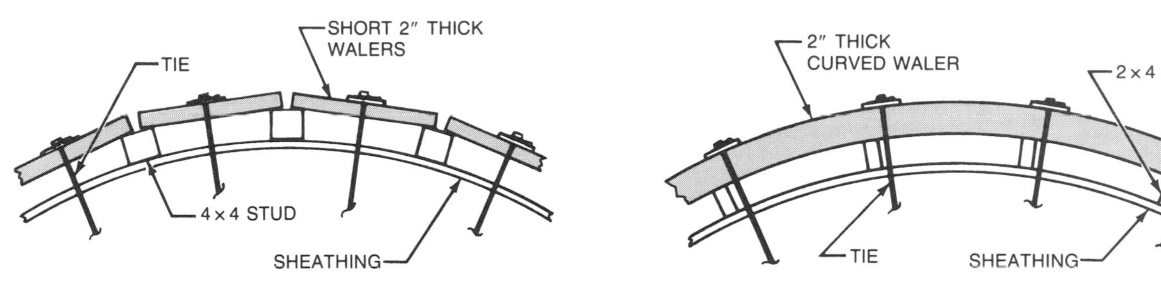

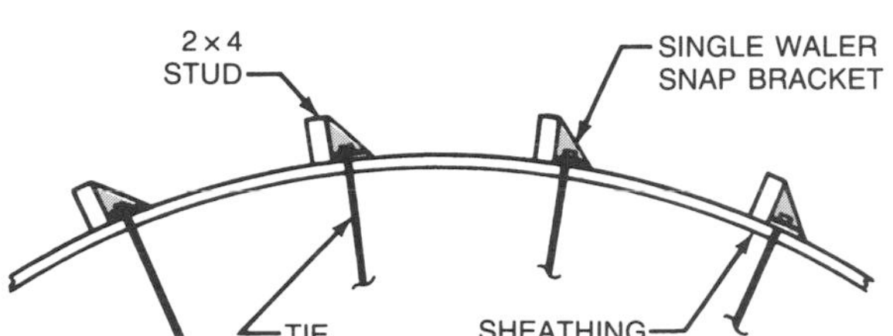

Figure 5-14. Various methods are used to secure and reinforce walers for curved wall forms.

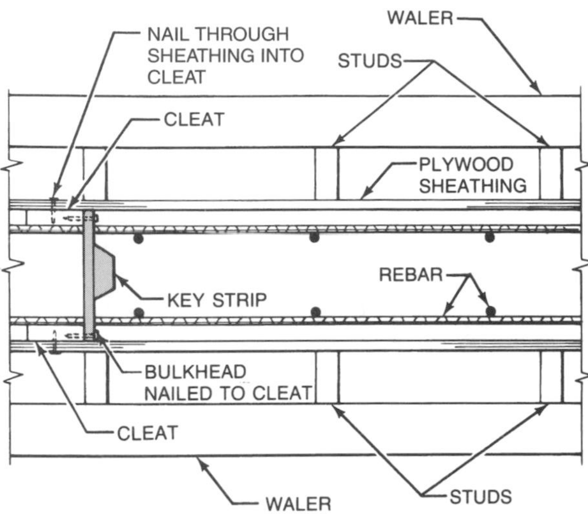

Figure 5-15. A vertical construction joint requires a bulkhead constructed of short boards.

Keyways may be required to prevent lateral movement between walls. Keyways are formed by attaching a tapered key strip to the bulkhead.

A water stop made of rubber, neoprene, polyvinyl chloride (PVC), or other plastic is installed to prevent water leakage at a vertical construction joint. Center-placed water stops (water stops placed at the center of the wall) are available in various designs, including single piece, split fin, labyrinth, and cellular. They are placed and attached to the bulkhead before the first placement of concrete. **See Figure 5-16.**

Horizontal Construction Joints. When constructing walls with climbing forms, the bottoms of the upper lift forms are secured toward the top of the lower lift of concrete. A row of tie rods or bolts is embedded 4″ below the top of the lower lift. The bottom row of ties for the upper lift should be placed 6″ above the joint. The bottom of the form panel should extend approximately 1″ below the joint of the lower lift. A greater overlap may result in concrete leakage because of unevenness in the surface of the lower wall. A compressible gasket may be placed beneath the lap to help prevent leakage of concrete. **See Figure 5-17.**

Control joints must be spaced properly. A general rule of thumb is to space joints (in feet) no more than 2 to 3 times the slab thickness (in inches). For example, a 4″ slab should have joints 8′ to 12′ apart (2 × 4 = 8; 3 × 4 = 12).

Patent Construction Systems

Construction joints and control joints help to prevent cracking in walls.

A 1 × 2 "pour" strip tacked toward the top of the lower wall form helps produce a straight horizontal construction joint. Wall forms should be filled until the concrete is slightly above the bottom of the strip. The strip is removed when the concrete sets. A horizontal construction joint can be made less apparent by nailing tapered pieces inside the wall form to create horizontal grooves at the joint.

Where form panels are raised from floor to floor, a waler rod attached to a J-bolt secures the bottom of the outside form wall. The J-bolt and coupling are embedded in the floor slab and a waler rod secures the form panel in place. The inside form wall rests on the floor slab. When the concrete has set, the waler rod is removed and the hole is grouted. **See Figure 5-18.**

Control Joints. Control joints control cracking in a wall. Control joints are usually required in walls with an architectural concrete finish in which the surface of the wall has a special texture or design. The location and spacing of control joints are shown in working drawings and are commonly part of the decorative texture or pattern. Control joints are formed by attaching a beveled strip of wood, metal, rubber, or plastic to the sheathing of the wall form. **See Figure 5-19.** The strips are removed after the concrete has set and the forms have been stripped. Control joints are usually caulked after the strips have been removed.

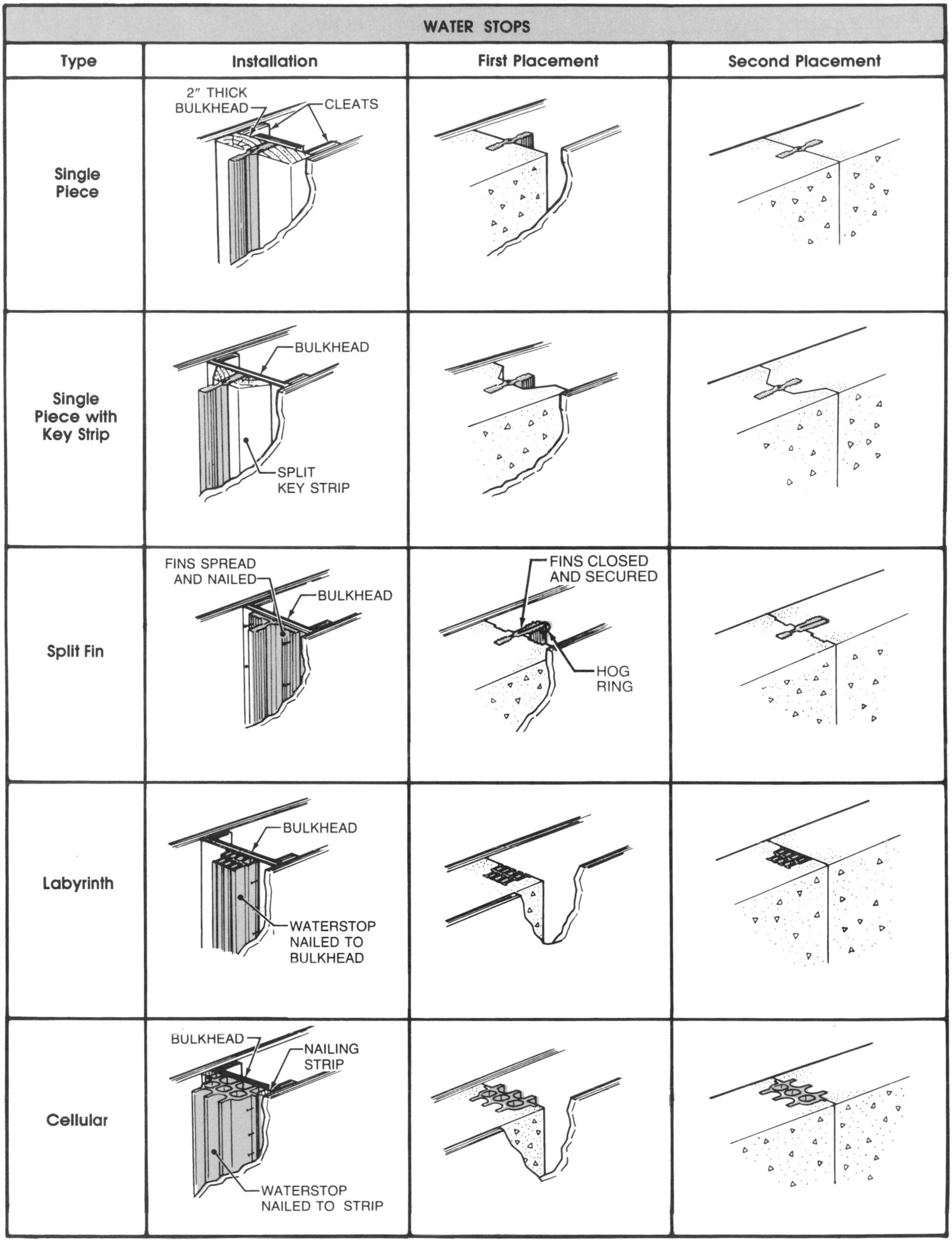

WATER STOPS			
Type	**Installation**	**First Placement**	**Second Placement**
Single Piece	2" THICK BULKHEAD CLEATS		
Single Piece with Key Strip	BULKHEAD SPLIT KEY STRIP		
Split Fin	FINS SPREAD AND NAILED BULKHEAD	FINS CLOSED AND SECURED HOG RING	
Labyrinth	BULKHEAD WATERSTOP NAILED TO BULKHEAD		
Cellular	BULKHEAD NAILING STRIP WATERSTOP NAILED TO STRIP		

Figure 5-16. Water stops prevent water leakage in vertical construction joints. The type of water stop used is determined by water pressure, wall thickness, and anticipated wall movement.

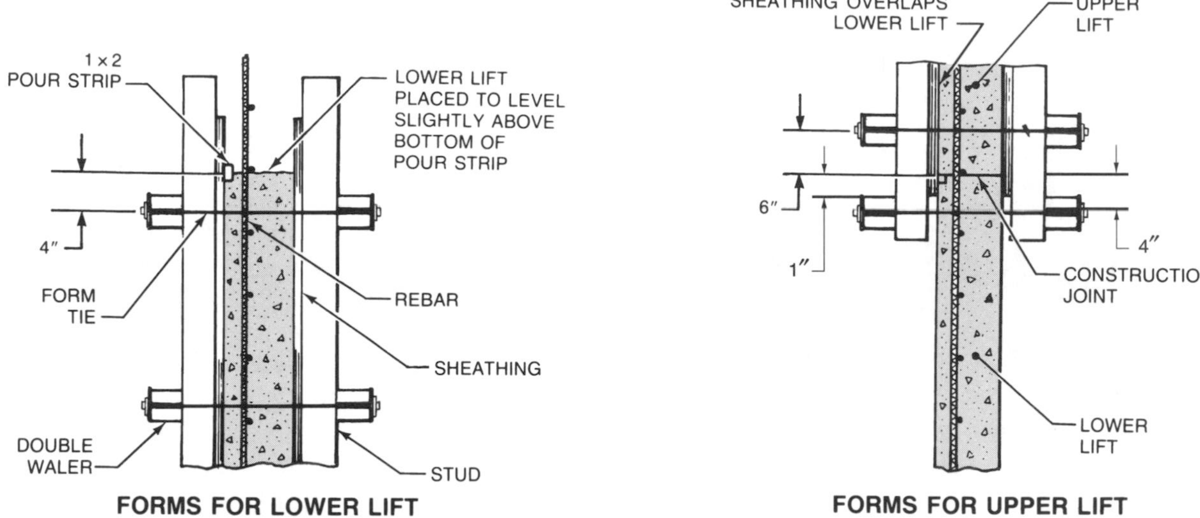

Figure 5-17. Forms for the upper lift of a concrete wall overlap 1″ past the lower lift. Tie rods are positioned 4″ from the top of the lower lift and 6″ from the bottom of the upper lift.

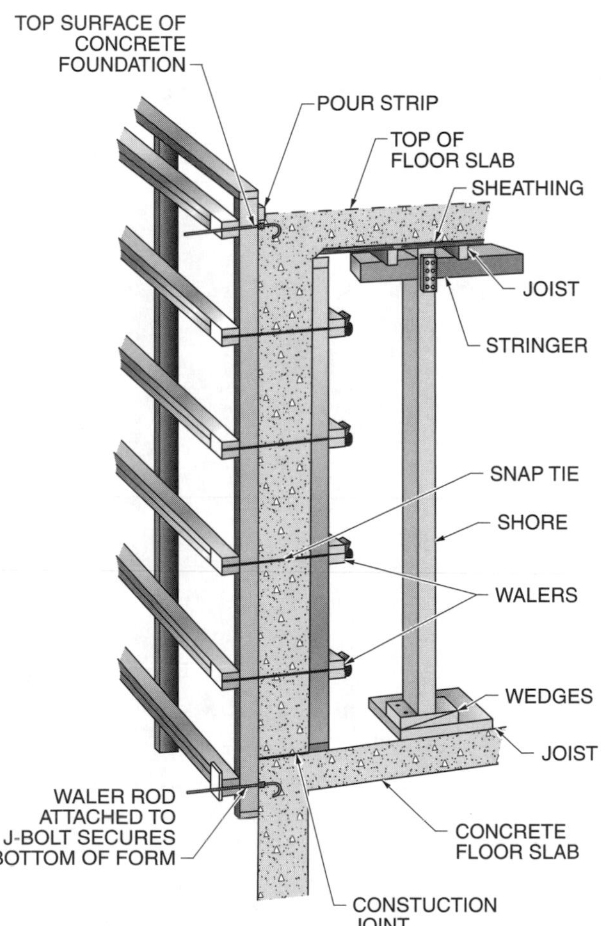

Figure 5-18. An outside wall form for a concrete wall at an upper level is attached using J-bolts and waler rods.

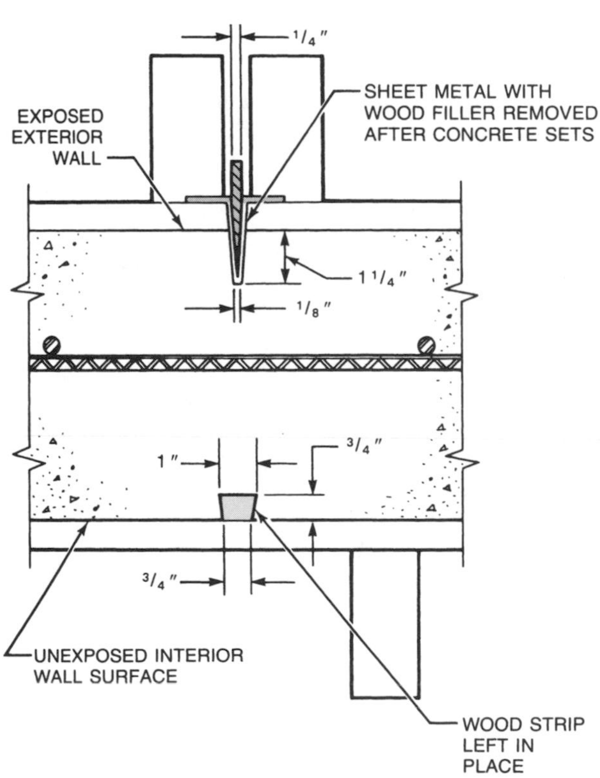

Figure 5-19. Control joints control cracking in a concrete wall. Control joints are formed with sheet metal, wood, rubber, or plastic strips.

APA—The Engineered Wood Association

Column forms must be properly braced to maintain plumb and level of the forms. Clamps or ties prevent collapse of the forms due to lateral pressure exerted by the concrete.

COLUMNS, GIRDERS, BEAMS, AND FLOOR SLABS

Columns, girders, beams, and floor slabs are combined to form integral structural units of a concrete building. Columns support girders and beams that hold up the floor slabs. Girders are heavy horizontal members that support beams and other bending loads. Beams are horizontal members that support a bending load over a span, such as from column to column. **See Figure 5-20.** Girders and beams are also lateral ties between the outside walls of the building. The columns, girders, beams, and floor slabs must be tied together and reinforced with rebar.

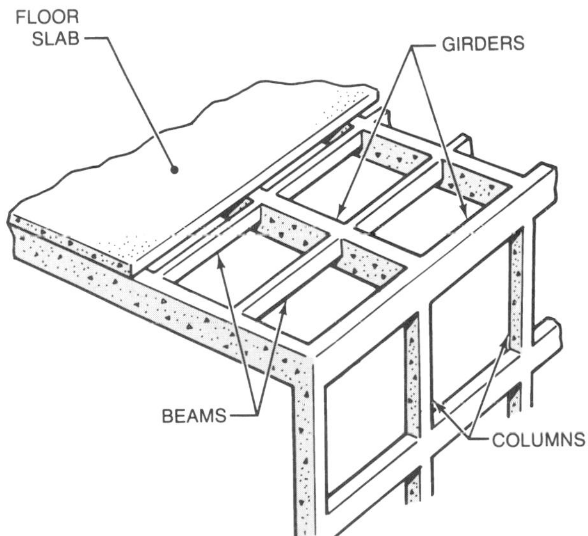

Figure 5-20. Columns, girders, beams, and floor slabs are combined to form integral structural units of a concrete building.

Columns, girders, and beams provide intermediate support for the floor slab when the perimeter of the floor is tied into and supported by the outside walls. Columns and girders may be formed in the perimeter of the building, with additional beams and girders providing interior floor support. The space between the girders and columns at the perimeter of the building is filled with panes of glass or curtain walls. A *curtain wall* is a light, non-load-bearing section of wall made of metal or precast lightweight concrete that is attached to the exterior framework of a building.

Flat slab and flat plate systems eliminate the use of beams and girders. In a flat slab system, the floor slab is directly supported by columns and drop panels (thickened area over a column). A capital (flared section at the top of a column) may also be used beneath the drop panel. In the flat plate system, the columns tie directly to the floor above without using drop panels or capitals.

One- and two-way joist and floor slab systems also eliminate beams and girders. The systems have thin slabs integrated with supporting girders, beams, and columns. One-way joist and floor slab systems (ribbed slabs) have cast-in-place joists running in one direction. Two-way joist and floor slab systems (waffle slabs) have joists running at right angles to each other. **See Figure 5-21.**

Column Forms

Column forms are subjected to greater lateral pressure than wall forms because of the small cross-sectional area in relation to the height. Column forms require tight joints, adequate bracing, and strong anchorage at the base of the form. A cleanout opening at the base of the column form is used to remove debris before the concrete is placed. A compressed air hose is lowered into the form and the debris is blown out. In high column forms, a pocket or window is placed midway in the height of the form to place and consolidate the concrete in the bottom section of the form. The pocket or window is nailed shut when the concrete reaches the bottom of the pocket or window.

Most columns are square, rectangular, or round. L-shaped or oval columns are less frequently used. **See Figure 5-22.** Square and rectangular columns are usually constructed with plywood, prefabricated metal-framed plywood forms, or all-steel custom-made forms. Most round columns are constructed with tubular fiber forms or all-steel custom-made forms.

FLAT-SLAB FLOOR WITH DROP PANEL AND CAPITAL OVER COLUMN

ONE-WAY JOIST SYSTEM

TWO-WAY JOIST SYSTEM

Portland Cement Association

Figure 5-21. Flat-slab and flat-plate construction methods eliminate the use of beams and girders. One- and two-way joist systems integrate floor slabs with supporting girders, beams, and columns.

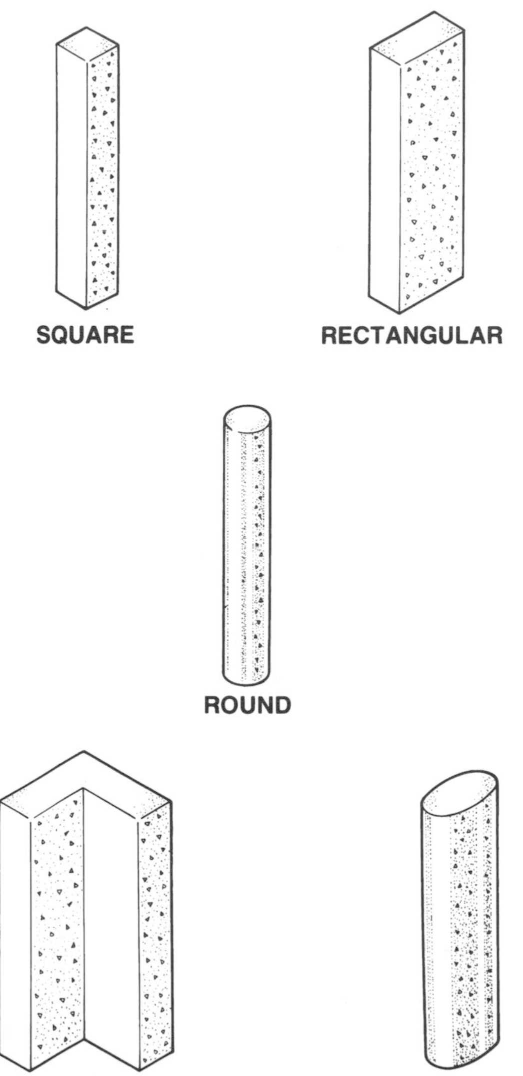

Figure 5-22. Typical column designs used in heavy construction include square, rectangular, round, L-shaped, and oval columns.

Gates & Sons

Metal devices are commonly used to tie together column forms.

Square and Rectangular Column Forms. When plywood is used to construct square or rectangular column forms, two sides are cut to the width of the column, and the other two sides are cut to the width plus twice the thickness of the plywood and twice the width of the cleats. Light column forms up to 12″ square may be stiffened using battens and ties, or adjustable metal clamps placed directly against the plywood. Heavy column forms are stiffened with vertical 2 × 4s nailed to the plywood. Hinged adjustable metal scissor clamps are placed around the stiffeners and tightened with wedges. **See Figure 5-23.**

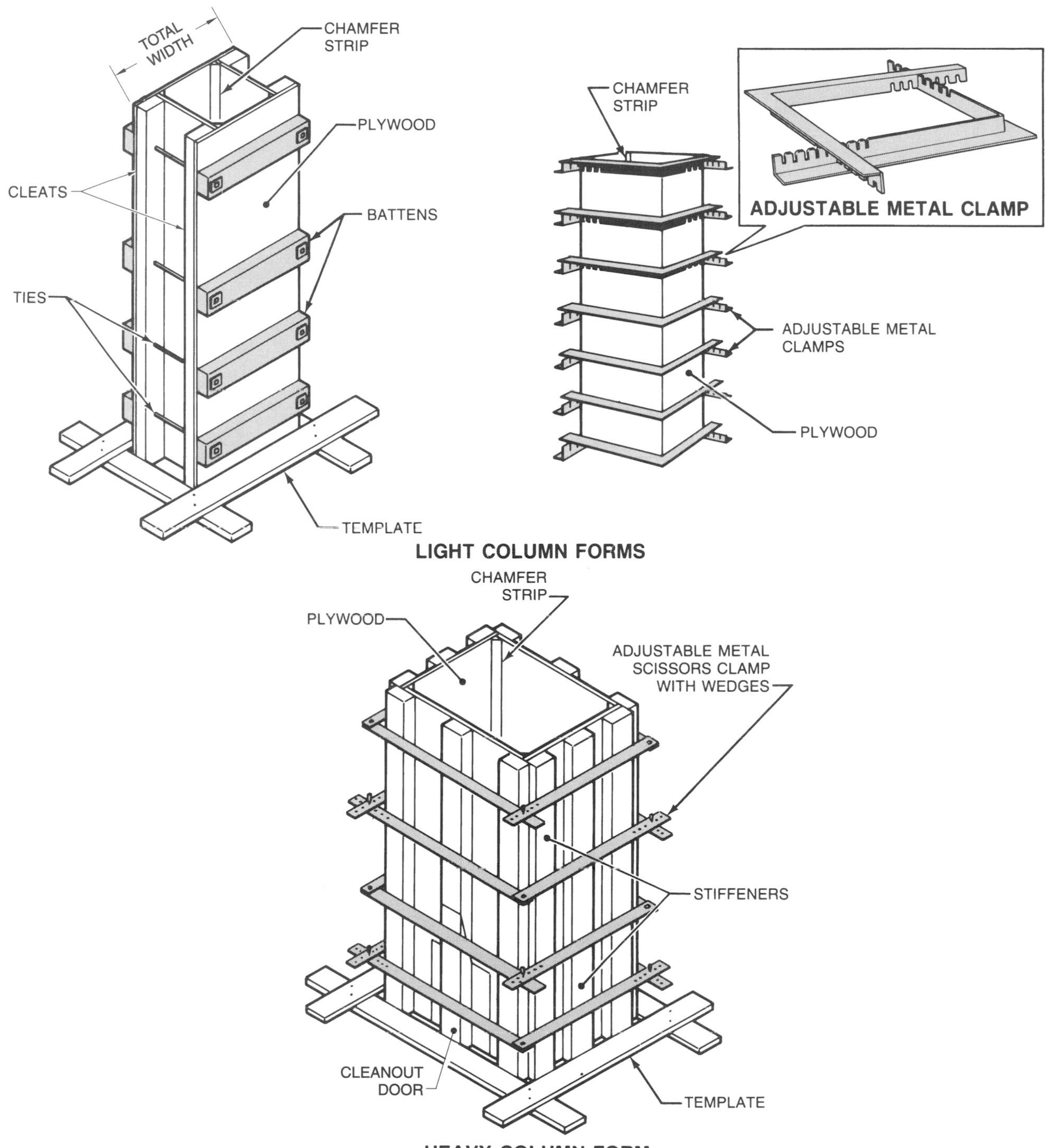

Figure 5-23. Light column forms are stiffened using battens and ties or adjustable metal clamps. Heavy column forms are stiffened with vertical 2 × 4s and adjustable metal scissor clamps.

A cleanout opening is cut out of one side of the column form before the form is assembled. The plywood and stiffener attached to the cleanout door should be cut at a 45° angle. The cleanout door is replaced and clamped in position after the debris inside the column form has been removed. Chamfer strips should be placed at all four corners of the form. A *chamfer strip* is a narrow strip of wood ripped at a 45° angle. The chamfer strips produce beveled corners for the finished concrete columns, making them less susceptible to chipping and other damage. When erecting a square or rectangular column form, the bottom is secured in position by a template aligned with centerlines marked on the slab or footing. **See Figure 5-24.**

If the rebar for a column is in place, three sides of the form are nailed together and set in position. The fourth side is then nailed into place and the column form is braced. If a rebar cage is used, the four sides of the column form are nailed together and the form is positioned. The rebar cage is then lowered into position with a crane and properly braced. When columns are required in the floor above, the rebar should project above the form to later tie into the rebar placed in the column above.

Since the rate of placing concrete in a column form is very high and the bursting pressure exerted on the form by the concrete increases directly with an increase in the placement rate, a column form must be properly constructed. Because the bursting pressure is greater at the bottom of the form than it is at the top, column clamps are placed closer together at the bottom.

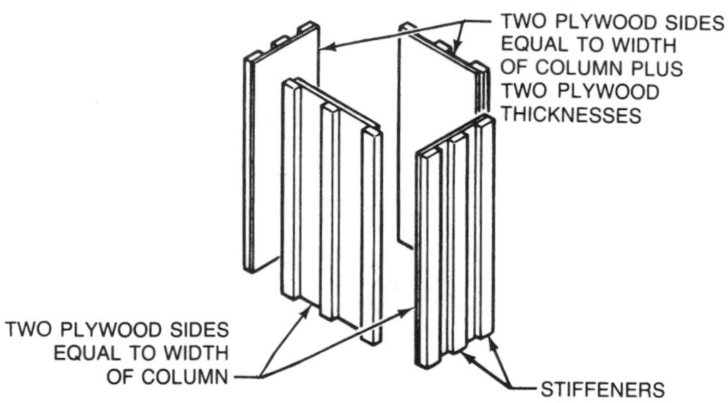

1. Cut four pieces of plywood to dimension. Nail 2 × 4 stiffeners to the plywood.

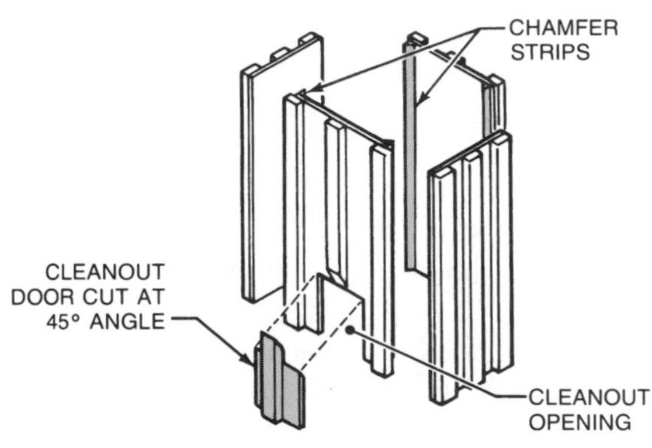

2. Nail chamfer strips to two sides. Cut a cleanout door in one of the sides.

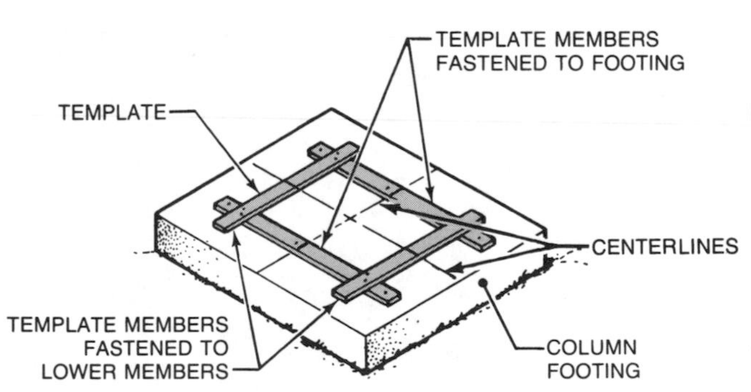

3. Establish center lines on the column footing. Fasten a template down to hold the column form bottom.

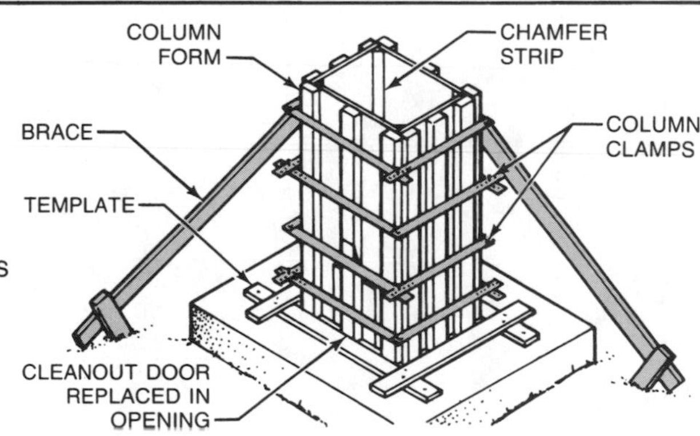

4. Nail column form together and position in template. Clean debris from form and replace door. Tie column forms together with column clamps. Plumb and brace form.

Figure 5-24. Square and rectangular column forms are constructed with plywood and 2 × 4 stiffeners. A cleanout door is provided for access to debris on the inside of the form.

Steel-Framed Plywood Column Forms. Steel-framed plywood column forms (Steel-Ply® or hand-set column forms) are commonly used to form standard size columns ranging from 8″ × 8″ to 24″ × 24″ in 2″ increments. Columns with odd dimensions are formed by adding filler pieces to the column form. Steel-framed plywood column forms do not require clamps or additional stiffeners and are often designed with hinged corners that facilitate erecting and stripping. **See Figure 5-25.** Steel-framed plywood column forms are also used for large columns. They are set in place by crane and may require additional ties and stiffeners.

Round Column Forms. Tubular fiber forms are used to form standard size round columns ranging from 6″ to 48″ in diameter and up to 18′ in length. Longer lengths are also available. Tubular fiber forms are made of spirally constructed fiber plies and are available with wax-impregnated inner and outer surfaces for weather and moisture protection.

Tubular fiber forms are positioned after the rebar cage is in place. **See Figure 5-26.** Small column forms can be lowered by hand or with a block and tackle. Large column forms may require a crane. The interior surface of the form should not be damaged when lowering the form. The base of the form is secured with a wood template, and the sides are plumbed and secured with braces nailed to a wood collar.

MFG Construction Products

Figure 5-26. The tubular fiber forms are positioned after the rebar cage has been set in place are used to form round columns.

Tubular fiber forms can also be used to construct oval-shaped columns. A rectangular column form is constructed and a tubular fiber form that is cut in half lengthwise is inserted at both ends. The edges of the tubular fiber form and rectangular form should be flush to ensure a smooth transition. **See Figure 5-27.**

Figure 5-25. Standard size columns are often constructed with prefabricated steel-framed panels with plywood facing.

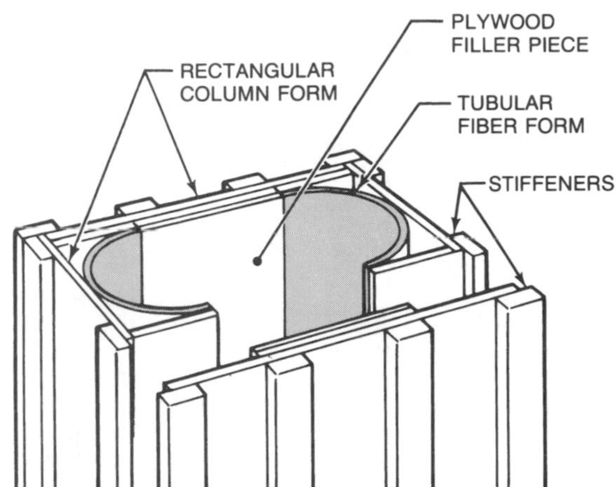

Figure 5-27. Oval column forms are constructed with tubular fiber forms combined with a rectangular column form.

A spiral concrete column is a round column that has a continuous spiral winding of lateral reinforcement encircling its core and vertical reinforcing bars to restrain expansion. A design engineer specifies the pitch of the spiral, the tie size, and the number of vertical bars for a spiral column.

Tubular fiber forms are notched with a saw for beam openings or utilities, such as light switches and electrical outlets. Wooden blocks are nailed to the inside of the form to form slots for beams or joists if required.

A circular saw or knife may be used to strip the forms after the concrete has set. Care must be taken not to mar the concrete column. A circular saw is adjusted to the thickness of the form and two vertical cuts are made the complete length of the form on opposite sides. The two sections of the form are then removed. When using a sharp knife, a $1/2''$ slit is made and a broad-bladed tool is then used to pry the form.

Round Steel and Fiberglass Column Forms. Round steel and fiberglass forms are used to construct large columns for heavy construction projects. Round steel column forms are available in diameters ranging from 14″ to 10′. The sections that make up the form range in length from 1′ to 10′. Bracing is an integral component of the round steel column forms so additional bracing is not required.

A fiberglass form is pulled apart and placed in position around previously installed rebar. The edges are then secured with bolts at closure flanges reinforced with a predrilled steel bar. The form is plumbed and secured with braces tied to a steel bracing collar. Fiberglass forms provide a smooth architectural finish and can be easily stripped. Fiberglass forms can be combined with a two-piece capital form to construct a column with a capital. **See Figure 5-28.**

Beam and Girder Forms

Beam and girder forms are constructed after the column forms are positioned and braced. In general, the concrete for beams, girders, and columns is placed monolithically at each floor level. Therefore, beam, girder, and column forms must be framed and tied to each other.

Two methods are commonly used to frame beam and girder forms to column forms. In one method, the beam or girder forms rest on or butt against the top of the column form. In the second method, the beam or girder forms frame into a pocket in the side of the column form. **See Figure 5-29.**

A bottom and two sides are the main components of beam and girder forms. Chamfer strips are used to produce beveled edges where the sides and bottom join. Beam and girder forms are supported by T-head shores, double-post shores, or other shoring placed beneath the bottom of the form.

A beam or girder form framing into a tubular fiber form is supported by a yoke or collar placed at the top of the column form. A half circle with a diameter equal to the column diameter must be cut at the ends of the bottom piece to facilitate construction. **See Figure 5-30.**

MFG Construction Products

Columns are often used on building facades to provide a decorative finish for an entrance.

POSITIONING FIBERGLASS FORM

PLACING CONCRETE IN FIBERGLASS FORM

STRIPPING FIBERGLASS FORM

FIBERGLASS FORM WITH CAPITAL FORM

MFG Construction Products

Figure 5-28. A fiberglass form produces a smooth finish with one vertical seam.

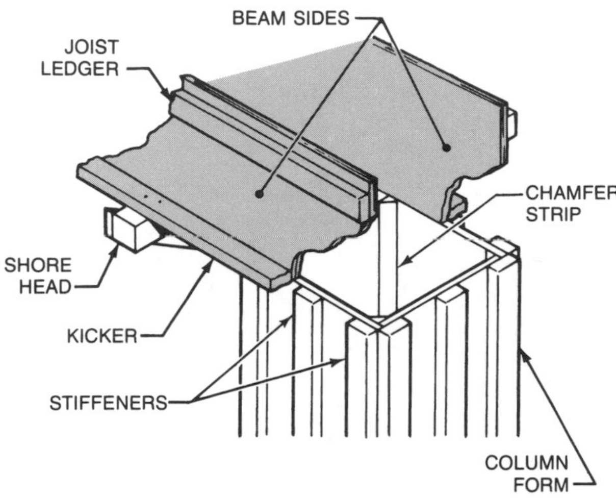

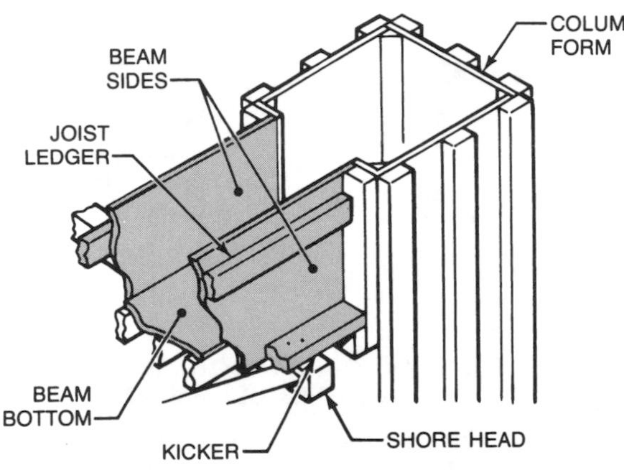

Figure 5-29. Beam and girder forms rest on or butt against a column form, or frame into a pocket cut into the side of the column form.

Beam and Girder Bottoms. Beam and girder bottoms are constructed of 2″ planks or plywood stiffened with 2 × 4s. The width of the bottom piece should be the same as the width of the finished soffit. The bottom piece should be long enough to butt against the column form or rest on top of the column form. If the beam or girder forms are designed to be stripped before the column form is stripped, the bottom piece should be butted against the column form. If the ends rest on the top of the column form, a 45° bevel is cut to provide a chamfer where the finished beam or girder meets the column. Joist ledgers are nailed against the columns and beneath the ends of the beam bottoms for additional support. **See Figure 5-31.**

Beam and Girder Sides. Plywood is commonly used to form beam and girder sides. The bottoms of the beam sides may be nailed against or set on top of the bottom piece. In either case, a kicker is nailed against the bottom of the sides to secure it in position. A joist ledger is nailed against the column form and below the ends of the beam bottoms for additional support. Blocking between the joist ledger and kicker may also be added to stiffen the sides and support the ledger.

Studs, bracing, walers, and ties are used to construct and reinforce the sides for large beams and girders. The height of the beam or girder sides is determined by the form framing method used. When the sides are nailed against the beam bottom and the slab sheathing rests on top of the beam side, the total height of the sides is equal to the height of the beam plus the thickness of the beam bottom and the width of the stiffeners, minus the concrete slab and slab sheathing thickness. The length of the beam or girder sides depends on whether the beam sides butt against or pass beyond the sheathing of the column. **See Figure 5-32.**

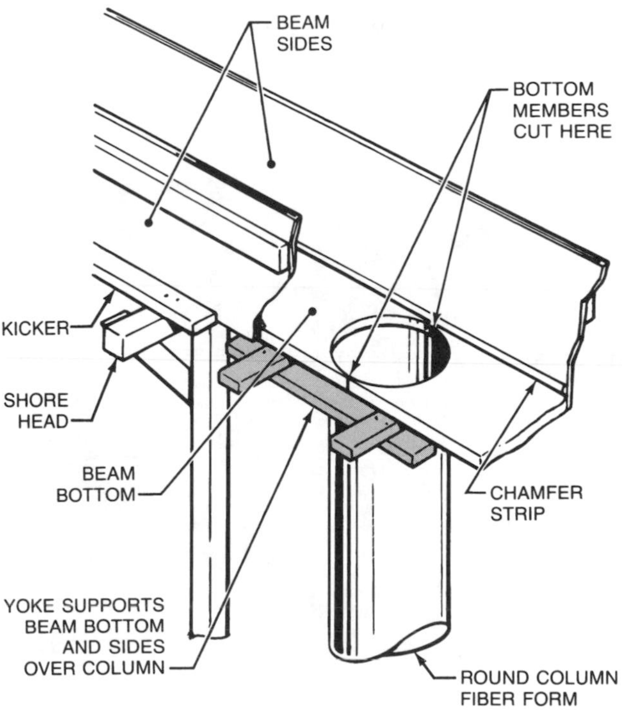

Figure 5-30. A yoke supports the beam sides and bottom over a round column fiber form.

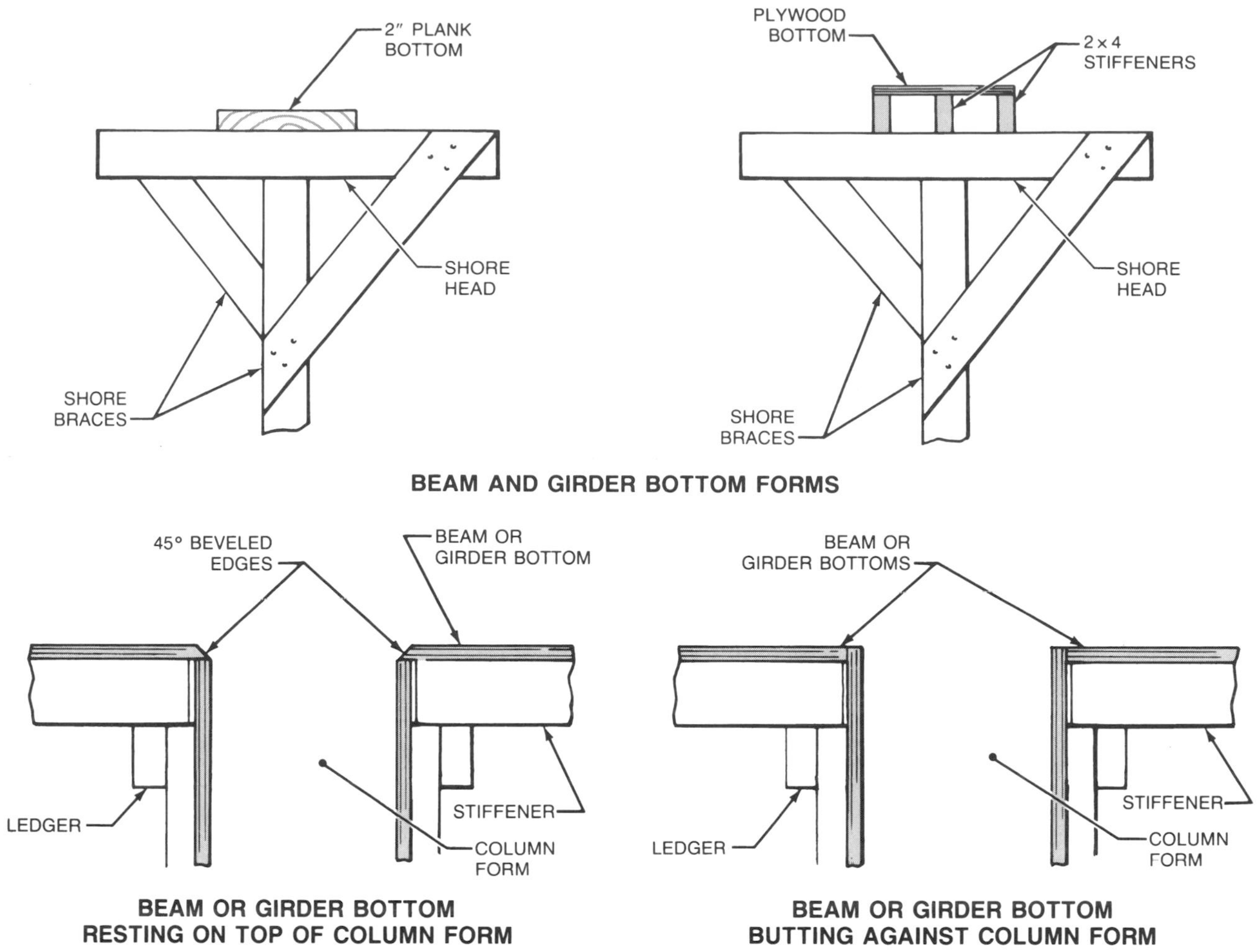

2" PLANK BOTTOM
SHORE HEAD
SHORE BRACES

PLYWOOD BOTTOM
2 x 4 STIFFENERS
SHORE HEAD
SHORE BRACES

BEAM AND GIRDER BOTTOM FORMS

45° BEVELED EDGES
BEAM OR GIRDER BOTTOM
LEDGER
STIFFENER
COLUMN FORM

BEAM OR GIRDER BOTTOM RESTING ON TOP OF COLUMN FORM

BEAM OR GIRDER BOTTOMS
LEDGER
STIFFENER
COLUMN FORM

BEAM OR GIRDER BOTTOM BUTTING AGAINST COLUMN FORM

Figure 5-31. Beam or girder bottoms are formed with 2" thick planks or plywood reinforced with 2 × 4s. The bottom rests on a column form or butts against it.

Symons Corporation

Beam and girder forms must be properly braced and supported until concrete is placed and has reached sufficient strength.

Constructing Beam and Girder Forms. Shores are braced horizontally between two columns or between a wall and column to support beam and girder forms. The beam and girder forms may be prefabricated on the ground and lifted in place or constructed on top of shores. When constructing the forms on shores, the bottom is positioned on the shores, and the sides are then attached. Studs are nailed to the form sides and a joist ledger is nailed to the studs. The sides are braced between the studs and the shore heads. **See Figure 5-33.** After the slab forms have been constructed, the beam or girder forms are adjusted to their correct height by raising or lowering the shores with wedges or other adjusting devices.

Framing Beam Forms to Girder Forms. Beam forms are framed to girder forms by cutting a beam pocket in the side of the girder form. If the sides and bottom of the beam form butt against the girder form, the pocket is cut to the size of the finished beam. If the sides and bottom of the beam form extend past the sheathing of the girder form, the pocket is cut to accommodate the thickness of the beam bottom and sides, plus a small allowance for easy fitting. The opening should be reinforced with cleats and a beam ledger. **See Figure 5-34.**

Spandrel Beams. Spandrel beams are located in the outer walls of a building and tie into the floor slab above. Spandrel beam forms consist of a bottom piece and two sides supported on extended shore heads that are supported with double posts. If a walkway is required, the shore head should extend beyond the outside wall. Kickers are nailed against the bottoms of the beam sides. A joist ledger is nailed toward the top of the inside wall to support the joists of the slab form. **See Figure 5-35.**

Symons Corporation

Spandrel beams are used in large commercial buildings to support the floors above or the roof.

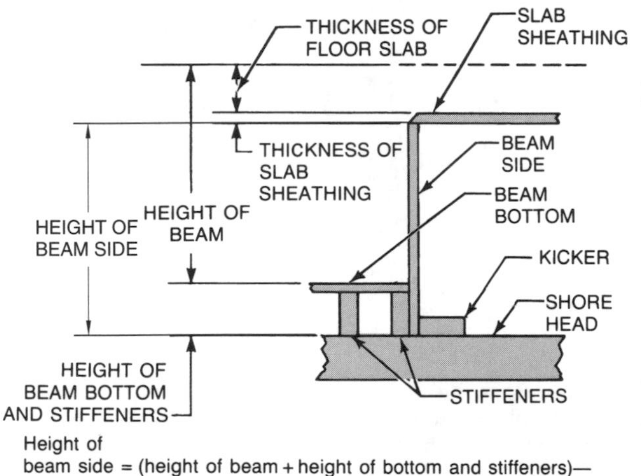

CALCULATING THE HEIGHT OF BEAM OR GIRDER SIDES

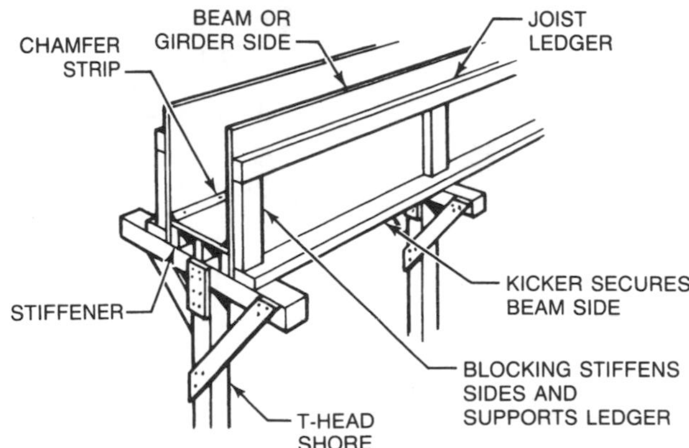

FORMING SMALL BEAM OR GIRDER SIDES

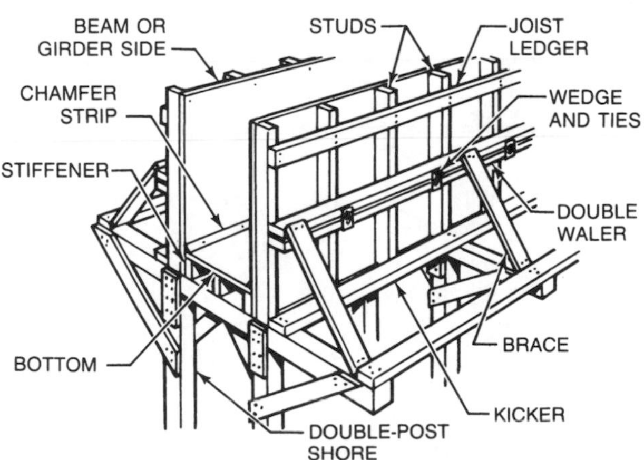

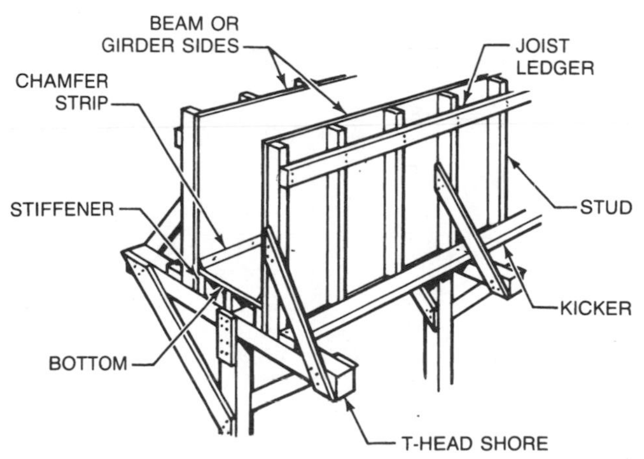

FORMING LARGE BEAM OR GIRDER SIDES

Figure 5-32. Beam and girder sides are reinforced with blocking, kickers, studs, walers, ties, and braces.

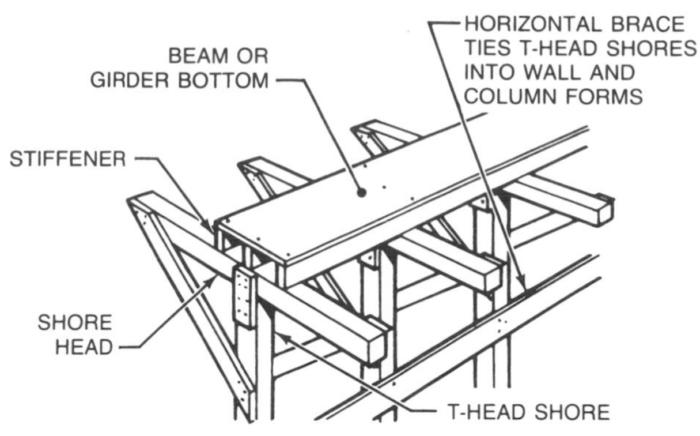

1. Position T-head shores and nail a horizontal brace across them. Nail the beam or girder bottom to the top of the shore head.

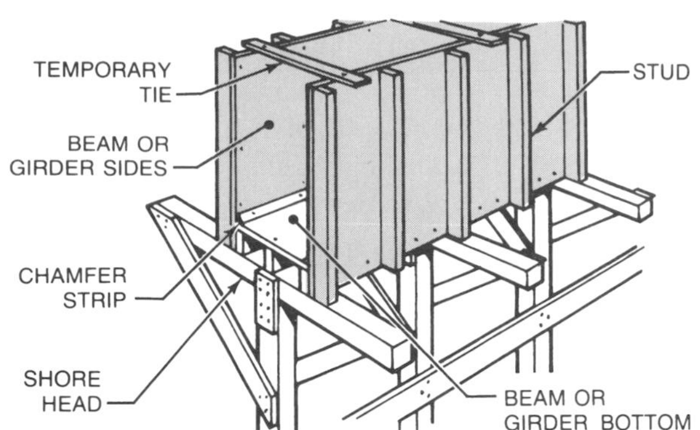

2. Nail the form sides to the bottom. Nail ties across the top of the form sides. Nail studs to the form sides. Nail chamfer strips to the beam or girder bottom.

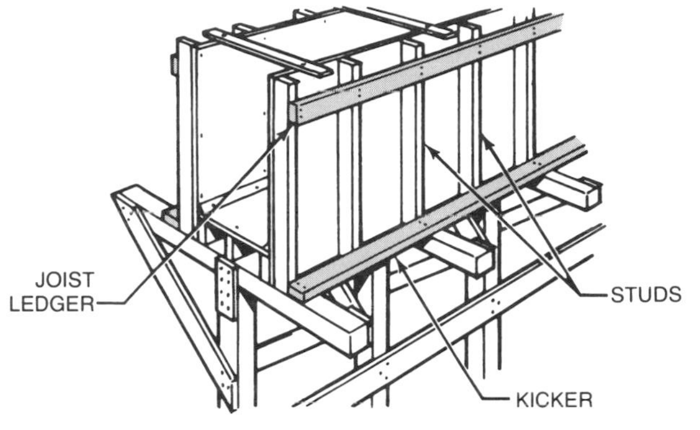

3. Nail a kicker and joist ledger against the studs.

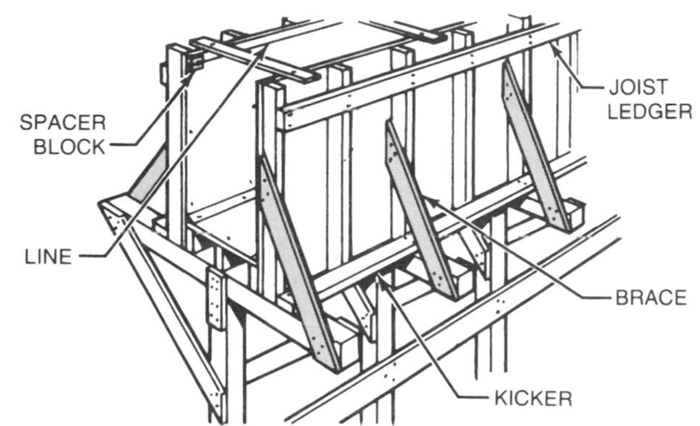

4. Set up a line held away from the form sides with a spacer block. Align and brace the sides.

Figure 5-33. Beam and girder forms are constructed on well-braced T-head shores. The sides of the beam or girder forms are braced to resist lateral pressure.

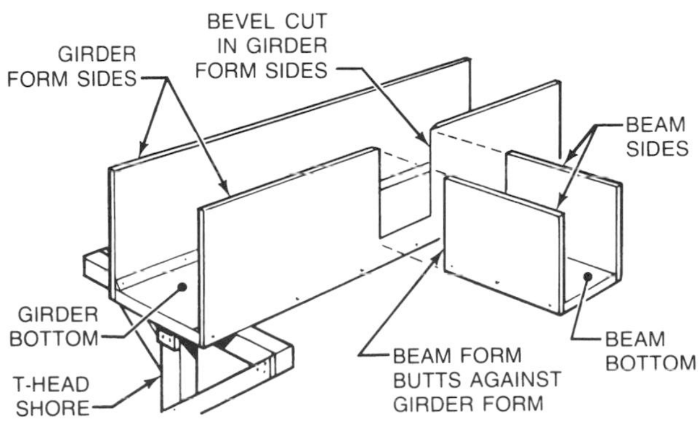

1. Lay out a beam pocket on the girder form side. Cut the pocket at a bevel.

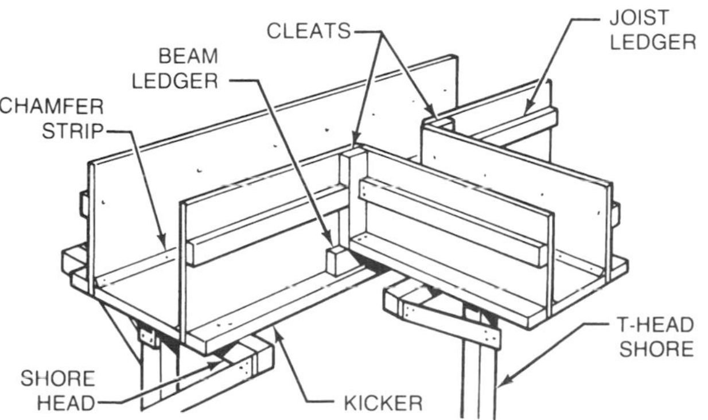

2. Nail a kicker across the shore heads. Frame the beam pocket with cleats and a beam ledger. Shore the beam form and nail into position.

Figure 5-34. Beam forms are framed to girder forms with cleats and a beam ledger.

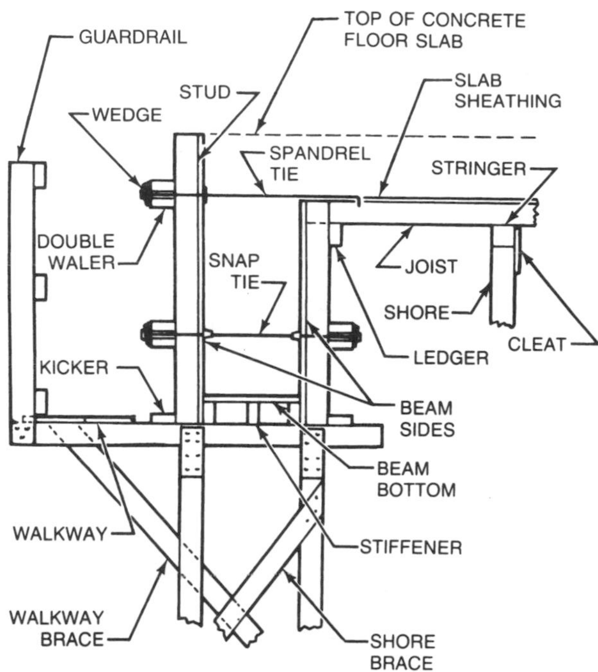

Figure 5-35. Spandrel beams are located in the outside walls of a concrete structure and tie into the floor slab above.

Spandrel beams often require intermediate ties and walers because of their depth. A spandrel tie is driven into the deck sheathing at the top of the form. A *spandrel tie* is a type of snap tie with a hooked end.

Suspended Forms. Suspended forms are used to form concrete slabs that are supported by steel beams or girders. Suspended forms are secured with U-shaped snap ties or coil hangers that slip over a steel beam or girder, eliminating the need for shores to support the forms. If the steel beam or girder is to be encased in concrete for fireproofing, the hangers support formwork consisting of a bottom member and two sides. The ends of the hangers extend through short walers placed beneath the formwork and are secured with wedges or bolts. The hangers for steel beams or girders not encased in concrete secure the joists supporting the slab form. **See Figure 5-36.**

Floor Form Construction

Floor forms are constructed after the column, beam, and girder formwork has been completed. Shores are used to support stringers. The ends of stringers must butt over the centers of shores and each end must be toenailed. Stringers provide support and a nailing surface for the joists. Stagger joints between joists and butt ends over the center of stringers.

The spacing of shores, stringers, and joists is based on the floor span and load to be carried by the form. After the shores, stringers, and joists are placed, the floor form sheathing, usually plywood panels, is fastened to the joists. The panels should be placed lengthwise across the joists and nailed at the corners. **See Figure 5-37.** The joists and sheathing can also be prefabricated in large panels and hoisted and set in place over the stringers.

Beam, Girder, and Slab Floor Systems. Concrete for beam, girder, and slab floor systems is commonly placed monolithically. Ledgers are nailed toward the top of the beam or girder sides to support one end of the floor form joists. If the slab sheathing rests on top of the sides, the ledgers are positioned down from the top of the beam or girder sides the width of the ledger. If the slab sheathing butts against the sides, the ledgers are positioned down the width of the ledger plus the thickness of the slab sheathing. When the span between beams or girders requires intermediate stringers, the top surface of the stringers must be level with the top of the ledgers. To simplify the stripping operation, only a few nails are used to secure the joists.

Block bridging is placed between the joists to prevent the joists from tipping. Block bridging may be eliminated by using 4 × 4s for joists. Plywood sheathing is positioned after all the joists are in place. The plywood sheathing may butt against or rest on top of the beam or girder sides. When the plywood sheathing rests on top of the sides, the edge of the plywood sheathing should not be allowed to form a groove in the concrete where the beam and floor intersect. The end of the plywood sheathing should be held back ¼″ to ½″ from the inside face of the beam or girder sides and beveled. **See Figure 5-38.**

Portland Cement Association

Floor slab systems are commonly used for large expanses, such as an office building or parking garage.

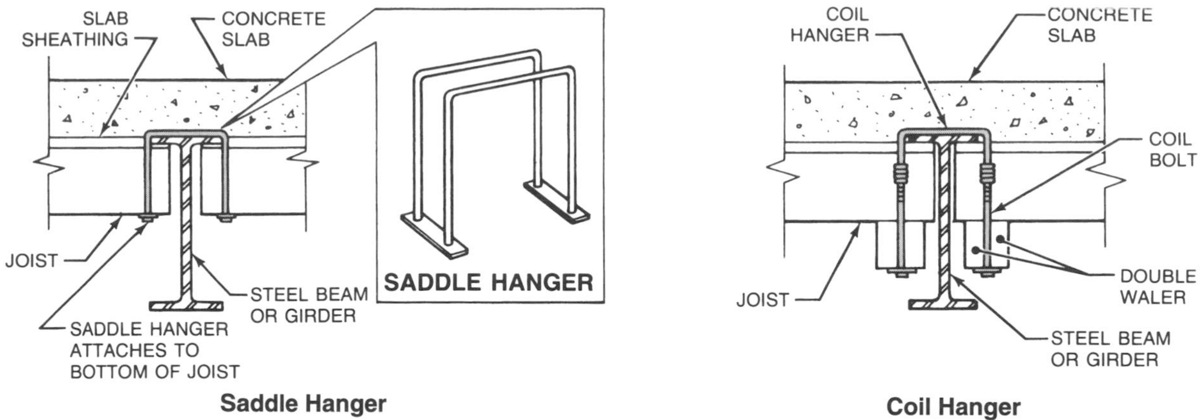

Saddle Hanger **Coil Hanger**

SLAB SUPPORT FORMS

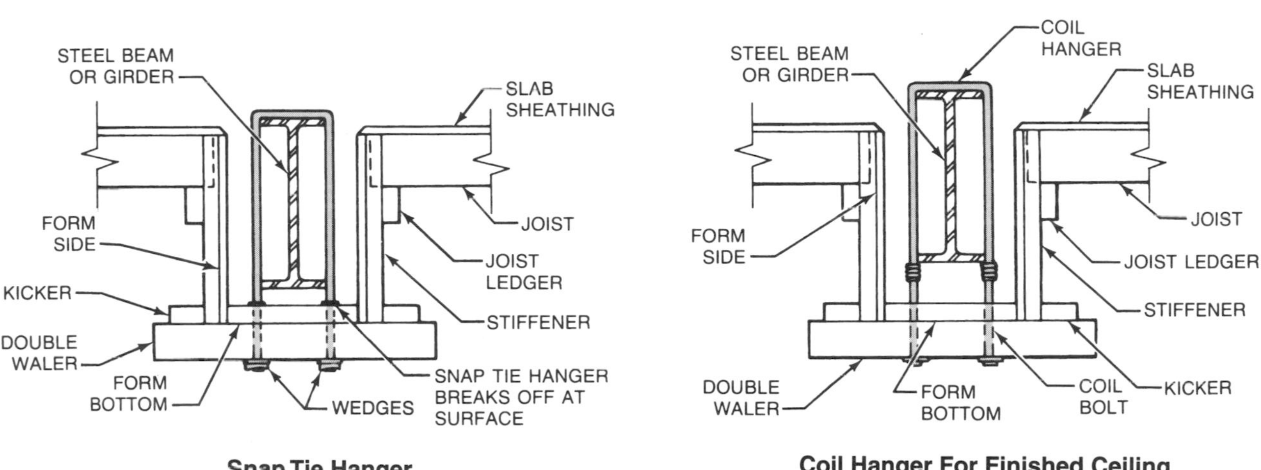

Snap Tie Hanger **Coil Hanger For Finished Ceiling**

STEEL BEAM OR GIRDER
COIL HANGER
SLAB SHEATHING
JOIST
JOIST LEDGER
FORM SIDE
STIFFENER
KICKER
FORM BOTTOM
DOUBLE WALER
BOLT

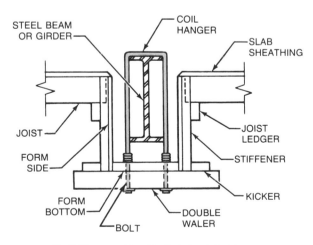

Coil Hanger For Unfinished Ceiling

BEAM ENCASEMENT FORMS

Figure 5-36. Suspended forms are used to form a concrete slab supported by steel beams or girders. The steel beams or girders may be encased in concrete or exposed.

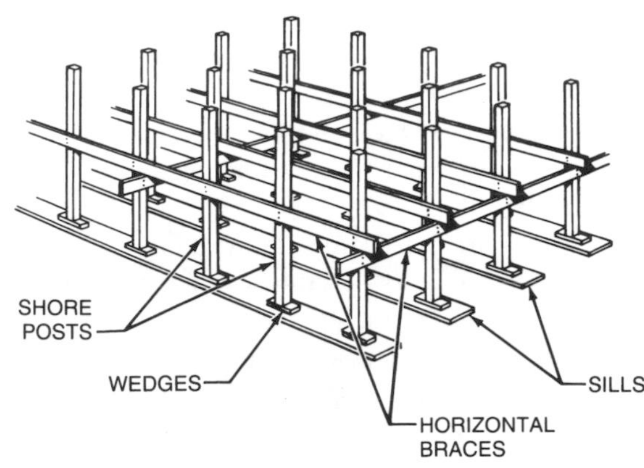

1. Place sills firmly over soil. Erect shore posts and tie together with horizontal braces.

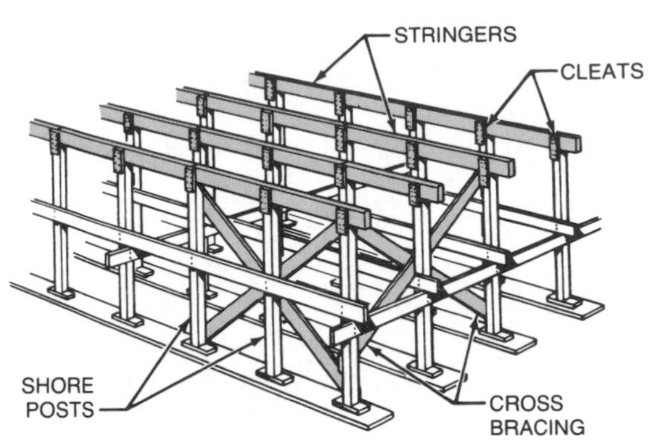

2. Fasten cross bracing to the shore posts. Secure stringers to the top of the shore posts with plywood cleats or steel angles.

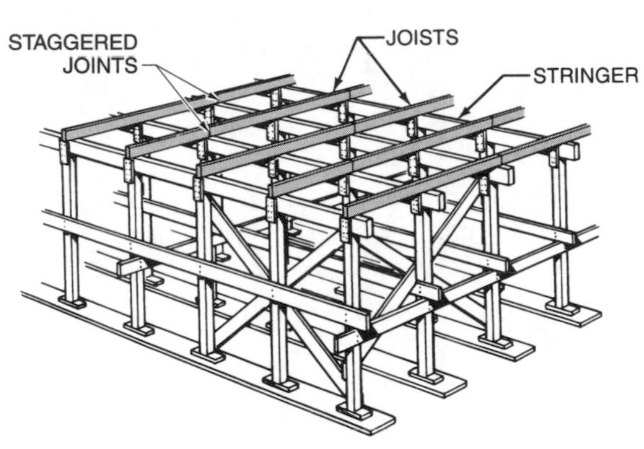

3. Place joists over the stringers, staggering the joints. Nail the joists to the stringers.

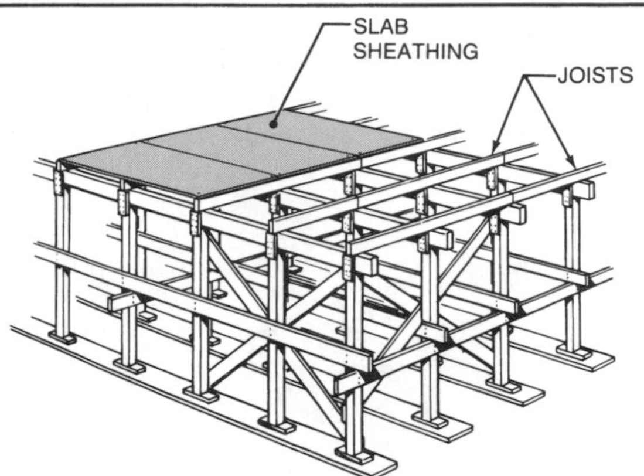

4. Place slab sheathing perpendicular to the joist direction. Nail the slab sheathing into place by nailing at the corners.

Figure 5-37. Floor forms are supported with stringers and shores. Plywood sheathing is fastened to joists positioned perpendicular to the stringers.

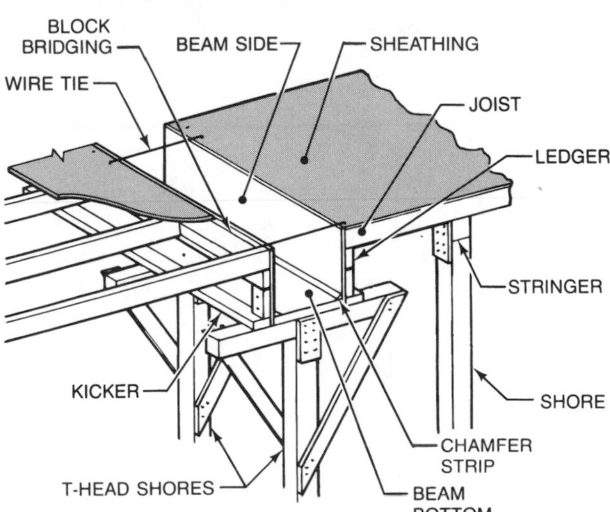

Figure 5-38. The sheathing for a floor form is secured to the beam or girder side when forming a monolithic beam or girder and floor slab system.

Flat Plate and Flat Slab Floors. Floor joists for flat plate and flat slab floor systems are supported entirely by stringers and shores. The formwork for flat plate and flat slab floors is similar to beam and slab formwork. Slab sheathing is cut to the shape of the column form, beveled, and held back 1/4″ to 1/2″ from the inside face of the column form.

The column forms and flat plate floor forms are tied to one another. For flat slab floors, a drop panel form is constructed over the column before the slab sheathing is positioned. The drop panel form is a plywood bottom piece with four sides. The bottom piece must be large enough to accommodate cleats behind the side pieces. An opening the size and shape of the column is cut out of the bottom piece to allow for rebar extending from the column. The drop panel form rests on joists placed on top of stringers. The edges of the slab sheathing are beveled and held back slightly when placed on top of the drop panel form sides. **See Figure 5-39.**

The drop panel form may also be prefabricated on the floor or ground and positioned on top of the joists. The slab sheathing may butt against or rest on top of the drop panel form sides.

Concrete Joist Systems. Concrete joist systems combine concrete joists with a concrete slab. The concrete for the joist systems is placed monolithically with beams, girders, and columns. Reusable prefabricated pans are placed at regular intervals on soffits to form the concrete joist system. The soffits are supported by stringers and shores. Steel or fiberglass pans are commonly predrilled for nailing to the soffit. Nail-down and adjustable pan designs are used to form the concrete joist systems.

Nail-down pans are the easiest to install because the flanges allow them to be nailed from the top side into the soffit. Standard pans for one-way joist systems are positioned with the flanges parallel to the soffit. Long pans are secured in place with the flanges perpendicular to the soffits. Long pans reduce the number of seams and produce a smooth exposed surface. The adjoining flanges clamp together to form the concrete joist. **See Figure 5-40.**

One-way joist pan forms are available in standard size widths of 20″, 30″, and 40″, and depths ranging from 8″ to 20″ in 2″ increments. The concrete joists formed range in size from 4″ to 8″. Concrete slabs incorporated with the joist systems are from 2½″ to 4½″ thick.

Forming costs can account for 30% to 50% of a concrete structure. It is more economical to reuse forms. It is also usually cheaper to use one column size rather than to vary column sizes.

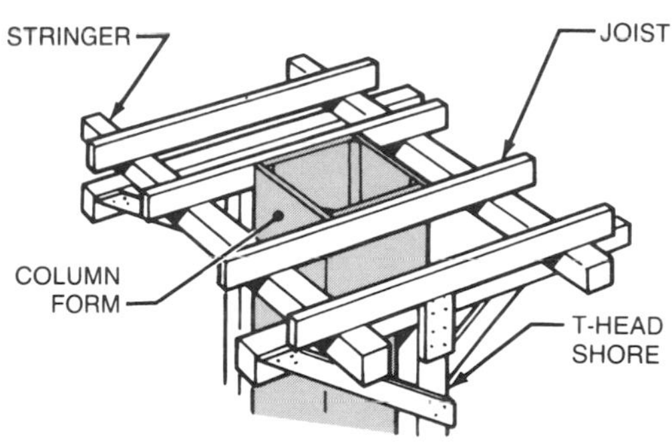

1. Place shores and stringers around column form. Place joists across the stringers.

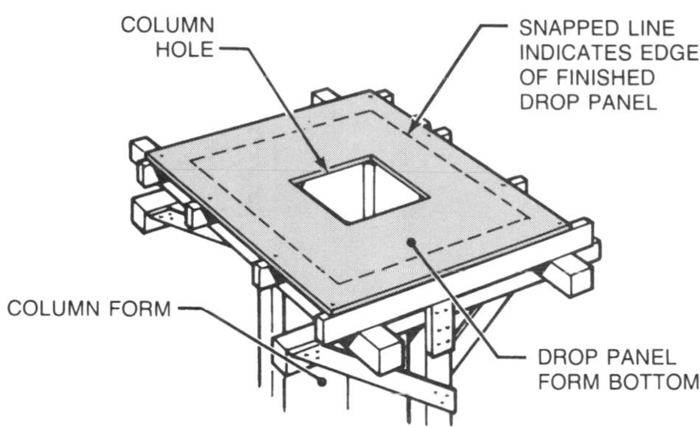

2. Cut a hole in the drop panel bottom and nail to the joists. Lay out and snap a line to indicate the edges of the completed drop panel.

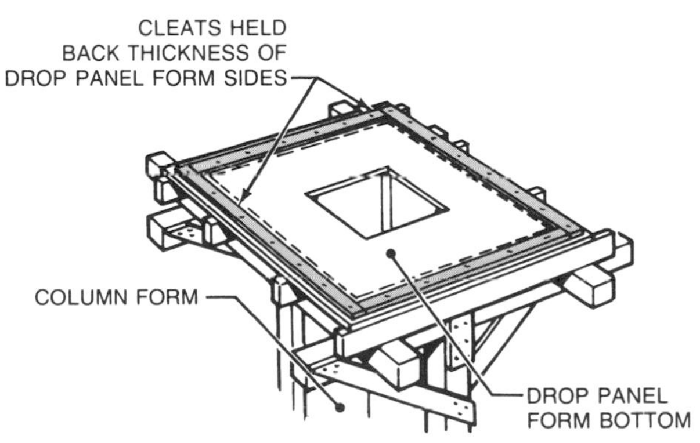

3. Nail cleats to the bottom, holding them back the thickness of the drop panel form sides.

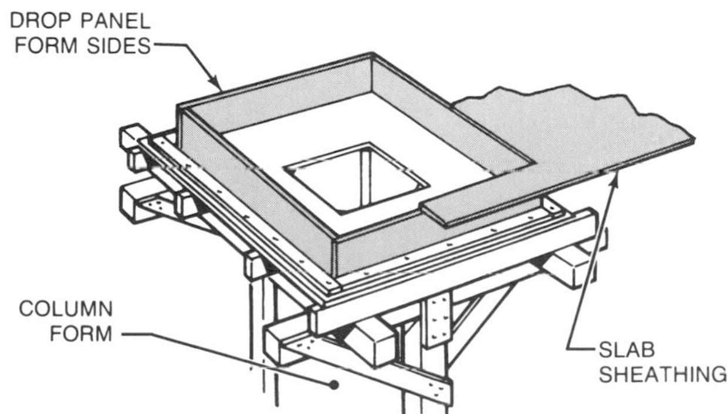

4. Nail the drop panel form sides to the cleats. Fit the slab sheathing to the inside edge of the drop panel form.

Figure 5-39. A drop panel form is positioned over a column form. Floor slab sheathing is fitted around the drop panel form.

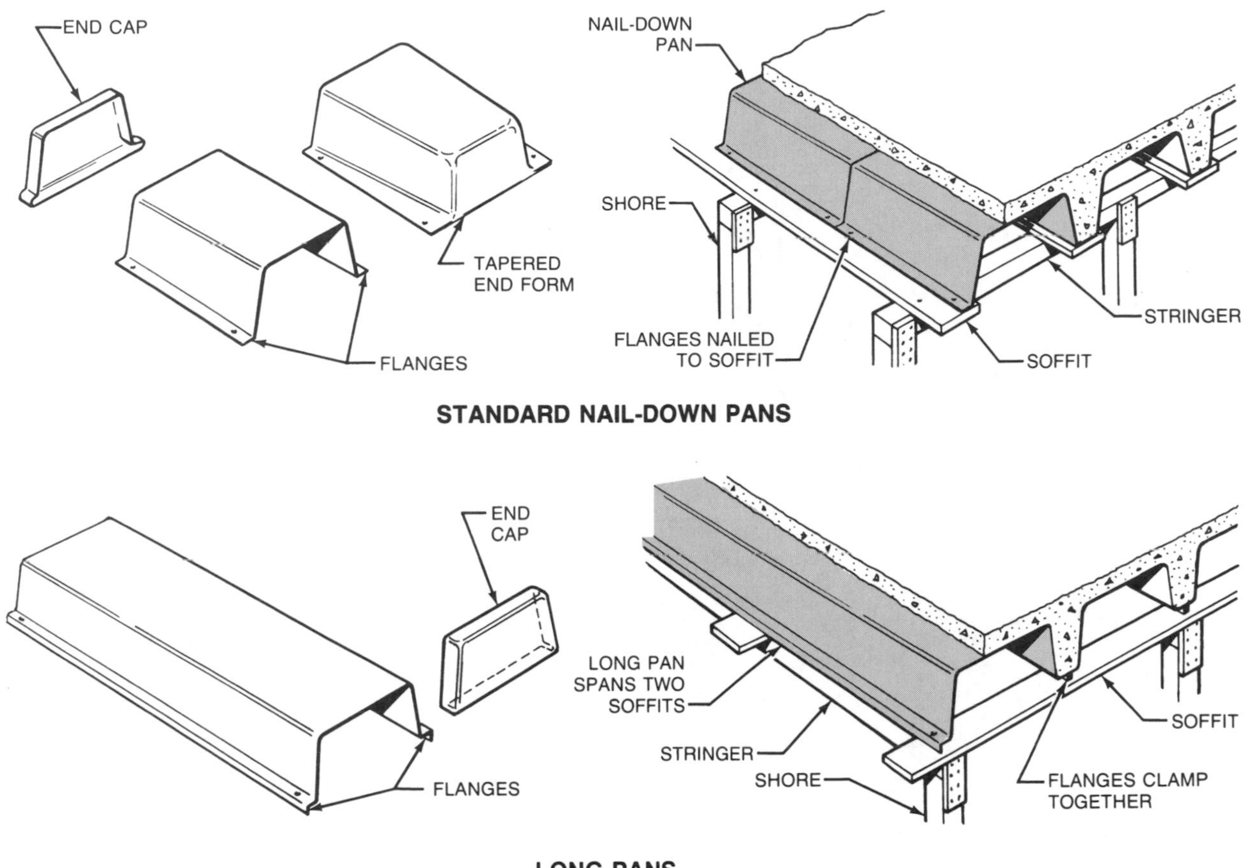

Figure 5-40. Standard nail-down pans and long pans are used to construct one-way joist systems.

Pan forms commonly frame into girder forms. Shores, stringers, and soffits are positioned and tapered end pans are placed against end caps. The pans are then placed from the ends and progress toward the center, with the pans overlapping 1″ to 5″. A filler piece is placed at the center to fill the open area. **See Figure 5-41.** Plywood sheathing may also be used as a base for nail-down pans. Plywood sheathing provides a convenient working surface and simplifies one-way joist system construction.

Adjustable pans for one-way joist systems are nailed to the sides of the soffit. Adjustable pans produce a smoother finish than nail-down pans because flange nail head impressions are not left in the exposed concrete. Adjustable pans are often used to form exposed ceilings.

Flanged and unflanged adjustable pans are also used to form one-way joist systems. When using unflanged adjustable pans, a soffit the width of the bottom of the joist is supported by a stringer. A template is used to hold the pans at their correct height while the pans are fastened to the soffit with double-headed nails.

When using flanged adjustable pans, stringers support a joist that is placed on edge. Soffits the width of the bottom of the concrete joist are positioned on top of the joists. Spreaders are placed between the joists, with ledgers nailed along the sides of the joists. A short spreader, equal to the distance between joists minus twice the flange width and pan thickness, is positioned on top of the bottom spreader to hold the pan to its correct width. **See Figure 5-42.**

When sizing individual floor members, it is more economical to use wider girders that are the same depth as the joists or beams they support than to use narrow, deeper girders. Wall pilasters, lugs, and openings should be kept to a minimum since their use increases forming costs. All members should be sized so that readily available standard forms can be used instead of custom job-built forms.

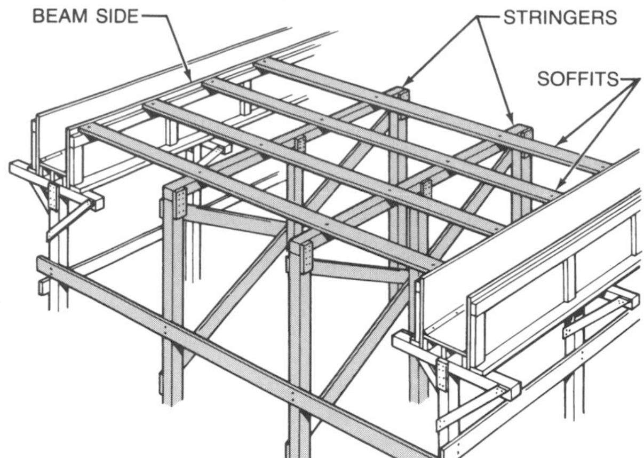

1. Set up and adjust the shores and stringers to the required height. Nail soffits to the top of the stringers.

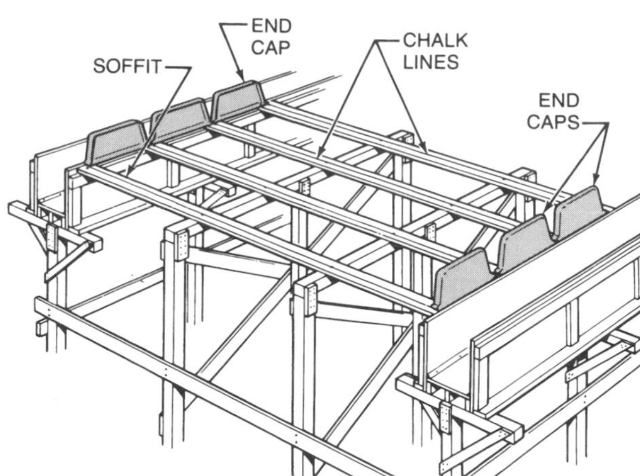

2. Snap chalk lines on the soffits. Nail end caps to the soffits at both ends.

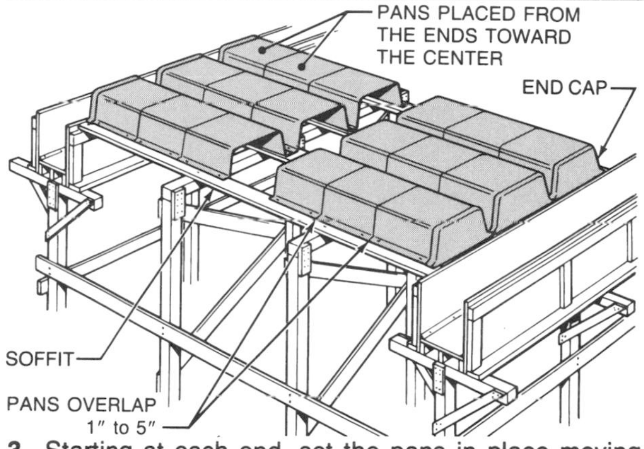

3. Starting at each end, set the pans in place moving toward the center of the span. Overlap the pans a few inches.

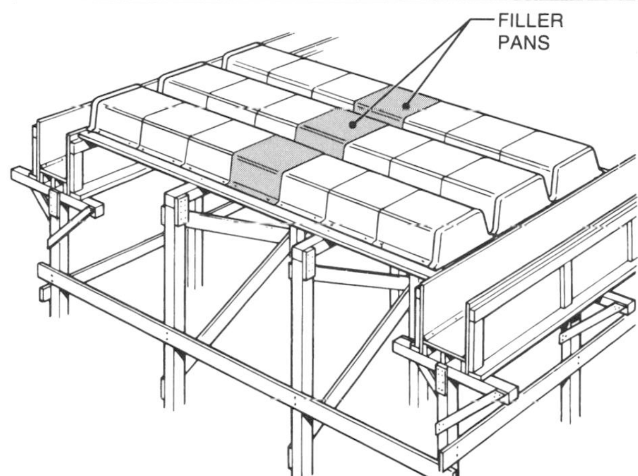

4. Set filler pans in position at the center of the span.

Figure 5-41. Pan forms are supported by soffits or solid deck sheathing. Pan placement starts at both ends with the pans overlapping 1″ to 5″.

Portland Cement Association

Two-way joist systems are produced by using dome pans placed on the soffit.

A two-way joist system is constructed in a similar manner to the nail-down pan method. Dome pans are placed on the soffits or plywood sheathing after the shores and stringers have been set up. A *dome pan* is a square prefabricated pan form nailed in position through holes in the flanges. Most dome pans are designed so the flanges butt together to produce the required joist size. If a wider joist is required, the pans are set to chalk lines snapped on the soffit or plywood sheathing. Dome pans are available in 2′, 3′, 4′, and 5′ widths and 8″ to 24″ depths. **See Figure 5-43.**

Two-way joist systems usually require a solid area around a column equal to the thickness of the slab and joist system. Solid deck sheathing is placed in this area and the dome pans are omitted.

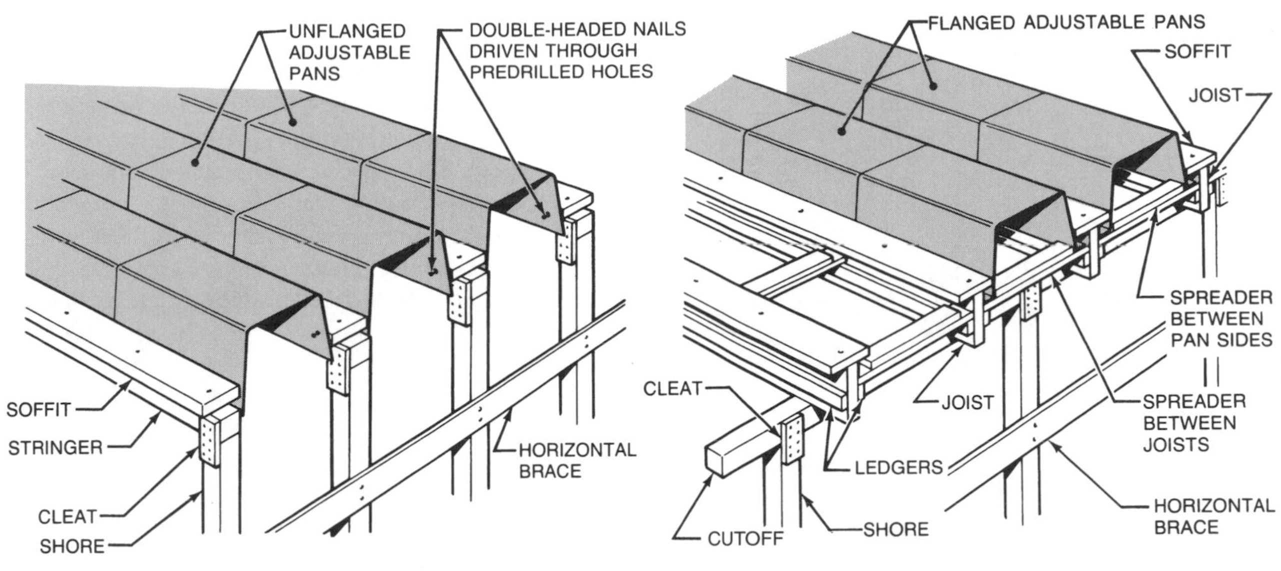

Figure 5-42. Adjustable pans are used to construct one-way joist systems. Unflanged adjustable pans are nailed to the sides of the soffits. Flanged adjustable pans are supported and secured by spreaders.

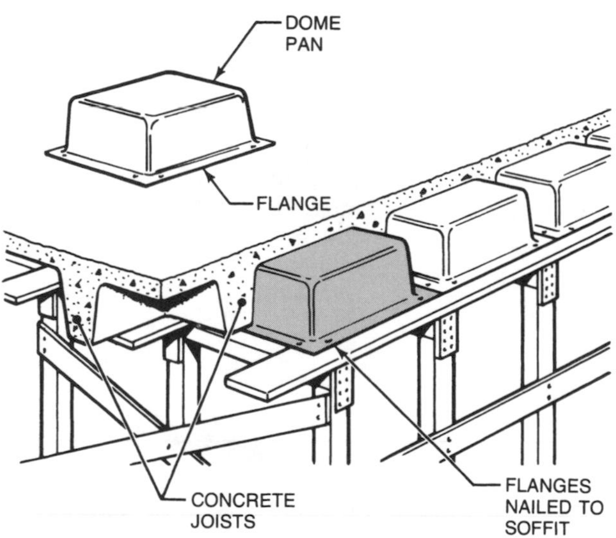

Figure 5-43. Dome pans are used to form a two-way joist system. The dome pans are placed on soffits or plywood sheathing.

Care must be taken to prevent damage to the plastic sheath covering post-tensioning tendons. Tendons within the sheath are coated their entire length with corrosion-inhibiting grease to protect the tendons from corrosion, weakening, and potential breakage. Do not make sharp angles with the tendons.

Post-Tensioning Concrete

Post-tensioning is a method designed to reinforce and strengthen concrete with high-strength steel strands, referred to as tendons. While post-tensioning is used for slab-on-grade floors, it is also widely used to reinforce concrete floor slabs and beams in multistory structures such as office and apartment buildings, parking garages, bridges, sport stadiums, and water tanks.

Concrete is very strong in compressive strength, but weak in tensile strength. *Compression* is stress caused by pushing together or a crushing force. *Compressive strength* is the maximum resistance of a concrete or mortar specimen to vertical loads. *Tension* is a pulling or stretching force. *Tensile strength* is the resistance of a material to forces attempting to pull it apart. Concrete may crack when compression or tension forces are higher than the strength of the concrete. **See Figure 5-44.** For example, the weight of a car resting on a floor slab in conventional concrete construction may cause the slab to deflect or sag. This could cause the bottom of the slab to slightly elongate, producing tensile forces and cracking.

Steel reinforcing bars (rebar) are passive reinforcement. Rebar is placed in the concrete as tensile reinforcement, which may be sufficient for the load requirements of a particular structure. Post-tensioning is considered active reinforcement, and is far more effective in resisting deflection and cracking under extremely heavy loads.

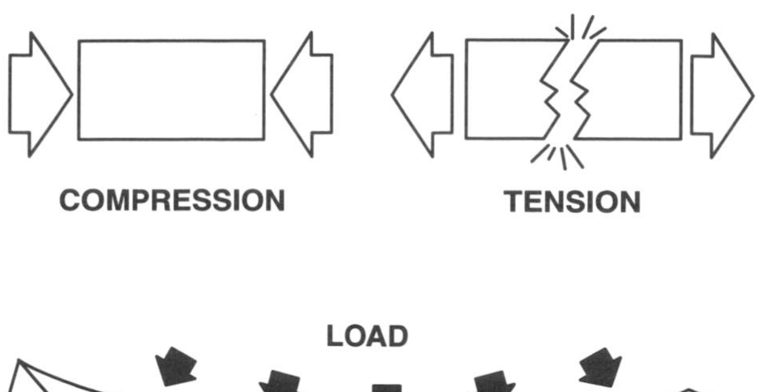

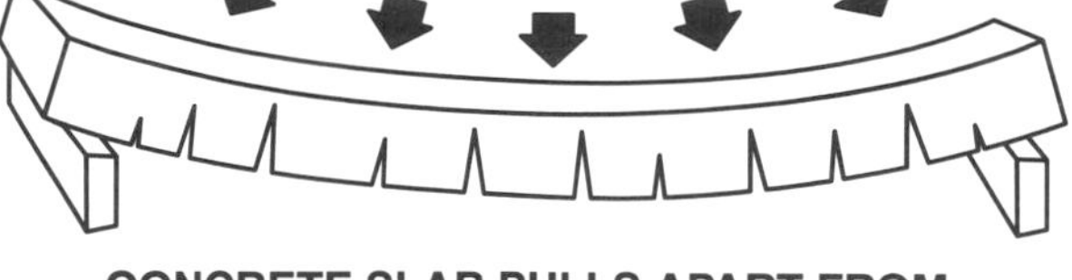

Figure 5-44. Tensile forces may cause a concrete slab to deflect or sag, which may result in cracking.

Advantages. There are many advantages to post-tensioning concrete. In the construction of a building, post-tensioning allows thinner slabs with longer spans without the need for intermediate support. Thinner slabs mean less concrete is needed, which lowers the overall building height while maintaining the same floor-to-floor height. Lower building height can result in substantial savings in mechanical systems and façade (face of building) costs. Post-tensioning results in a considerable reduction of weight as compared to a conventional concrete building with the same number of floors. This lessens the load imposed on the foundation of a building, which is an important consideration in more active seismic (earthquake) areas.

Post-Tensioning Priniciples. Horizontal post-tensioning is commonly applied to floor slabs and beams. Vertical post-tensioning is used in walls and columns. The main components of a post-tensioning system are very high strength steel strands called tendons enclosed in a plastic or metal tubing. The tendons are coated along their entire length with corrosion-inhibiting grease to protect the tendons from corrosion, weakening, and potential breakage. A typical tendon used in post-tensioning has a tensile strength of 270,000 pounds per square inch (psi). Conventional rebar has a tensile strength of 60,000 psi.

The height and spacing of the tendons in a slab is determined by engineering calculations. To ensure that the tendons stay at the required height in the slab or wall, plastic or cement supports (chairs) are placed and wired to the tendons where needed. One end of the tendons is securely anchored. The opposite end is inserted through an anchoring device attached to a hydraulic jack or other type of tensioning device. **See Figure 5-45.**

Concrete is placed and must cure (harden) sufficiently to withstand the pressure of the tensioning. Curing commonly takes between 3 days and 5 days, after which the concrete is tested to ensure that it has reached the necessary strength. The next step of the procedure is to tension the tendons. A strong force, typically up to 27,000 psi, is then applied to the tendons by the stressing equipment.

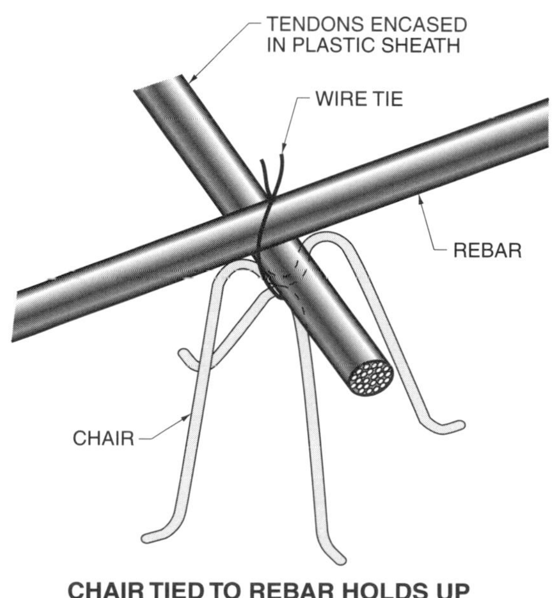

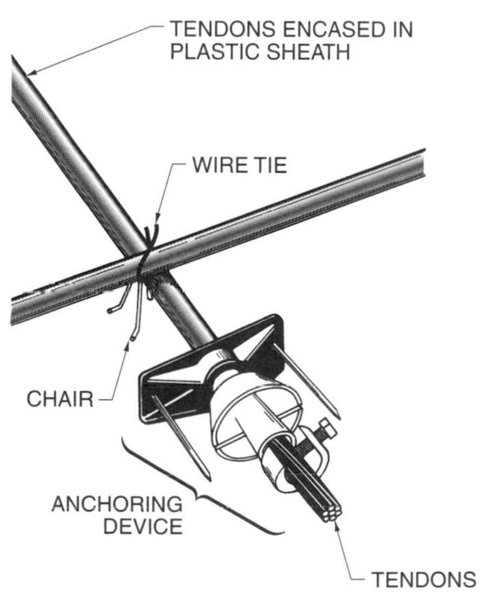

Figure 5-45. Tendons are encased in tubing and anchored at both ends.

A number of methods are used to hold the tendons under pressure after they have been released. In one method, the ends are locked into place using an anchoring device that remains in the concrete. Another method uses a tapered tube containing steel wedges that grip the tendons. Because of the tapered shape of the tube, the wedges grip more tightly as tension is applied.

Vertical Shoring

Vertical shoring provides the main support for beam, girder, and slab formwork. The vertical and lateral pressures that occur during and after concrete placement, and the weight of the form materials, machinery, and workers are considered in designing shoring systems. Shoring system design should adhere to guidelines established by the American Concrete Institute (ACI) and the U.S. Occupational Safety and Health Administration (OSHA). (See ACI and OSHA Shoring Standards in Appendices C and D.) Shoring systems may consist of wood or metal shores, or metal scaffold shoring. Traditionally, wood was used entirely for shoring purposes. Wood shoring is still used, but usually for smaller projects. Metal shores and scaffolding are now common on most larger construction projects. The shoring system used should be designed by a structural engineer. The shoring system and scaffolding design should include the spacing, size, and types of wood or metal shoring to be used. **See Figure 5-46.**

Symons Corporation

Figure 5-46. Vertical shoring is placed beneath stringers that support the joists for floor forms. Diagonal and horizontal bracing is used to secure the shores.

Wood Shoring. All wood shoring members must be straight and true. Cuts made at the bearing ends and splices of the shoring members should be square. Vertical shores must be placed in a plumb position and secured with braces so they cannot tilt. Inclined shores must be securely braced to prevent slippage. Shoring systems must be properly braced to ensure the integrity of the shoring. Horizontal braces tie the shores together. Diagonal braces prevent sway or lateral movement of the shores.

Wood shores are used to support girders, beams, and slab forms for multistory structures with an average height (approximately 10′) between floors. Wood shores are constructed with 3 × 4s, 4 × 4s, or 6 × 6s. The load and unsupported height of the shore are used to determine the cross-sectional dimensions of the shore required.

A T-head shore supports beam and girder forms. The head of a T-head shore is centered on top of a vertical post. Cleats are nailed to the post and head. An L-head shore is commonly used under spandrel beam forms. The head of the L-head shore is offset and braced with 1 × 4s or 1 × 6s extending from the head to the post. The heads of wood shores are attached to the posts using plywood cleats or metal angle brackets.

Types of wood shores used for vertical shoring are the single-post, double-post, and two-piece adjustable wood shores. A single-post wood shore is a single vertical member placed beneath stringers supporting floor slab forms. A double-post wood shore consists of a head placed over two vertical posts. Cross bracing may be used to reinforce the shore. Double-post wood shores support heavy girder loads, spandrel beams, and drop panels. A two-piece adjustable wood shore has two overlapping wood posts held in place with a post clamp (Ellis clamp). The post clamp is nailed to the lower post. The upper post is then raised into position with a portable jack. The two posts are held in place by friction against the post clamp. **See Figure 5-47.**

Mudsills should be placed beneath wood shores positioned over the ground to spread the load over a large area. A single 2 × 10 may be used as a mudsill on good load-bearing soil. Two or three 3 × 10 mudsills placed next to one another with a 4 × 4 or 4 × 6 plate placed at a right angle may be used as a mudsill in poor soil conditions. The shore post is toenailed to the plate.

Wood shores are generally cut shorter than required to accommodate a pair of wedges beneath the posts. The wedges are used to adjust the height of the shores when aligning the formwork. When shores are required over concrete surfaces, a wood sill is placed beneath each row of shores to facilitate driving the wedges and to provide a nailing base for the post bottom and wedges.

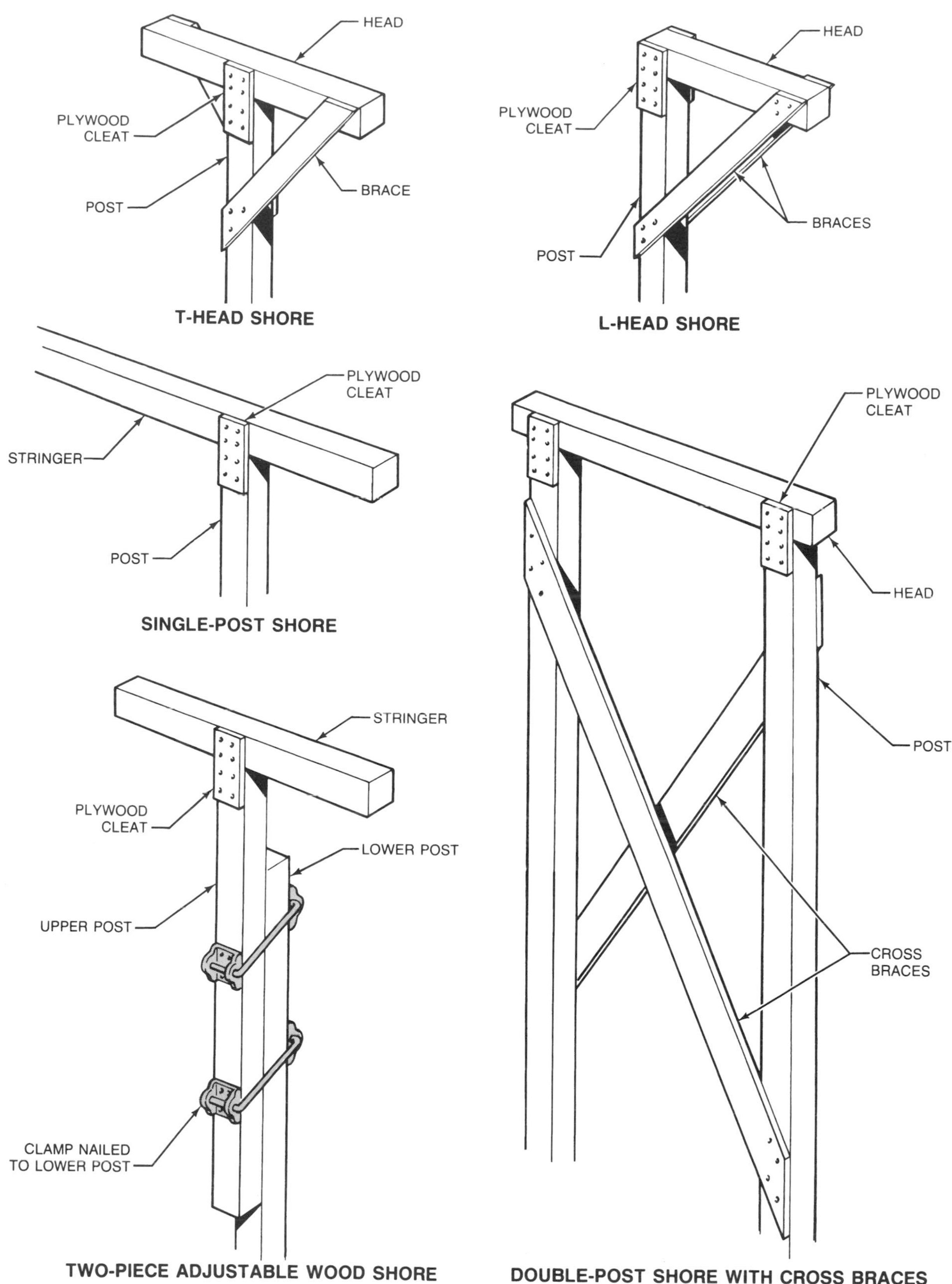

Figure 5-47. Wood shores are used to support beam, girder, and floor slab formwork. Heads or stringers are fastened to the tops of the posts to support formwork.

Wood shores are fastened securely to the stringers or any other member they directly support. A plywood cleat is nailed to the side of the post and stringer to secure the stringer. Various types of metal brackets are also used to secure the stringer by screwing or nailing them to the stringer and post.

Spliced wood shores are commonly used to cut material costs. Two-inch lumber or 5/8″ plywood is used as splicing cleats and is fastened to all sides of the post. Splicing cleats should be as wide as the shore post and extend a minimum of 12″ past each side of the splice. Unbraced shores should not be spliced at midheight or midway between horizontal supports. Shore jacks are used to adjust the height of wood shores without using wedges. A metal fitting slips over a 4 × 4 or 6 × 6 post and is nailed into place. The jack is then adjusted to the final height. **See Figure 5-48.**

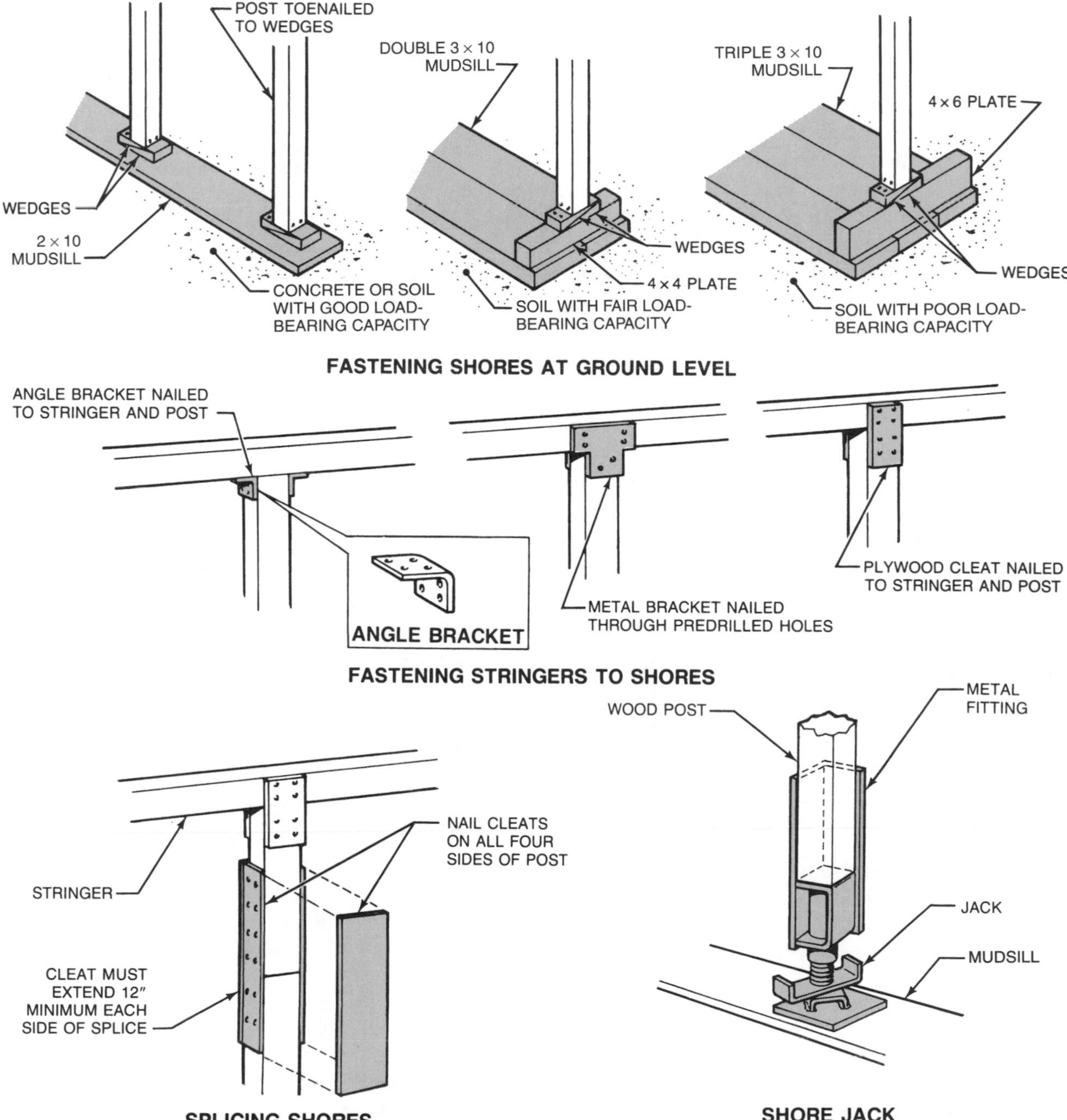

Figure 5-48. The load-bearing capacity of the soil determines the method used to fasten shores at ground level. Metal or angle brackets, or plywood cleats secure stringers to posts. Metal shore jacks are used to adjust the height of wood shores, eliminating the need for wood wedges.

Reshores and Permanent Shores. Reshoring takes place during the construction of multistory buildings. It consists of placing shores beneath beams and floor slabs immediately after the formwork has been stripped away. The shores must be placed immediately because the floor slab and beams may not have cured sufficiently to adequately support materials and equipment to be used in the construction of the floors above.

Metal shores and 4 × 4 or 6 × 6 wood shores are used for reshoring. While reshoring beneath a recently placed structural member, construction loads should not be placed on the floor level above. Reshores are placed directly above one another because they may be required to stay in place until the entire structure has been completed. Reshores should not be wedged or jacked up to lift or crack the concrete above. In order to keep the load distribution from changing, reshores are tightened in position uniformly and only enough to hold them securely. **See Figure 5-49.** Improper reshoring can result in deflection of recently placed concrete, causing cracking or collapse.

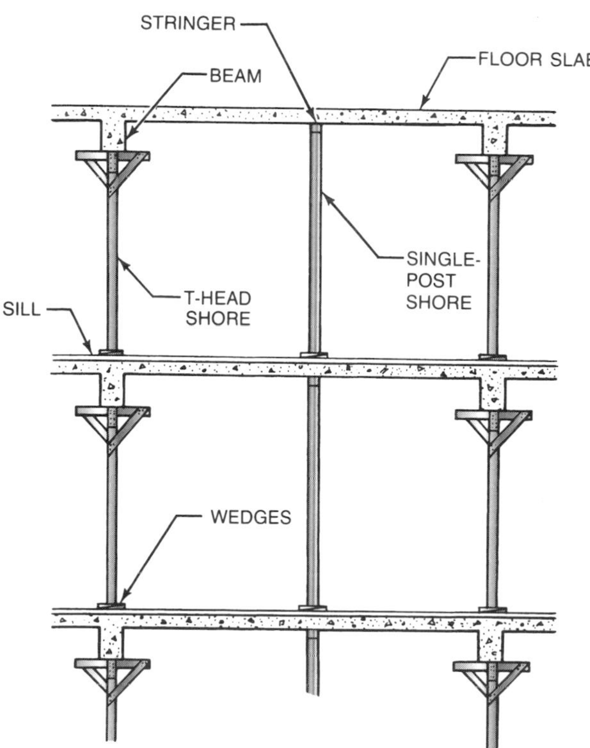

Figure 5-49. Reshores are single-post shores placed under structural members after the original shoring is removed. Reshores are placed directly above one another in successive levels.

Permanent shores are erected with the top of the stringers flush with the top of the floor slab sheathing. Ledgers are held down the thickness of the floor slab sheathing from the tops of the stringers and bolted into position. The floor slab sheathing is then placed on top of the ledger. After the concrete has set, the intermediate shores, floor slab sheathing, and ledgers are removed, leaving the permanent shores and stringers in place. **See Figure 5-50.**

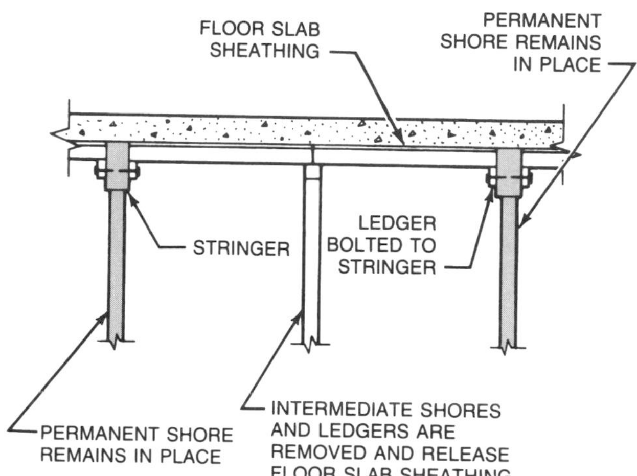

Figure 5-50. Permanent shores remain in place after the floor slab forms have been removed.

The main advantage of permanent shores is that the cost of reshoring is avoided and the deck and beam forms can be stripped sooner. Although called permanent shores, they will be removed at a later time.

As each floor level has gained sufficient strength to carry loads, the reshores or permanent shores are removed. The removal begins at the upper floor level and proceeds toward the lower levels of the building.

Adjustable Metal Shoring. Adjustable metal shores, used for vertical shoring, are constructed of tubular steel. The tubular steel is open at both ends to prevent accumulation of water and rust. The upper tube is adjusted to the approximate height required and a locking pin is inserted through a hole in a slot above the adjustment collar. The adjustment collar is turned to make the final adjustments. Flanges secure the top and bottom of the adjustable metal shore to a stringer and pad or mudsill. Braces may be attached to the shores by using nailing brackets or other devices. Like wood shores, horizontal and diagonal braces must be attached to the metal shores. One method uses metal clamps designed for this purpose. **See Figure 5-51.**

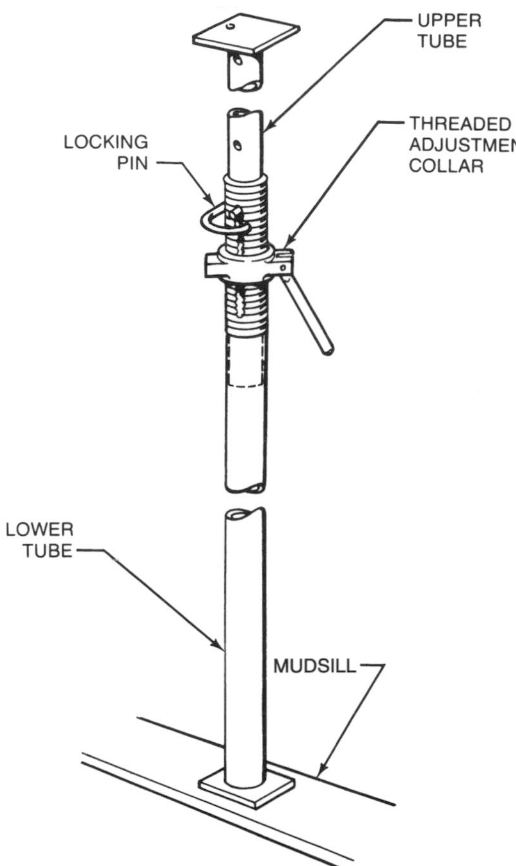

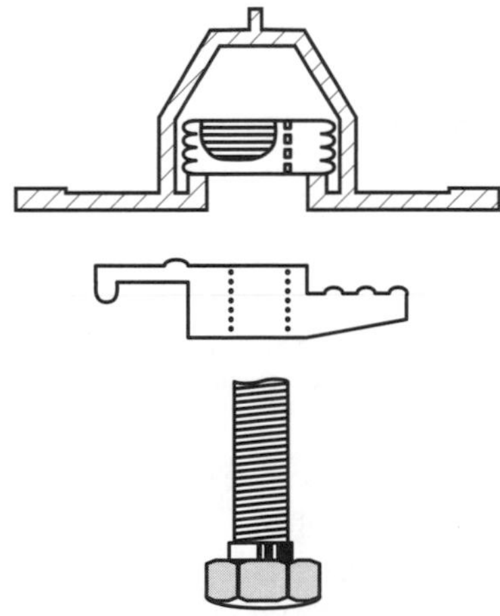

TYPICAL ATTACHMENT CLAMP
FOR SECURING BRACES
TO STEEL SHORES

Figure 5-51. Adjustable metal tubular shores can be adjusted to different heights. Metal clamps are designed to attach horizontal and diagonal braces.

Horizontal and Diagonal Bracing. Horizontal and diagonal bracing ensure the stability and safeness of a shoring system. Bracing reduces the possibility of form collapse resulting from overloading forms and lateral pressure caused by wind, movement of heavy weight, and disturbance of the forms caused by crane booms or other equipment.

Horizontal bracing is placed at the midpoint of the shores and extends in two directions. Shores over 10 high may require two or more rows of horizontal bracing. Diagonal bracing should also be installed in two directions. At ground level, bracing may be extended from the outside row of shores and fastened to stakes driven into the ground. **See Figure 5-52.**

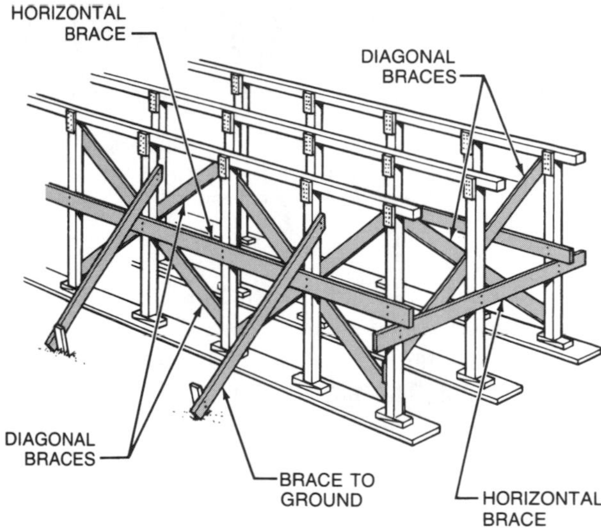

Figure 5-52. Shores supporting formwork for floor systems require extensive horizontal and diagonal bracing.

Metal-Framed Shoring. In metal-framed shoring, sections of tubular steel frames are assembled to the required heights. Metal-framed shoring is used to support slab and beam, flat plate and flat slab floor forms. Metal-framed shoring is also used in structures with high beam and slab soffits, and bridge and overpass construction. **See Figure 5-53.**

Metal-framed shoring is erected according to a layout plan. Safe working loads for metal-framed shoring range from 4000 lb to 25,000 lb per leg, depending on the type of frame, bracing, and height of the scaffold. Greater working loads are possible with specially designed towers. Typical frame sections are 2′ to 4′ wide and 4′ to 6′ high.

Symons Corporation

Figure 5-53. Metal-framed shoring is used to support flat plate and flat slab floor forms.

Individual frames are made up of two tubular steel uprights joined by horizontal members. Cross braces extending across the uprights of individual frames provide lateral stability. The individual frames are mounted on top of each other and secured with coupling insert pins. Opposite frames are fastened to each other with diagonal cross braces. The assembled metal-framed shoring is supported by adjustable swivel screw jacks that rest on metal base plates. The plates rest on mudsills similar to those used for wood shores. **See Figure 5-54.**

Metal-framed shoring supporting heavy loads over soil is placed on thick timbers to distribute the load over a large area. A beam clamp is installed at the top of the shoring to hold and support the stringers. A *beam clamp* is an adjustable U-shaped device.

Loaded shoring equipment, including cross braces, shall not be released or removed until the approval of a qualified engineer has been received. Premature releasing or stripping of forms can cause failure. A qualified engineer must decide when and how stripping is to proceed. Weather conditions, variations in different parts of the structure, and the setting qualities of the concrete all affect the stripping time schedule.

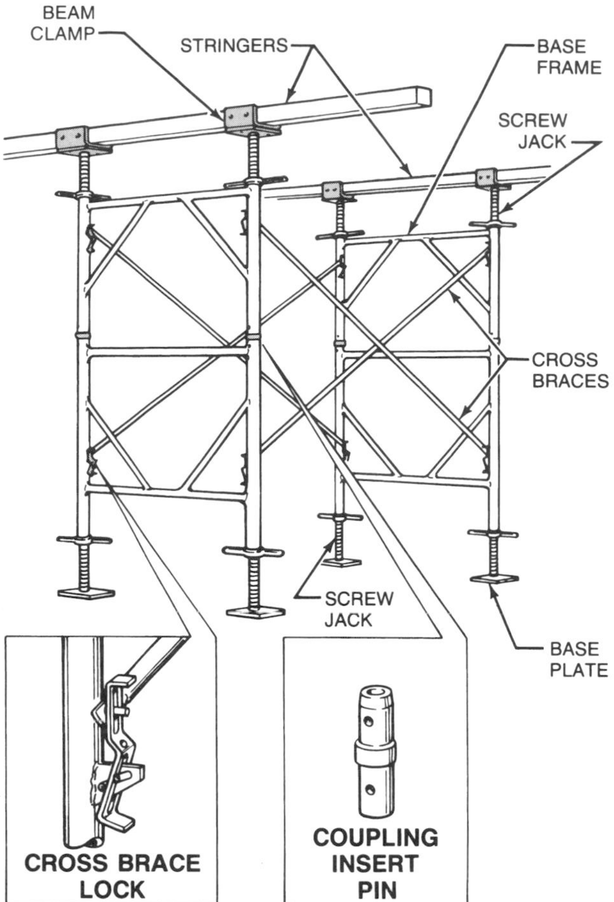

Figure 5-54. Metal-framed shoring consists of a frame secured together with insert coupling pins. A cross brace lock secures cross braces extending from the frames.

SLIP FORMS

Slip forms were first used in the construction of curved concrete structures such as silos and towers. Today, slipforming methods are used to construct rectangular buildings as well as building cores, dams, caissons, underground shafts, communication towers, and other types of concrete structures. **See Figure 5-55.** One advantage of slipforming is the savings in labor and material costs.

Typical slip forms are designed with 4′ high inner and outer walls. The walls are usually constructed of 3/8″ to 3/4″ Plyform panels backed with 2 × 4 studs and 2 × 6 walers. To reduce the amount of drag as the forms are being raised, the inner and outer form walls are tapered outward 1/8″ per foot. Steel cross beams and yoke legs hold the walls together. The cross beams fasten the opposite yokes together and provide a base for the hydraulic jacks. **See Figure 5-56.**

Portland Cement Association

Figure 5-55. High-rise buildings may be constructed using slip forms.

The steel yoke legs are spaced approximately 6′ apart along the entire length of the slip form. The tops of the yoke legs are fastened to cross beams, and the yokes extend down and fasten to the walers. The slip forms are raised by pneumatically powered hydraulic jacks that are mounted on the cross beams. The jacks climb the jackrods that extend into the form.

The lifting operation requires perfect coordination of the hydraulic jacks to ensure accurate alignment of the form. The hydraulic jacks must lift at the same time and the same rate of speed. The slip forms move up the jackrods as the concrete is being placed at rates ranging from 2″ to 70″ per hour. The climbing speed is based on the type of structure and the rate of concrete placement. Other factors affecting climbing speed are how quickly rebar and built-ins can be placed. *Built-ins* are the frames (bucks) for the door and window openings, and beam pockets where required. Additional provisions must be made for the positioning of plumbing and electrical recesses, brackets, and anchors.

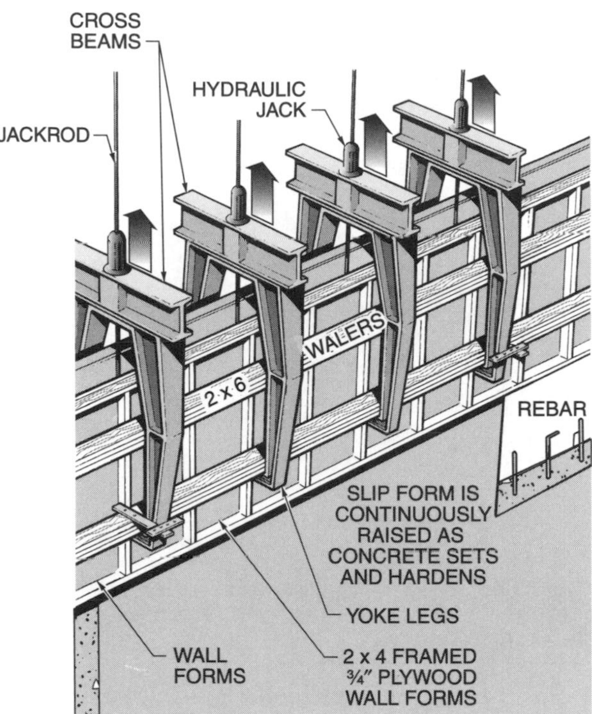

Figure 5-56. A typical slip form design includes cross beams, yoke legs, hydraulic jacks, and jackrods.

FLYING FORMS

A *flying form* is an engineered prefabricated form that consists of a wood deck and an aluminum frame system. **See Figure 5-57.** Aluminum trusses on each side of the assembly and beams resting on the trusses support the wooden deck. Adjustable jacks are used to raise the flying form to its final position. Flying forms are usually used for the placement of concrete for floor slabs and beams. One- or two-way pan systems may be installed on the top of flying forms on the ground and lifted into position. The forms are set into position by crane.

Cast-in-place or prefabricated walls and/or columns are constructed under the flying form. After the concrete has been placed in the flying form and sufficiently cured, the entire flying form is removed and raised to the next level. The use of flying forms can greatly increase productivity on multistory buildings where all the floor levels are identical.

CONCRETE STAIRWAYS

Concrete stairways are classified as exterior or interior stairways. Exterior stairways are built over sloping soil and receive their main support from the earth. Interior stairways are built over an open area and receive their main support from the reinforced soffit below the steps.

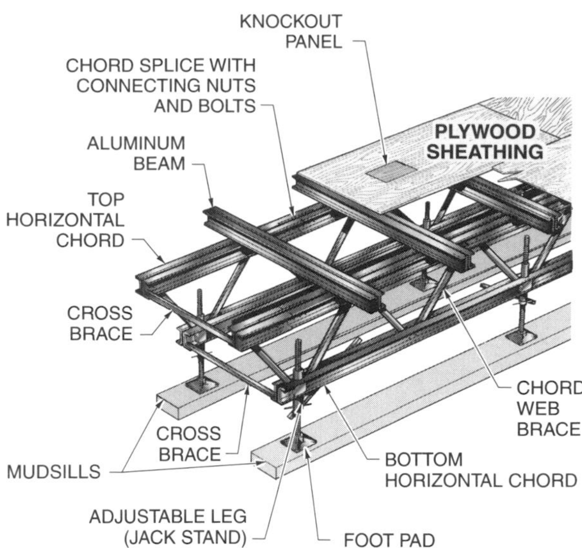

Figure 5-57. Flying forms can increase productivity in the construction of multistory buildings where all floor levels are identical.

Interior stairways are further classified as open or closed. Open stairways are not enclosed by walls and rely entirely on built-in-place forms and shoring to support the concrete. Closed stairways are enclosed by walls on two sides. A stairway constructed with a wall on one side is a combination of the two designs. **See Figure 5-58.**

Most concrete stairways are variations of a straight flight design. A straight flight stairway extends from one floor to the next without a change in direction. L-shaped and U-shaped stairways have landings which change the direction of the stairway.

Riser and Tread Layout

Riser and tread layout establishes the height of the individual risers (vertical surface) and the depth of the individual treads (horizontal surface). The riser height (unit rise) is determined by dividing the total rise of the stairway by the number of risers. *Total rise* is the vertical distance from one floor to the floor above. The tread depth (unit run) is determined by dividing the total run by the number of treads. *Total run* is the horizontal length of a stairway measured from the foot of the stairway to a point plumbed down from where the stairway ends at a floor or landing above.

Calculating Unit Rise and Unit Run. Many building codes specify minimum and maximum riser heights and minimum tread depth. For commercial stairways, the International Building Code (IBC) specifies a minimum riser height of 4″ and a maximum of 7″. Minimum tread depth is 11″. The IBC also permits a ⅜″ variation in the depth of adjacent treads or height of adjacent risers. The code having jurisdiction in the area should be consulted before determining unit rise and unit run.

Unit rise and unit run are calculated by converting the total rise and total run to inches and dividing by the number of risers and treads. For example, the unit rise for a stairway with a total rise of 10′-2″ is determined as follows:

Step 1: Convert total rise to inches.

Total rise (inches)
= *(no. of feet* × 12) + no. *of inches*
= (10′ × 12) + 2″
= 120 + 2
Total rise = **122″**

Step 2: Determine the number of risers by dividing the total rise by the minimum desired riser height.

No. of risers = total rise ÷ minimum riser height
= 122 ÷ 7
No. of risers = **17.428**

If the answer contains a decimal value, a fraction of an inch is added to each 7″ riser to ensure that the 17 risers are equal height.

Step 3: Determine the exact riser height by dividing the total rise by the number of risers. Calculate the answer to three decimal places.

Riser height = total rise ÷ no. of risers
= 122″ ÷ 17
Riser height = **7.176″**

Step 4: Convert the decimal value of the riser height to sixteenths of an inch by multiplying the remainder by 16.

Fractional equivalent (16ths)
= *decimal value* × 16
= .176 × 16
Fractional equivalent (16ths) = **2.816**

Section 1009, Stairways and Handrails, *of the International Building Code provides information regarding stairway construction, including minimum and maximum riser heights, minimum tread depths, uniformity of tread and riser size and shape, and stair profile for commercial structures. Minimum riser height is 4″ and maximum riser height is 7″, with a minimum tread depth of 11″. The greatest tread depth and riser height must not exceed the smallest by more than ⅜″. The front edge of the tread must not have a radius greater than ½″.*

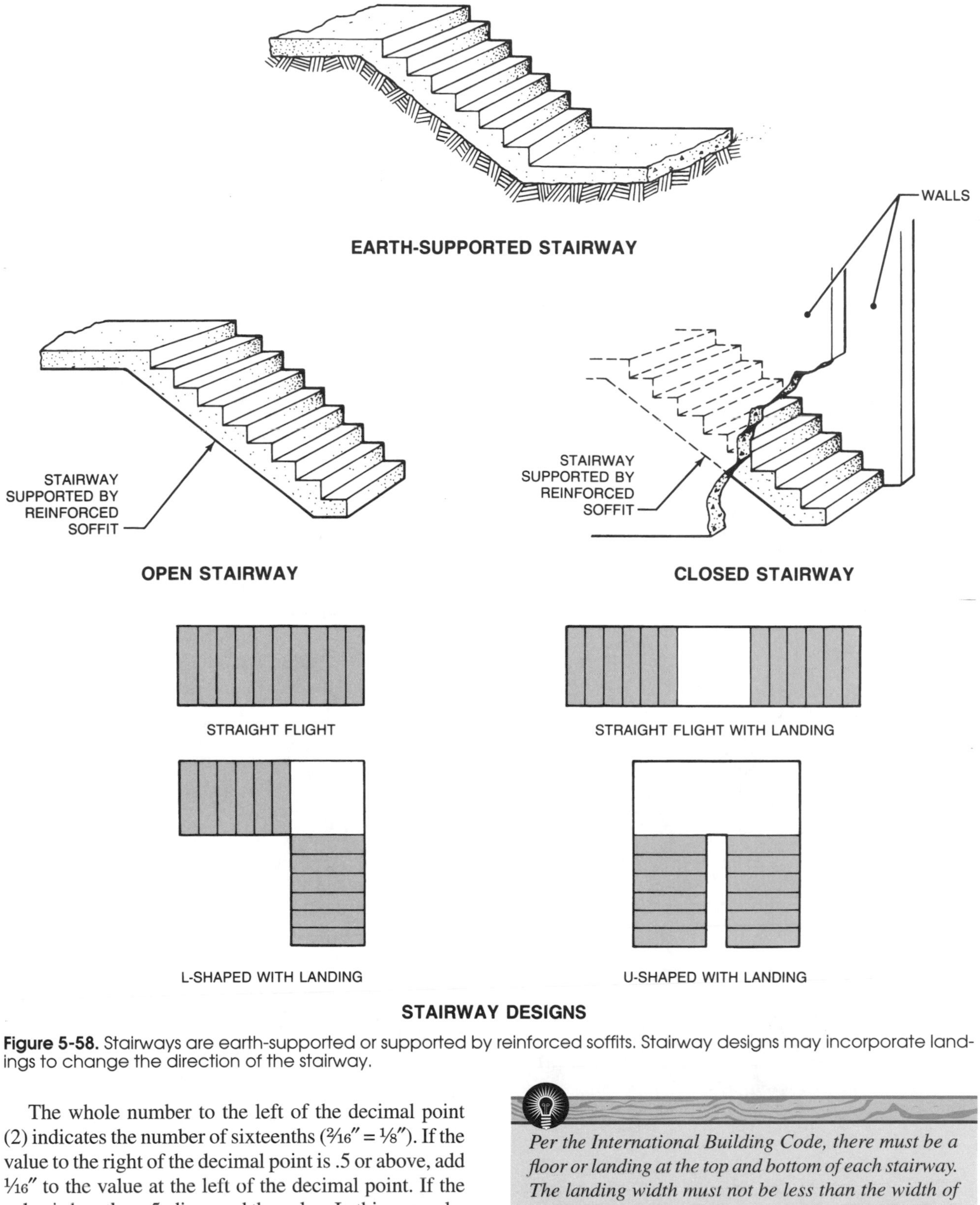

Figure 5-58. Stairways are earth-supported or supported by reinforced soffits. Stairway designs may incorporate landings to change the direction of the stairway.

The whole number to the left of the decimal point (2) indicates the number of sixteenths (2⁄16″ = 1⁄8″). If the value to the right of the decimal point is .5 or above, add 1⁄16″ to the value at the left of the decimal point. If the value is less than .5, disregard the value. In this example, the value 2.816 is converted to 3⁄16″. The whole number in step 3 (7″) is added to the fractional equivalent obtained in step 4 (3⁄16″) to equal the exact riser height (7 3⁄16″).

Per the International Building Code, there must be a floor or landing at the top and bottom of each stairway. The landing width must not be less than the width of the stairway it serves. The length of the landing, measured in the direction of travel, must be at minimum equal to the stairway width.

When calculating the tread depth, the total run is converted to inches and divided by the total number of treads. The total number of treads (total run) is always one less than the number of risers. For example, the tread depth of a stairway with 17 risers and a total run of 16′-0″ is determined as follows:

Step 1: Convert the total run to inches.

Total run (inches)

= *(no. of feet* × 12) + *no. of inches*

= (16′ × 12) + 0″

Total run = **192″**

Step 2: Determine the tread depth by dividing the total run by the number of treads.

Tread depth = *total run* ÷ *no. of treads*

= 192″ ÷ 16

Tread depth = **12″**

If the answer has a decimal value, use the same procedure used for determining fractions when calculating riser height.

The riser and tread dimensions are laid out on a pair of skirt boards using a steel square. The skirt boards are part of the stair form to which the riser form boards are fastened. The riser height is marked on the tongue of a steel square and the tread depth is marked on the blade. Square gauges may be used to hold the square in position along the edge of a skirt board. The steel square is positioned along the top edge of the skirt board and the steps are laid out. **See Figure 5-59.**

Constructing Stairway Forms

Concrete stairway forms require accurate layout to ensure accurate finish dimensions for the stairway. Interior or exterior stairways are reinforced with rebar that tie into the floors and landings.

Concrete stairways are formed monolithically with floor slabs by framing the stairway forms into the floor slab forms. Concrete stairway forms may also be constructed after the concrete for the floor slabs has set. Concrete stairways formed after the floor slab has set are anchored to a wall or beam by tying the stairway rebar to rebar projecting from the wall or beam, or by providing a keyway in the wall or beam. Step supports are commonly placed monolithically with the exterior wall. Rebar projecting from the walls may also be used to tie into rebar in the stairway form in closed stairways. **See Figure 5-60.**

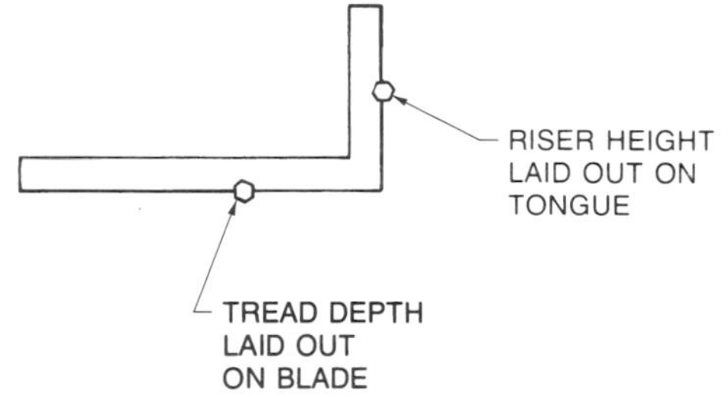

1. Clamp square gauges on the tongue of the steel square at the riser height and on the blade at the tread depth.

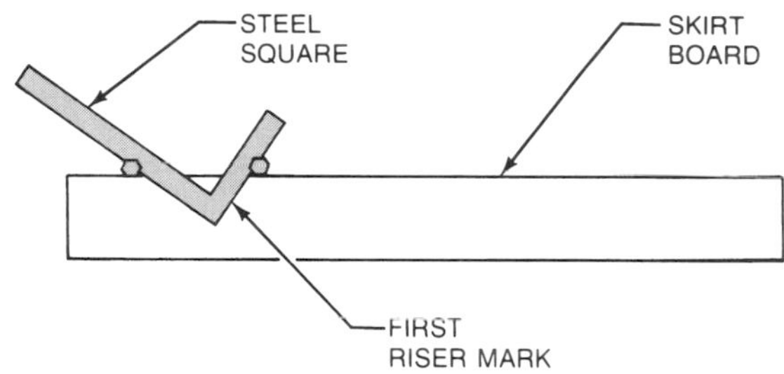

2. Position the steel square along the top edge of the skirt board and lay out the first riser.

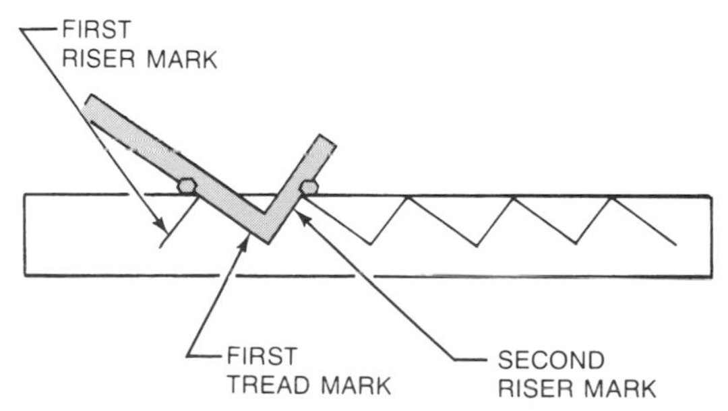

3. Slide the steel square until the tread square gauge aligns with the riser mark. Lay out the first tread and second riser. Continue sliding the steel square and laying out the treads and risers.

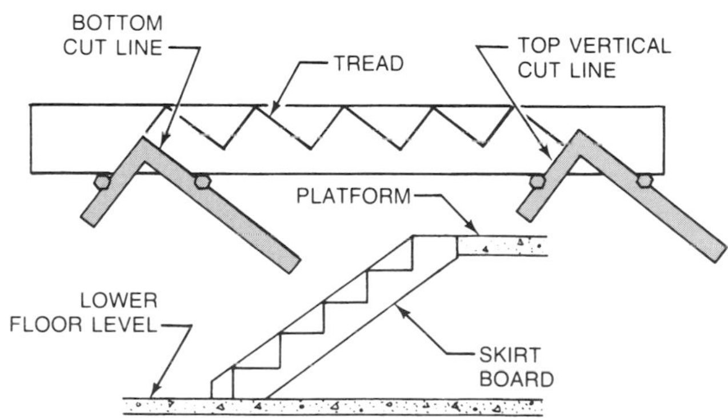

4. Place the steel square along the bottom edge of the skirt board and lay out the bottom and top cut lines.

Figure 5-59. Riser and tread dimensions are laid out on a skirt board. The skirt board is secured in position and riser form boards are fastened to it.

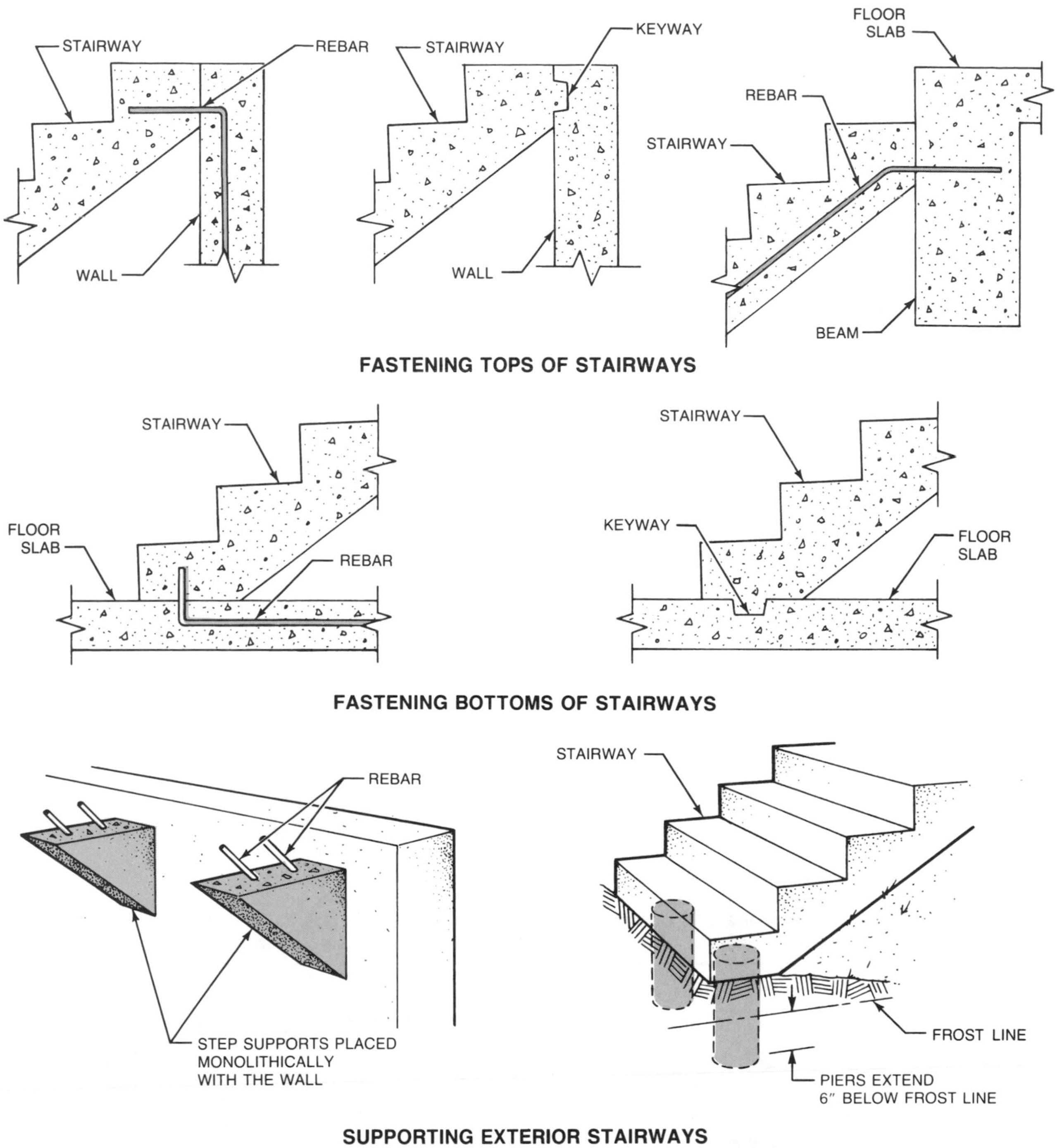

FASTENING TOPS OF STAIRWAYS

FASTENING BOTTOMS OF STAIRWAYS

SUPPORTING EXTERIOR STAIRWAYS

Figure 5-60. The tops and bottoms of concrete stairways are secured to walls or beams and floor slabs using rebar or keyways. Exterior stairways may require step supports or piers.

The bottom of an interior concrete stairway may be secured by rebar projecting from the floor slab or a keyway cut in the floor slab. The area beneath the bottom step of an exterior stairway resting on the soil should be excavated to load-bearing soil. For exterior stairways, small piers are placed below the first step if a frost line or unstable soil conditions are present.

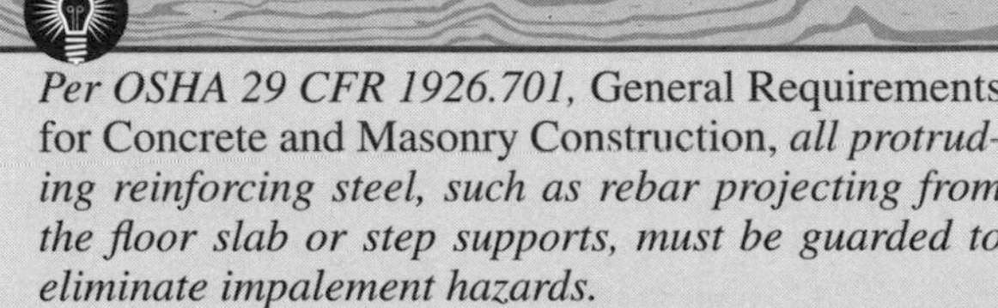

Per OSHA 29 CFR 1926.701, General Requirements for Concrete and Masonry Construction, *all protruding reinforcing steel, such as rebar projecting from the floor slab or step supports, must be guarded to eliminate impalement hazards.*

The riser of the stairway form is laid out with a ¾″ to 1″ slope to create a nosing at the front of each step. The tread slopes ⅛″ to ¼″ from front to back. **See Figure 5-61.** The bottom edge of the riser form boards should be beveled to facilitate troweling and finishing the treads.

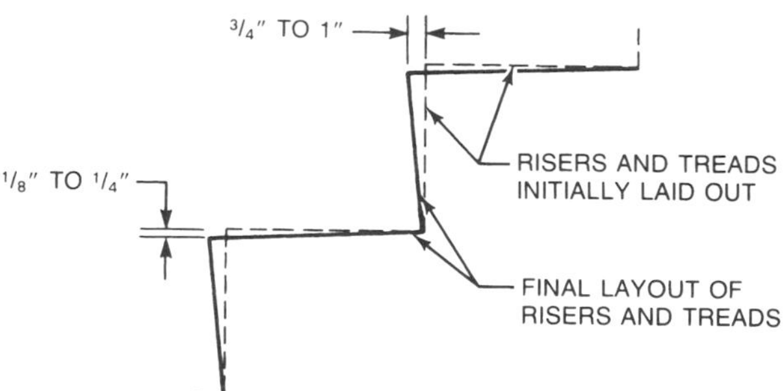

Figure 5-61. The riser is sloped to create a nosing at the front of the tread. The tread is sloped toward the front to facilitate water drainage.

Constructing Stairway Forms over Sloping Ground. Concrete stairways constructed over sloping ground receive their main support from the earth. Preliminary groundwork includes stepping the slope to prevent the fresh concrete from sliding and placing a well-compacted layer of gravel over poorly draining soil.

Two skirt boards (planks or plywood) are braced and staked to the ground. Riser form boards are secured by cleats nailed to the skirt board. The riser form boards may be end nailed through the skirt boards if 1½″ form boards are used. For wide stairs, braces are placed 4′-0″ OC between the skirt boards. **See Figure 5-62.**

Per the International Building Code, the riser angle of a stair must not be greater than 30° from the vertical.

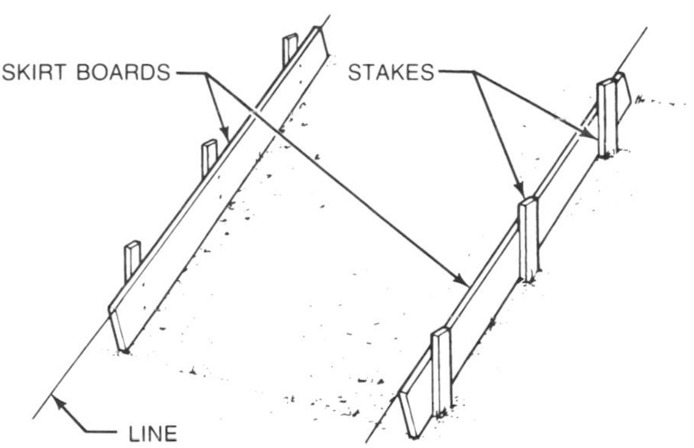

1. Drive stakes to support skirt boards to a line. Lay out the position of the top edge of the skirt boards on the stakes. Nail the skirt boards in position.

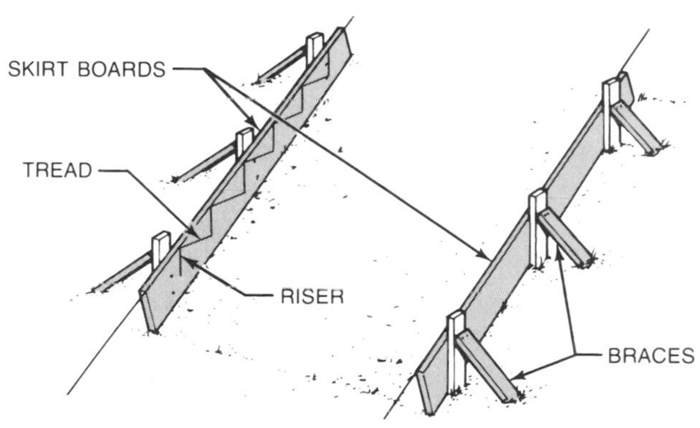

2. Align the skirt boards and brace the stakes. Lay out the treads and risers on the skirt boards.

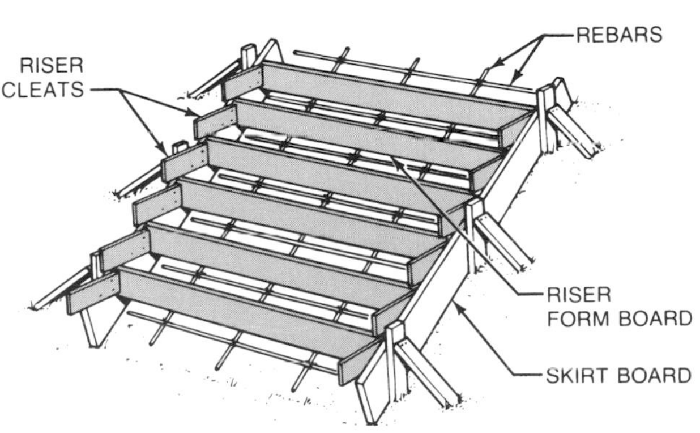

3. After rebar have been placed, nail riser cleats in position. Nail the riser form boards to the cleats.

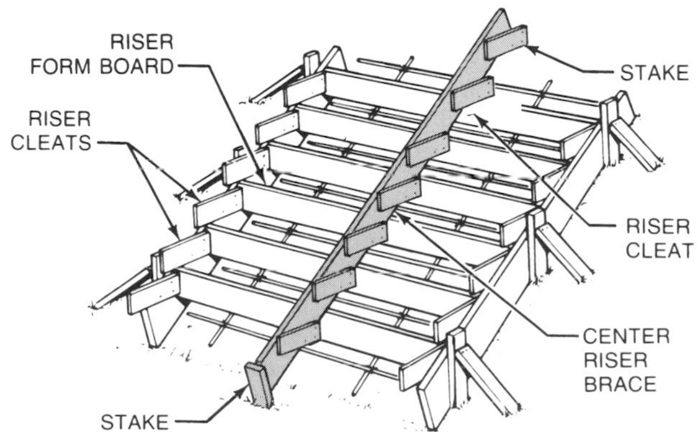

4. On wider stairways, reinforce the riser form boards with a center riser brace and cleats.

Figure 5-62. Two skirt boards are required to form a stairway over sloping ground. For wide stairways, a center riser brace and cleats are used to prevent distortion.

Constructing Open Stairways. Open stairways are commonly constructed inside a structure to provide access from one level to another. Where landings are required, a landing form is integrally constructed with the stairway form. An open stairway is constructed by setting up temporary panels along the form construction area. The treads and risers and slab thickness are laid out on the panels. The slab thickness is laid out perpendicular to the stairway angle. The thickness of the soffit panel sheathing, width of the joists, and width of the stringers are laid out. The width of the skirt board is determined by measuring from the end of a tread to the top of the joists. The length of the shores is determined by subtracting 3″ (1½″ sill thickness + 1½″ wedge thickness) from the total shore length.

The formwork for an open stairway includes plywood soffit panels supported by joists, stringers, and shores. The stringers are positioned and braced beneath the soffit panels, and the steps are formed with riser form boards secured by cleats. Bottoms of riser form boards are beveled to permit troweling and finishing the treads. **See Figure 5-63.**

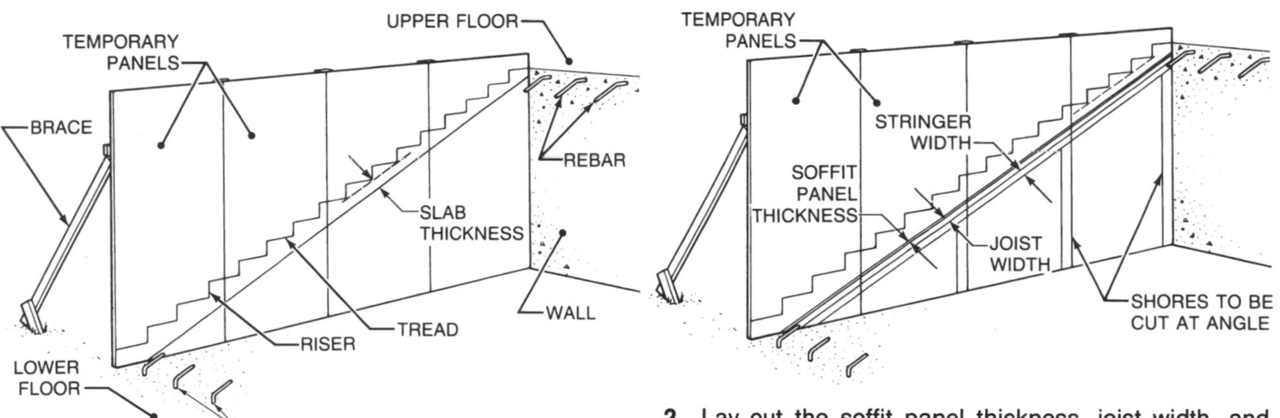

1. Set up and brace temporary panels along the stairway construction area. Lay out treads and risers on the panels. Measure the slab thickness at a right angle to the slope of the stairway and snap a line.

2. Lay out the soffit panel thickness, joist width, and stringer width and snap lines. Determine the shore length by measuring the distance from the lower stringer line to the floor and subtracting sill and wedge thickness. Determine the side form width by measuring at a right angle to the slope of the stairway.

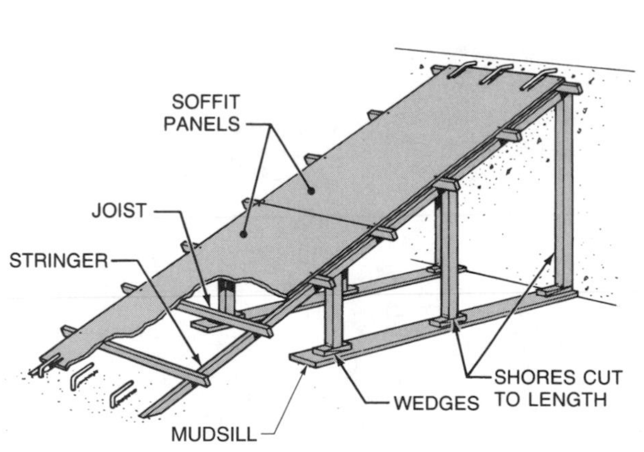

3. Cut shores to length and secure in position. Nail stringers to the tops of the shores. Nail joists to the tops of the stringers. Nail the soffit panels in position. Remove the temporary panels.

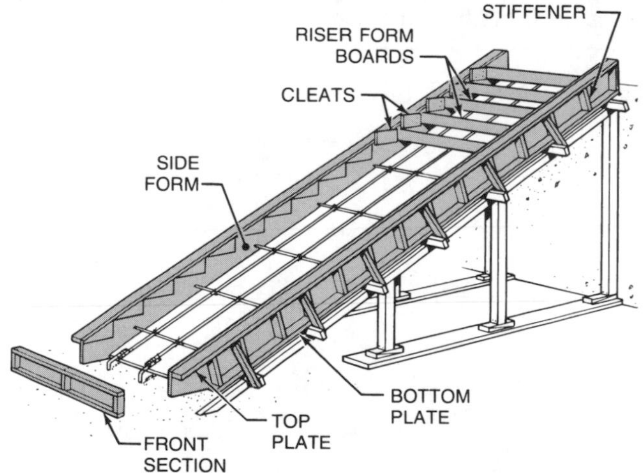

4. Lay out treads and risers on the side form. Nail top and bottom plates and stiffeners through the side forms. Fasten side forms to the top of the joists. Align and brace side forms. After the rebar have been placed, fasten cleats and riser form boards to side forms. Nail the front section into place.

Figure 5-63. Temporary panels are set up along the stairway construction area for the layout of open stairways. After the soffit panels are positioned, side forms are secured in position and riser form boards are nailed to the side forms.

Constructing Closed Stairways. Closed stairway construction is similar to open stairway construction. A soffit is supported by joists, stringers, and shores. However, the stringers and other form components are laid out on the two enclosing walls. The skirt boards are fastened to the walls and the riser form boards are secured with cleats nailed to the plank. See Figure 5-64.

HIGHWAY AND BRIDGE CONSTRUCTION

Highway construction includes the construction and maintenance of highway systems. Most of the paving and curbing of road surfaces is accomplished with mechanical slip form paving and finishing equipment. However, form construction is required for bridges, ramps, approaches, and overpasses.

Bridges are a means to cross over natural barriers such as rivers and canyons, and artificial barriers such as railroads and highways. Ramps and approaches allow vehicular traffic to merge smoothly onto the highway. Overpasses are bridges that are an integral part of highway systems that provide crossovers for traffic and entrance and exit ramps at intersections. **See Figure 5-65.** Formwork procedures and materials, such as ties, reinforcement, bracing, and shoring used in the construction of highway bridges and overpasses, are similar to other types of concrete structures.

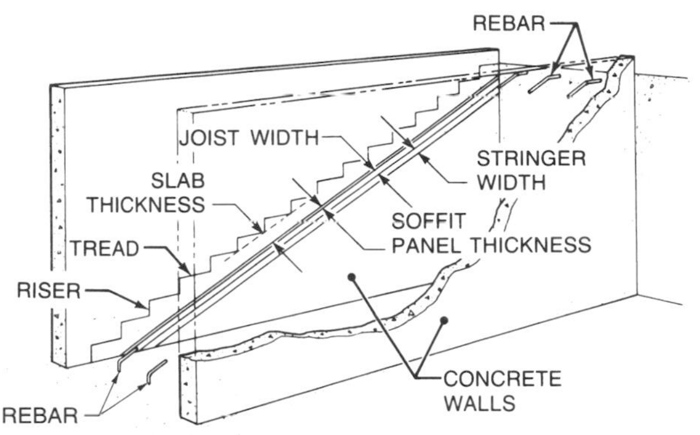

1. Lay out treads and risers on the concrete walls. Lay out and snap lines for the slab thickness, soffit panel thickness, joist width, and stringer width.

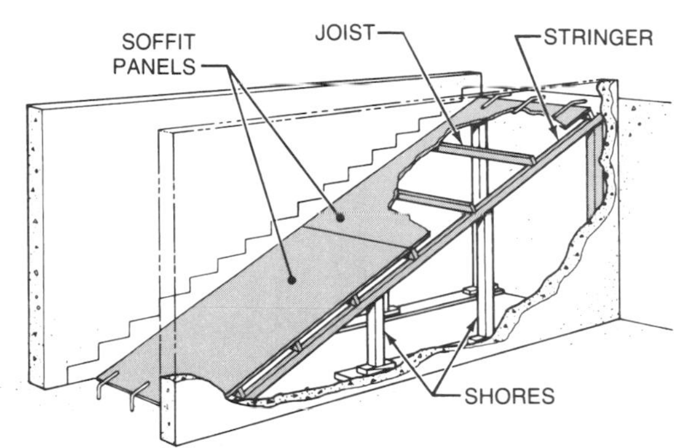

2. Lay out and cut shores to length. Construct the stairway soffit between the concrete walls.

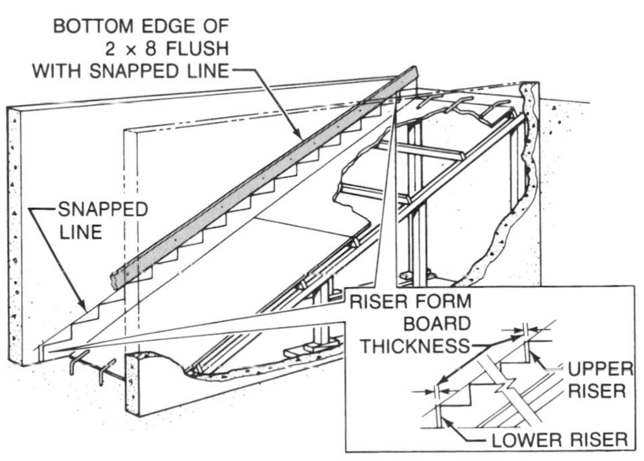

3. Lay out the riser form board thickness along the upper and lower risers. Snap a line from the outside corners of the riser form boards. Nail a 2 × 8 to the concrete walls with the bottom edge flush with the line.

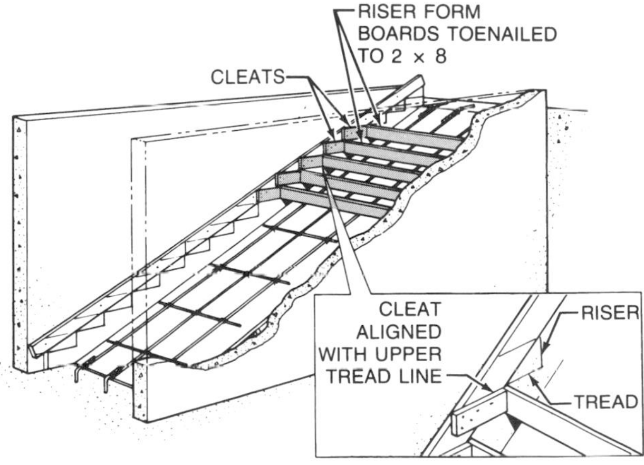

4. Mark risers and treads on the 2 × 8. Nail form board cleats to the riser and tread marks on the 2 × 8. After the rebar have been placed, toenail the riser form boards to the 2 × 8 and against the cleats.

Figure 5-64. Risers and treads are laid out on the enclosing walls when laying out a closed stairway.

Portland Cement Association

Figure 5-65. Bridges, ramps, and overpasses are an integral part of a highway system.

The structural components of bridges and overpasses are classified as part of the substructure or superstructure. The *substructure* is the footings, piers, pier caps, and abutments that support bridges, ramps, or overpass decks. The *superstructure* is the bridge deck, sidewalks, and parapets (low walls formed along the edges of the deck) of a highway system. **See Figure 5-66.**

Precast concrete members are commonly used in highway construction. Precast members are prefabricated in casting yards and delivered by truck to the job site. They are lifted in position by crane and tied together with steel dowels and/or welding plates.

Substructure

The footings of a substructure are located beneath the piers and abutments and provide a solid foundation for the bridge. An *abutment* is the end structure that supports the beams, girders, and deck of a bridge or arch. Piers extend from the footings and support the pier cap and superstructure. The pier cap directly supports and provides a larger bearing surface for the superstructure. **See Figure 5-67.**

Wing walls are commonly placed monolithically with the abutments. A *wing wall* is a short section of wall at an angle to the abutment used as a retaining wall and to stabilize the abutment. Grouped piles or friction piles are driven into solid bedrock where poor soil conditions exist. A pile cap is then constructed over the piles to transmit loads to the piles. Rebar extending from the piles are tied to the rebar in the pile cap.

Cofferdams are constructed to restrain water when constructing footing forms in rivers, lakes, and other bodies of water. A *cofferdam* is a large, rectangular, watertight enclosure constructed of interlocking sheet piling. The interlocking sheet piling is driven into the river bottom around the work area, and water is pumped out of the enclosure to permit access for the formwork.

Pier forms are constructed over footings after the concrete for the footings has set. Pier forms consisting of sheathing, studs, and/or walers secured with ties are similar to column forms for buildings. Tall piers may require climbing forms that move vertically for successive lifts. Single or multiple piers may be used to support the pier caps. Some pier designs extend the full width of the superstructure. Pier designs include round, square, rectangular, battered on two or four sides, and inverted-batter. **See Figure 5-68.** Some pier designs incorporate a tie beam (tie strut) midway between the footing and pier cap to tie two piers together. Piers may rest on and be joined to crash walls. Crash walls prevent structural damage caused by moving vehicles.

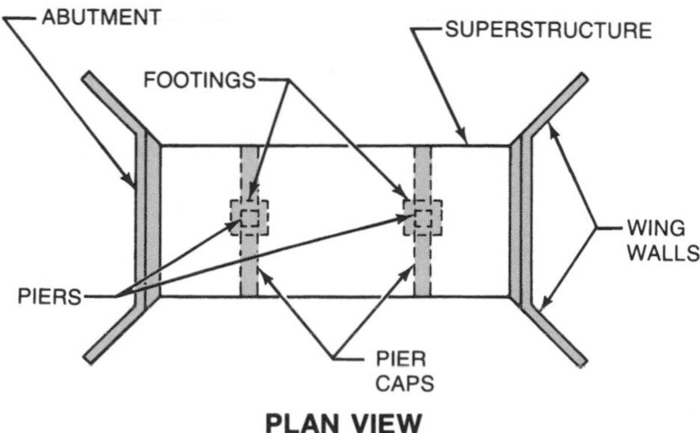

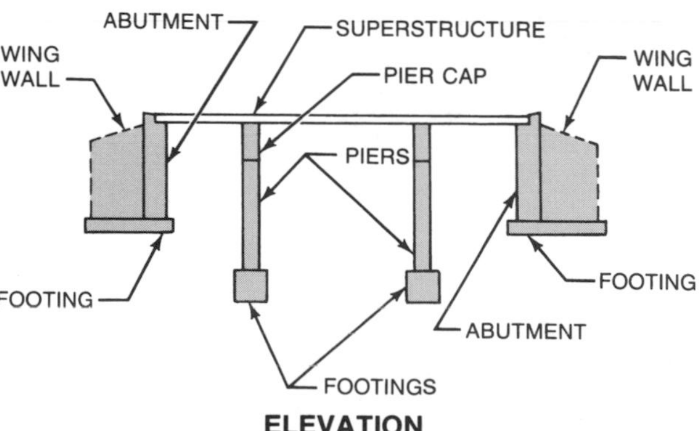

Figure 5-66. The major components of a bridge are classified as the substructure or superstructure. The substructure includes footings, piers and pier caps, abutments, and wing walls.

Figure 5-67. Piers support the pier cap, which directly supports and provides a larger bearing surface for the superstructure.

The type of forms used to form piers depends on the shape and height of the piers, and the number of similar piers to be formed. Custom-made forms are used to form multiple piers with the same design. Custom-made forms are constructed by form manufacturers to the exact dimensions of the piers. Custom-made forms are constructed of steel, metal-framed plywood panels, or all wood, and may be reused many times.

Piers and footings provide the main support for the bridge superstructure. Abutments and wing walls acting as earth-retaining walls support the ends of bridges. Abutments and wing walls are commonly constructed over large spread footings. The formwork for abutments and wing walls is similar to that used for buildings. The concrete for an abutment is placed in two lifts. The first lift forms the stem wall, and the second lift forms the head wall. **See Figure 5-69.** After the stem and head walls are constructed, concrete bridge seats are placed on top of the stem wall to support the superstructure girders.

Pier cap forms may be constructed over the completed piers or the concrete for the pier caps can be placed monolithically with the piers. Pier caps support the full width of the bridge superstructure. The hammerhead pier cap (T-cap) rests on a single round or rectangular pier. Pier caps serve as tie beams when they are supported by two or more piers. **See Figure 5-70.**

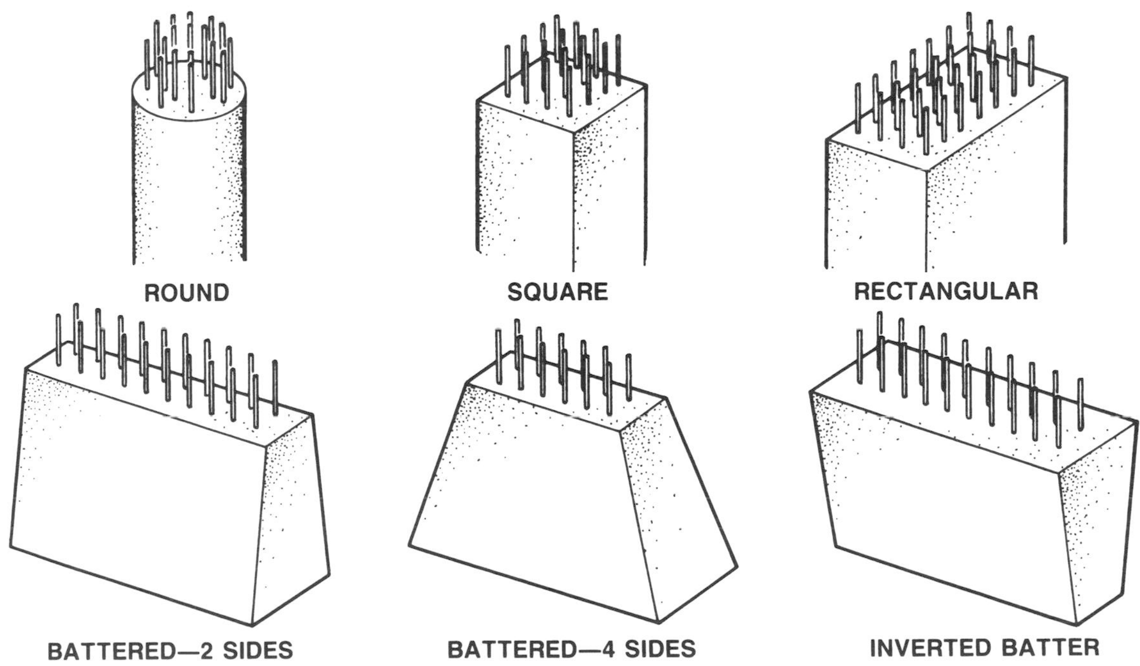

Figure 5-68. Common bridge pier designs are round, square, rectangular, battered on two or four sides, and inverted batter. Rebar extends from the top of the pier tie into the pier cap.

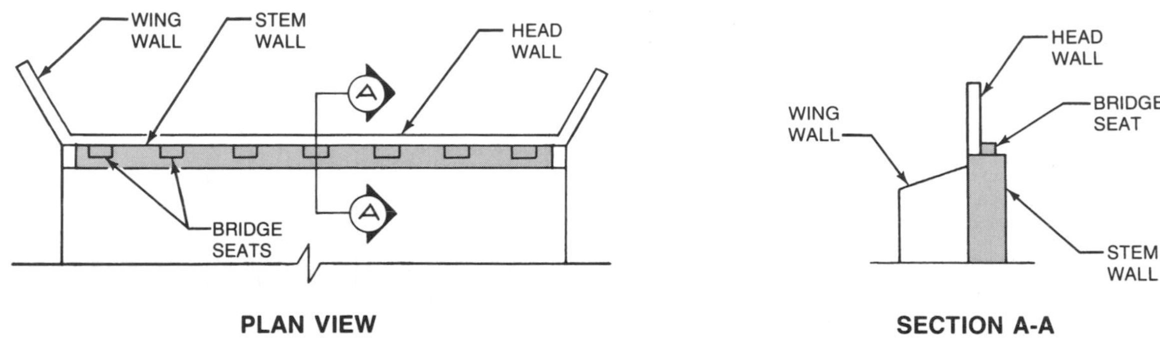

Figure 5-69. An abutment consists of a head wall and a back wall. Abutments and wing walls are earth-retaining walls that support the ends of bridges.

Symons Corporation

Figure 5-70. Pier caps are secured to the top of the piers and support the full width of the bridge superstructure.

Pier cap forms consist of two sides, a soffit, and bulkheads at the ends. The formwork is reinforced with studs and/or walers with internal disconnecting ties extending between the opposite walls. Lift plates are secured to the top of the forms for the crane attachment. Large pier cap forms are fitted with metal walkway brackets to support a work platform. Wood uprights are inserted into the ends of the walkway brackets and a safety railing is nailed to the uprights.

Friction clamps or screw jack support brackets support pier cap formwork over the piers. Friction clamps support small pier cap formwork placed over round columns. A screw jack support bracket supports pier cap formwork for a rectangular pier. Pier cap formwork may also be supported by the existing formwork on the pier. **See Figure 5-71.** Heavy pier cap formwork is supported by steel beams secured to wide-flange beam inserts that are embedded in the pier. Scaffold or timber shoring may also be required for additional support.

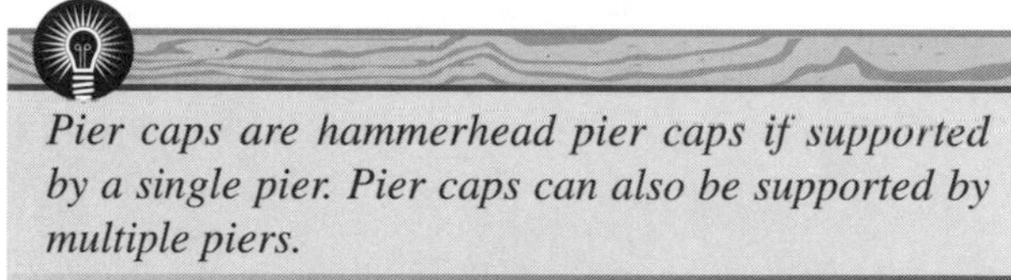

Pier caps are hammerhead pier caps if supported by a single pier. Pier caps can also be supported by multiple piers.

Superstructure

The superstructure of a bridge includes the bridge deck, curbs, sidewalks, and parapets. The *deck* is the concrete surface that supports the traffic load. Curbs form the edge of the deck and are used to direct the flow of rainwater. Sidewalks commonly border the curbs and are used to support pedestrian traffic. *Parapets* are short walls that act as a safety barrier along the edge of the superstructure.

A bridge deck for short spans consists of a concrete slab resting directly on the pier caps. Long spans may require slabs to be reinforced by concrete or steel girders. Bridge decks are cast-in-place or precast concrete and are supported by steel or precast concrete girders.

Cast-in-place bridge deck construction requires formwork attached to the pier caps and/or girders. Forms for bridge decks may be built in place in a manner similar to constructing formwork for a floor slab. Plywood or metal panels are used for deck sheathing. The deck sheathing is supported by joists and stringers secured in place by wood or metal shores. The joists and deck sheathing must extend beyond the deck width to allow for edge forms, bracing, walkways, and guardrails. **See Figure 5-72.** The edge form should be at the same elevation as the top of the slab if low parapets are constructed with slipforming machines. Wall forms are not required when using slipforming equipment and the concrete is placed over rebar extending from the bridge deck. When constructing high parapets, an outside wall form is used as the edge form.

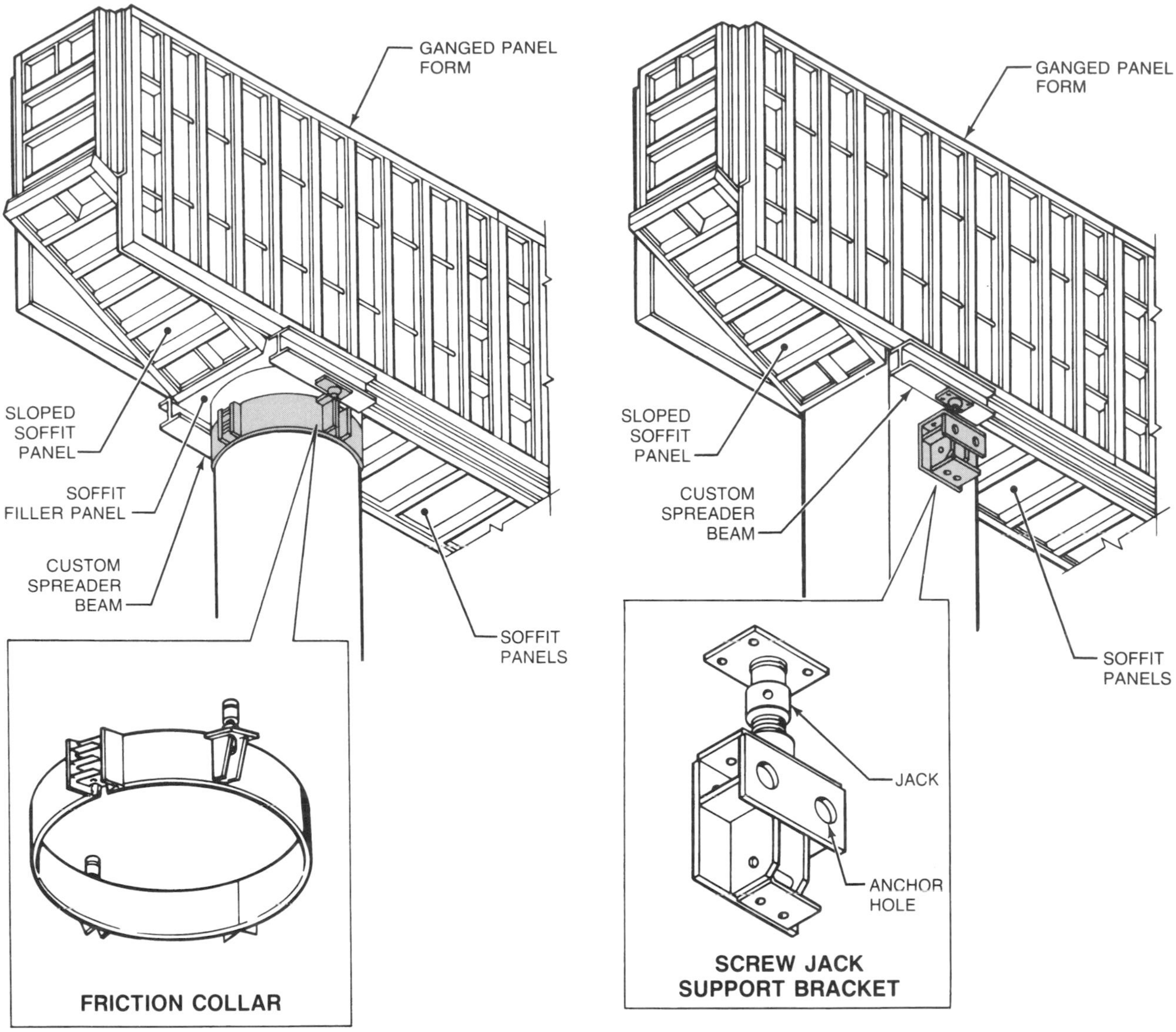

SUPPORT SYSTEMS FOR PIER CAP FORMS

Figure 5-71. Friction collars, screw jack support brackets, and existing pier formwork are used to support pier cap forms.

Symons Corporation

Friction collars are installed before the pier cap to provide additional support for the pier cap form.

Shoring for bridge deck formwork is similar to shoring floor slabs for upper levels in concrete buildings. In many cases, bridge deck shoring extends to greater heights and must be designed for traffic movement below. Wood and metal scaffold shoring placed over sills is used to support high bridge formwork. The shoring is braced horizontally and diagonally in both directions at every tier.

Cast-in-place bridge decks are frequently placed over steel or precast girders. The girders are spaced close together and rest on pier caps. Temporary wooden joists are secured between the girders, and the deck panels are placed over the joists. After the rebar is positioned, the concrete is placed over the plywood panels.

Portland Cement Association

Figure 5-72. Bridge deck formwork is supported by pier caps or girders. Bridge deck sheathing must extend beyond the deck width to support walkways and guardrails.

A more recent development is all-steel bridge deck-forming systems. The steel forms are adjustable for width and are placed by crane between the girders. The forms are then secured with adjustable pipe struts that rest on the bottom flanges of the steel or concrete girders. The forms are also adjustable for different girder depths and types. This method eliminates the cost of fitting joists and plywood between the girders, and the steel form can be used over and over again.

OSHA 29 CFR 1926.704, Requirements for Precast Construction, *provides information regarding lifting inserts and associated hardware. Lifting inserts that are embedded or otherwise attached to precast members must be capable of supporting at least four times the maximum intended load.*

Precast, Prestressed Concrete Bridges

In a precast, prestressed concrete bridge, all the superstructure components of the bridge are formed in a casting plant and delivered to the job site. Precast methods have largely replaced cast-in-place concrete for short- to medium-span bridges (20′ to 130′) on low-volume roads. Precast bridges can be constructed much faster and the work can proceed during cold weather. However, cast-in-place concrete has advantages when erecting larger and more complex bridge structures with curves, flares, and multiple spans.

Western Forms

Precast bridge components, including piers and pier footings, are cast at a casting plant and delivered to the job site.

There are a variety of designs for prestressed bridge components. One very successful method was originally developed for rural areas in the Pacific Northwest and has now spread to other parts of the country. This system of standard bridge sections was jointly developed by the American Association of State Highway and Transportation Officials (AASHTO) and the Precast/Prestressed Concrete Institute (PCI).

The basic design of a standard bridge section consists of an integral deck combined with a prestressed concrete beam. The sections are tied together with weld plates and intermediate diaphragms. **See Figure 5-73.** A *diaphragm* is a short wall placed between the sections. The opposite ends of the bridge are supported by abutments or pilings. The abutments may be entirely cast in place, or may be a precast unit resting on cast-in-place footings.

In a more recent development, the sections are tied together with post-tensioned cables placed in tubes provided in the sections. This method supports more weight and allows a higher volume of traffic than the sections joined with weld plates and diaphragms.

Precast concrete piers are a recent development. Load-bearing capacity of precast concrete is comparable to cast-in-place concrete, but less forming, placement, and form removal time is required.

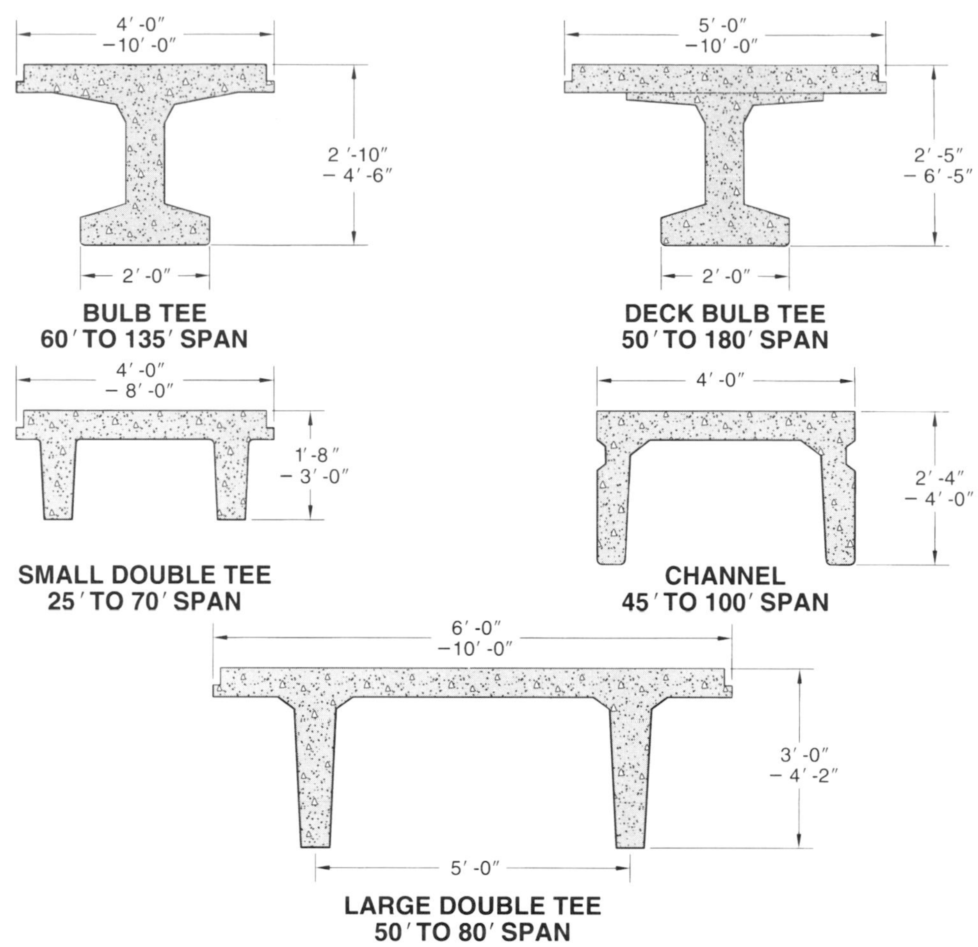

Figure 5-73. Precast, prestressed concrete sections with integral decks are tied together with weld plates and diaphragms.

Heavy Construction

Review Questions

5 CHAPTER

Name ______________________________ **Date** ______________

Completion

__________ **1.** T-foundations and ___ beam and pile foundations support heavy concrete structures.

__________ **2.** Mat and raft foundations are types of ___ foundations.

__________ **3.** A(n) ___ beam is a reinforced beam running along the surface of the ground that does not rest on supporting piers or piles.

__________ **4.** ___ tower cranes are set in position during foundation construction and move upward as the height of the building increases.

__________ **5.** A(n) ___ is a long structural member that penetrates deep into soil.

__________ **6.** The tip of a concrete pile is protected from damage with a pile ___ while it is being driven into the ground.

__________ **7.** A(n) ___ pile is driven completely through unstable soil layers and rests on firm, load-bearing soil.

__________ **8.** In the ___ method of cast-in-place piles, a metal casing is driven into the ground and remains in place while the concrete is being placed.

__________ **9.** The pile ___ is the upper surface of a precast pile in its final position.

__________ **10.** ___ are used in heavy construction projects where the building design and/or soil conditions make pile driving difficult or inadequate.

__________ **11.** A(n) ___ caisson provides a large bearing area at the base of the caisson.

__________ **12.** ___ piles are placed beneath a concrete cap and are a base for load-bearing columns.

__________ **13.** A(n) ___ is the concrete placed between horizontal construction joints.

__________ **14.** ___ panel forms are constructed by bolting many small panels together.

__________ **15.** Short-radius curves are formed by using two or more layers of ___″ plywood.

__________ **16.** Large panel forms and ganged panel forms are types of ___ forms.

__________ **17.** A(n) ___ joint is formed when the concrete of a wall section is placed on top of a previously placed concrete wall section.

__________ **18.** ___ consists of placing shores beneath beams and floor slabs immediately after formwork has been stripped away.

__________ **19.** A(n) ___ is embedded in a concrete wall to prevent water leakage at a vertical construction joint.

__________ **20.** The wall form panel for an upper lift should overlap the hardened concrete for the lower lift approximately ___″.

__________ **21.** ___ joints are shallow grooves made in a concrete wall to control cracking.

__________ **22.** ___ support beams and other bending loads.

__________ **23.** A(n) ___ wall is a light non-load-bearing section of a wall made of metal or precast lightweight concrete.

__________ **24.** The floor slab in a flat-slab system is directly supported by columns and ___ panels.

__________ **25.** A(n) ___ is a flared section at the top of a column.

__________ **26.** Columns tie directly into the floor slab above without using capitals or drop panels in a flat-___ system.

__________ **27.** ___ strips are placed in the corners of column forms to produce columns with beveled edges.

__________ **28.** Round columns are formed using tubular ___ forms.

__________ **29.** ___ are the frames for the door and window openings, and beam pockets where required.

__________ **30.** The sheathing material for beam and girder side forms is ___.

__________ **31.** A(n) ___ is nailed against the bottoms of beam and girder side forms to secure them in position.

__________ **32.** A(n) ___ tie is a snap tie with a hooked end that is driven into deck sheathing at the top of the form.

__________ **33.** A(n) ___ pan is secured with its flanges nailed to the top of the soffit.

__________ **34.** A(n) ___ is an engineered prefabricated form that consists of a wood deck and an aluminum frame system.

__________ **35.** ___ are placed under structural members after the original shoring has been removed.

__________ **36.** ___ stairways are enclosed by walls on both sides.

__________ **37.** The vertical surface of a step is the ___.

__________ **38.** The ___ of a bridge includes the footings, piers, pier caps, and abutments.

__________ **39.** A(n) ___ is constructed to restrain water when constructing footing forms in rivers, lakes, and other bodies of water.

__________ **40.** A bridge ___ is the concrete surface that directly supports the traffic load.

Multiple Choice

__________ **1.** Two-way joist systems are constructed with ___ pans.
A. long
B. adjustable
C. dome
D. all of the above

__________ **2.** A ___ is placed on top of a pile head to receive the pile-driving rig's blows and protect the head from damage.
A. driving shoe
B. pile cutoff
C. driving head
D. pile foot

__________ **3.** ___ steel piles are used as a foundation support and for shoring around deep excavations.
A. Bearing
B. H-shaped
C. Tubular
D. Friction

__________ **4.** When placing a concrete wall in two lifts, a row of tie rods is embedded ___″ below the top of the lower lift.
A. 4
B. 6
C. 8
D. 10

__________ **5.** When placing a concrete wall in two lifts, the bottom row of tie rods is embedded ___″ above the joint.
A. 4
B. 6
C. 8
D. 10

__________ **6.** Flying forms are usually used for the placement of concrete for ___ and ___.
A. beams; walls
B. floor slabs; walls
C. floor slabs; beams
D. beams; piers

__________ **7.** ___ pans for one-way joist systems are nailed to the sides of the soffit.
A. Dome
B. Nail-down
C. Adjustable
D. Long

__________ **8.** ___ pans are used to form one-way joist systems.
A. Flanged adjustable
B. Long
C. Nail-down
D. all of the above

__________ **9.** A ___ is used to secure the two members of a two-piece adjustable wood shore.

A. cleat
B. post clamp
C. brace
D. stake

__________ **10.** Wood shores placed over poor load-bearing soil should rest on ___.

A. 3/4″ plywood
B. 2 × 10 planks
C. 3 × 10 mudsills reinforced with 4 × 4 or 4 × 6 plates
D. 2 × 6 plates

__________ **11.** Splicing cleats for wood shores extend a minimum of ___″ on each side of a splice.

A. 6
B. 8
C. 10
D. 12

__________ **12.** ___ shores remain in place as formwork around them is removed.

A. Permanent
B. T-head
C. Scaffold
D. Adjustable metal

__________ **13.** In a stairway, the ___ is the vertical distance from a floor to the floor above.

A. total run
B. total rise
C. unit run
D. unit rise

__________ **14.** The ___ of a bridge includes the bridge deck, sidewalks, and parapets.

A. superstructure
B. substructure
C. abutment
D. none of the above

__________ **15.** A(n) ___ supports the ends of a bridge deck or arch.

A. pier
B. wing wall
C. abutment
D. cofferdam

__________ **16.** A ___ wall is a retaining wall used to stabilize an abutment.

A. head
B. stem
C. back
D. wing

__________ **17.** Slip forms move up jackrods as the concrete is being placed at rates ranging from ___″ to ___″ per hour.

A. 2; 20
B. 2; 70
C. 4; 45
D. 4; 70

__________ **18.** ___ are low walls that act as a safety barrier along the edge of a bridge superstructure.

A. Wing walls
B. Parapets
C. Head walls
D. Bridge seats

__________ **19.** ___ support pier cap formwork for bridge piers.

A. Friction clamps
B. Screw jack support brackets
C. Pier forms
D. all of the above

__________ **20.** ___ are placed on top of the stem wall to support the bridge superstructure girders.

A. Crash walls
B. Head walls
C. Bridge seats
D. Pier caps

Identification

__________ **1.** Beam bottom

__________ **2.** Temporary tie

__________ **3.** Shore head

__________ **4.** Stud

__________ **5.** Stiffener

__________ **6.** Beam side form

__________ **7.** T-head shore

__________ **8.** Brace

__________ **9.** Chamfer strip

__________ **10.** Joist ledger

__________ **11.** Kicker

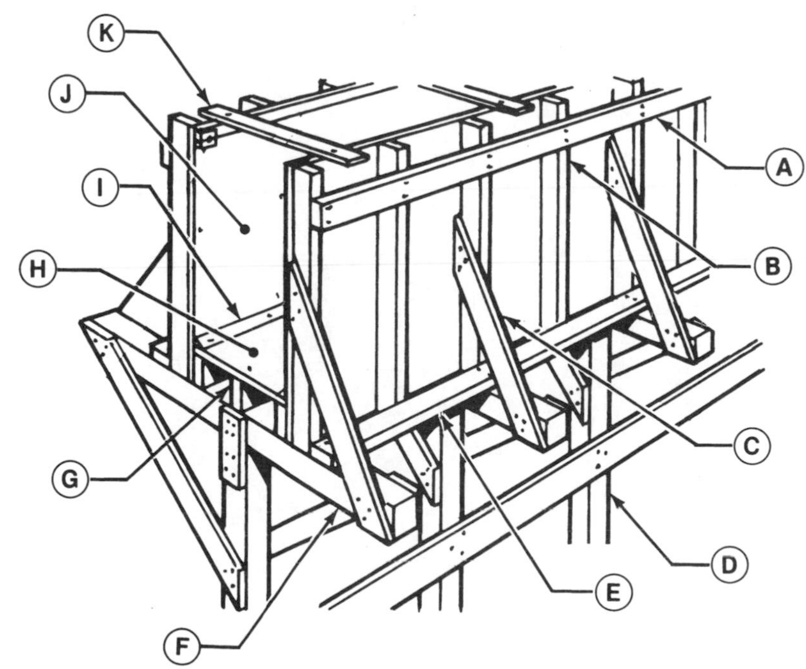

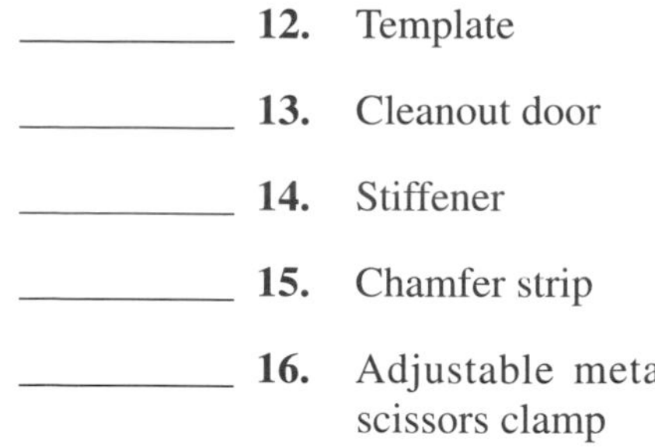

_________ **12.** Template

_________ **13.** Cleanout door

_________ **14.** Stiffener

_________ **15.** Chamfer strip

_________ **16.** Adjustable metal scissors clamp

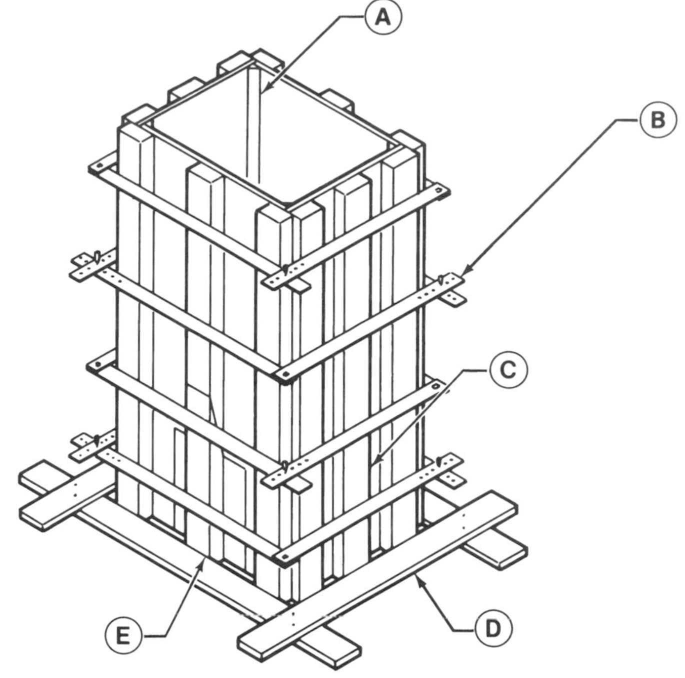

_________ **17.** Pile head

_________ **18.** Pile shoe

_________ **19.** Tip

_________ **20.** Foot

_________ **21.** Driving head

_________ **22.** Butt

_________ **23.** Pile cutoff

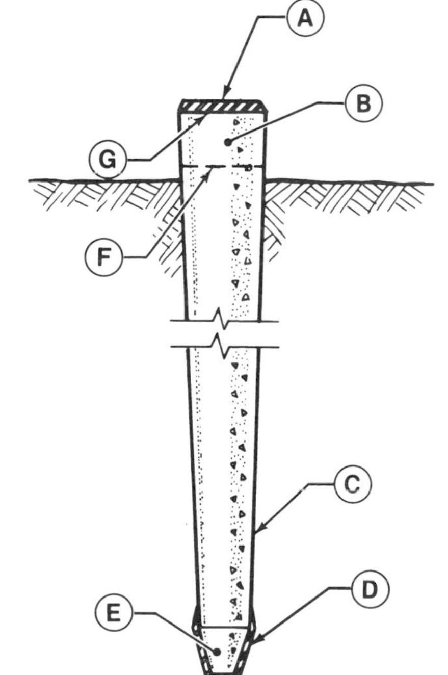

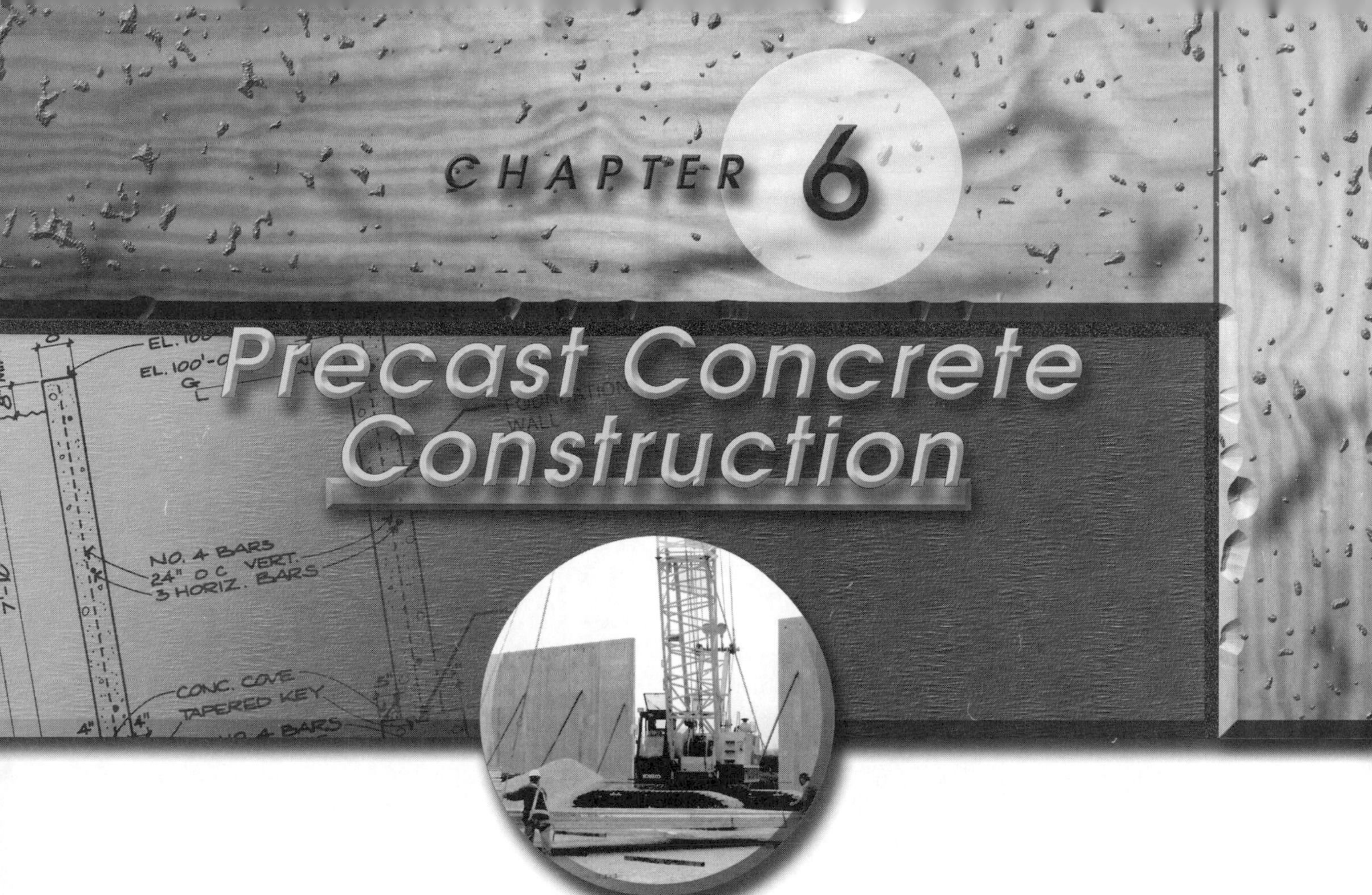

CHAPTER 6 Precast Concrete Construction

Precast concrete systems have gained wide acceptance in recent years. Many modern concrete structures are partially or completely constructed with precast concrete members. Precast concrete members are also used in the construction of bridges, tunnels, and wharves.

Precast concrete construction is advantageous in many situations. Formwork costs are reduced because fewer and simpler forms are required, less scaffolding is used to support the precast members, and production schedules are not affected as much as schedules for cast-in-place concrete.

Precast structural members are fabricated in a factory and transported to the job site by truck, or fabricated on the job site. The design and quantity of precast members, location of the casting factory, and cost of transportation are considered when determining whether to cast the precast members at a factory or on the job site.

Precast members are fabricated by placing reinforcement and concrete into forms constructed over casting beds. When the concrete has set, the precast members are raised and positioned by crane. The precast members are braced in position and tied together using steel dowels and/or welding plates.

Rebar or prestressed steel cables are used to reinforce precast concrete members. Prestressed steel cables are used to reinforce precast concrete members by pretensioning or post-tensioning.

PRESTRESSED CONCRETE

Prestressed concrete is concrete in which internal stresses are introduced to such a degree that tensile stresses resulting from service loads are counteracted to the desired degree. The prestressed concrete is held in a state of compression by stressing the concrete with high tensile steel cables. Prestressed concrete members may be precast or cast-in-place. **See Figure 6-1.** Two methods used to prestress concrete are pretensioning and post-tensioning. *Pretensioning* is a method of prestressing in which the steel cables are tensioned before the concrete is placed in the casting bed. *Post-tensioning* is a method of prestressing in which the steel cables are tensioned after the concrete has been placed.

Pretensioned concrete is commonly used for factory-produced members. Powerful jacks stretch the cables until they are under the required tension for a particular structural member. Concrete is then placed in the form. When the concrete has set to its specified strength, the tension from the jack is released. As the cables return to their original state, the concrete member is placed under compression, resulting in greater resistance to lateral loads and pressures than conventionally reinforced concrete. Prestressed members require less concrete and are lighter than members reinforced with rebar.

Post-tensioned members are commonly formed at the job site. Concrete is placed around unstressed cables that are enclosed in flexible metal or plastic ducts. The cables are stressed and anchored at both ends after the concrete has been placed and set. Post-tensioning is also used with cast-in-place concrete.

Portland Cement Association

Post-tensioning cables can be used in cast-in-place concrete and are commonly set with the rebar in the concrete slab before concrete is placed.

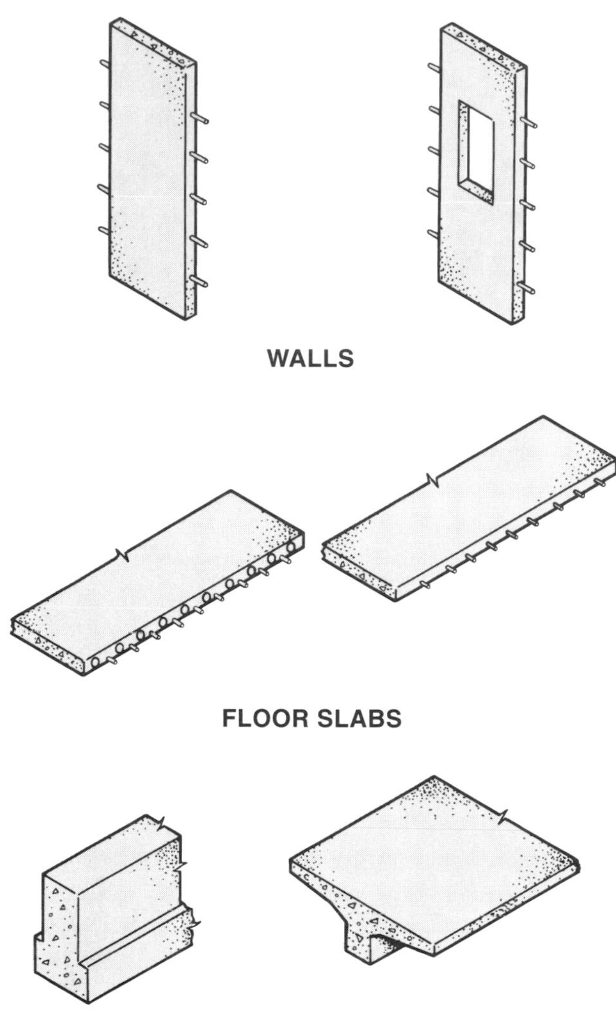

Figure 6-1. Precast members for precast concrete structures include walls, floor slabs, and beams.

Job Site Precast Construction

Job site precast construction lowers costs and increases productivity when producing multiple precast units of a design. The time required for precast formwork is considerably less than for cast-in-place concrete because the forms are constructed on casting beds built at ground level. Positioning of rebar, and placing, consolidating, and finishing the concrete are easier. Precasting of structural members is done during the construction of the foundation. Precast members can carry full loads, thereby eliminating the need for temporary shoring. **See Figure 6-2.**

Job site precasting is used when there is adequate space for a casting yard within or adjacent to the job site, and when lifting equipment (cranes or hoists) is available. A casting yard must be well organized for efficient assembly line production. Materials used to build the forms, the reinforcing steel, and an area for stockpiling completed precast members should be nearby. If concrete is mixed on the job site, an adjacent area for storing cement, aggregate, and water is required.

CONCRETE PLACED IN CASTING BEDS

PANELS RAISED AND SET IN PLACE BY CRANE

Portland Cement Association

Figure 6-2. Entire structures may be constructed of precast wall panels and floor slabs that are cast at the job site. When the concrete has set, they are raised into position with a crane.

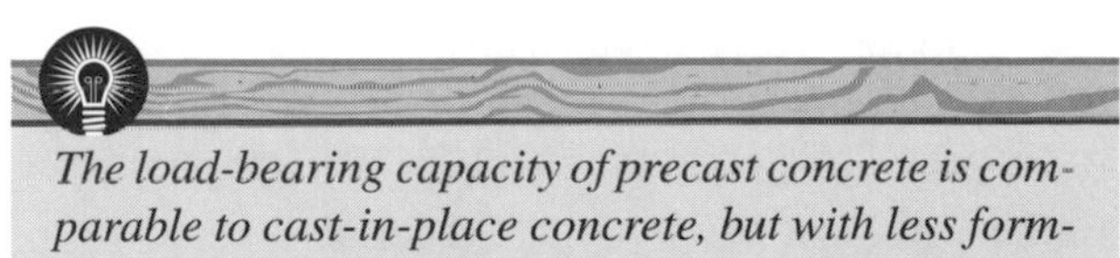

The load-bearing capacity of precast concrete is comparable to cast-in-place concrete, but with less forming, placement, and form removal time required.

Casting Beds and Forms

Casting beds are a base and support for the forms. The surface of a casting bed is smooth, level, and free from defects. Casting beds must be rigidly supported to prevent deflection from the weight of the concrete. A casting bed is constructed of plywood panels supported by timbers or a concrete slab. Precast wall and floor panels require edge forms around the perimeter of the casting bed. Wood casting bed forms are commonly lined with plastic or hardboard to facilitate stripping and increase the life of the form. **See Figure 6-3.**

Prefabricated metal and plastic forms are also available for standard structural units, or they can be custom-made to different dimensions. Prefabricated metal and plastic forms have a long life expectancy and can be reused many times.

REBAR POSITIONED PRIOR TO PLACING CONCRETE

FINISHING CONCRETE—BOLTS POSITIONED IN INSERTS

Figure 6-3. Prefabricated casting beds are commonly used to form precast concrete members. Reinforcement is placed in the beds prior to the concrete placement.

Inserts, anchors, and lifting units are placed in precast members and used for the crane attachment. **See Figure 6-4.** Threaded or coiled inserts are positioned below the surface of thin panels. Plastic caps are screwed into the insert while the concrete is placed. After the concrete has set, the plastic caps are removed and replaced with bolts to hold the lift plate. Anchors are used to raise the precast panels when the concrete surface is not visible. The anchors are embedded in the concrete with the opening for the crane attachment protruding from the surface. After the panel is raised, the anchor is left in place. Lifting units are used on heavy panels and usually have a steel bar that is recessed in the precast member. The lifting unit has a flexible vinyl piece that, when removed from the precast member, creates a semicircular impression. A clutch head engages the embedded steel bar to a shackle that is hooked to the crane attachment point. After the precast member is positioned and released by the crane, the lift plates and bolts are removed and the bolt holes are patched with a sand and cement grout mixture. Precast members and tilt-up panels must be adequately supported to prevent overturning or collapse before connections are completed.

Precast forms are designed for easy stripping. The side sections are hinged or bolted so they can be folded down or removed easily. The forms must also be oiled or treated with a release agent to facilitate stripping. After the concrete has set, the precast member is lifted out of the form by crane.

External vibrators are used for precast concrete applications. Vibrators should be spaced as equally and symmetrically as possible over the formwork to ensure even distribution of the vibration. A test run should be performed to determine the exact quantity and location of external vibrators. More small vibrators are recommended over a few large ones.

Connecting Methods. Precast members must be connected, or anchored, to adjoining parts of the structure after they are positioned. Steel plates or angles, dowels, tensioning bars, and bolts are commonly used to connect precast members. A column resting on a footing is often secured with bolts placed in the footing. The bolts align and extend through steel plates attached to the column base. Beams resting on the top of columns are tied together by welding steel angles to tensioning bars extending from the ends of the beams. Beams are also secured with steel dowels projecting from the column below. The steel dowels slip into steel tubes embedded in the beam and the tubes are filled with grout.

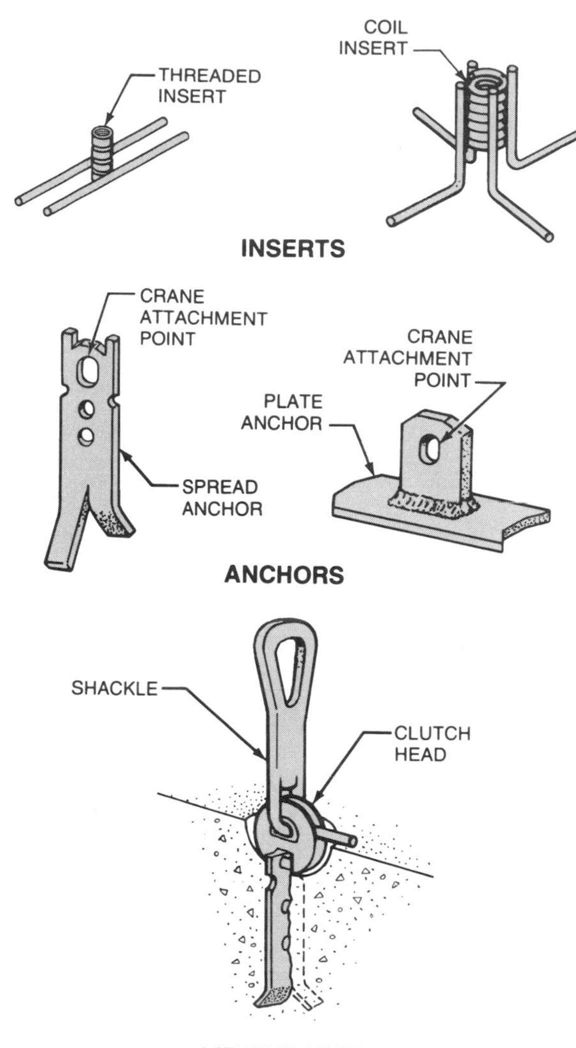

Figure 6-4. Inserts, anchors, and lifting units are embedded in precast members and used for the crane attachment.

Precast members can also be connected with a post-tensioning cable that is stressed and secured into position with a washer after placing grout between the column and beam end. Floor slabs are commonly anchored to precast beams or walls with welding plates. Steel angles that are embedded in the precast beam are welded to the plates. The plates are welded to rebar running through the floor slab. Steel plates can also be embedded in the slabs and welded to a steel connection plate placed across the precast beam. **See Figure 6-5.**

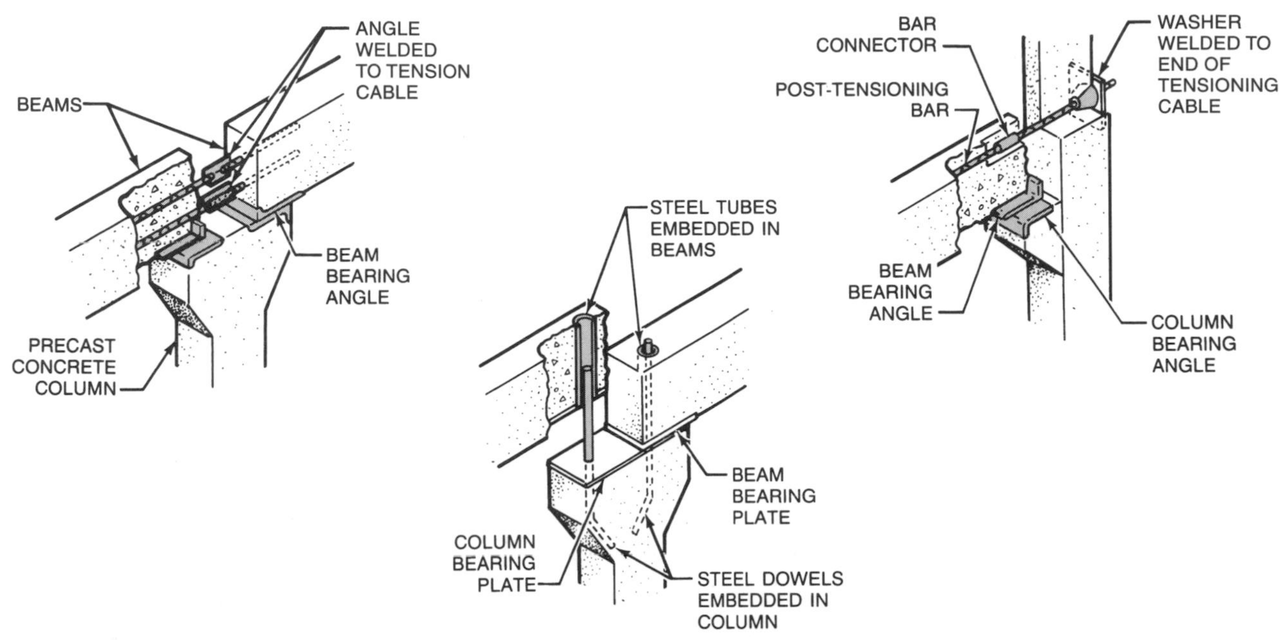

PRECAST BEAM AND COLUMN CONNECTIONS

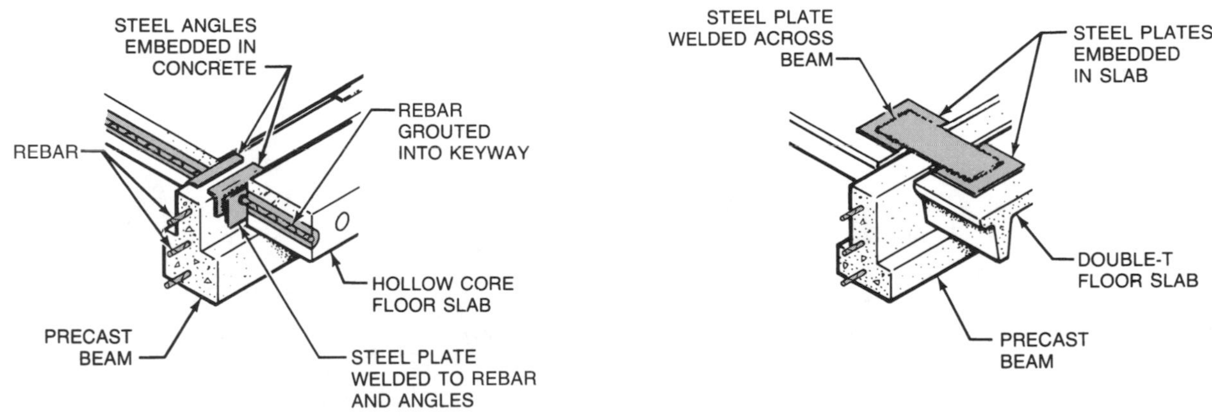

PRECAST FLOOR SLAB AND BEAM CONNECTIONS

Figure 6-5. Steel welding plates, angles, dowels, and tensioning cables are used to connect precast members.

TILT-UP CONSTRUCTION

Tilt-up construction is a major branch of precast concrete construction. Over 10,000 buildings are constructed annually because of the low cost, low maintenance, and speed of construction of tilt-up, and because of the long-lasting durability of the building. These advantages apply mainly to buildings greater than 10,000 sq ft with 20′ or higher side walls.

Tilt-up planning begins with an evaluation of the building site, which includes factors such as slab placement and the movement of materials on or around the slab. Proper engineering is crucial for the floor and wall designs, and should be designed by a qualified engineer.

Floor Slabs

In tilt-up construction, a concrete slab serves as the casting bed for wall panels and is often the ground floor slab of the building. However, if conditions require that the panels must be cast outside of the building, a temporary concrete slab, wood platform, or well-compacted fill is used as a casting bed.

Floor slabs used as casting beds must be level and have a smooth trowel finish. The slab, as well as the compacted sub grade below, must be strong enough to support material trucks and mobile cranes. In some cases, loads may exceed building occupancy loads. Slabs 5″or 6″ thick are most often used unless structural requirements require a thicker one. Pipe or utility openings in the slab should be temporarily filled with sand and topped off with a ¾″ layer of concrete.

Form Release Agents. A bond breaking agent is sprayed on the floor slab before the wall panels are constructed to ensure a clean lift when they are raised. Chemical, resin, and wax-based compounds that act as a bond-breaking agent and also as a curing compound are commonly used. A second coat of compound is applied after the edge forms have been constructed but before rebar and inserts are placed.

Tilt-up Formwork

The first step in constructing the perimeter forms is to lay out the edges of the wall and snap lines on the floor slab. The perimeter forms will be set to these lines. Tilt-up formwork consists of 2″ thick members placed on edge and fastened to the floor slab. The width of the planks varies with the thickness of the wall. The edge form planks can be secured by laying other planks behind the edge form and bolting or pinning the flat planks to the floor slab. The bottom of the perimeter form plank is then nailed to the flat plank, and the top of the perimeter form is secured with short 1 × 4 braces. Another method is to brace the edge forms with triangular plywood braces nailed to short 2 × 4 plates that are secured to the floor slab. Metal brackets are also available for securing edge forms. Manufactured metal units are also used for edge forms. They are designed with predrilled nail holes that allow the edge forms to be temporarily nailed into the base concrete slab. **See Figure 6-6.**

After the edge forms for the concrete panels have been built and secured to the floor slab, window and door bucks are fastened into position. Rebar is then positioned. The size and spacing of the rebar are based on the dimensions of the wall and the anticipated vertical and lateral loads. The areas around door and window openings and the edges of the wall are more heavily reinforced than other parts of the wall panel. Wall panels containing larger openings also require additional reinforcement to withstand the strain at the time of lift. Electrical conduit, outlet boxes, and inserts for crane attachments are positioned after the rebar is positioned. **See Figure 6-7.**

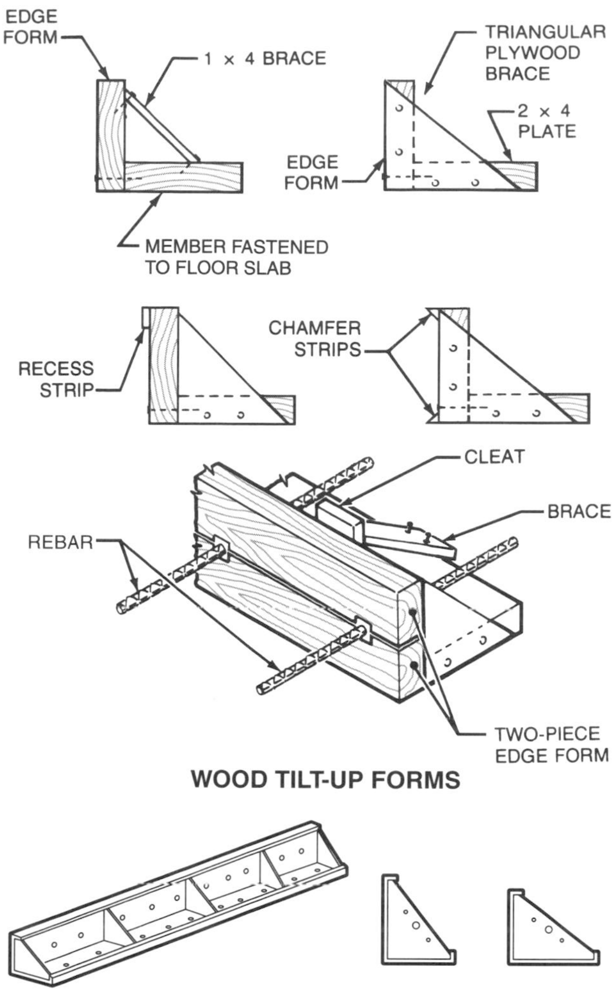

Figure 6-6. Edge forms for tilt-up construction are constructed with 2″ thick planks. Offset and beveled edges are formed by using recess or chamfer strips. Manufactured metal units are also used.

Meadow Burke Products, Inc.

Figure 6-7. Forms for tilt-up wall panels are constructed on the floor slab. Window and door bucks, rebar, and inserts are positioned prior to placing the concrete.

Avoid disturbing the form release materials on the floor slab during the vibrating and working of the concrete. After the concrete has set and reached the required strength, the wall panels are raised and set into position over the foundation footings. Formwork and shores must not be removed until the employer determines that the concrete has gained sufficient strength to support its own weight and the weight of imposed loads.

Lifting Systems

Before placing concrete for a tilt-up wall, inserts must be set in place and secured. This is usually done by tying the inserts to rebar at determined lifting points of the wall. Plastic caps are temporarily screwed into the inserts to protect the coil threads from concrete, sand, grit, and water. After the concrete has set, the plugs can be removed. There are a variety of insert designs, each of which provides crane attachments for lifting the walls. Some of the more common inserts are coil inserts and remote release systems.

Coil Inserts. Coil inserts provide a threaded coil that will receive bolts. Coil inserts are designed to be placed on the flat plane of the wall, or to permit edge lifting of the panel. Coil inserts are also placed in the walls to anchor wall braces. Lift plates bolted to the inserts provide crane attachments. **See Figure 6-8.**

Remote-Release Systems. Remote-release systems are commonly used to raise wall panels when the lifted edge of the wall is not to be permanently visible. Workers can release the lifting units from the ground, eliminating the safety hazard of working at heights and decreasing erection time of the wall panels. Remote-release systems consist of an insert embedded in the wall panel and a lifting unit. The insert is supported by a base and positioned at a predetermined lift point prior to concrete placement. After the concrete has set, lifting units are attached to the inserts. The clutch-type and encasement ball inserts are common remote-release systems for tilt-up construction. **See Figure 6-9.**

A *clutch-type insert* is an insert that consists of a T-bar anchor and a recess former supported by a base. The T-bar anchor provides a hook point for the lifting unit. The recess former fits over the top of the T-bar anchor and creates a void for the clutch of the lifting unit. Locator antennae extend from the top of the recess former to indicate the position of the insert after the concrete has been placed.

An *encasement ball insert* is a one-piece insert that rests on a base. A plastic cap with locator antennae is secured in the top of the insert to prevent concrete from accumulating inside. After the concrete has set, the recess former or plastic cap is removed and the lifting unit is attached.

Lifting units for remote-release systems are attached to inserts after the concrete has obtained its required strength. **See Figure 6-10.**

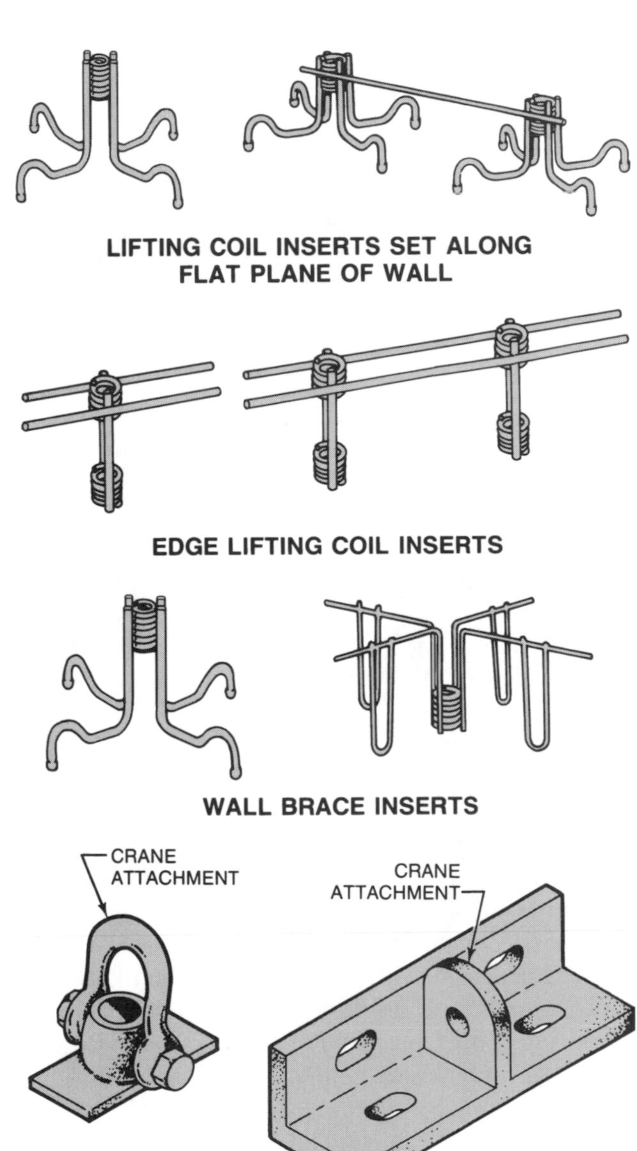

Figure 6-8. Single or double inserts are embedded in the precast panels. Lift plates are bolted to the inserts for the crane attachment.

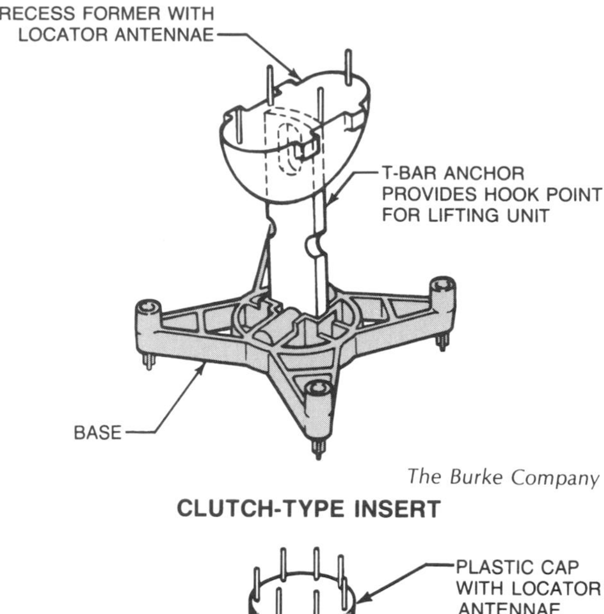

CLUTCH-TYPE INSERT

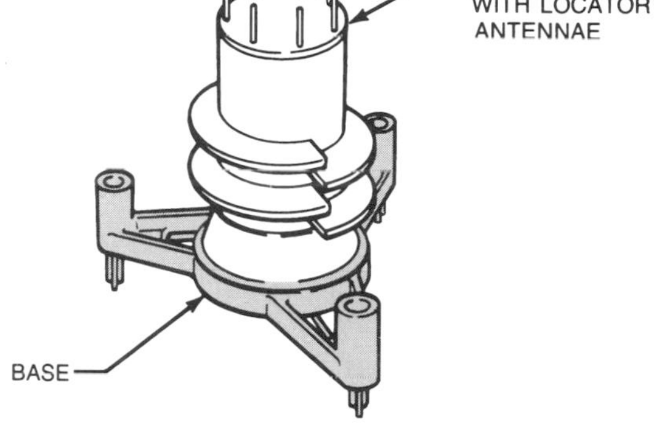

ENCASEMENT BALL INSERT

Figure 6-9. Inserts for remote-release systems are embedded in tilt-up wall panels. Lifting units are attached to the inserts and the wall panels are raised into position.

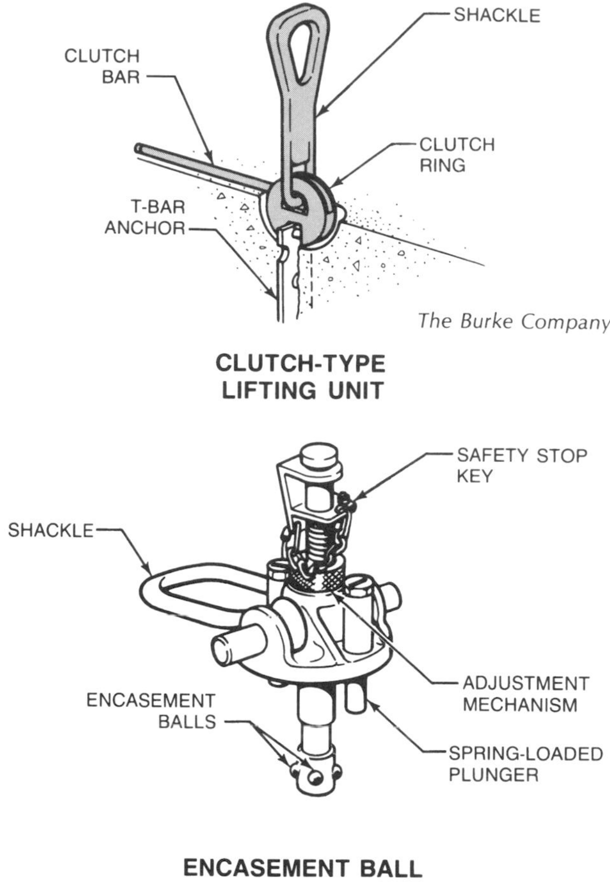

ENCASEMENT BALL LIFTING UNIT

Figure 6-10. Lifting units for remote-release systems are fastened to the embedded inserts. The lifting units are released from ground level.

The lifting unit for a clutch-type insert consists of a clutch ring mechanism and a shackle. The clutch ring is lowered into the preformed void and attached to the T-bar anchor by pushing the clutch bar against the wall panel. The shackle is hooked to a crane and the wall panel is raised into position. After the wall panel is raised and braced, a lanyard attached to the clutch bar is pulled and the lifting unit is released.

An *encasement ball lifting unit* is a lifting unit that consists of a shaft containing the encasement balls and an adjustment mechanism, two spring-loaded plungers, and a shackle. The encasement balls are forced against the sides of the insert as the adjustment mechanism is screwed into the lifting unit. As the lifting unit is positioned in the insert, the spring-loaded plungers are depressed. A safety stop key is placed against the adjustment mechanism to secure the encasement balls in position. The shackle is hooked to a crane and the wall panel is raised into position. After the wall panel is set and braced, a lanyard attached to the safety stop key is pulled to release the pressure between the encasement balls and the insert. The spring-loaded plungers eject the lifting unit from the insert, leaving the insert in the wall panel.

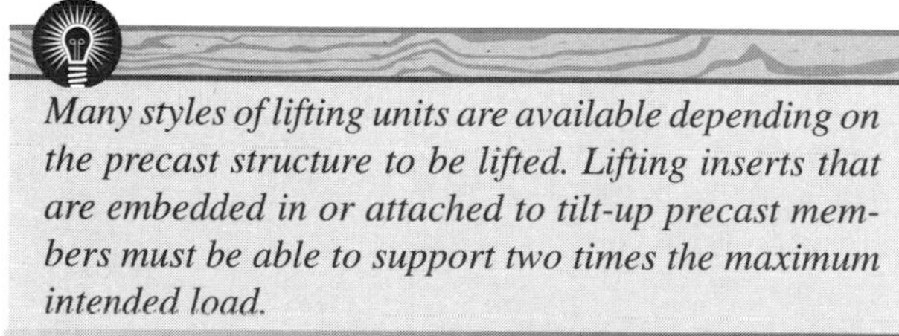

Many styles of lifting units are available depending on the precast structure to be lifted. Lifting inserts that are embedded in or attached to tilt-up precast members must be able to support two times the maximum intended load.

Raising and Bracing Wall Panels

When precasting a wall panel on a floor slab, the formed panels are cast next to each other in a row. After the concrete has set and gained sufficient strength, the wall panels are raised and set into place in one continuous operation. Wall panels may also be stack cast. Stack casting consists of panels cast on top of each other and then raised and placed in position.

Cranes used for lifting tilt-up wall panels are equipped with horizontal spreader bars. Steel cables are attached to the lift plates or lifting units of the wall panels and threaded over pulleys fastened to the spreader bars.

The lift points (placement of the lift plates or lifting units) must be positioned carefully to equalize the lifting force when the wall panels are raised. **See Figure 6-11.** The layout of the lift points should be determined by a qualified engineer. An engineer often uses a computer to detail the lift points based on the following:

- weight and dimensions of wall panel
- concrete strength at time of lift
- type of concrete used
- location and dimensions of openings
- preferred rigging configuration

After the lift points are determined, the crane size is selected based on the weight of the panels, lift point positions, position of panels on the job site, and potential obstructions during lifting. Wood timbers may be used as strongbacks for thin wall panels or wall panels containing numerous openings. The timbers are temporarily bolted to the wall panels to prevent structural damage from lifting stress.

Meadow Burke Products, Inc.

Figure 6-11. The lift points for tilt-up wall panels must be positioned to equalize the lifting force when raising the panels into place.

Bracing Panels. A wall panel must be temporarily braced after it is raised into position. The braces must be able to resist all lateral stress, including wind stress. Telescoping steel braces are commonly used as temporary braces. A typical telescoping brace has a top and bottom shoe for attaching to inserts in the wall and floor. A screw jack is located at the lower end of the brace for making final adjustments. Inserts that will receive brace attachments must be laid out accurately so that the wall inserts align with inserts embedded in the floor slab. If bracing to the inside of a wall to the floor slab is not possible or convenient, the bottom end of a brace may also be fastened to a "deadman", which is a concrete block buried in the ground. **See Figure 6-12.**

Wedge Bolts. Wedge bolts are a type of floor anchoring device that can be installed in the floor after the concrete has set. After drilling a hole completely through the concrete floor slab, a wedge bolt with a bottom lip is placed into the hole. A wedge is then driven next to the bolt to secure it in position and engage the bolt lip to the bottom of the slab. A brace shoe is then fastened to the bolt. **See Figure 6-13.** Braces must remain in place until permanent roof structures and columns are in place. When the bracing is no longer required, the wedges are pried up and the bolts are removed. The holes are then filled with a sand and cement grout mixture. Expansion bolts are not recommended for anchoring braces as they may not withstand the tension and shear forces exerted by the wall. OSHA recommended safety precautions that should be followed when erecting tilt-up panels are as follows:

- employers shall maintain programs for frequent inspection of the job site, materials, and equipment
- employers shall instruct employees in how to recognize and avoid unsafe conditions
- employers shall comply with all precast concrete construction requirements
- employers shall ensure that tilt-up panels are properly braced
- employers shall use only certified welders when welding steel joists to embeds and inserts.

The layout of lift points for tilt-up construction is determined by an engineer based on the requirements of the concrete. Proper safety procedures should always be followed when lifting tilt-up wall panels.

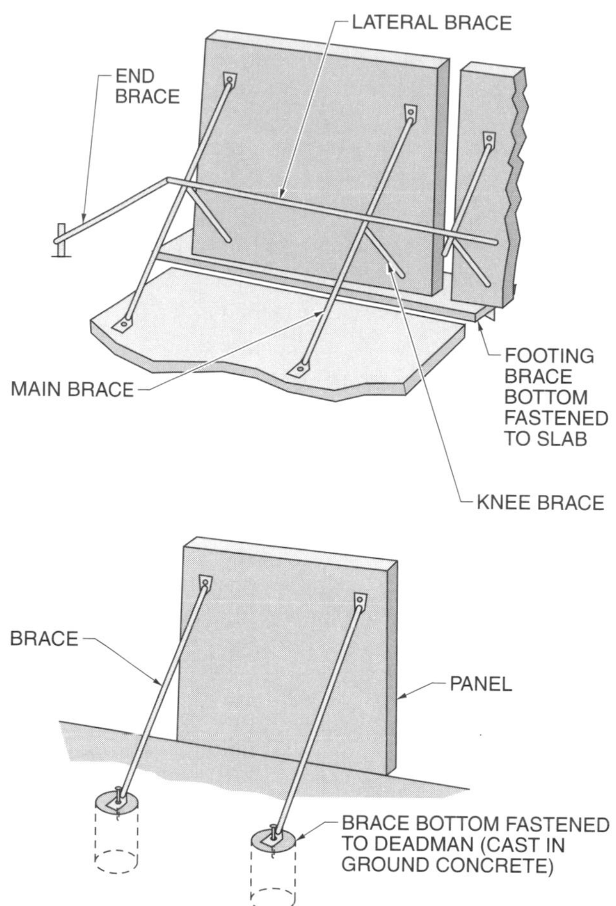

Figure 6-12. Panels must be secured with temporary braces. The tops are secured to inserts in the wall. The bottoms are fastened to inserts located in the floor or to a concrete "deadman" cast in the ground.

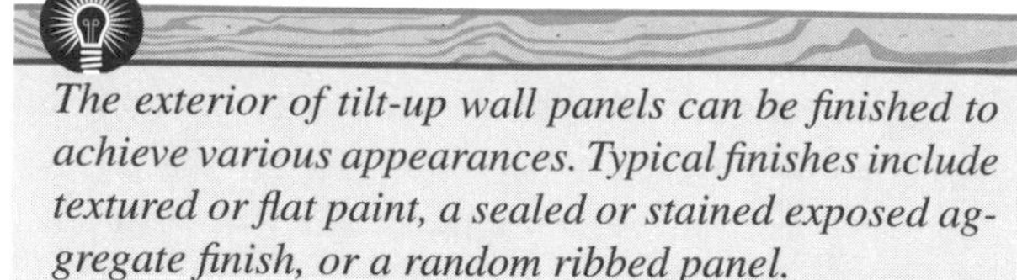

The exterior of tilt-up wall panels can be finished to achieve various appearances. Typical finishes include textured or flat paint, a sealed or stained exposed aggregate finish, or a random ribbed panel.

Securing Wall Panels. After wall panels have been raised, they are permanently fastened into position. Wall panels can be fastened by placing the panels directly on grout pads placed on top of the foundation footing. Rebar extending from the wall is welded or tied to rebar extending from the floor slab. A backer rod is positioned under the wall panel before raising a tilt-up panel. A *backer rod* is foam material used to prevent moisture from seeping between the wall panel and the footing. The area between the bottom of the wall panels and the foundation is then filled with grout or concrete. Concrete is placed between the wall and the floor slab. **See Figure 6-14.** Other methods of securing the wall panels include placing the bottom of the wall in a slot at the top of the foundation wall, slipping the walls over steel dowels extending vertically from the foundation, and securing the wall panel bottoms with steel welding plates.

Various methods are used to tie vertical edges of wall panels together, and new devices continue to become available. An older but still used method is to form and pour cast-in-place columns between the wall columns. Another method features independent precast columns positioned before the wall panels are placed.

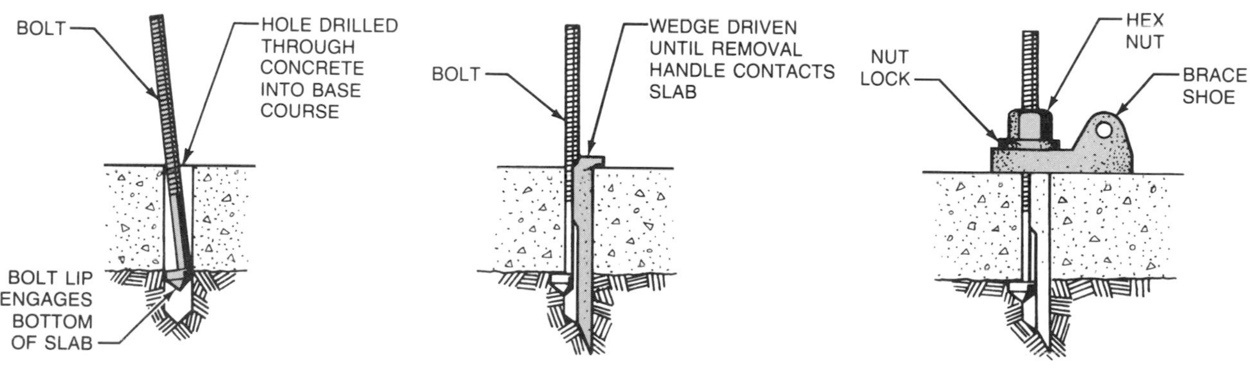

1. Insert the bolt through a predrilled hole in the floor slab. The bolt lip must be opposite the brace shoe and engage the bottom of the slab.

2. Insert the wedge opposite the bolt lip and drive it into the slab until the removal handle touches the slab.

3. Slide the brace shoe over the bolt. Tighten a hex nut on top of the brace shoe and secure with a nut lock.

Figure 6-13. Wedge bolts are used to fasten the brace shoe to the floor slab.

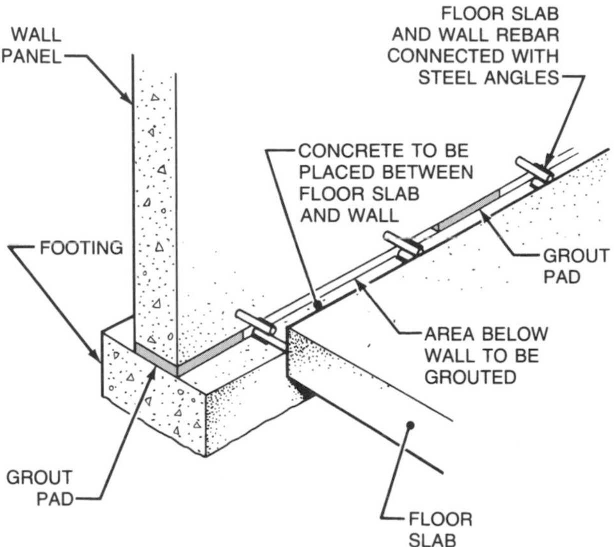

Figure 6-14. Tilt-up wall panels are placed on top of the footing. Rebar extending from the floor slab and the wall panel are tied together and concrete is placed between the wall and slab.

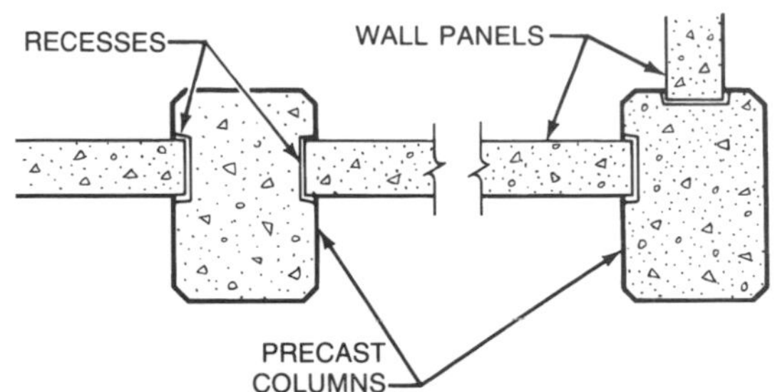

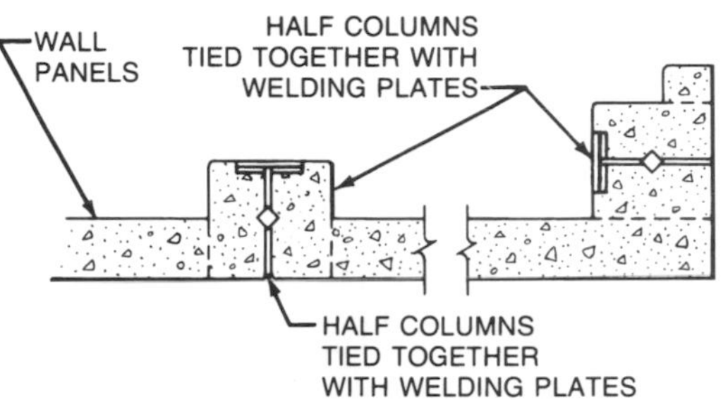

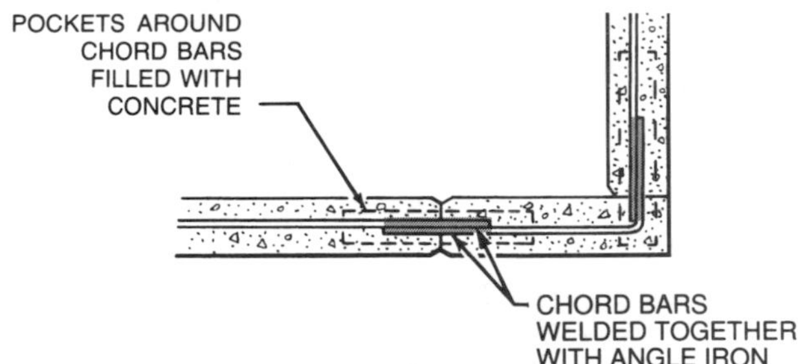

Figure 6-15. Tilt-up wall panels are tied together with precast columns or chord bars.

Independent precast columns are formed with oversize recesses that accommodate the edges of the wall panel. The wall panels are secured to the columns with steel welding plates. When columns and wall panels are cast monolithically, one-half of a column is formed at each end of the wall panel. The half column is tied to a half column extending from an adjoining wall panel. Another tilt-up method uses heavy chord bars that extend horizontally through the wall panels. Chord bars are heavy rebar that resist lateral pressure exerted on the wall panels and tie the wall panels together. When forming the wall panels, a small pocket around the ends of the chord bars is blocked out. After the wall panels have been raised and positioned, the exposed bars are welded together and the pocket is filled with concrete. **See Figure 6-15.**

The exterior wall panels for tilt-up structures that are at least two stories high extend the entire height of the structure. The floor slabs are suspended and cast in place. The floor slab forms are supported by stringers held in position by wood or metal scaffold shoring or by wood or metal joists or trusses. The joists or trusses are secured into position with metal brackets or steel angles that are bolted to the exterior wall panels. The floor slab sheathing is placed over the joists or trusses and the concrete is placed in the forms using buggies or pumps.

Roofs for tilt-up structures are commonly constructed with open web steel joists, glulam timbers, or trusses, and sheathed with plywood, OSB, or other approved material.

TILT-UP SANDWICH PANELS

Sandwich panels are a more recent and rapidly growing development in the tilt-up industry. While there may be some variation in system design among different manufacturers, all sandwich panels consist of board insulation placed between two concrete wall slabs. **See Figure 6-16.** Sandwich panels are used in both residential and commercial construction.

Sandwich panels offer a number of advantages. Concrete by itself does not offer very good insulation against heat and cold. Therefore, the insulation placed between the slabs greatly increases the thermal insulation of the walls and eliminates the need for insulation materials added to the outside surfaces of the concrete walls. R-values of finished walls using sandwich panels can reach as high as R-41.5, exceeding that of wood or steel walls. Sandwich wall systems may produce an average energy savings of 20% to 60%. Sandwich panels also greatly reduce sound penetration through the walls.

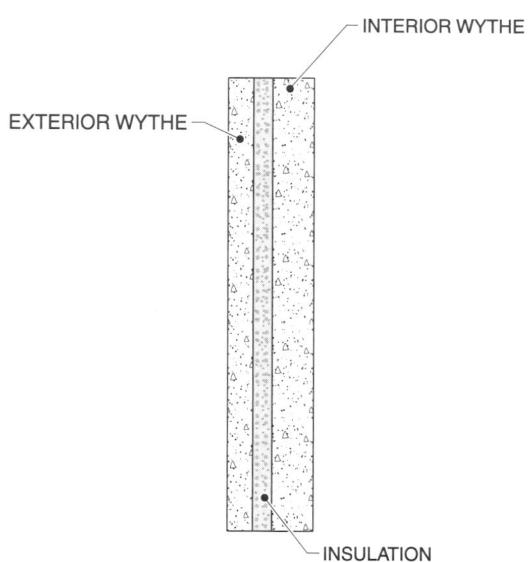

Figure 6-16. Tilt-up sandwich panels have concrete interior and exterior wythes, with insulation in between. The interior wythe is thicker than the outer wythe, as it is considered the main bearing section of the wall.

Construction Procedures

The form construction for sandwich wall panels is very similar to solid tilt-up walls. Higher side forms are needed because the insulation materials will add to the total thickness of the walls. Total wall thicknesses range from 5″ to 8″ and in some cases up to 14″. Insulation layers range in thickness from 1″ to 10″ in ½″ increments. However, a minimum thickness of 1½″ or 2″ is recommended. The majority of tilt-up sandwich designs have 2½″ to 3″ thick inside and outside concrete coverings, referred to as wythes. For example, a sandwich wall with 2″ insulation and 3″ concrete wythes will have a total wall thickness of 8½″.

In a typical construction procedure, the outer concrete wythe is placed in the forms. The insulating material is then placed over the outer wythe, and the inner concrete wythe is placed over the insulation.

Reinforcement and Connectors. The wythes of sandwich walls are usually reinforced with welded wire reinforcement, commonly 6 × 6—W2.9 × W2.9. A vital component of these walls are the connectors that tie the wythes and the insulation material together. A variety of connector designs are available; and they are usually made of stainless steel or fibrous materials. After the insulation board is placed over the bottom layer of concrete the connectors are inserted through pre-drilled holes in the insulation board. Another system features wires that pierce the insulation core and are welded to the wire mesh in the concrete. **See Figure 6-17.**

Sandwich panels are prepared with the same type of inserts and lifting plates used for solid tilt-up panels. They are raised into place by crane and joined in the same manner as solid panels.

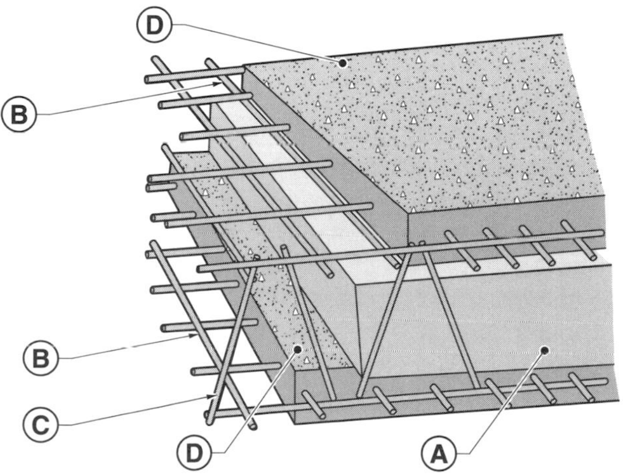

Figure 6-17. Sandwich tilt-up panels usually use welded wire reinforcement and metal or fibrous connectors. Truss wires run through the insulation and are welded to the outside mesh layers.

Name ______________________ Date ______________

Completion

______ **1.** Precast members are reinforced with rebar or ___ steel cables.

______ **2.** Two methods used to prestress concrete are ___ and ___.

______ **3.** ___ beds act as a base and support for precast forms.

______ **4.** The lift points for tilt-up wall panels must be positioned to ___ the lifting force when raising the panels into place.

______ **5.** A(n) ___ insert is a one-piece insert that rests on a base and has a plastic cap with locator antennae.

______ **6.** Prestressed ___ cables are placed in the form when pretensioning concrete.

______ **7.** Concrete is placed under ___ when pretensioning occurs.

______ **8.** Advantages of job site precasting include ___ costs and increased ___.

______ **9.** In addition to wood, ___ and ___ may be used for precast forms.

______ **10.** Forms should be ___ or ___ before the concrete is placed to facilitate stripping.

______ **11.** Precast wall and floor section panels require ___ forms around the perimeter of the casting bed.

______ **12.** Casting beds must be ___ to avoid deflection.

______ **13.** ___ are heavy rebar that resist lateral pressure exerted on wall panels, tie the wall panels together, and extend horizontally through the wall panels.

______ **14.** Roofs for tilt-up structures are commonly constructed with ___, ___, or ___.

______ **15.** ___ precasting is used when adequate space is available for a casting yard.

Multiple Choice

______ **1.** Tilt-up construction is ___.
A. a factory precast method
B. a method for constructing high-rise buildings
C. primarily used in the construction of one- and two-story buildings
D. not a widely accepted method of construction

______ **2.** An engineer uses a computer to detail the lift points based on ___.
A. type of concrete used
B. weight and dimensions of wall panel
C. concrete strength at time of lift
D. all of the above

_________ **3.** Precast members are connected, or anchored, to adjoining parts of the structure using ___ for connections.

A. steel plates

B. dowels

C. tensioning bars

D. all of the above

_________ **4.** When placing rebar for tilt-up wall panels, the areas around the door and window openings require ___ reinforcement than other parts of the wall.

A. less

B. the same amount of

C. more

D. no

_________ **5.** When constructing tilt-up sandwich panels the first step in the construction process is to ___.

A. place the outer concrete wythe in the forms

B. place the inner concrete wythe in the forms

C. attach the insulation to the outer concrete wythe

D. none of the above

_________ **6.** A ___ strip is a wood piece nailed to the edge forms to provide beveled corners at the vertical ends of the wall panel.

A. recess

B. angle

C. chamfer

D. vertical

_________ **7.** Inserts for tilt-up wall panels are ___.

A. placed along the flat plane of the wall

B. placed for edge lifting

C. protected by plastic caps during the placement of concrete

D. all of the above

_________ **8.** Stack casting is a procedure ___.

A. commonly utilized when casting panels on the floor slab of the building

B. in which a series of panels are cast on top of each other and then lifted into place

C. used most often in tilt-up construction

D. none of the above

_________ **9.** After raising a wall panel, the first step is to ___.

A. tie it to a column

B. securely brace it

C. remove the lift plates

D. patch the insert holes

_________ **10.** Grout pads are used to ___.

A. level the bottom of the wall panel

B. fill in the space between the wall bottom and the foundation

C. secure the wall to the foundation

D. waterproof the joint between the wall and the foundation

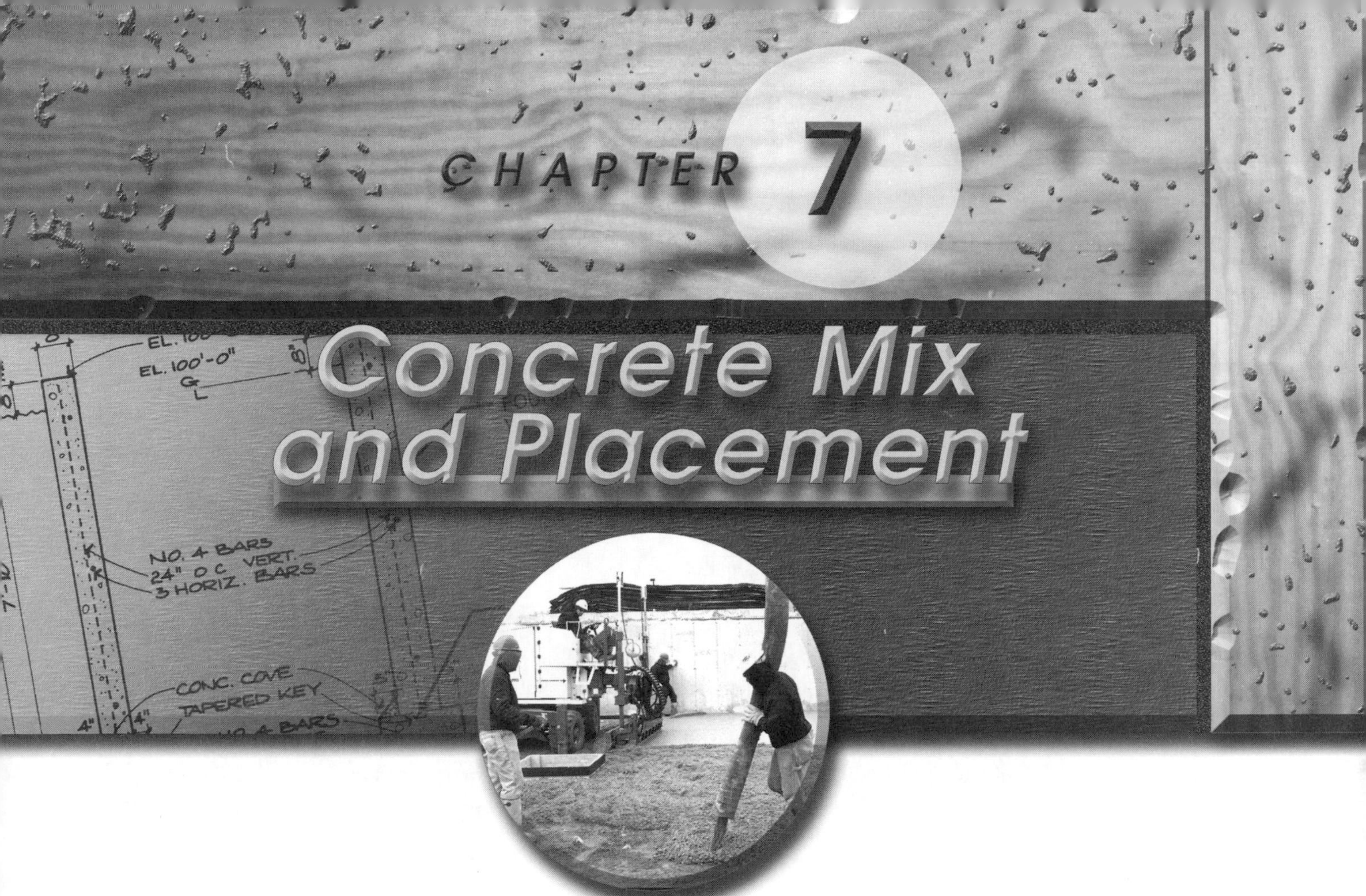

CHAPTER 7

Concrete Mix and Placement

Concrete is one of the most durable materials used in modern construction. Concrete is strong and fireproof, and resists decay. Concrete is often the principal construction material used in buildings, highways, and other heavy construction projects.

Concrete is composed of portland cement, coarse and fine aggregate, and water. When water is added to these ingredients, concrete in a plastic state is produced. Hydration, a chemical reaction between cement and water, occurs and the concrete begins to set and solidify.

Modern techniques used to manufacture portland cement have been used for over 160 years with minor modifications. An Englishman, Joseph Aspdin, obtained a patent for the manufacturing process in 1824. Portland cement refers to rocks and limestone quarried from deposits on the island of Portland, off the coast of England.

Quality control of concrete and its placement is essential to ensure its final strength and appearance. Proper mixing techniques and proportions must be used to produce concrete with the proper amount of slump and the desired compressive strength. Proper mixing, transportation, and placement methods must be used to prevent segregation. *Segregation* is the separation of sand and cement ingredients from the coarse aggregate in a concrete mix.

The forms are removed (stripped) from the concrete member after the concrete has set and it is hard enough to resist damage. Built-in-place forms are removed from the concrete in the sequence opposite to that in which they were erected. Large panel forms and ganged panel forms are stripped using a crane.

COMPOSITION OF CONCRETE

Concrete is composed of fine and coarse aggregate, cement, and water. Fine and coarse aggregate make up the greatest part of the concrete mixture. Fine aggregate is sand and coarse aggregate is gravel or crushed stone. Coarse aggregate ranges in size from ¼″ to 1½″ in diameter.

The size of coarse aggregate chosen for a concrete mixture is based on the spacing between the rebar placed in the form and the distance between opposite form walls. Walls with many rebar or walls that are narrow in width require a concrete mixture with small gravel or stone. In general, the maximum size coarse aggregate permitted should be used in the mixture. However, the maximum size should not be larger than one-fifth the narrowest dimension between form walls or not larger than three-fourths the minimum distance between rebar.

Cement constitutes the smallest portion of the concrete mixture. Cement acts as a paste that binds the aggregate in the concrete mixture when water is added. Most cement products are derived from limestone, which is obtained by digging into the Earth's surface or from mining beneath the ground. It may also be dredged from deposits covered by water.

The first step in manufacturing cement is to reduce the size of the stone. The stone is then mixed with other raw materials and ground to powder and blended. This raw mixture is then burnt and converted to cement clinkers. Gypsum is combined with the clinkers and ground up to complete the process. **See Figure 7-1.**

Portland Cement Association

Fine and coarse aggregate material are mixed at the batch plant to the specifications of a particular concrete application.

When water is added to cement, a chemical reaction called hydration occurs. *Hydration* is a chemical reaction between cement and water that produces hardened concrete. The rate and degree of hydration directly affect the final strength of concrete. The water used in a concrete mixture should be clear and free of oils, alkalis, and acids. The amount of water combined with the cement and aggregate is also important to the strength of the concrete. Too much water dilutes the cement and causes the aggregate to separate, producing weak concrete. Too little water results in poor mixing action of the cement and aggregate and also produces weak concrete.

Portland cement can be stored indefinitely in dry conditions without losing any of its properties. However, bagged portland cement stored in warehouses may develop "warehouse pack," which is compaction of the cement that causes the cement to become lumpy. Warehouse pack can be corrected by rolling the cement bags on the floor to break up the lumps.

The Concrete Mix

The concrete mix is the proportion of cement, fine and coarse aggregate, and water in a batch of concrete. As the size of the coarse aggregate increases, the proportion of coarse aggregate increases also. **See Figure 7-2.** Information about the structure being built must be determined before the concrete mix is proportioned, including shape and size of structural members (walls, slabs, beams, columns, etc.) and their required design strength. Exposure of concrete members to weather conditions and other environmental factors must also be considered.

The concrete mix affects the compressive strength of concrete. Compressive strength is the amount of force the concrete can withstand 28 days after it has been placed. Twenty-eight days is the average period of time required for concrete to gain its full strength. Compressive strength is expressed in pounds per square inch (psi).

Water-Cement Ratio. The water-cement (w/c) ratio of a concrete mix is the major factor affecting compressive strength of concrete. The *water-cement ratio* is the amount of water used in a concrete mix in relation to the amount of cement. A low water-cement ratio produces strong and dense mixtures. Therefore, the water-cement ratio selected should be the lowest value possible to meet the design requirements of the structure. **See Figure 7-3.**

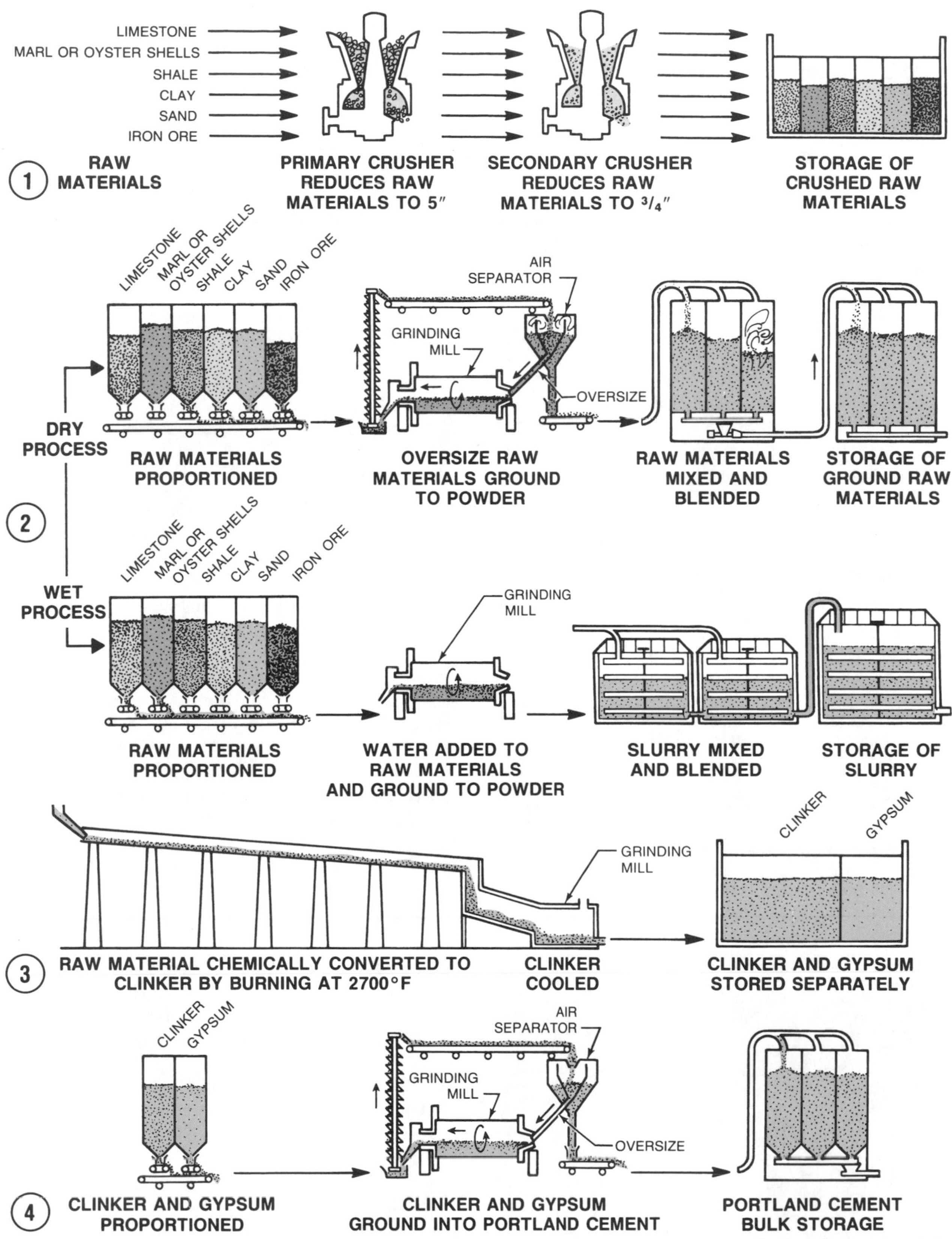

Portland Cement Association

Figure 7-1. Raw materials are crushed, blended together, burned to partial fusion, and ground to a fine consistency to obtain portland cement.

CONCRETE PROPORTIONS				
Coarse Aggregate				
Size (max.)	(cu ft)	Cement (cu ft)	Sand (cu ft)	Water (gal.)
3/8″	1 1/2	1	2 1/2	1/2
1/2″	2	1	2 1/2	1/2
3/4″	2	1	2 1/2	1/2
1″	2 3/4	1	2 1/2	1/2
1 1/2″	3	1	2 1/2	1/2

Figure 7-2. Various concrete proportions are used for different sizes of coarse aggregate.

Water-cement ratio is determined by dividing the weight of the water by the weight of the cement contained in a cubic yard (cu yd) of concrete. For example, if the weight of the water used in 1 cu yd of concrete is 8 lb and the weight of the cement is 18 lb, the water-cement ratio is .44 (8 ÷ 18 = .44).

COMPARISON OF COMPRESSIVE STRENGTH TO WATER-CEMENT RATIO		
Compressive Strength at 28 days (psi)	Water-Cement Ratio, by Weight	
	Non-Air-Entrained Concrete	Air-Entrained Concrete
2000	0.82	0.74
3000	0.68	0.59
4000	0.57	0.48
5000	0.48	0.40
6000	0.41	—

American Concrete Institute

Figure 7-3. The compressive strength of concrete is directly related to the water-cement ratio of the concrete mix. Air-entrained concrete requires a lower water-cement ratio than non-air-entrained concrete.

Samples for compression tests are taken throughout concrete placement and sent to the lab for testing.

In heavy construction work, proper proportions of the concrete mix are determined by an engineer or concrete field specialist and are included in the print specifications. Other information may include minimum cement content in relation to the lowest water-cement ratio, size and amount of coarse aggregate, and type and amount of admixtures. Local building codes may furnish the required information for an acceptable concrete mix for small projects such as residential foundations. Local batch plants that produce ready-mixed concrete may also provide additional information.

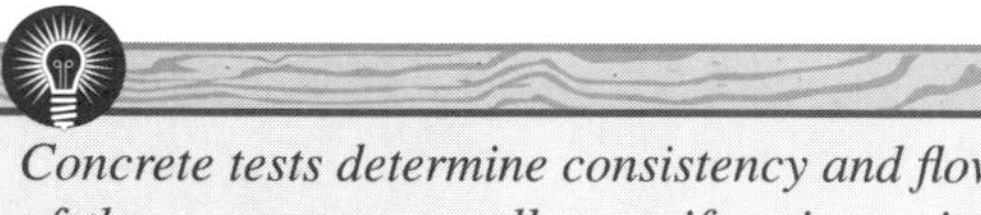

Concrete tests determine consistency and flowability of the concrete, as well as uniformity, unit weight, and air content.

Job Site Testing

Job site testing of concrete is often required for heavy construction projects where concrete is being continuously placed. Two tests commonly conducted on concrete are the slump test and the compression test.

A *slump test* is a test that measures the consistency, or slump, of concrete. *Consistency* is the ability of concrete, in its plastic form, to flow as it is being placed into the form. A stiff mix has a higher proportion of aggregate and poorer consistency than a fluid mix. The material cost of stiff mixes is less than fluid mixes. However, they are more difficult to place because of their inability to flow around reinforcement and other areas of the form. Separation of the concrete materials or bleedwater may also occur. *Bleedwater* is excess water that collects on the surface of concrete as aggregate material sinks in the concrete mixture.

Slump tests are conducted before or during placement of concrete in the forms. Concrete samples are taken from stationary or truck mixers. Variations in the results of slump tests indicate that changes have occurred in the grading or proportions of the aggregate, or in water content. Corrections are made immediately to ensure correct and uniform consistency of the concrete.

A *slump cone* is a cone made of galvanized metal that is 8″ in diameter at the bottom, 4″ in diameter at the top, and 12″ high and is used to perform slump tests. The cone is dampened immediately before use and placed on a smooth, nonabsorbent surface. **See Figure 7-4.** Three layers of concrete are placed in the cone and each layer is rodded with a 5/8″ × 24″ rod. Rodding is done with an up-and-down motion. After rodding is completed the cone is carefully lifted from the concrete, without tilting or jarring the cone.

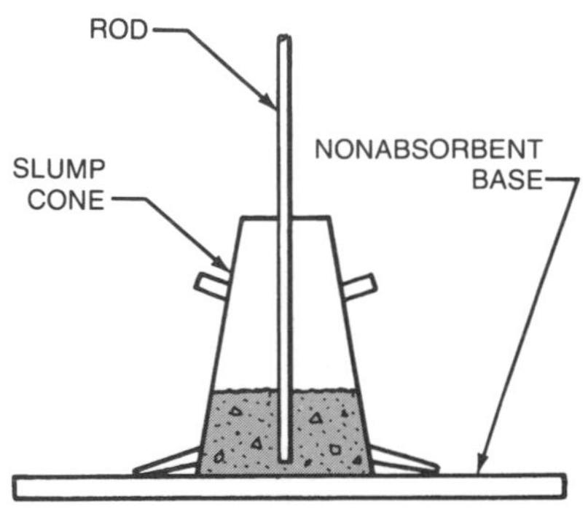

1. Fill one-third of the cone with concrete and rod 25 times.

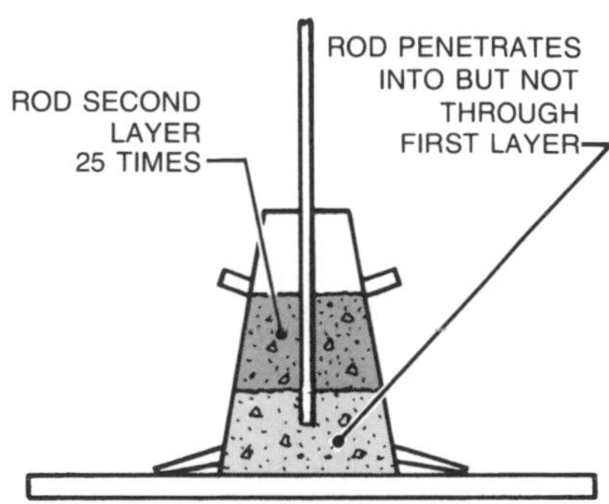

2. Fill two-thirds of the cone and rod the second layer 25 times.

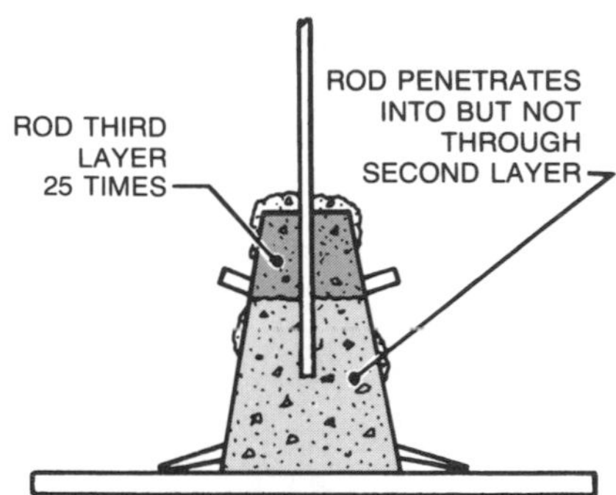

3. Fill the cone to overflow and rod 25 times.

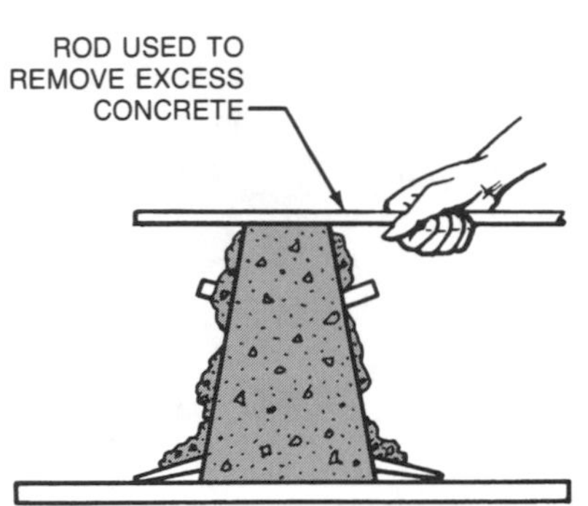

4. Remove the excess from the top of the cone and at the base.

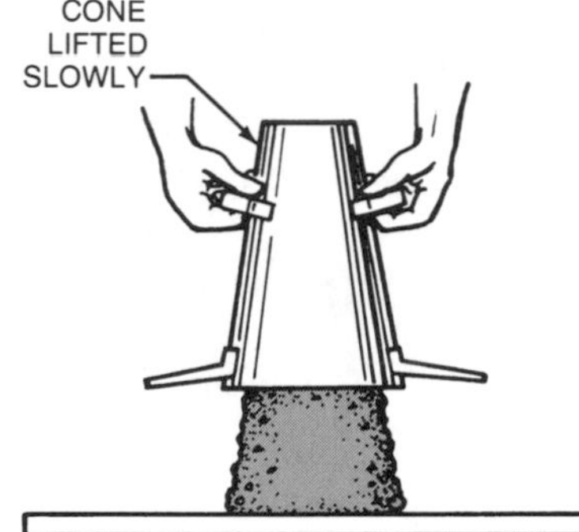

5. Lift the cone vertically with a slow, even motion.

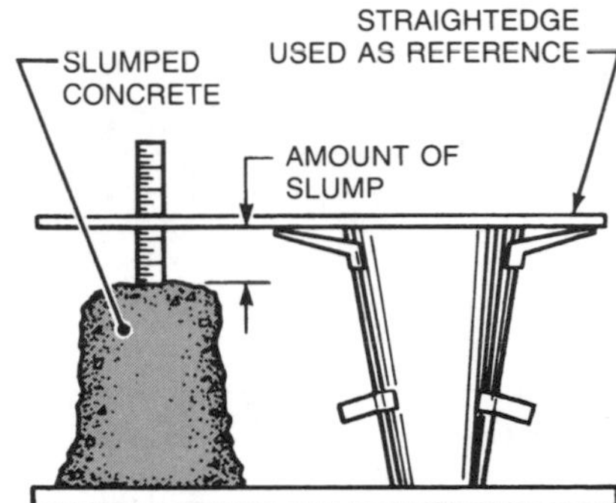

6. Invert the cone and place next to the concrete. Measure the distance from the top of the cone to the top of the concrete.

Figure 7-4. A slump test is used to measure the consistency of the concrete.

The amount of slump is determined by measuring down from a straightedge placed on top of the inverted cone to the top of the concrete slump pile. For example, if the top of the slump pile is 3″ below the top of the cone, the concrete has a 3″ slump. Slump ranges are allowed for different types of construction work. Narrow, congested forms erected for reinforced walls, beams, and columns require more fluid concrete mixtures than wide forms. Fluid concrete mixtures consist of small coarse aggregate combined with more water and cement, and produce a greater amount of slump. Wide, less congested forms such as footings and slabs allow a stiffer mix with less slump. **See Figure 7-5.**

A *compression test* is a test that measures the compressive strength of concrete. Compressive strength is the force the concrete can withstand after 28 days. The required compressive strength of concrete for a structure is determined by exerting force on a specimen that has set. A major factor determining the compressive strength of concrete is the water-cement ratio.

ALLOWABLE SLUMP

Concrete Construction	Slump	
	Max.* †	Min. †
Reinforced foundation walls and footings	3	1
Plain footings, caissons, and substructure walls	3	1
Beams and reinforced walls	4	1
Building columns	4	1
Pavements and slabs	3	1
Mass concrete	2	1

* may be increased 1″ for consolidation by hand methods such as rodding and spading

† in in.

Portland Cement Association

Figure 7-5. Allowable slump is determined by the type of structural member.

ASTM C172, Standard Practice for Sampling Freshly-Mixed Concrete, *and ASTM C31,* Standard Practice for Making and Curing Concrete Test Specimens in the Field, *detail the process for compression testing.*

Samples of concrete for a compression test must be taken at three or more regular intervals throughout the discharge of the concrete batch. Samples should not be taken at the very beginning or end of the discharge. The concrete specimens are placed in a watertight metal or nonabsorbent cylindrical mold 6″ in diameter and 12″ long. Each mold is filled in three layers, and each layer is rodded approximately 25 times with a ⅝″ × 24″ rod. When the rodding has been completed for the last layer of concrete, the surface is leveled off and covered with a glass or metal plate to prevent evaporation.

The cylinders are stored on the job site and must be protected from jarring. After 24 hours, the cylinders are taken to a laboratory where the concrete sample is removed from the cylinder and allowed to set in a "moisture room" where temperature and humidity are controlled. Tests are performed after 7 days, 14 days, and 28 days. Before each test, a thin layer of capping compound is applied to the specimen. After the cap sets, the concrete specimen is placed in a compression testing machine. Pressure is exerted until the specimen breaks. A dial on the machine indicates the pressure required to break the concrete. **See Figure 7-6.**

ELE International, Inc.

Figure 7-6. A concrete sample is placed in a compression testing machine. Pressure is exerted on the sample until it breaks.

Concrete Admixtures

A *concrete admixture* is a material other than cement, aggregate, or water that is added to a batch of concrete immediately before or during the mixing process to modify the concrete's properties. Admixtures increase the effectiveness of concrete under various conditions. Admixtures can reduce the overall cost of a concrete project, impart certain desirable properties to concrete, and help to ensure the quality of concrete under poor weather conditions. Properties such as frost resistance, workability, increased strength, and retardation or acceleration of setting time can be altered by using an admixture or combination of admixtures. While certain advantages are gained from admixtures, special consideration should be given to their use and their impact upon other physical characteristics of concrete.

It may be necessary to combine admixtures in a concrete mixture to achieve the desired properties. Admixtures should be tested with job materials and other admixtures before they are used in an actual construction project. While most admixtures are compatible in a mixture, admixtures should not be combined prior to adding them to mix water or aggregate.

All admixtures should be tested before use. Concrete samples should be prepared using the same cement, aggregate, water, and other admixtures that will be used on the job. Testing determines the approximate amount of water to be used, as well as the effects of the admixtures on air entrainment, hardening rate, and strength development. The appropriate quantity of admixture to be used is determined by these tests.

Whenever a new shipment of admixture is received from a supplier, close attention must be paid to the performance of the concrete with the new admixture. If water requirements, slump, hardening rate, or other properties change, adjust the amount of admixture accordingly. The types of admixtures used most often are air-entraining agents, accelerators, water-reducing retarders, and pozzolans, each of which affects the characteristics of concrete in a different way. **See Figure 7-7.**

Air-Entraining Admixtures. An *air-entraining admixture* is a foaming substance used to add microscopic air bubbles to concrete. Normal concrete contains small capillaries through which water can migrate to the surface. During freeze/thaw cycles, water expands and contracts in the capillaries, resulting in damage to the concrete. Water that penetrates through a concrete surface expands and breaks off pieces of the concrete surface when it freezes. Air-entraining admixtures increase the durability of concrete during freeze/thaw cycles. The system of air bubbles in air-entrained concrete allows water to expand and contract without damaging the concrete surface. Concrete exposed to cold environments must be air-entrained. Concrete exposed to repeated freeze/thaw cycles should contain 6% to 8% entrained air. In exposed flatwork, such as walks and driveways, air-entrainment produces concrete that is highly resistant to severe frost and the effects of salt used for snow and ice removal. Non-air-entrained concrete can be used for basement slabs or projects where concrete is not exposed to freezing, thawing, or de-icers.

Air-entrained concrete requires less water per cubic yard than non-air-entrained concrete with the same amount of slump. The amount of sand in the mix may be reduced by an amount equal to the volume of the entrained air. Air-entrainment increases the durability of concrete, particularly in cold climates. Air-entraining admixtures also improve the workability of concrete and decrease segregation of concrete mixture ingredients.

Another way in which air entrainment protects concrete from freeze/thaw cycles occurs while concrete is still plastic. Aggregate, which is heavier than concrete paste, tends to settle after concrete has been placed. As it settles, the aggregate leaves tube-like voids. Bleedwater travels up these voids to the surface of the slab. This increases the water-cement ratio of the surface paste, resulting in a weak and porous surface on the cured slab. A slab with a weak and porous surface flakes, scales, and spalls after repeated freezing and thawing. The microscopic bubbles created by air-entrainment block the capillaries and stop bleedwater from traveling up to the surface of the slab. The air bubbles make mixtures more placeable or workable.

Although air-entrained concrete is very workable, it is difficult to finish. Magnesium floats are more effective than wood floats since they do not tear the surface. Troweling is more difficult and requires a smoother motion to bring out a hard, dense surface.

Accelerating Admixtures. An *accelerating admixture (accelerator)* is a substance added to a concrete mixture to reduce setting time and improve the early strength of concrete. Accelerators are especially useful in cold weather conditions because they help concrete set before it freezes, which allows the concrete to gain early strength slowly. Accelerators can be replaced by using high-early-strength cement, increasing the cement content, or providing a longer curing period. Accelerators should not be used as a substitute for proper curing and frost protection techniques.

Accelerators such as sodium silicate, triethanolamine, and high alumina cement produce rapid setting and considerable strength within a few hours. These accelerators are typically used to seal water leaks in hardened concrete or to grout construction joints in concrete dams. Concrete in which these accelerators are used has a lower-than-normal strength at later ages. Workers who place and finish concrete containing accelerating admixtures should be aware that these materials are in a mixture, because concrete containing accelerators can set very quickly, making the mixture difficult to place and finish properly.

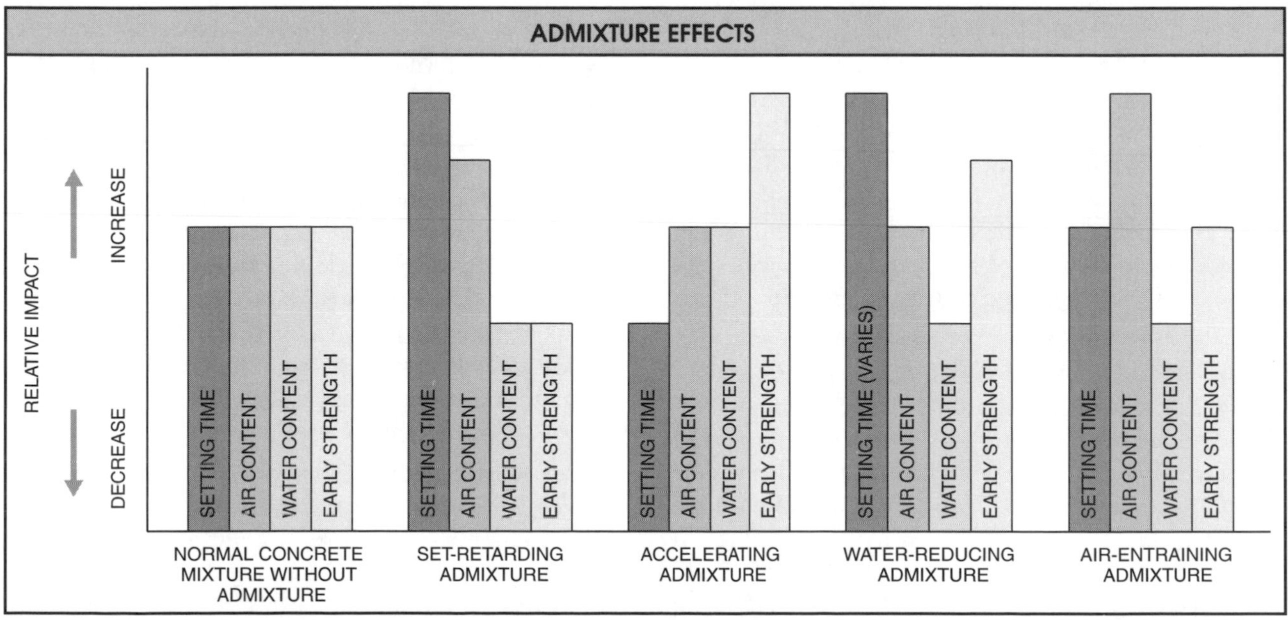

Figure 7-7. Admixtures affect the characteristics of concrete in different ways, depending on the admixture used.

Calcium chloride is one of the most commonly used accelerators. *Calcium chloride* is a crystalline solid accelerating admixture. Calcium chloride increases the strength of concrete at the age of 1 day to 7 days. Calcium chloride is not an antifreeze. When it is added to a concrete mixture in proper proportions, calcium chloride does not decrease the freezing point of concrete to any great extent. It does, however, reduce setting time of concrete and decrease the likelihood of freezing damage.

Calcium chloride is added to concrete after it has been dissolved in water to achieve better mixing. The amount of calcium chloride used should never exceed more than 2% of the weight of cement. More than this amount may cause reinforcement corrosion and rapid stiffening. If greater amounts are added, the concrete stiffens quickly, making placement and finishing difficult. In addition, use of too much calcium chloride could cause flash set or increase shrinkage, which results in cracking, corrosion of reinforcement, and weakening and discoloration of concrete. Never add calcium chloride to concrete in its dry form as it might not dissolve properly. If it is not completely dissolved, calcium chloride could cause popouts and dark spots in concrete as well as affect air-entraining admixtures in the mixture.

Do not use calcium chloride when aluminum is present, such as when aluminum forms are used. The aluminum and calcium chloride react and create heat and gas bubbles. Do not use calcium chloride if concrete is in permanent contact with galvanized steel. Do not use calcium chloride if concrete is exposed to soil or water containing sulfates. Do not use calcium chloride in nuclear-shielding concrete or in massive placements of concrete.

Do not use calcium chloride with Type V (sulfate-resistant) cement. Instead, use a mixture that has a greater amount of cement. In addition, consider using Type III cement to accelerate the set before deciding to use calcium chloride. When using calcium chloride, do not use polyethylene film to cure concrete, as this discolors the concrete.

Water-Reducing Admixtures. A *water-reducing admixture* is a substance added to a concrete mixture to reduce the amount of water needed to produce a desired mixture. They also slow down the set of concrete during hot weather and delay early stiffening of concrete placed under different conditions. Water-reducing admixtures may entrain some air in the concrete. Since water-reducing admixtures may have varying effects on concrete, technical advice is recommended.

Concrete strength is increased using water-reducing admixtures, provided that water content in the mixture is reduced and cement content and slump remain constant. Even though water content has been reduced, concrete containing some water-reducing admixtures shrinks more when drying.

A *plasticizer* is a type of water-reducing admixture that provides concrete with increased workability with less mix water. Plasticizers typically reduce the amount of water needed by 10% to 15%, with only a slight increase in slump. When plasticizers are added to concrete, its strength increases because of the lower water-cement ratio. In addition, concrete mixtures with lower water-cement ratios can be placed and vibrated easily with decreased risk of voids in the concrete. Plasticizers allow very stiff mixtures to be placed and finished. Plasticizers can also be accelerating or set-retarding admixtures. Some plasticizers also entrain air.

Water-reducing admixtures and plasticizers are usually added to a mixture just before it is discharged. These admixtures are used in small amounts, generally less than .3% of the weight of the cement. Excessive amounts of admixtures severely retard the setting time of concrete.

Set-Retarding Admixtures. A *set-retarding admixture (retarder)* is a substance added to concrete to extend its setting time. Set-retarding admixtures are useful in hot weather conditions when concrete sets so rapidly that it cannot be finished properly. They are also useful when more time is needed to place concrete. For example, it may be necessary to delay the set in some applications, such as when placing concrete in a large foundation or when more time is needed to complete a finishing operation such as patterning, texturing, and coloring. **See Figure 7-8.** Most retarders also reduce the amount of water needed in a mixture and are commonly referred to as water-reducing retarders. Water-reducing retarders decrease setting time while making concrete more plastic to allow for proper placement.

Generally, there is some reduction in strength of concrete for the first three days when a set-retarding admixture is used. Retarders may entrain some air into concrete. Therefore, determine the possible extent of air-entrainment and compensate when determining the amount of air-entraining admixture to be used, if any.

Water-Reducing, Set-Retarding Admixtures. A *water-reducing, set-retarding admixture* is a substance that allows less mix water to be used to produce concrete of a desired slump while retarding the set of concrete. Mix water can be reduced 4% to 15% for a given slump, resulting in an increase in strength.

Increte Systems

Figure 7-8. Set-retarding admixtures are used when more time is needed to complete a finishing operation such as patterning, texturing, and coloring.

Water-reducing, set-retarding admixtures do not provide the same results with all cements or mixes or at all temperatures. For example, the quantity of admixture added to a particular mixture to achieve proper retardation in hot weather will likely cause too long a delay in initial hardening in cold weather.

Pozzolan. A *pozzolan* is a fine substance that chemically reacts with calcium hydroxide, which is produced in the hydration process of cement. By itself, pozzolan adds little value to concrete. When combined with calcium hydroxide, it improves the workability and plasticity of concrete mixtures and is commonly used in pumped concrete. Common types of pozzolans include flyash, volcanic glass, calcined shales and clays, blast furnace slag, and ground brick. In the presence of moisture and under normal temperature conditions, pozzolans react with other cement materials to form compounds with cementing properties.

As a general rule, concrete containing pozzolans requires more water to produce the same slump. Therefore, the set concrete has a higher rate of contraction, resulting in a greater tendency to crack. Pozzolans, when used as a partial replacement for cement, decrease the early strength of concrete, but the concrete usually develops higher strength at a later stage. Pozzolans may be used as a partial replacement for cement in mass concrete operations, such as dam construction, where early strength development is not critical.

Dampproofing Admixtures. A *dampproofing admixture* is a substance added to a concrete mixture to improve the impermeability (resistance to water penetration) of hardened concrete. Dampproofing admixtures are used in concrete for tanks, pipes, swimming pools, and other vessels or structures that must retain or transport liquids.

Coloring Admixtures. A *coloring admixture* is a substance that imparts a desired color to concrete. Coloring admixtures are used in concrete slabs and in portland cement-stucco mixtures. Coloring admixtures should be stable in ultraviolet light and should not have adverse effects on the other ingredients in the mixture.

Specialized Admixtures. Some applications may require the use of specialized admixtures to fulfill certain design requirements. A *gas former* is an admixture that facilitates expansion setting and is used in nonshrink grouts. Gas formers typically contain aluminum powder, which generates gas in a concrete mixture to cause expansion. An *air detrainer* is an admixture that decreases air content in concrete mixtures so hardeners may be cast on a wet slab and incorporated into the surface. Air detrainers should not be used in cold environments.

Hot and Cold Weather Concreting

Placing concrete during extremely hot or cold conditions creates problems that affect the hydration of concrete. The problem encountered during hot weather is the rapid rate of water evaporation. This can cause an early slump loss and rapid setting, which may result in strength loss and the possibility of cracking. The problem encountered in cold weather is the slow rate of hydration of the concrete. The ideal setting temperature for concrete is 70°F.

Hot Weather Concrete Placement. Hot weather concrete placement is required when the ambient temperature is 90°F or higher. High temperatures increase the hydration rate, resulting in lower long-term strength of the hardened concrete. The evaporation rate of mix water depends on air temperature, concrete temperature, wind speed, and relative humidity. **See Figure 7-9.**

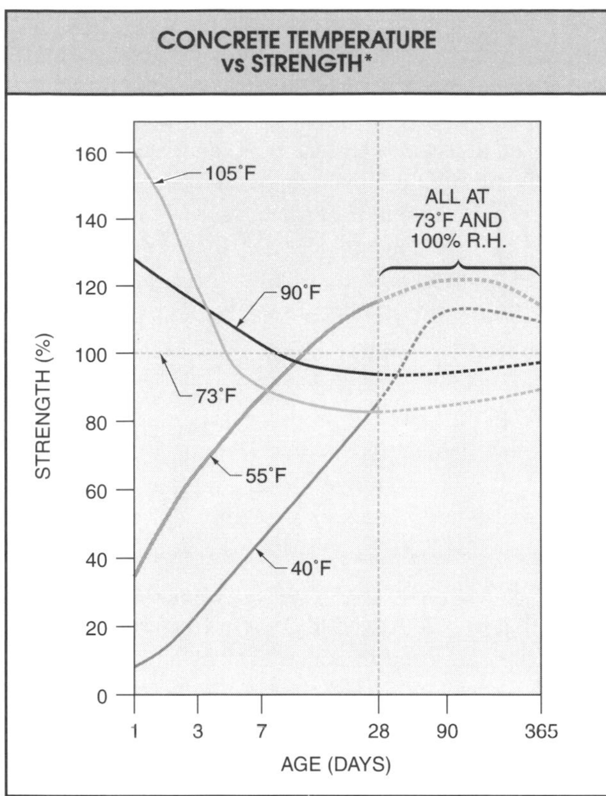

Figure 7-9. The evaporation rate of mix water depends on air temperature, concrete temperature, wind speed, and relative humidity.

Rapid evaporation of the mix water results in high plastic shrinkage, crazing, and excessive loss of workability. Rapid evaporation of mix water also slows down hydration, resulting in inadequate strength development. High temperatures are also damaging to large concrete volumes because greater temperature differential exists between batch placements of the mass. These temperature differences create tensile stresses that commonly result in thermal cracking.

When the ambient temperature reaches or exceeds 85°F, measures must be taken to control the temperature of the concrete. The most effective way to maintain a low concrete temperature is to keep the concrete ingredients as cool as possible before mixing. The easiest and most effective methods for cooling concrete are to cool the mix water, cool the aggregate, or cool the cement. The easiest way to cool concrete is to cool the mix water. The water used in the mix can be cooled before it is added to the mix by refrigerating it or adding liquid nitrogen or crushed ice. Care must be taken to ensure that the ice is completely melted before placement to prevent weak spots in the concrete. When mixing concrete at the job site, aggregate should be stockpiled in a shady place before use and kept moist by sprinkling water on it. Aggregate cooling is commonly accomplished by shading aggregate stockpiles, or by sprinkling stockpiles with water to allow the release of heat through evaporation. Other techniques used to reduce concrete placement temperature include spraying the formwork prior to placement, locating the water tanks in shaded areas, placing concrete in the cooler evening hours, and/or painting the storage mix water tanks white, which reflects sunlight better than darker colors.

Placing equipment such as mixers, chutes, buckets, buggies, and pump lines should be shaded or covered with wet burlap. Wall forms and reinforcing steel should be wetted down. The subgrade below concrete slabs should be thoroughly soaked the night before placing the concrete. The curing of the slabs should be started as soon as possible. Curing is the process of retaining the moisture in freshly placed concrete to ensure proper hydration.

Excessive evaporation of mix water after placement must be prevented. Evaporation rates greater than .1 psf (pound per square foot) of exposed concrete surface per hour must be avoided to ensure satisfactory curing and prevent plastic cracking. Concrete should also be protected from the sun during hot weather. White-colored blankets should be used to reflect sunlight, reducing heat gain.

Cold Weather Concrete Placement. Cold weather concrete procedures are required when the mean daily temperature is less than 40°F. At temperatures 0°F or below, no hydration occurs. To ensure proper hydration in concrete placed during freezing weather, the concrete mix should be heated prior to being placed in the forms. In most cases, heating of the mixing water is all that is required, provided the aggregate is free of ice and snow. If it is necessary to heat the aggregate, steam coils may be placed in the storage pile or holder bins, or steam can be injected into the pile. After the concrete has been placed, the forms can be covered with tarpaulins or sheets of plastic film to retain heat. If necessary, an enclosed area can be further warmed with gasoline-powered heaters.

The American Concrete Institute (ACI) recommends specific concrete temperatures for cold weather concrete placement. **See Figure 7-10.** Concrete must not be placed on a frozen subgrade. A frozen subgrade acts as a heat sink to drain core temperature from the concrete mixture. When the subgrade thaws, settling occurs, causing structural cracks.

RECOMMENDED COLD WEATHER CONCRETE TEMPERATURES*					
	Air Temperature†	Minimum Dimension or Section†			
		Less than 12	12 to 36	36 to 72	More than 72
Minimum concrete temperature as placed and maintained	Below 40	55	50	45	40
Minimum concrete temperature as mixed for indicated air temperature	Above 30	60	55	50	45
	0 to 30	65	60	55	50
	Below 0	70	65	60	55
Maximum concrete temperature drop permitted in first 24 hr after protection		50	40	30	20

* in °F
† in in.

RECOMMENDED COLD WEATHER AIR-ENTRAINED CONCRETE PROTECTION TIMES*							
Cement Type	Service Category/Frost Protection				Service Category/Safe Strength Level		
	No Load, No Exposure	No Load, Exposure	Partial Load, Exposure	Full Load, Exposure	No Load, No Exposure	No Load, Exposure	Partial Load, Exposure
Types I and II	2	3	3	3	2	3	6
Type III	1	2	2	2	1	2	4

* in days

RECOMMENDED COLD WEATHER FULLY LOADED CONCRETE PROTECTION TIMES*					
Cement Type	Temperature†	Percent of 28-Day Strength			
		50	65	85	95
Type I	50	6	11	21	29
	70	4	8	16	23
Type II	50	9	14	28	35
	70	6	10	18	24
Type III	50	3	5	16	26
	70	3	4	12	20

* in days
† in °F

Figure 7-10. The American Concrete Institute (ACI) recommends specific concrete temperatures for cold weather concrete placement.

After placement, concrete must be kept at the proper temperature for a specified time period while it is green. *Green concrete* is concrete that has been placed, but has not yet reached full strength. Enclosures are often constructed with a heat source to produce an environment suitable for proper curing. The heat source should be selected and positioned so there are no localized hot spots. The concrete temperature should be monitored during the curing process. Monitoring the temperature reveals the need to adjust the heat source or insulation to provide an even temperature within the enclosure. For example, changing variables such as wind conditions can cause heat fluctuations within the enclosure.

If concrete is allowed to freeze before it has set, the mix water is changed to ice, increasing the overall volume of concrete. When mix water is changed to ice, the hydration process and hardening of concrete is delayed because there is no longer any water available for hydration to occur. When thawed, the concrete sets and hardens in its expanded state, causing a large volume of pores and consequently low strength. If freezing takes place after the concrete has set but before it has reached reasonable strength, expansion from ice formation causes a permanent loss of strength. Concrete that has reached a certain strength can withstand stresses caused by freezing mix water. Concrete should not be allowed to freeze before it has reached a compressive strength of at least 500 psi. This critical strength can be achieved with most mixes in 48 hr if the concrete temperature is kept above 49°F.

The amount of water in a concrete mixture is a primary factor in determining the quality and strength of the concrete.

Mixing and Transporting Concrete

Concrete is usually mixed at a batch plant and delivered by truck to the job site. For isolated projects located too far from a batch plant, large onsite stationary mixers are used. **See Figure 7-11.**

Concrete must be mixed thoroughly to ensure consistency and uniform distribution of the ingredients. Mixing time can vary; however, common manufacturer recommendations specify 1 min of mixing for 1 cu yd of mix plus 15 sec for each additional cubic yard. The mixing period is measured from the time all ingredients are in the mixer. Approximately 10% of the required mixing water should be placed in the drum before the cement and aggregate are deposited. Water is then added along with the dry materials. The last 10% of water is added after all the materials are in the drum.

Ready-Mixed Concrete. Ready-mixed concrete is prepared at a batch plant and then delivered to the job site. The concrete ingredients are proportioned according to specifications for the particular job.

Most batch plants are highly automated, ensuring fast and accurate batching. *Batching* is the measuring and proportioning of the concrete mix. The type of truck most often used to transport ready-mixed concrete from the batch plant to the job site is the ready-mixed truck or transit mixer truck. Agitating trucks are also used to transport ready-mixed concrete.

BATCH PLANT

STATIONARY MIXER

Portland Cement Association

Figure 7-11. Large quantities of concrete are mixed at batch plants or in on-site stationary mixers.

A ready-mixed truck is equipped with a large revolving drum operated by an auxiliary engine. **See Figure 7-12.** Truck sizes vary, with drum capacities ranging from 1 cu yd to 12 cu yd. A typical ready-mixed truck also has a water tank so water can be added en route to the job site, if necessary. A number of methods are used to mix and transport concrete with ready-mixed trucks. A common method of mixing concrete is dry batching. *Dry batching* is a procedure in which the dry concrete ingredients are placed in the truck and then mixed with water on the way to the job. Another commonly used method of mixing concrete is shrink mixing. *Shrink mixing* is mixing all the concrete ingredients (including water) for approximately 30 sec at the batch plant, depositing the mix in the drum of the ready-mixed truck, then mixing en route to the job.

Care must be taken with all mixing methods to avoid loss of slump and plasticity while the concrete is transported. Concrete should be delivered and discharged from the truck mixer within 1½ hr after water has been added to the cement and aggregate.

Agitating trucks can also be used to transport concrete instead of ready-mixed trucks. An agitating truck has an open-top body with a paddle mechanism. The paddle mechanism, located at the bottom of the body, maintains the proper plasticity of the concrete during transportation. Concrete delivered by agitating trucks must be premixed at the batch plant. Agitating trucks can only be used for short hauls (30 min to 45 min) or to transport concrete from on-site concrete mixers. The concrete is discharged from the rear by tilting the body of the truck.

READY-MIXED TRUCK

AGITATING TRUCK

Portland Cement Association

Figure 7-12. Ready-mixed and agitating trucks are used to transport ready-mixed concrete.

Special dump trucks with sealed seams may be used for short hauls in moderate weather conditions. Other transportation methods include small rail cars used on special jobs such as concrete tunnel liners, where large quantities of concrete are required and other standard methods of transportation are not possible. Boats and barges may be used to deliver concrete to waterbound projects. Helicopters may be used to carry concrete that is placed in buckets and flown to inaccessible mountain areas.

CONCRETE PLACEMENT

Concrete placement is the transfer of concrete from a concrete truck or other transporting means into the forms. Concrete must be placed properly to prevent segregation. Segregation results in loss of strength of the hardened concrete.

Proper placement procedures are essential to guarantee uniformity of the batch of concrete. Concrete is deposited into the forms as quickly as possible after it has been transported to or mixed on the job site. A delay in placing concrete may result in slump loss, thus affecting the consistency and workability of the concrete. Segregation should be avoided when transferring concrete from a truck to chutes, buckets, and buggies.

The subgrade should be well compacted and graded to its proper elevation when placing concrete for slabs directly over a subgrade. The subgrade is dampened to a depth of 4″ to 6″ to prevent rapid loss of water from the freshly placed concrete. If the subgrade is frozen, all snow and ice should be removed from the placement area.

When placing concrete on rock or existing concrete, the surface of the rock or concrete must be clean, rough, and damp. This allows fresh concrete to firmly grip the surface. If it is necessary to cut away rock or old concrete, all cut surfaces should be horizontal and vertical (not sloping), and all loose fragments and dust should be removed.

Wall forms receiving concrete should have a final check for accurate dimensions as well as proper bracing. Sheathing joints must be tight to prevent mortar loss, and holes in the sheathing must be plugged. Sawdust, nails, and any other debris within the form walls are removed. The inside of the forms is then thoroughly wetted down with water before the concrete is placed. Rebar and inserts placed in the form should be free of rust, oil, mud, and mortar.

Placing Equipment

After concrete is delivered to or mixed on the job site, it is placed in the forms. Various methods and equipment are used and are chosen based on the type of construction project. Equipment most commonly used for transferring and placing concrete includes chutes, buckets, and manual or motorized buggies. In addition, pumping concrete by hose is often used on construction jobs. **See Figure 7-13.**

Chutes. Chutes are used to place concrete directly from trucks into the forms if the truck can be maneuvered close enough to the location. Chutes are also used to transfer concrete to buckets or buggies that deliver the concrete to the forms. Extended chutes may be used to transfer concrete from one elevation to another. The most efficient type of chute has a round bottom constructed of metal, or one equipped with a metal liner. The sides of the chute must be high enough to avoid concrete overflow. The chute should be sloped enough to allow the concrete to flow continuously down the chute without segregation.

Buckets. Buckets are one of the more efficient and flexible methods of moving concrete on the job. Concrete is deposited from the truck or stationary mixer into a bucket ranging in capacity from ½ cu yd to 4 cu yd. The bucket is then lifted by crane and moved over the placement area. A discharge gate is opened at the bottom of the bucket and the concrete flows into the form.

Various attachments may be used to facilitate the flow of concrete from the bucket to the form. A heavy rubber boot or elephant trunk may be attached below the discharge gate to direct the concrete into forms. Another convenient attachment is a side dump chute. The side dump chute provides flexibility when positioning the bucket over wall and slab forms.

CHUTE

CONCRETE BUCKET

MOTORIZED BUGGY

CONCRETE PUMP

Figure 7-13. Concrete is placed using chutes, buckets, motorized buggies, and pumps.

Manual and Motorized Buggies. Concrete buggies are similar in appearance to two-wheeled wheelbarrows. Concrete is deposited in the buggies by chute or bucket. Smooth, rigid runways are constructed to allow the concrete to be transported. Small amounts of concrete can be moved with buggies to points of placement that are not easily accessible by chute or bucket. Manual buggies are pushed by hand and are not as common as the motorized types. The recommended maximum delivery distance is 1000′ for motorized buggies and 200′ for manual buggies.

In addition to concrete placing tools, a variety of equipment and hand tools are also required for concrete placement. Equipment and tools are exposed to concrete, and are subjected to adverse weather conditions that affect their use. All equipment and tools on a job site should be checked for safety and usability prior to use. Tools should be cleaned after each concrete placement to prevent corrosion and buildup of concrete.

Concrete Pumps. Pumping concrete is a method of conveying concrete by pressure through a rigid pipe or flexible hose and depositing it at the placement area. This procedure is widely used for placing concrete for highway decks, foundations, low-rise buildings, and lower levels of tall structures. Pumping concrete eliminates the need for buckets, buggies, and other placement equipment. Pumping rates vary from 10 cu yd to 90 cu yd per hour, depending on the equipment used. An effective range for pumping is from 300′ to 1000′ horizontally and 100′ to 300′ vertically.

The pumping apparatus consists of a mobile unit containing a pump, a boom and cables, and the transport lines that deposit the concrete. The pump supplies the pressure necessary to transport the concrete. The boom and cables support the transport lines, which may be rigid pipes or flexible hoses. Rigid pipes made of steel, aluminum, or plastic are available in 3″ to 8″ diameters. Flexible hoses made of rubber, spiral-wound flexible metal, or plastic are available in 3″ to 5″ diameters.

The concrete is discharged from a concrete truck into a hopper on the pumping apparatus. Thc transport line supported by the boom and cables is moved into position at the placement area. The concrete is then conveyed to the forms. **See Figure 7-14.**

Figure 7-14. Concrete is pumped into the placement area through a flexible transport line.

Placing and Consolidating Concrete

Concrete should be placed as close as possible to its final location within the forms to prevent excessive movement. Excessive movement of concrete within the forms results in segregation and poor consolidation. This occurs because the sand-cement paste in the mix flows ahead of the coarse aggregate materials.

Concrete must be worked and compacted while it is being placed in order to consolidate each fresh layer of concrete with the layer below. Proper consolidation reduces or eliminates rock pockets or honeycombs. A *rock pocket* is a porous void in hardened concrete that consists primarily of coarse aggregate and open voids with little or no mortar. Rock pockets are typically visible on the surface of the concrete after it has set. A *honeycomb* is a void left in concrete due to mortar not effectively filling the spaces among the coarse aggregate. **See Figure 7-15.**

Figure 7-15. Surface defects such as honeycombing are a result of improper consolidation methods.

In small foundation forms, compaction and consolidation can be accomplished by spading the concrete with a wood rod or spading tool. The rod or spading tool must be thin enough to pass between the rebar and form walls and reach the bottom of the form. In heavy construction work, immersion vibrators (or internal vibrators) and external vibrators are used to consolidate the concrete. **See Figure 7-16.** An *immersion vibrator (internal vibrator)* is a tool that consists of a motor, a flexible shaft, and an electrically or pneumatically powered metal vibrating head that is dipped into and pulled through concrete. When activated by a motor, the steel spring causes the metal head to vibrate. The vibrating head is immersed in the concrete and produces rapid consolidation. Internal vibrators are powered by electricity, gas, or compressed air. An *external vibrator* is a vibrator that generates and transmits vibration waves from the exterior to the interior of the concrete. External vibrators are used primarily in precast concrete construction. External vibrators are attached along the outside of the casting forms at strategic positions and are vibrated using a pneumatic or electric power supply.

Portland Cement Association

Figure 7-16. Internal vibrators are used to consolidate larger masses of concrete. The vibrator is immersed vertically into the concrete at 18″ to 30″ intervals for 5 sec to 15 sec.

Floor Slabs. Concrete for floor slabs may be placed directly on a subgrade or on elevated floor forms. The concrete is deposited at the far end of the slab area so that each new batch of concrete is discharged against the face of the concrete already in place. As the concrete is placed, it is spread and consolidated. Immersion vibrators are often used to consolidate concrete for slabs covering large areas. After consolidation, cement finishers strikeoff the concrete with the screeds already set up. The screeds and screed supports are removed and the concrete is finished.

Concrete is placed and distributed in an even layer, rather than placed in piles and leveled off. When placing concrete on a sloping surface, the placement begins at the base of the slope so that compaction is increased by the added weight of each batch. A baffle should be positioned at the end of a chute when working with a sloping surface to prevent segregation and to keep the concrete on the slope. **See Figure 7-17.**

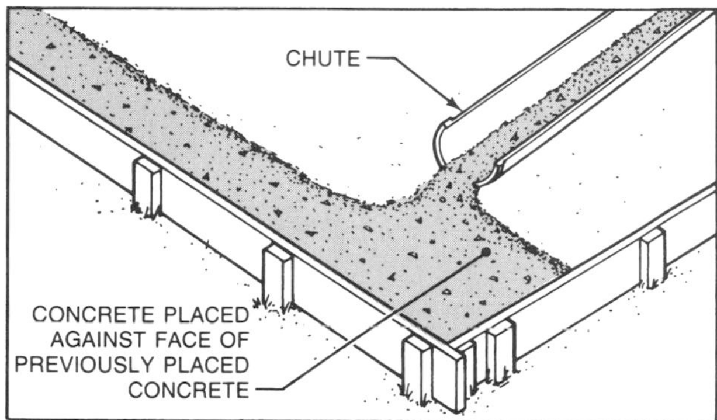

PLACING A CONCRETE FLOOR SLAB

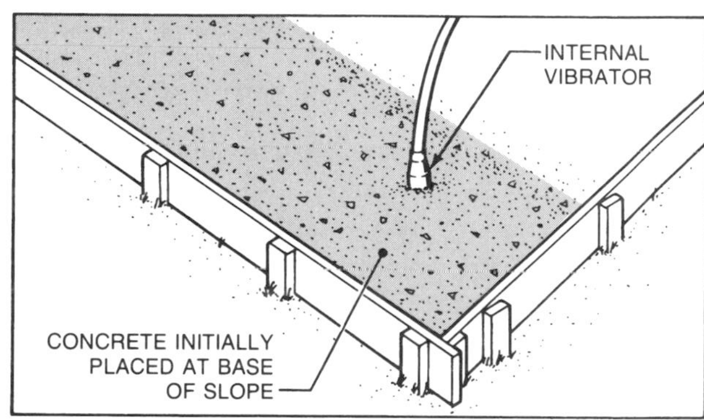

PLACING A SLOPED CONCRETE SURFACE

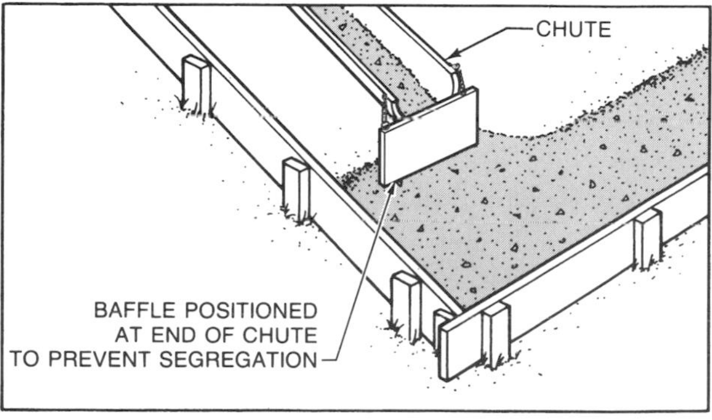

PLACING A SLOPED CONCRETE SURFACE USING A BAFFLE

Figure 7-17. Proper concrete placement ensures consolidation and uniformity of the concrete slab.

Walls. Concrete placed in wall forms should first be placed at the ends of the forms and progress toward the center. Working from the ends toward the center prevents water from being trapped in the corners of the form, which would result in voids in the concrete. Concrete in wall forms should always be placed in level courses from 12″ to 20″ thick. Thinner courses are recommended in forms containing heavy, closely spaced rebar, and when placing concrete from great heights.

After a course is placed, it is immediately consolidated with the course directly below. A spading tool is often used to consolidate the concrete in small and low forms, and an immersion vibrator is used on heavy construction projects. When an immersion vibrator is used, it is held vertically and should pass through the top course, penetrating a few inches into the course below. The vibrator is immersed at regular intervals into the course for periods of 5 sec to 15 sec. Overvibration should be avoided because water and paste may flow to the surface.

To avoid segregation in wall forms, the drop of the concrete must be as near to vertical as possible. Hoppers and flexible drop chutes should be used, particularly in walls with closely spaced rebar. Segregation resulting from concrete striking rebar and form walls above the placement level should be avoided. The free-fall distance of concrete within the wall form should be limited to 4′ to 6′.

In high walls and columns, openings (ports) are commonly cut at intervals in the side of the forms to eliminate long vertical drops. A pocket built along the outside of the opening will momentarily stop the concrete and allow it to flow slowly and evenly into the form. When the concrete level reaches the bottom of the opening, the opening is closed off and the concrete chute is moved to the next higher opening. If concrete is placed into the form by pump, the pipe or hose should be lowered close to the point of discharge. **See Figure 7-18.**

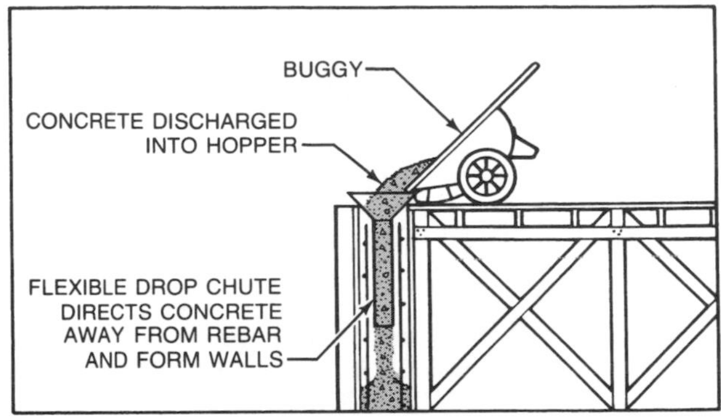

PLACING CONCRETE IN NARROW WALL FORMS

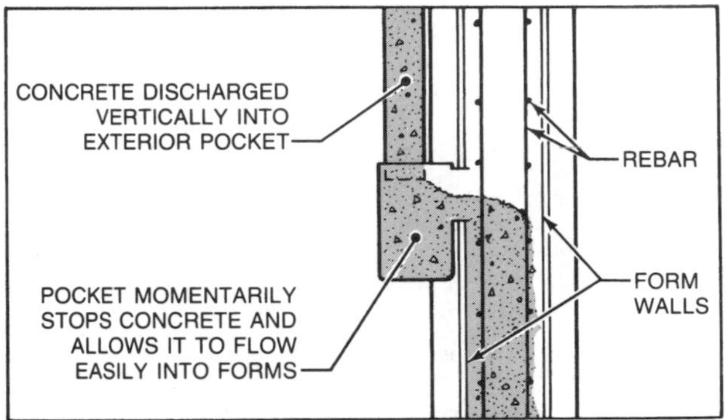

PLACING CONCRETE THROUGH WALL OPENINGS

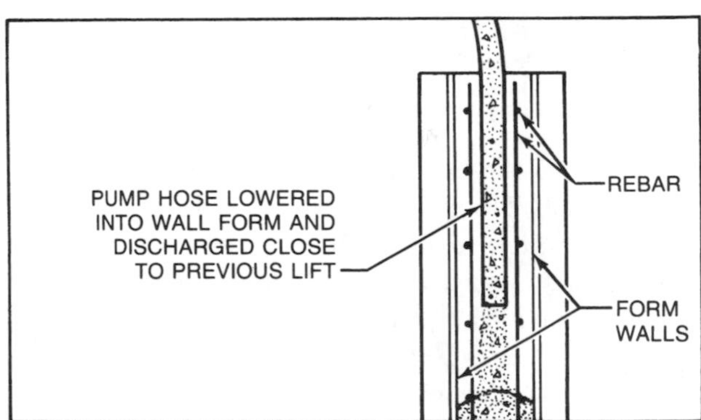

PLACING PUMPED CONCRETE INTO WALL FORMS

Figure 7-18. Considerations must be made for placement of concrete in wall forms.

Placement Rate and Form Pressure. Wall forms must be designed and constructed to withstand the lateral pressure of the concrete as it is being placed into the forms. Lateral pressure of concrete is pressure the concrete exerts against the forms. Lateral pressure of concrete is primarily affected by the rate and height of the placement. Other factors affecting lateral pressure of concrete are internal vibration, temperature of the concrete, weight of the concrete, type of cement, concrete slump, and admixtures.

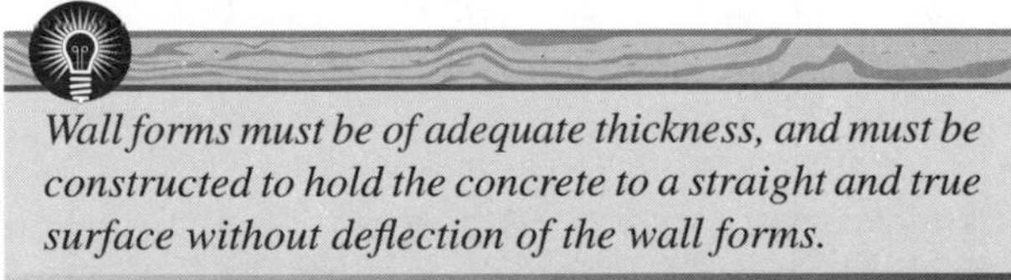

Wall forms must be of adequate thickness, and must be constructed to hold the concrete to a straight and true surface without deflection of the wall forms.

Before concrete sets, it acts like a liquid and exerts force against the form walls in all directions. The amount of pressure at any given point in a form is directly affected by the weight and height of the concrete above that point. Wall thickness does not affect the pressure. Concrete in a plastic state normally exerts a pressure of 150 psf at a placement rate of 1 foot per hour. If, for example, the placement rate is 3 feet per hour, the pressure at the bottom level is 450 psf (3 × 150 = 450). A placement rate of 5 feet per hour produces a pressure at the bottom level of 750 psf (5 × 150 = 750). **See Figure 7-19.**

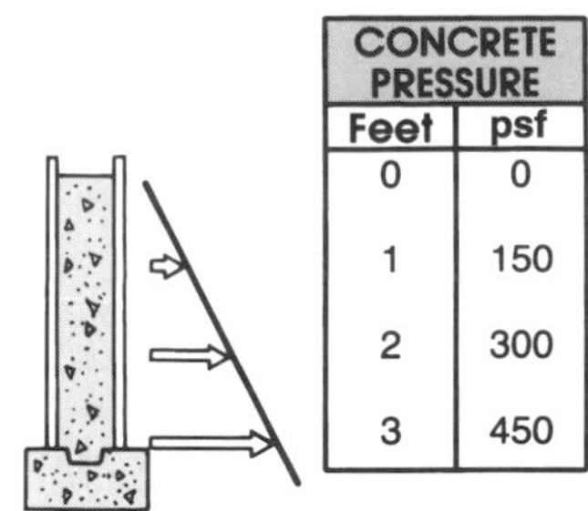

CONCRETE PRESSURE
Feet psf
0 0
1 150
2 300
3 450

CONCRETE PRESSURE AT PLACEMENT RATE OF 3 FT/HR

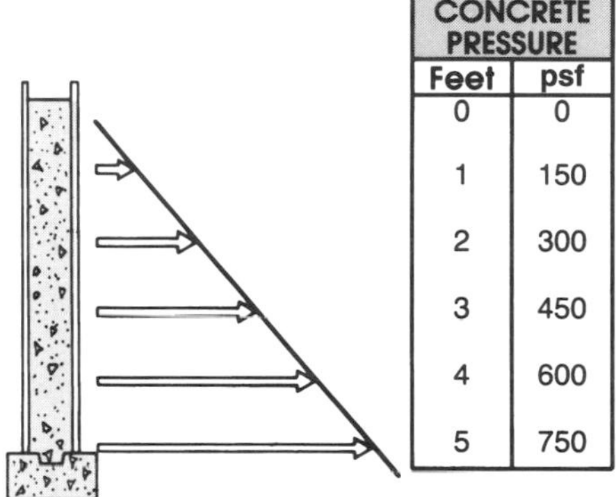

CONCRETE PRESSURE
Feet psf
0 0
1 150
2 300
3 450
4 600
5 750

CONCRETE PRESSURE AT PLACEMENT RATE OF 5 FT/HR

Figure 7-19. Lateral concrete pressure is directly affected by the placement rate and height of the concrete in a form. Wall thickness does not affect the lateral pressure exerted on the form.

Once concrete sets, it does not exert pressure against the form walls. The time it takes concrete to set is directly affected by its temperature. Most form designs are based on an assumed air and concrete temperature of 70°F. Concrete should be placed at a temperature ranging between 60°F and 100°F. Under these conditions the concrete will set in approximately 1 hr. Concrete takes longer to set at lower temperatures; therefore, the placement rate must be decreased or the concrete heated to maintain acceptable lateral form pressure.

CURING CONCRETE

Curing is maintaining proper concrete moisture content and temperature long enough to allow hydration of the concrete and development of the desired properties. Hydration is the chemical reaction between water and cement, at which time the cement becomes the bonding agent of the concrete mix. Hydration begins as soon as the water and cement are combined and continues as long as there is water in the mixture and temperature conditions are favorable. If the water in the concrete mix evaporates too quickly, the hydration process will end before the concrete attains its required design strength. Rapid water loss also results in the concrete shrinking and cracking.

In the initial curing stage, the temperature of the concrete should be maintained at approximately 70°F and the concrete should be kept thoroughly moist for a minimum of 3 days. This is the most critical period in concrete curing. The cement and water combine rapidly and the concrete is most vulnerable to permanent damage. Concrete attains about 70% of its design strength after 7 days, and about 85% of its design strength after 14 days. Full strength is reached after approximately 28 days.

Curing Methods

Wall forms are kept in place during the critical drying period to facilitate the curing of the concrete. The forms retain moisture in the walls and can be sprinkled with water and kept damp during hot, dry weather. Some types of wall forms can be loosened to allow water to run inside the forms.

A concrete slab presents a curing problem because its large exposed surface area allows a great amount of moisture to escape. However, various methods can be used to maintain moisture content. The slab can be sprayed with water by placing a pipe with a series of spray nozzles across the center of the slab. Fogging, a method similar to spraying, produces a mist-like effect.

To retain moisture, the concrete slab can be covered with waterproof paper or polyethylene film, which also protects the slab from frost, direct sun, traffic, and debris. Waterproof paper, available in 18″ to 96″ widths, is laid on the slab and anchored with sand or planks. White polyethylene film is recommended because the white pigment reflects heat. Water-saturated burlap material spread over the slab surface is another effective method used to retain moisture. **See Figure 7-20.** Chemical sealing compounds are also available and are applied by manual sprayers or automatic self-powered sprayers. The compounds are available in clear, black, and white finishes.

Concrete curing times are determined by the type of cement and the temperature during placement. Typical curing times are, for Type I cement, 7 days (at 50°F to 70°F) or 5 days (above 70°F); Type II cement, 14 days (at 50°F to 70°F) or 7 days (above 70°F); Type III cement, 3 days (at 50°F or higher).

POLYETHYLENE FILM

WATER-SATURATED BURLAP

CHEMICAL SEALING COMPOUND

Portland Cement Association

Figure 7-20. Moisture in a concrete slab is retained by covering the slab with polyethylene film or water-saturated burlap, or spraying the slab with a chemical sealing compound.

Concrete placed during hot weather is subject to excessive water evaporation. Because of the heat, curing should begin as soon as possible. Water should be fogged or sprayed for 12 hr before using other curing methods. A white, heat-reflective membrane or polyethylene may be applied later. The curing period should also be extended 7 days or longer during hot weather.

During cold weather, freezing of the concrete must be prevented during the curing period. A minimum concrete temperature of 50°F to 70°F must be maintained at the time of placement. A number of methods may be used to maintain the temperature, including heating the concrete, covering the concrete, or providing a heated enclosure.

STRIPPING FORMS

Stripping forms is the removal of forms after the concrete has set and achieved its required design strength. The concrete must also be hard enough to ensure that its surface will not be damaged when stripping the forms.

Stripping and Removal Schedules

On many building projects, forms must be stripped and removed as soon as possible in order to reuse the form materials. Stripping schedules for low foundation forms can be obtained from local building codes. The period of time required before form removal on heavy construction projects may require the approval of an architect or engineer. This information is often included in the print specifications. On construction projects where stripping specifications are not given, the American Concrete Institute (ACI) has established a stripping schedule that applies to concrete placed under normal conditions. **See Figure 7-21.**

In the case of suspended forms (arch centers and joist, beam, or girder soffits), the forms must remain in place for a longer period where the design live load is less than the dead load. Under these conditions, a larger percentage of the design load is included in the dead load.

Stripping and Removal Methods

Forms should be designed for safe and convenient removal. Panels and other form materials must be stripped carefully to avoid damage so they can be reused. The sequence of stripping is determined by the original assembly of the form.

ACI FORM STRIPPING SCHEDULE		
Structure	**Removal Time**	
Walls*	12 hr	
Columns*	12 hr	
Sides of beams and girders	12 hr	
Pan joists forms† 30″ wide or less Over 30″ wide	 3 days 4 days	
Where Design Live Load is:	**Greater Than Static Load**	**Less Than Static Load**
Arch centers	14 days	7 days
Joist, beam, or grider soffits		
Under 10′ clear span between structual supports	7 days‡	4 days
10′ to 20′ clear span between structual supports	14 days‡	7 days
Over 20′ clear span between structual supports	21 days‡	14 days
One-way floor slabs		
Under 10′ clear span between structual supports	4 days‡	3 days
10′ to 20′ clear span between structual supports	7 days†	4 days
Over 20′ clear span between structual supports	10 days†	7 days

* where such forms also support formwork for slab or beam soffits, the removal time of the later should govern
† of the type which can be removed without disturbing forming or shoring
‡ where forms may be removed without disturbing shores, use one-half of values shown but not less than three days

Figure 7-21. The American Concrete Institute (ACI) has established a stripping schedule that applies to concrete placed under normal conditions.

When stripping a wall form, the tie clamps or wedges are removed and the walers are pried off. In a panel system, the panel section and studs can be removed as a unit. In a built-in-place form, the studs and/or walers are pried off, followed by the removal of the plywood sheathing.

Column forms should be constructed so the sides can be pried off and removed without disturbing the adjoining beam or girder forms. Beam or girder forms should be constructed so the side panels can be stripped before the beam bottoms. The floor soffit forms are removed by releasing the supporting shores and stringers. If it is necessary to allow a section of floor slab form to fall free, a platform or other support should be constructed to reduce the distance of drop.

Cranes, using two lines, are used to strip larger panels or ganged panel forms. One line is attached at the top of the form for the upward pull. The second line is attached at a lower point to exert an outward pull. When stripping large panels and ganged panel forms, a few ties should remain connected until the crane lines are securely attached.

Metal stripping bars should not be used to pry form panels directly from concrete surfaces, as this results in damage to the concrete. Wooden wedges are placed against the concrete surface and the stripping bar is then used to pry the form.

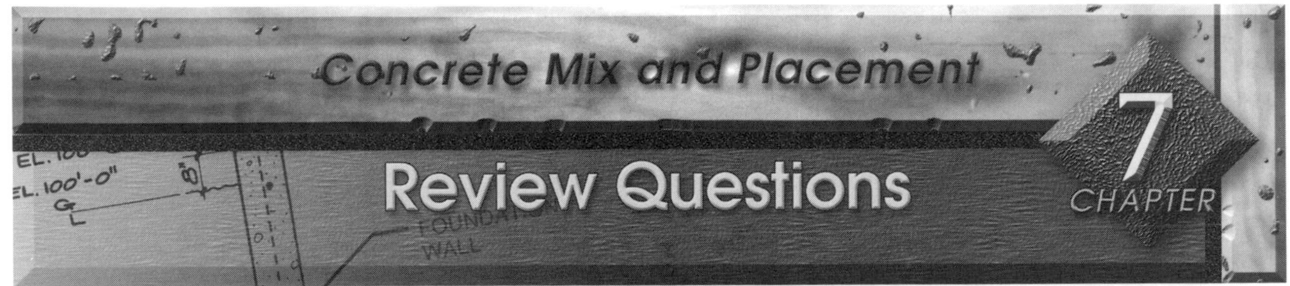

Name ______________________________ Date ______________

Completion

__________ **1.** The basic ingredients of concrete are fine and coarse ___, ___, and ___.

__________ **2.** ___ is the source of most cement products.

__________ **3.** ___ is the chemical reaction between water and cement that produces concrete.

__________ **4.** The smallest proportion in a concrete mix is ___.

__________ **5.** If the weight of water in a cubic yard of concrete is 10 lb, and the weight of the cement is 20 lb, the water-cement ratio is ___.

__________ **6.** A measure of the consistency of concrete is determined with a(n) ___ test.

__________ **7.** A compression test measures the ___ strength of concrete.

__________ **8.** ___ concrete is concrete that has been placed, but has not yet reached full strength.

__________ **9.** Concrete exposed to repeated freeze/thaw cycles should contain ___% to ___% entrained air.

__________ **10.** A(n) ___ is a type of water-reducing admixture that provides concrete with increased workability with less mix water.

__________ **11.** As a general rule, concrete containing ___ requires more water to produce the same slump.

__________ **12.** When the ambient temperature reaches or exceeds ___°F, measures must be taken to control the temperature of the concrete.

__________ **13.** Concrete gains its full strength ___ days after placement.

__________ **14.** The durability of concrete in freezing temperatures is increased by adding ___ admixtures.

__________ **15.** ___ admixtures are effective in slowing down the setting process of concrete during hot weather.

__________ **16.** Rapid water evaporation is a problem encountered when placing concrete during ___ weather.

__________ **17.** ___°F is the ideal setting temperature for concrete.

__________ **18.** A slow ___ rate is encountered when placing concrete during cold weather.

__________ **19.** ___ concrete is mixed at a batch plant.

__________ **20.** A revolving ___ contains the concrete in a ready-mixed truck.

__________ **21.** Concrete should be discharged from ready-mixed trucks within ___ hr after water has been added to it.

__________ **22.** ___ is the separation of the sand and cement ingredients from the coarse aggregate in a concrete mix.

__________ **23.** ___ loss is the result of segregation during concrete placement.

__________ **24.** The ___ ratio is the amount of cement in relation to the amount of water in a batch of concrete.

__________ **25.** ___ crane line(s) should be attached to the forms when stripping ganged panel forms.

Multiple Choice

__________ **1.** A ___ is used to discharge concrete directly from a ready-mixed truck into forms.
A. buggy
B. bucket
C. chute
D. all of the above

__________ **2.** A concrete bucket is ___.
A. a long, arch-shaped tube
B. lifted by crane and carried to the placement area
C. moved by motorized buggies
D. only used on small concrete projects

__________ **3.** The maximum recommended delivery distance for concrete using a motorized buggy is ___′.
A. 200
B. 400
C. 1000
D. 2000

__________ **4.** The effective range for pumping concrete is ___ horizontally and ___ vertically.
A. 100′ to 300′; 300′ to 1000′
B. 250′ to 1000′; 150′ to 300′
C. 300′ to 1000′; 100′ to 300′
D. none of the above

__________ **5.** When placing concrete, it should be ___.
A. placed and moved from one end of the form
B. allowed to free-fall 10′ to 12′ in the form
C. compacted as little as possible
D. placed as close as possible to its final location in the form

__________ **6.** A(n) ___ is used to consolidate concrete in heavy construction work.
A. spade
B. rod
C. immersion vibrator
D. consolidator

_______ **7.** Concrete in wall forms is placed in level lifts ranging from ___″ to ___″ deep.
A. 6; 12
B. 8; 16
C. 12; 20
D. 20; 30

_______ **8.** When placing concrete for walls, segregation can be avoided by ___.
A. dropping the concrete vertically
B. using hoppers and chutes
C. limiting the free-fall to 4′ to 6′
D. all of the above

_______ **9.** Lateral concrete pressure against a form wall at the time the concrete is being placed ___.
A. depends on the thickness of the wall
B. is affected only by the concrete temperature
C. is affected by the rate of pour and the concrete temperature
D. depends on the method of consolidation

_______ **10.** The most critical period in the curing and hydration process of concrete is ___ days.
A. during the first 3
B. during the first 28
C. after the first 3
D. after the first 7

_______ **11.** An effective method for curing a concrete slab is to ___.
A. leave the edge forms in place
B. cover the slab with boards
C. place a layer of sand over the slab
D. spread waterproof paper or plastic film over the slab

_______ **12.** A cubic yard of concrete containing 9.5 lb of water and 22 lb of cement has a water-cement ratio of ___.
A. .33
B. .37
C. .43
D. .46

_______ **13.** The American Concrete Institute recommends that wall forms not supporting beam soffits may be stripped after ___ hours.
A. 3
B. 6
C. 12
D. 24

_______ **14.** The maximum allowable slump for concrete used for beams and reinforced walls is ___″.
A. 1
B. 2
C. 3
D. 4

_________ **15.** ___ is excess water that collects on the surface of concrete as aggregate material sinks in the concrete mixture.

A. Mix water
B. Bleedwater
C. Segregation water
D. Hydrated water

_________ **16.** ___ is one of the most commonly used accelerators.

A. Calcium chloride
B. Pozzolan
C. Sodium chloride
D. Permicite

_________ **17.** Hydration stops when the ambient temperature is ___°F or below.

A. 15
B. 10
C. 0
D. –10

_________ **18.** The lateral pressure at the base of a wall form with a placement rate of 4 feet per hour is ___ psf.

A. 300
B. 400
C. 500
D. 600

_________ **19.** Concrete attains approximately ___% of its design strength after 14 days.

A. 55
B. 70
C. 85
D. 100

_________ **20.** During the initial curing stage of concrete, the concrete should be kept moist for ___ days.

A. three
B. four
C. five
D. six

Chapter 8
Concrete Formwork Computations

A form builder encounters applications on the job that involve math and printreading skills. The applications range from calculating area and volume to determining placement of forms for foundation walls and footings.

Section 1, Math Fundamentals, includes information regarding conversions of decimal values to foot and inch equivalents and foot and inch values to decimal equivalents, area and volume calculations, and tread and riser calculations. Conversions and area and volume calculations are used to determine the surface area of a wall form or volume of concrete required for a structure. Tread and riser calculations determine the number of treads and risers and their respective dimensions.

Section 2, Printreading Exercises, includes questions related to five prints: Plot Plan, Crawl Space Foundation, Full Basement Foundation, Slab-on-Grade Foundation, and Heavy Construction Foundation. The prints contain information regarding foundation construction that is presented in chapters 3 through 5 of the text.

Section 3, Form Materials and Concrete Quantity Takeoff, provides a step-by-step procedure for calculating the amount of form materials and volume of concrete required for a small residential structure. The Review Questions contain exercises for estimating the amount of form materials and volume of concrete for a full basement foundation.

SECTION 1

MATH FUNDAMENTALS

A form builder must have an understanding of basic math concepts to perform estimating and formwork operations. Mixed numbers and fractions, such as 4⅝″ or ½″, are routinely used in form construction. Decimal numbers, such as .55 or 101.8′, are used when referring to ratios or elevations. A form builder must convert decimal numbers to mixed numbers and fractions, and mixed numbers and fractions to decimal numbers, to calculate area, volume, and tread and riser dimensions.

Decimal Foot to Inch Equivalents

Elevations on a plot or foundation plan are commonly expressed in decimal numbers, such as 45.2′ or 10.2′. A form builder must convert decimal numbers to their foot and inch equivalents for use with a standard tape or wood rule.

Mathematical conversion of decimal numbers to inches is accomplished by first determining the number of inches and then the fraction. The answers are combined to obtain the inch equivalent.

Example

Convert .83′ to an inch equivalent.

Solution

1. Multiply the decimal value by 12 to determine the number of inches and decimal part of an inch.

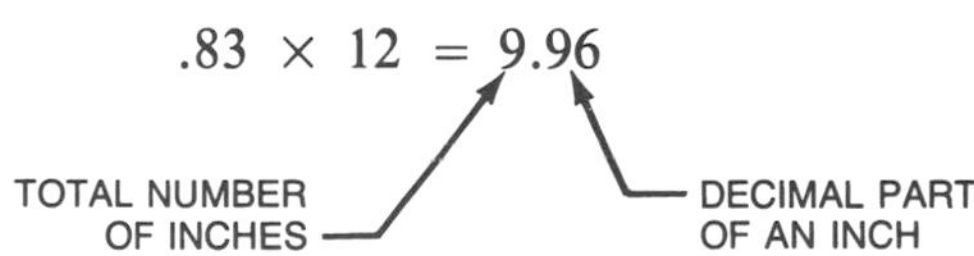

2. Multiply the decimal part of the inch by a commonly used denominator (for example 4, 8, or 16). A larger denominator value results in greater accuracy.

.96 × 16 = 15.36

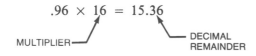

MULTIPLIER — DECIMAL REMAINDER

3. If the decimal remainder is .5 or greater, round up the value preceding the decimal point. If the decimal remainder is less than .5, round the value down. Use the rounded value as the numerator and the commonly used denominator in step 2 as the denominator. Reduce the fraction to lowest terms if possible.

15.36 rounded down to 15

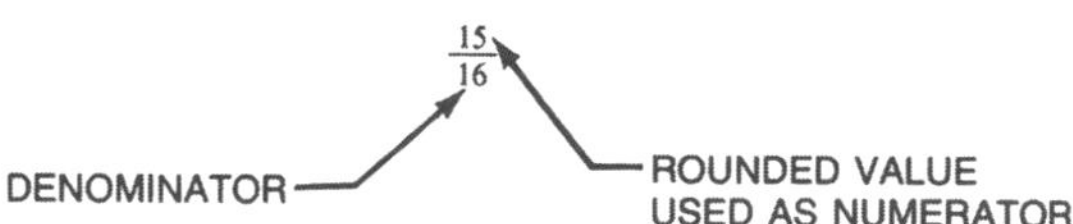

4. Combine the answers from steps 1 and 3 to obtain the inch equivalent.

9″ + 15⁄16″ = 9 15⁄16″

.83′ = 9 15⁄16″

When converting an elevation to an inch equivalent, the number preceding the decimal point is the total number of feet and the number following the decimal point is converted to inches.

Example

Convert 10.9′ to a foot and inch equivalent.

Solution

1. Multiply the value after the decimal point by 12 to determine the number of inches and decimal part of an inch.

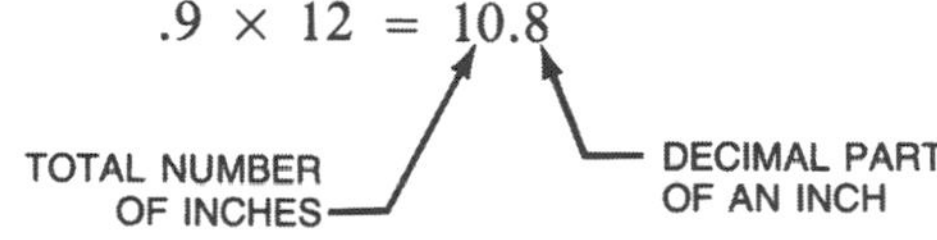

2. Multiply the decimal part of an inch by a commonly used denominator (for example 4, 8, or 16). A larger denominator value results in greater accuracy.

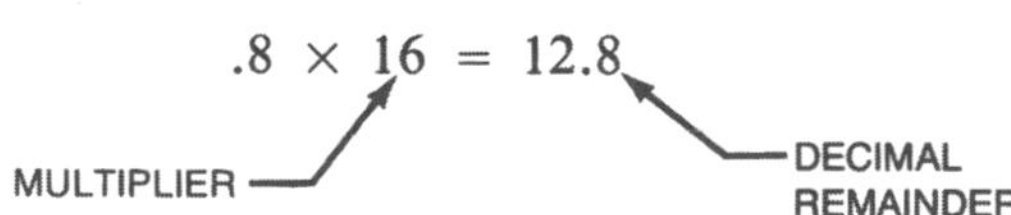

3. If the decimal remainder is .5 or greater, round up the value preceding the decimal point. If the decimal remainder is less than .5, round the value down. Use the rounded value as the numerator and the commonly used denominator in step 2 as the denominator. Reduce the fraction to lowest terms if possible.

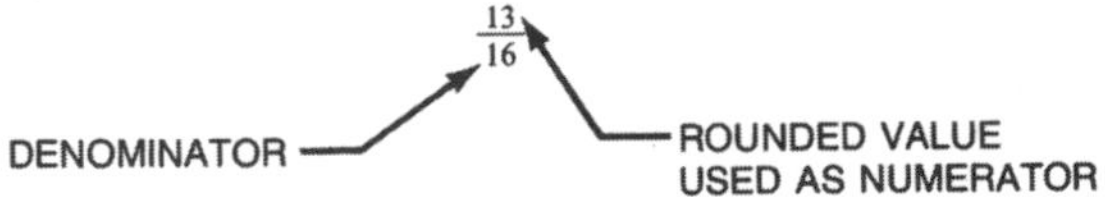

4. Combine the foot value (number preceding the decimal point in the example) with inch values.
   $10' + 10'' + \frac{13}{16}'' = 10'\text{-}10\frac{13}{16}''$
   $10.9' = 10'\text{-}10\frac{13}{16}''$

A conversion table is also used to convert decimal feet to inches. (See Appendix A for Conversion Table.) The decimal foot value to be converted is located in the table and the number of inches is read from the horizontal row above. The fractional part of an inch, in eighths, is read from the vertical column to the left. If a decimal foot value is not located in the conversion table, the mathematical conversion is used.

Example

Convert .22′ to an inch equivalent.

Solution

		Inches		
		0	1	2
8th of an Inch	0	.00	.08	.17
	1	.01	.09	.18
	2	.02	.10	.19
	3	.03	.11	.20
	4	.04	.13	.21
	5	.05	.14	.22
	6	.06	.15	.23
	7	.07	.16	.24

$.22' = 2\frac{5}{8}''$

Inch to Decimal Inch and Foot Equivalents

Fractions and mixed numbers are converted to decimal inch and foot equivalents for use in applications such as determining stair riser and tread dimensions. Decimal inch equivalents express the given value in terms of 1″, such as $\frac{3}{4}'' = .75''$. Decimal foot equivalents express the given value in terms of 1′, such as $\frac{3}{4}'' = .06'$.

A decimal inch equivalent is determined mathematically or by using a conversion table. (See Appendix A for Conversion Table.) To convert a fraction to a decimal inch equivalent, the numerator is divided by the denominator. A decimal point is placed before the equivalent.

Example

Convert $\frac{7}{8}''$ to a decimal inch equivalent.

Solution

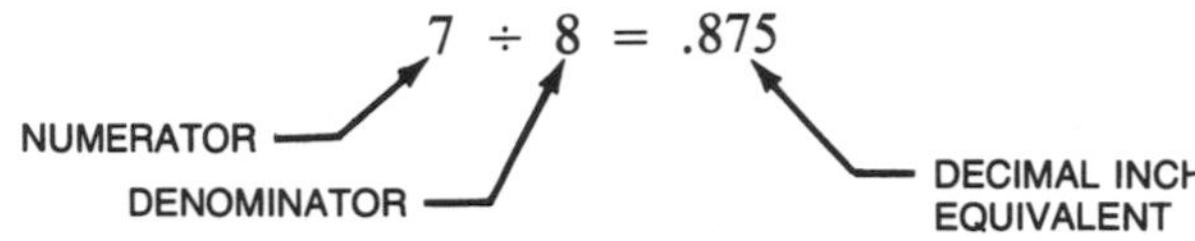

$\frac{7}{8}'' = .875''$

A mixed number is converted to a decimal inch equivalent in a similar manner. The whole number remains the same and the fraction is converted to a decimal inch equivalent as in the previous example.

Example

Convert $8\frac{1}{4}''$ to a decimal inch equivalent.

Solution

$8\frac{1}{4}'' = 8.25''$

A decimal foot equivalent is determined by converting a fraction, whole number, or mixed number to a decimal inch equivalent. The decimal inch equivalent is divided by 12 to determine the decimal foot equivalent.

Example

Convert 10″ to a decimal foot equivalent.

Solution

$10 \div 12 = .833$
$10'' = .833'$

Example

Convert $8\frac{1}{2}''$ to a decimal foot equivalent.

Solution

1. Convert $8\frac{1}{2}''$ to a decimal inch equivalent.
   $8\frac{1}{2}'' = 8.5''$
2. Convert 8.5″ to a decimal foot equivalent.
   $8.5 \div 12 = .708$
   $8\frac{1}{2}'' = .708'$

Area Calculation

Area is a measurement of space that a two-dimensional plane or surface occupies. Area calculations determine the amount of form sheathing material required or the area within building or property lines. Area is expressed in units such as square feet (sq ft) or square inches (sq in.). The area for various shapes is calculated by different formulas, but in general is determined by multiplying the length by the height or width. The area of a circle is determined by multiplying π (3.14) by the radius squared or .7854 times the diameter squared.

Squares, Rectangles, and Parallelograms. The area of a horizontal or vertical square or rectangular surface, such as a building site or form wall, is determined by multiplying the two outside dimensions. The area of a horizontal surface is determined by multiplying the width by the length, and a vertical surface is determined by multiplying the length by the height.

Example

Determine the area (A) of the rectangle.

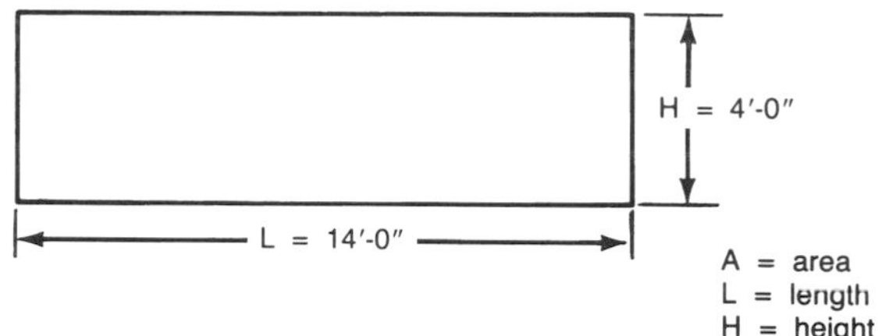

Solution

$A = L \times H$

$A = 14'\text{-}0'' \times 4'\text{-}0''$

$A =$ **56 sq ft**

Example

Determine the area (A) of the parallelogram.

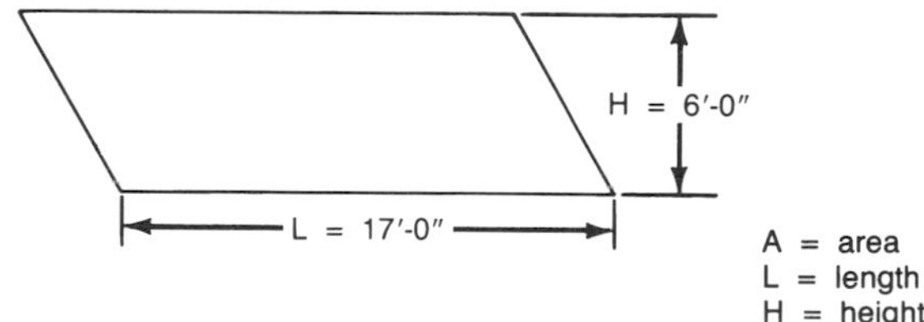

Solution

$A = L \times H$

$A = 17'\text{-}0'' \times 6'\text{-}0''$

$A =$ **102 sq ft**

Triangles. The area of a triangle is determined by multiplying the base (b) dimension by the altitude (a) and dividing by 2.

Example

Determine the area (A) of the triangle.

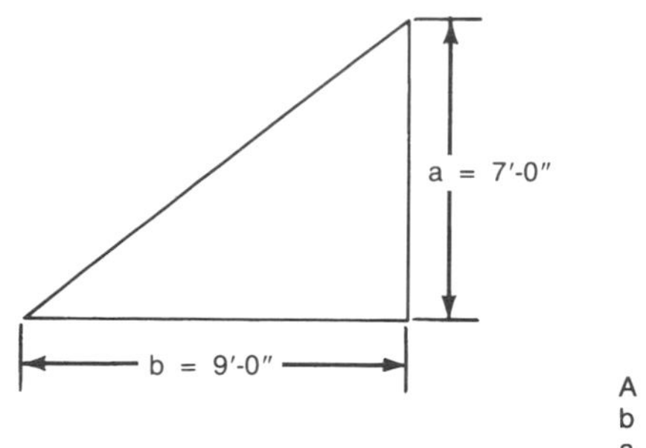

Solution

$$A = \frac{ba}{2}$$

$$A = \frac{9'\text{-}0'' \times 7'\text{-}0''}{2}$$

$$A = \frac{63}{2}$$

$A =$ **31.5 sq ft**

Trapezoids. A trapezoid is a geometric shape having four sides in which two of the sides are parallel. Battered foundation walls or tapered pier footings are common designs incorporating a trapezoid shape. To determine the area of a trapezoid, add the length of the top and bottom and multiply the sum by the height. Divide the results by 2 to obtain the area.

Example

Determine the area (A) of the trapezoid.

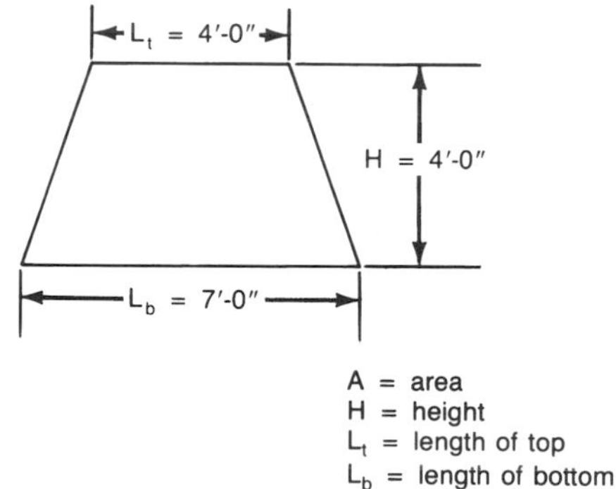

Solution

$$A = \frac{H(L_t + L_b)}{2}$$

$$A = \frac{4'\text{-}0''(4'\text{-}0'' + 7'\text{-}0'')}{2}$$

$$A = \frac{4'\text{-}0''(11'\text{-}0'')}{2}$$

$A =$ **22 sq ft**

Circles. The area of a circle is determined by using either the radius or the diameter of the circle. The radius is one-half the diameter and is measured from the center point to an edge of a circle. The diameter is the distance from one edge of a circle to the other, passing through the center point.

The area of a circle is calculated by multiplying π (3.14) by the radius squared (radius × radius). The area may also be determined by multiplying .7854 by the diameter squared (diameter × diameter).

Example

Determine the area (A) of the circle using the radius of the circle.

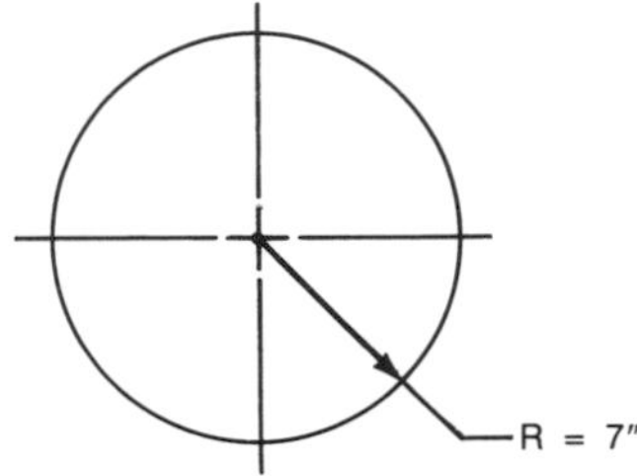

Solution

$A = \pi R^2$

$A = 3.14 \times (7'' \times 7'')$

$A = 3.14 \times 49$

$A =$ **153.86 sq in.**

Example

Determine the area (A) of the circle using the diameter of the circle.

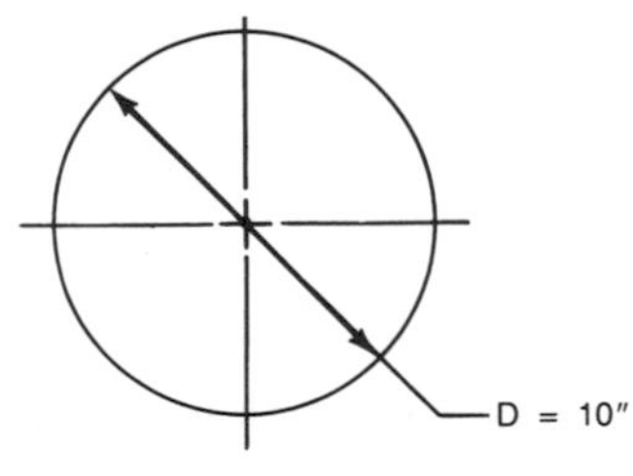

Solution

$A = .7854\ D^2$

$A = .7854 \times (10'' \times 10'')$

$A = .7854 \times 100$

$A =$ **78.54 sq in.**

Volume Calculation

Volume is the amount of space that a three-dimensional figure or object occupies. Volume calculations are used to determine the amount of fill required for a building site or the amount of concrete required to form a footing or wall. Volume is expressed in cubic units, such as cubic feet (cu ft) or cubic yards (cu yd).

Rectangular Solids and Cubes. A rectangular solid is a six-sided solid object with a rectangular base. A cube is a solid object with six equal square faces. Examples of a rectangular solid or cube are a foundation wall, square pier footing, or square or rectangular column. The volume of a rectangular solid or cube is determined by multiplying its thickness, length, and height. The volume of a square column is determined by multiplying the width squared by the height.

Example

Determine the volume (V) of the rectangular solid.

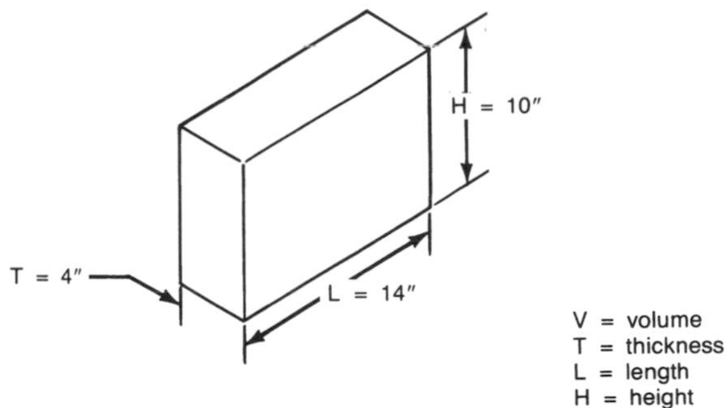

Solution

$V = T \times L \times H$

$V = 4'' \times 14'' \times 10''$

$V =$ **560 cu in.**

When calculating large amounts of material, volume is commonly expressed in cubic feet or cubic yards. One cubic foot equals 1728 cubic inches and 1 cubic yard equals 27 cubic feet. When calculating volume for large amounts of material, thickness, length, and height are converted to decimal feet and then multiplied to obtain cubic feet. The result is divided by 27 to obtain cubic yards.

Example

Determine the volume (V) of the foundation wall.

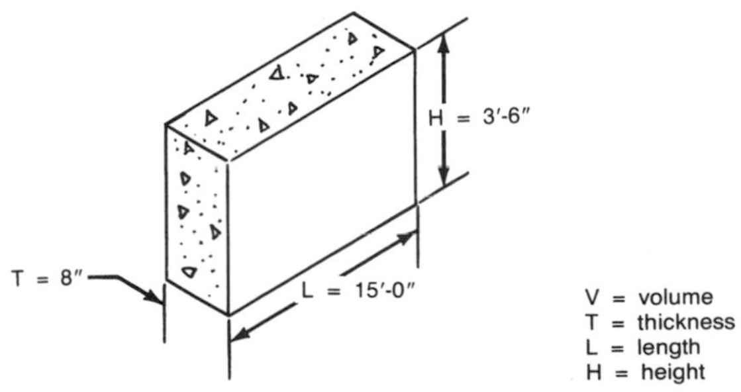

Solution

1. Convert the thickness (T), length (L), and height (H) to decimal foot equivalents.
   $T = 8'' = .67'$
   $L = 15'\text{-}0'' = 15.0'$
   $H = 3'\text{-}6'' = 3.5'$
2. Calculate the volume of the foundation wall.
   $V = T \times L \times H$
   $V = .67' \times 15.0' \times 3.5'$
   $V = \mathbf{35.18\ cu\ ft}$
3. Convert cubic feet to cubic yards.
   $cu\ yd = cu\ ft \div 27$
   $cu\ yd = 35.18 \div 27$
   $cu\ yd = \mathbf{1.3}$

Frustums of Pyramids. A frustum is a pyramid cut off parallel to its base. The sides of a frustum, such as a tapered pier footing, are trapezoids. The volume of a frustum with one battered side, such as a battered foundation wall, is determined by multiplying the average thickness by the height and length. The approximate volume of a frustum with four battered sides is determined by adding the areas of the top and bottom and dividing by 2 and then multiplying by the height.

Example

Determine the volume (V) of the battered foundation wall.

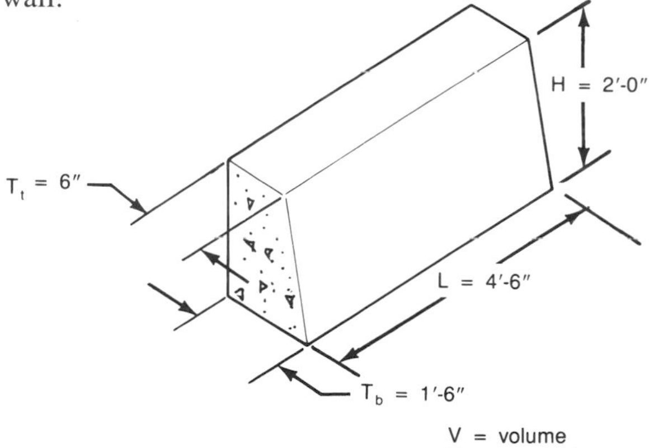

Solution

1. Convert the top thickness (T_t), bottom thickness (T_b), height (H), and length (L) to decimal foot equivalents.
   $T_t = 6'' = .5'$
   $T_b = 1'\text{-}6'' = 1.5'$
   $H = 2'\text{-}0'' = 2.0'$
   $L = 4'\text{-}6'' = 4.5'$
2. Calculate the volume of the battered foundation wall.
   $V = \dfrac{T_t + T_b}{2} \times H \times L$
   $V = \dfrac{.5' + 1.5'}{2} \times 2.0' \times 4.5'$
   $V = 1.0' \times 2.0' \times 4.5'$
   $V = \mathbf{9\ cu\ ft}$
3. Convert cubic feet to cubic yards.
   $cu\ yd = cu\ ft \div 27$
   $cu\ yd = 9 \div 27$
   $cu\ yd = \mathbf{.33}$

Example

Determine the volume (V) of the tapered pier footing.

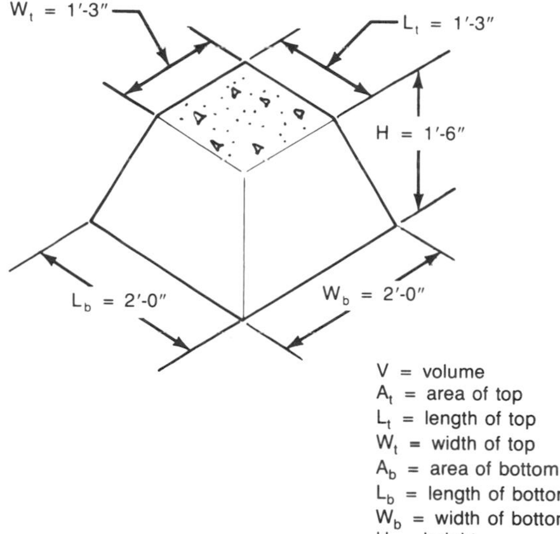

Solution

1. Convert dimensions to decimal foot equivalents.
   $W_t = 1'\text{-}3'' = 1.25'$
   $L_t = 1'\text{-}3'' = 1.25'$
   $W_b = 2'\text{-}0'' = 2.0'$
   $L_b = 2'\text{-}0'' = 2.0'$
   $H = 1'\text{-}6'' = 1.5'$
2. Determine the areas of the top (A_t) and bottom (A_b).
   $A_t = L_t \times W_t$
   $A_t = 1.25' \times 1.25'$
   $A_t = 1.56$ sq ft
   $A_b = L_b \times W_b$
   $A_b = 2.0' \times 2.0'$
   $A_b = 4.0$ sq ft

3. Determine the volume of the tapered pier footing.

$$V = \frac{A_t + A_b}{2} \times H$$

$$V = \frac{1.56 + 4.0}{2} \times 1.5'$$

$$V = 2.78 \times 1.5'$$

$$V = 4.17 \text{ cu ft}$$

4. Convert cubic feet to cubic yards.

cu yd = *cu ft* ÷ 27

cu yd = 4.17 ÷ 27

cu yd = **.15**

Cylinders. A cylinder is a solid object with a circular cross-sectional area. Round columns and piers are cylinders. The volume of a cylinder is determined by multiplying the cross-sectional area by the height. The cross-sectional area is determined by using the diameter or radius.

Example

Determine the volume (V) of the round column.

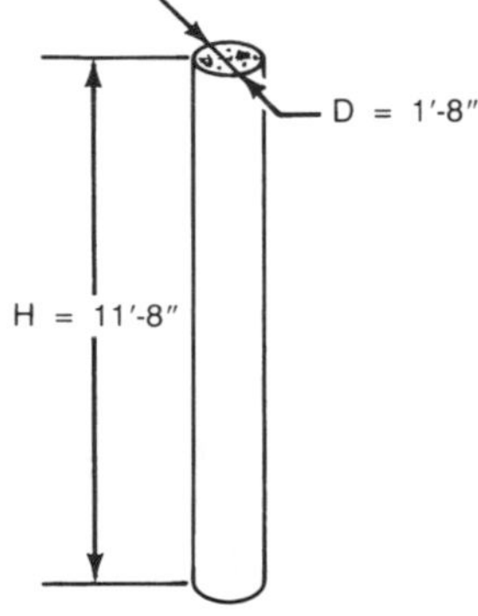

V = volume
D² = diameter²
H = height

Solution

1. Convert the diameter (D) and height (H) to decimal foot equivalents.

$D = 1'\text{-}8'' = 1.67'$

$H = 11'\text{-}8'' = 11.67'$

2. Determine the volume of the round column.

$V = .7854 \times D^2 \times H$

$V = (.7854 \times 1.67^2) \times 11.67'$

$V = (.7854 \times 2.79) \times 11.67'$

$V = 2.19 \times 11.67'$

$V = 25.56$ cu ft

3. Convert cubic feet to cubic yards.

cu yd = *cu ft* ÷ 27

cu yd = 25.56 ÷ 27

cu yd = **.95**

Example

Determine the volume (V) of the circular pier footing.

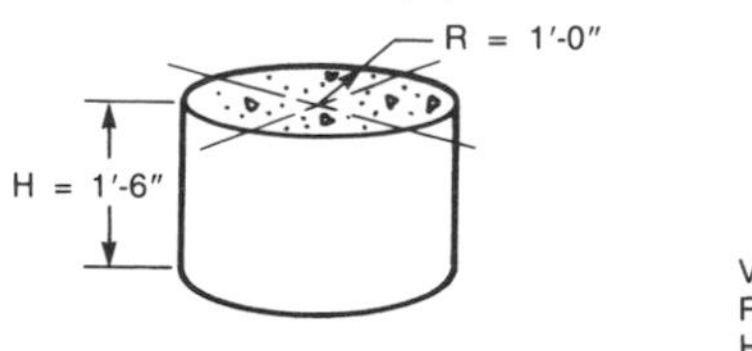

V = volume
R² = radius²
H = height

Solution

1. Convert the radius (R) and height (H) to decimal foot equivalents.

$R = 1'\text{-}0'' = 1.0'$

$H = 1'\text{-}6'' = 1.5'$

2. Determine the volume of the circular pier footing.

$V = \pi R^2 \times H$

$V = (3.14 \times 1.0^2) \times 1.5'$

$V = (3.14 \times 1.0) \times 1.5'$

$V = 3.14 \times 1.5'$

$V = 4.71$ cu ft

3. Convert cubic feet to cubic yards.

cu yd = cu ft ÷ 27

cu yd = 4.71 ÷ 27

cu yd = **.17**

Curved Walls and Footings. Curved walls and footings are used in the construction of storage tanks and silos. When calculating the volume of a curved wall or footing, the surface area is multiplied by the height.

Example

Determine the volume (V) of the curved wall and footing.

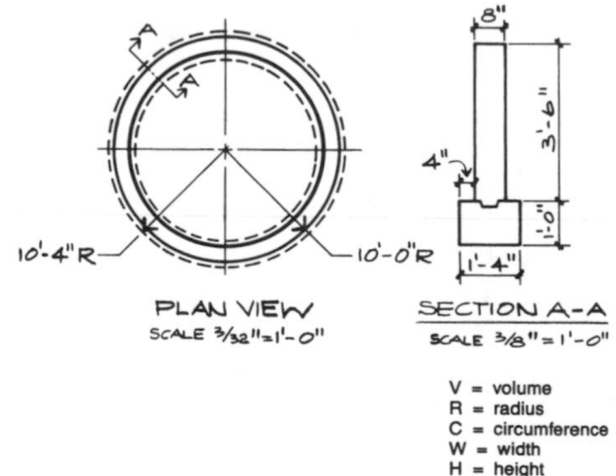

V = volume
R = radius
C = circumference
W = width
H = height

Solution

1. Convert the curved wall dimensions to decimal foot equivalents.

$R = 10'\text{-}0'' = 10.0'$

$H = 3'\text{-}6'' = 3.5'$

$T = 8'' = .67'$

2. Determine the area of a circle corresponding to the outside of the curved wall (A_o).
$A_o = \pi R^2$
$A_o = 3.14 \times 10.0' \times 10.0'$
$A_o = 314'$

3. Determine the area of a circle corresponding to the inside of the curved wall (A_i).
$C_i = \pi R^2$
$C_i = 3.14 \times 9.33' \times 9.33'$
$C_i = 273.3'$

4. Determine the surface area of the top of the curved wall.
$surface\ area = A_o - A_i$
$surface\ area = 314 - 273.3$
$surface\ area = 40.7'$

5. Determine volume of the curved wall.
$V = surface\ area \times H$
$V = 40.7 \times 3.5'$
$V = 142.5$ cu ft

6. Convert cubic feet to cubic yards.
cu yd = cu ft ÷ 27
cu yd = 142.5 ÷ 27
cu yd = **5.3**

7. Convert the footing dimensions to decimal foot equivalents.
$R = 10'\text{-}4'' = 10.33'$
$H = 1'\text{-}0'' = 1.0'$
$W = 1'\text{-}4'' = 1.33'$

8. Determine the area of a circle corresponding to the outside of the footing (F_o).
$F_o = \pi R^2$
$F_o = 3.14 \times 10.33' \times 10.33'$
$F_o = 335$ sq ft

9. Determine the area of a circle corresponding to the inside of the footing (F_i).
$F_i = \pi R^2$
$F_i = 3.14 \times 9' \times 9'$
$F_i = 254.3$ sq ft

10. Determine the surface area of the top of the footing.
$surface\ area = F_o - F_i$
$surface\ area = 335 - 254.3$
$surface\ area = 80.7$ sq ft

11. Determine the volume of the footing.
$V = \text{surface area} \times H$
$V = 80.7 \times 1.0'$
$V = 80.7$ cu ft

12. Convert cubic feet to cubic yards.
cu yd = cu ft ÷ 27
cu yd = 80.7 ÷ 27
cu yd = **3.0**

Tread and Riser Dimensions

Tread and riser dimensions are determined before the stairway forms are constructed. The tread is the horizontal surface of a step. The riser is the vertical member between two steps. Riser height is determined by dividing the total rise of a stairway by the number of risers. Tread depth is determined by dividing the total run of the stairway by the number of treads. The number of treads in a stairway is one less than the number of risers.

Example

Determine the number of treads and risers and the riser height and tread depth of a stairway. The total rise is 6′-7″ and the total run is 8′-6″. The riser height should be between 7″ and 7½″ and the tread depth should be 10″ minimum.

Solution

1. Convert the total rise to inches.
Total rise = (*no. of feet* × 12) + *no. of inches*
Total rise = (6′ × 12) + 7″
Total rise = 79″

2. Determine the number of risers by dividing the total rise by the minimum desired riser height. Disregard the decimal remainder.
No. of risers = *total rise* ÷ *minimum riser height*
No. of risers = 79 ÷ 7
No. of risers = 11.28
No. of risers = 11

3. Determine the exact riser height by dividing the total rise by the number of risers.
Riser height = *total rise* ÷ *no. of risers*
Riser height = 79 ÷ 11
Riser height = 7.18″

4. Convert the decimal inch value to a fractional equivalent.
7.18″ = 7³⁄₁₆″

5. Convert the total run to inches.
Total run = (*no. of feet* × 12) + *no. of inches*
Total run = (8′ × 12) + 6″
Total run = 102″

6. Determine the tread depth by dividing the total run by the number of treads.
Tread depth = *total run* ÷ *no. of treads*
Tread depth = 102″ ÷ 10
Tread depth = 10.2″

7. Convert the decimal inch value to a fractional equivalent.
10.2″ = 10³⁄₁₆″

Name ______________________ Date ______________

Convert the decimal foot values to foot and inch equivalents.

_______ **1.** .47′
_______ **2.** .20′
_______ **3.** 2.04′
_______ **4.** .79′
_______ **5.** 1.56′
_______ **6.** .09′
_______ **7.** 11.35′
_______ **8.** .98′
_______ **9.** .63′
_______ **10.** 3.16′

Convert the inch values to decimal inch equivalents.

_______ **11.** ½″
_______ **12.** 3⁄16″
_______ **13.** 1⅝″
_______ **14.** 4¼″
_______ **15.** 1⁄16″
_______ **16.** 10⅞″
_______ **17.** ¾″
_______ **18.** 7⁄16″
_______ **19.** 3⅛″
_______ **20.** ⅜″

Convert the foot and inch values to decimal foot equivalents.

_______ **21.** 9″
_______ **22.** 3′-6″
_______ **23.** 1′-4¾″
_______ **24.** 1½″
_______ **25.** 5′-9¼″
_______ **26.** 6′-0″
_______ **27.** 5′-8″
_______ **28.** 6′-2½″
_______ **29.** 1′-0¾″
_______ **30.** 12′-6⅜″

Determine the area of the geometric shapes. Express the answer in square feet.

_______ **1.** Height = 12″, Length = 4½″
_______ **2.** Height = 1′-6″, Length = 4′-0″
_______ **3.** Height = 5′-2″, Length = 5′-2″
_______ **4.** Height = 5′-0″, Length = 34′-0″
_______ **5.** Height = 2′-6″, Length = 2′-6″
_______ **6.** Height = 1′-8″, Length = 3′-6″

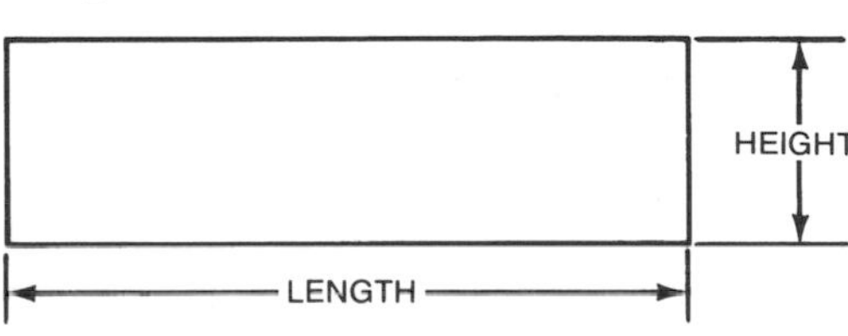

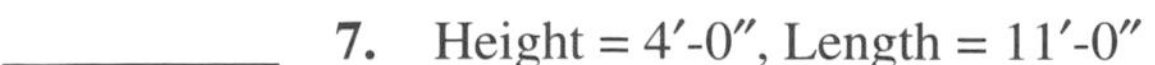

_________ 7. Height = 4′-0″, Length = 11′-0″

_________ 8. Height = 5′-6″, Length = 14′-0″

_________ 9. Height = 8′-2″, Length = 10′-6″

_________ 10. Height = 13′-4″, Length = 15′-10″

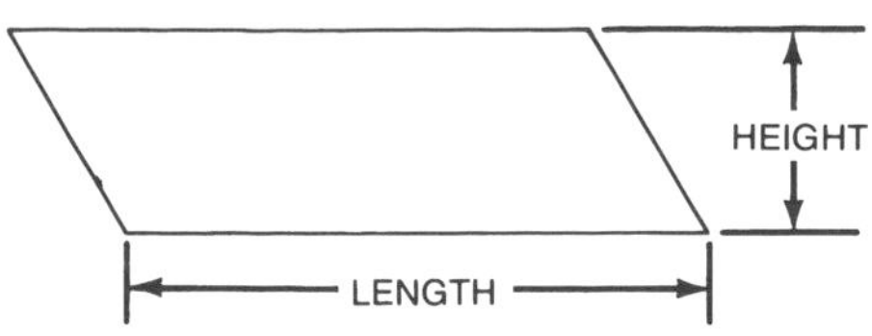

_________ 11. Base = 5′-0″, Altitude = 6′-0″

_________ 12. Base = 10′-0″, Altitude = 15′-6″

_________ 13. Base = 9′-2″, Altitude = 10′-6″

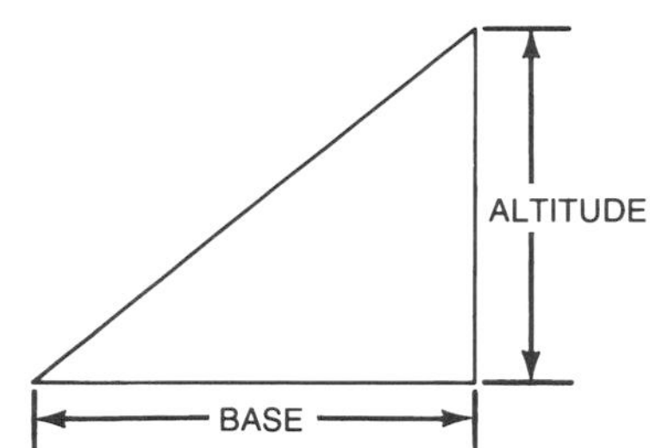

_________ 14. Height = 4′-0″, Length of top = 8′-0″, Length of bottom = 10′-0″

_________ 15. Height = 6′-0″, Length of top = 16′-4″, Length of bottom = 24′-8″

_________ 16. Height = 4′-6″, Length of top = 13′-10″, Length of bottom = 21′-8″

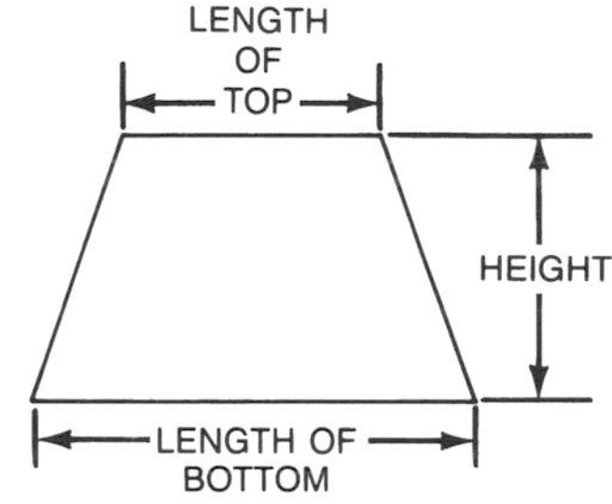

_________ 17. Diameter = 2′-6″

_________ 18. Radius = 5′-0″

_________ 19. Diameter = 3′-4″

_________ 20. Radius = 8′-6″

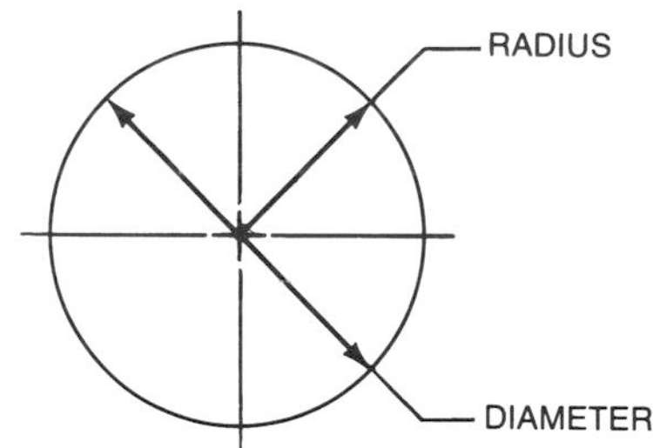

Determine the volume of the geometric solids. Express the answer in cubic yards.

_________ 1. Thickness = 1′-0″, Height = 5′-0″, Length = 12′-0″

_________ 2. Thickness = 7″, Height = 5′-10″, Length = 19′-0″

_________ 3. Thickness = 8″, Height = 3′-6″, Length = 10′-6″

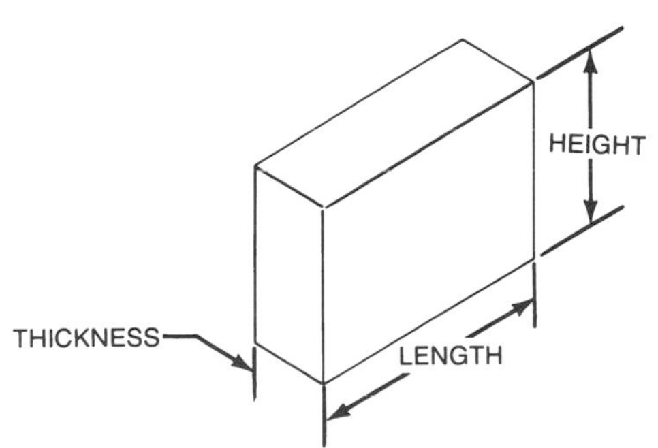

_________ **4.** Width = 10″, Height = 12′-0″

_________ **5.** Width = 1′-4″, Height = 9′-0″

_________ **6.** Width = 2′-6″, Height = 15′-6″

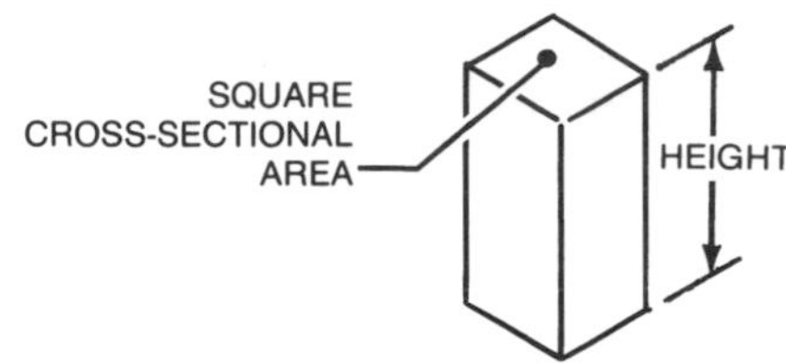

_________ **7.** Top thickness = 8″, Bottom thickness = 1′-4″, Height = 3′-0″, Length = 10′-0″

_________ **8.** Top thickness = 6″, Bottom thickness = 1′-0″, Height = 2′-0″, Length = 15′-0″

_________ **9.** Top thickness = 8″, Bottom thickness = 1′-8″, Height = 4′-0″, Length = 18′-0″

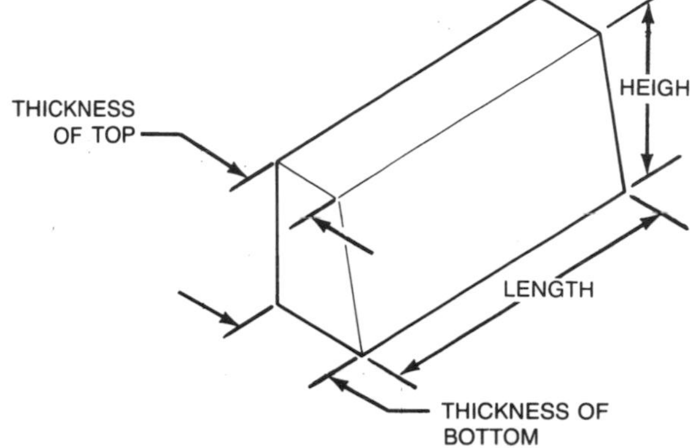

_________ **10.** Area of top = 2.25 sq ft, Area of bottom = 4.0 sq ft, Height = 2′-0″

_________ **11.** Area of top = 4.0 sq ft, Area of bottom = 6.25 sq ft, Height = 3′-6″

_________ **12.** Area of top = 6.25 sq ft, Area of bottom = 9.0 sq ft, Height = 3′-0″

_________ **13.** Area of top = 9.0 sq ft, Area of bottom = 12.25 sq ft, Height = 4′-0″

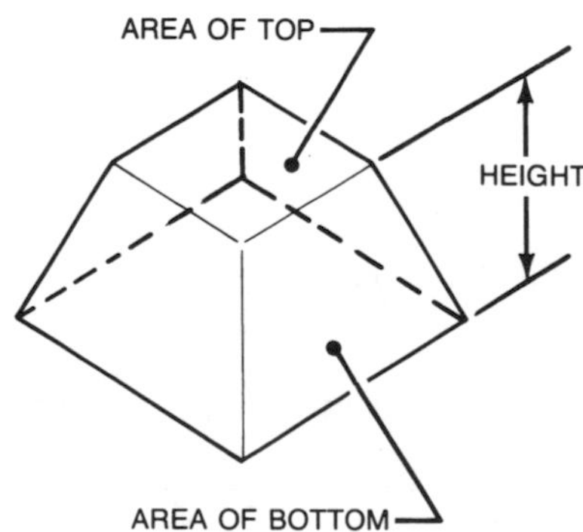

_________ **14.** Diameter = 2′-0″, Height = 10′-0″

_________ **15.** Radius = 1′-3″, Height = 8′-0″

_________ **16.** Diameter = 1′-6″, Height = 10′-3″

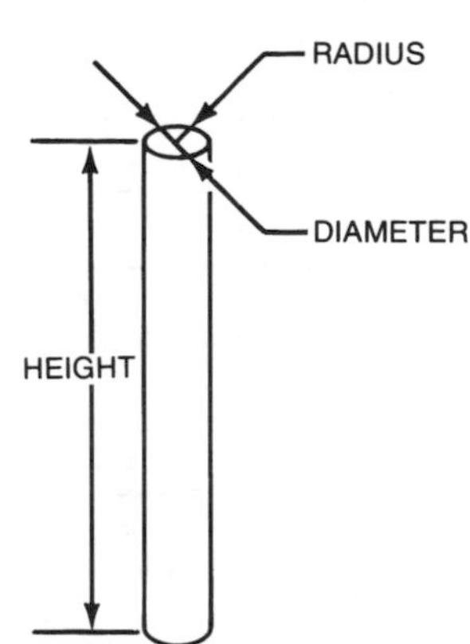

Determine the volume of the concrete required to form the secondary clarification tank walls and footing. Express the answers in cubic yards.

__________ **17.** Footing

__________ **18.** 8″ wall

__________ **19.** 12″ wall

__________ **20.** Total volume of concrete

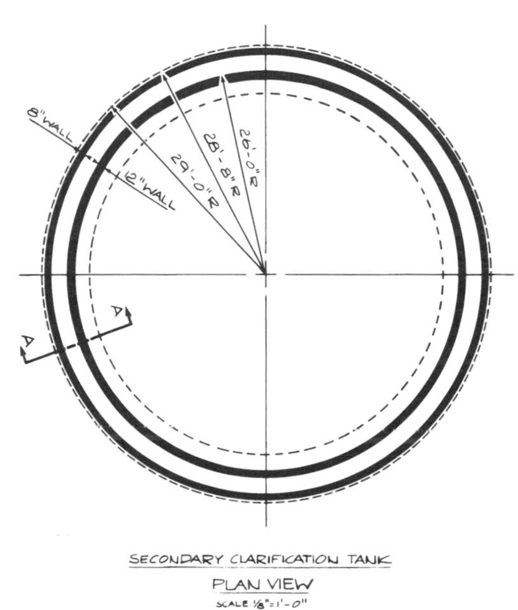

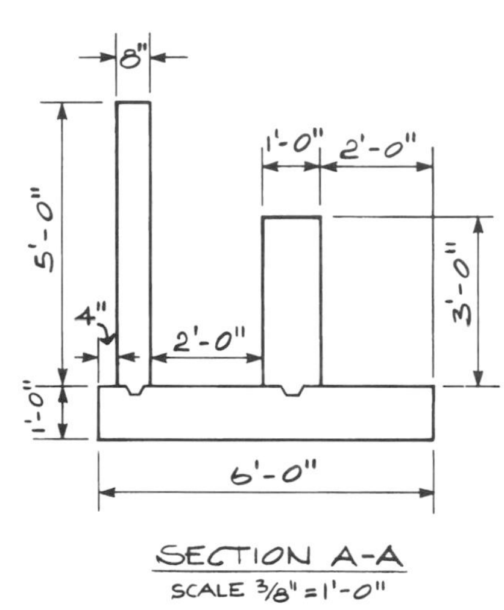

Symons Corporation

Determine the number of risers and treads and the riser height and tread depth for the stairways based on the following code requirements.

The rise of every step in a stairway shall not exceed 7½″. The tread depth shall not be less than 10″. The largest riser height or tread depth within a stairway shall not exceed the smallest by more than ¼″.

21.

TOTAL RISE	TOTAL RUN	NUMBER OF RISERS	HEIGHT OF RISERS	NUMBER OF TREADS	DEPTH OF TREADS
3′-9″	4′-2″				
4′-8″	5′-11¾″				
6′-1¾″	8′-7½″				
4′-1⅞″	5′-0″				
8′-6⅜″	12′-8¾″				
5′-6″	6′-11″				
10′-3¼″	14′-11″				
7′-3¾″	9′-2¾″				
11′-4 9/16″	15′-11¼″				
7′-5¼″	10′-5⅛″				

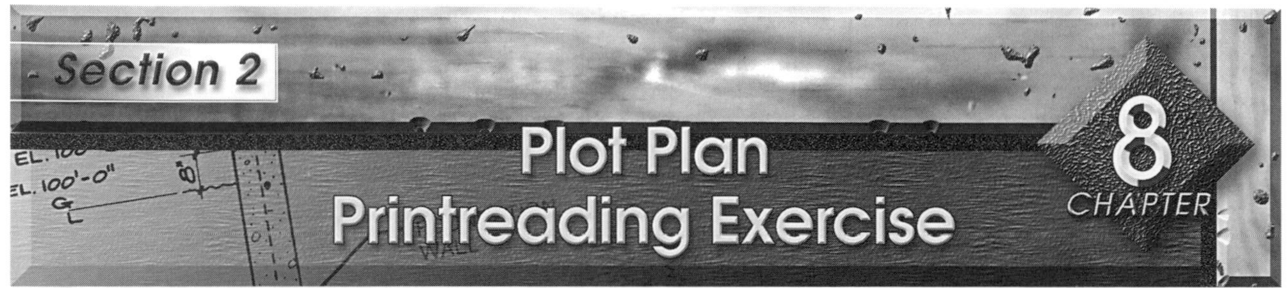

Name ______________________________ **Date** ______________

A plot plan is the main source of information regarding preliminary site work. A plot plan shows the location of a building, finish grade levels and elevations, utilities, streets and roads, sidewalks, easements, and locations of trees, bushes, and shrubs. Benchmarks are also identified and used as a reference for elevations.

Refer to the Plot Plan on page 273 to answer the following questions.

Completion

__________ **1.** The benchmark is located in the ___ corner of the building site.

__________ **2.** The finish floor elevation is ___′.

__________ **3.** The overall east to west dimension of the building site is ___.

__________ **4.** The difference in height between the benchmark and the grade at the northwest corner of the lot is___′.

__________ **5.** The distance from the curb line to the front sidewalk is ___.

__________ **6.** The widest north to south dimension of the building is ___.

__________ **7.** The front setback of the building is ___.

__________ **8.** The gas line and water main are located below the surface of the ___.

__________ **9.** The driveway is ___ wide.

__________ **10.** The grade level at the northeast corner of the building is ___′.

Multiple Choice

__________ **1.** The driveway slopes ___′ from the north end to the southeast corner.
A. .4
B. .5
C. .6
D. .8

__________ **2.** The sidewalk is ___ wide.
A. 3′-0″
B. 4′-6″
C. 6′-0″
D. 7′-5″

__________ **3.** The west property line is ___ long.
A. 80′-0″
B. 101.8′
C. 103.0′
D. 140′-0″

__________ **4.** The building site slopes ___′ from the northeast to northwest corner.
A. 1.0
B. 2.0
C. 3.4
D. 4.0

__________ **5.** The grade level at the southwest corner of the building is ___′.
A. 101.0
B. 101.8
C. 102.2
D. 102.4

__________ **6.** The front property line is ___ from the sidewalk.
A. 2′-0″
B. 4′-0″
C. 6′-0″
D. 15′-0″

__________ **7.** The east property line is ___ from the side of the building.
A. 15′-0″
B. 18′-0″
C. 19′-6″
D. 20′-0″

__________ **8.** The driveway is sloped ___′ from the southwest to southeast corner.
A. .1
B. .2
C. 1.1
D. 1.4

__________ **9.** The lowest grade level of the building site is ___′.
A. 99.8
B. 100.0
C. 100.2
D. 102.8

__________ **10.** The building site is sloped ___′ from the southwest corner of the building to the southwest corner of the lot.
A. .2
B. .4
C. .6
D. .9

__________ **11.** The finish floor elevation is ___′ higher than the southwest corner of the lot.
A. 3.2
B. 4.3
C. 5.0
D. 6.2

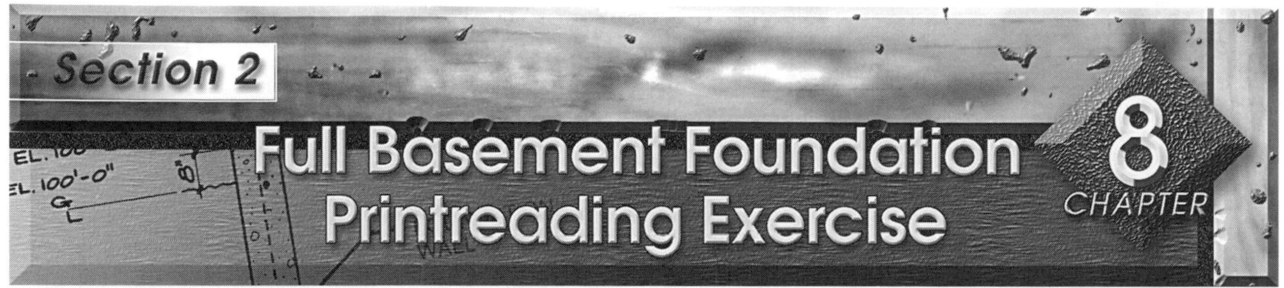

Name ______________________________ Date ______________

A full basement foundation provides an area below the superstructure for living or storage space. The basement area is commonly below ground level. A set of prints for a full basement foundation includes a plan view and related section view drawings. A plan view provides elevations and overall dimensions of a building and locations and dimensions of structural components such as pier footings and stairways. The size and spacing of first floor joists and size and location of supporting beams and posts are also included in the plan view.

Section view drawings indicate the thickness and height of foundation footings and walls, type and spacing of reinforcement, and thickness of the basement floor slab. Information regarding specially formed features, such as pockets or shoulders, is also provided in the section view drawings.

Refer to the Full Basement Foundation print on pages 275 and 276 to answer the following questions.

Completion

__________ **1.** The north foundation wall is ___ long.

__________ **2.** The east garage wall is ___ from the east wall of the house.

__________ **3.** The back stoop is ___ wide × ___ long.

__________ **4.** The center of the basement window in the southeast corner of the south foundation wall is ___ from the southeast corner of the building.

__________ **5.** The dimensions of the beam pocket are ___ × ___ × ___.

__________ **6.** There are ___ treads and ___ risers in the stairway.

__________ **7.** There are ___ No. 4 rebar running continuously in the foundation footings.

__________ **8.** The east foundation wall is ___ long.

__________ **9.** A(n) ___″ wide-flange, ___ lb beam is supported by the pipe columns.

__________ **10.** ___ × ___ floor joists are supported by the wide-flange beam.

__________ **11.** The front stoop is ___ wide × ___ long.

__________ **12.** The column footings are ___″ × ___″ × ___″.

__________ **13.** The stair risers are ___″ high and the treads are ___″ deep.

__________ **14.** The concrete floor slab is ___″ thick.

__________ **15.** The floor joists are spaced ___″ OC.

__________ **16.** The garage opening is ___ wide.

__________ **17.** The wide-flange beam is ___ long.

__________ **18.** The north foundation wall is ___″ thick.

__________ **19.** The footing supporting the south foundation wall is ___″ wide.

__________ **20.** Three No. ___ rebar are spaced ___″ OC vertically in the foundation walls.

__________ **21.** The elevation at the garage door opening is ___.

__________ **22.** The footing beneath the garage door opening extends ___″ beyond the foundation wall.

__________ **23.** The elevation at the top of the front stoop wall is ___.

__________ **24.** The wide-flange beam has a ___″ clearance at the end.

__________ **25.** The footing under the north foundation wall is ___″ wide.

Multiple Choice

__________ **1.** The pipe columns are ___″ in diameter.
A. 2
B. 4
C. 6
D. 8

__________ **2.** The stairway dimensions are ___.
A. 3′-0″ × 9′-0″
B. 3′-0″ × 12′-8″
C. 4′-0″ × 9′-0″
D. 5′-0″ × 12′-8″

__________ **3.** The front stoop wall is ___″ thick.
A. 4
B. 5
C. 6
D. 8

__________ **4.** The distance between the column centers is ___.
A. 9′-11″
B. 12′-0″
C. 12′-8″
D. 38′-0″

__________ **5.** The footing under the south foundation wall is ___″ high.
A. 6
B. 8
C. 10
D. 12

_________ **6.** The front and back stoop walls extend ___″ below the frost line.

A. 2
B. 4
C. 6
D. 10

_________ **7.** All foundation walls extend a minimum of ___″ above the grade line.

A. 4
B. 6
C. 8
D. 10

_________ **8.** The foundation footing extends ___″ beyond the west garage wall.

A. 4
B. 6
C. 8
D. 10

_________ **9.** Basement windows in the north foundation wall are ___ center-to-center.

A. 7′-8″
B. 9′-11″
C. 11′-8″
D. 12′-0″

_________ **10.** The top of the north foundation footing is ___ below the grade line.

A. 8″
B. 2′-6″
C. 7′-2″
D. 7′-10″

_________ **11.** The rebar in the north foundation footing are placed ___″ from the bottom of the footing.

A. ¾
B. 1
C. 1½
D. 2½

_________ **12.** The footing below the garage door opening is ___″ wide.

A. 10
B. 16
C. 20
D. 24

_________ **13.** The back stoop wall is ___ high.

A. 2′-6″
B. 3′-0″
C. 3′-6″
D. 4′-0″

__________ **14.** A concrete ___ is formed along the outside of the north section of the west foundation wall.
A. keyway
B. shoulder
C. pocket
D. core

__________ **15.** The distance from the southwest corner of the building to the garage door opening is ___.
A. 2′-0″
B. 2′-3⅝″
C. 16′-4¾″
D. 100′-8″

__________ **16.** The foundation walls under the brick veneer are ___″ thick.
A. 6
B. 8
C. 10
D. 12

__________ **17.** The foundation wall elevation at the garage opening is ___″ lower than the adjacent walls.
A. 6
B. 8
C. 10
D. 12

__________ **18.** The stairway has ___ risers.
A. 12
B. 13
C. 14
D. 15

Section 2

Crawl Space Foundation Printreading Exercise

8 CHAPTER

Name ______________________________ **Date** ______________

Crawl space foundations provide a crawl space beneath the floor joists for access to plumbing and other utilities. The prints for a crawl space foundation include a plan view and related section view drawings required to construct the foundation. The plan view provides overall dimensions of the foundation and locations and dimensions of foundation walls, footings, and pier footings. The size and direction of floor joists, posts, and beams are also included in the plan view.

Section view drawings show the thicknesses and heights of foundation walls and footings, and type and size of reinforcement. Section view drawings also provide information regarding the minimum clearance between the ground and floor joists.

Refer to the Crawl Space Foundation print on pages 277 to 279 to answer the following questions.

Completion

__________ **1.** The west foundation wall is ___ long.

__________ **2.** The minimum depth of excavation below the floor joists is ___.

__________ **3.** A(n) ___″ I beam extends across the crawl space area.

__________ **4.** The beam pockets are ___″ × ___″ × ___″.

__________ **5.** The foundation footing beneath the west wall is ___″ × ___″.

__________ **6.** The concrete piers in the crawl space area are ___″ in diameter.

__________ **7.** The fireplace footing is ___ × ___ × ___.

__________ **8.** The east foundation wall is ___ long.

__________ **9.** The fireplace extends ___ outside the east foundation wall.

__________ **10.** The centers of the piers in the crawl space area are ___ from the north foundation wall.

__________ **11.** The west foundation wall is ___″ thick × ___ high.

__________ **12.** The elevation of the garage door pockets is ___.

__________ **13.** The distance from the outside of the west garage wall to the east side of the west crawl space wall is ___.

__________ **14.** The south foundation wall is ___″ thick × ___ high.

__________ **15.** The pier footings are ___″ × ___″ × ___″.

__________ **16.** The difference in elevation between the garage floor and the top of the foundation walls is ___.

__________ **17.** The west garage wall is ___ from the center of the west door pocket.

__________ **18.** There are ___ No. 4 rebar required in the foundation footings.

__________ **19.** The center of the pier footing in the garage is ___ from the north foundation wall.

__________ **20.** No. 4 rebar are placed ___″ OC each way in the fireplace footing.

__________ **21.** The elevation at the top of the concrete pier is ___.

__________ **22.** A(n) ___″ concrete slab is placed below the fireplace.

__________ **23.** The hearth extends ___″ in front of the fireplace.

__________ **24.** The north foundation wall is ___″ thick × 3′-0″ high.

__________ **25.** The sill plates are fastened to the top of the foundation wall with ___″ × ___″ anchor bolts.

Multiple Choice

__________ **1.** The north foundation wall is ___ long.
A. 24′-0″
B. 26′-0″
C. 50′-0″
D. 65′-0″

__________ **2.** A ___″ concrete slab is placed in the garage.
A. 3
B. 4
C. 5
D. 6

__________ **3.** The center-to-center distance between the piers in the crawl space area is ___.
A. 8′-8″
B. 10′-6″
C. 13′-9″
D. 18′-8″

__________ **4.** The elevation of the door pocket in the north foundation wall is ___.
A. 95′-6″
B. 96′-6″
C. 99′-8″
D. 100′-0″

_________ **5.** The outside of the east foundation wall is ___ from the center foundation wall.
A. 8′-8″
B. 13′-9″
C. 24′-3″
D. 26′-0″

_________ **6.** The cavity between the walls under the fireplace is ___ wide.
A. 1′-4″
B. 1′-6″
C. 1′-10″
D. 2′-8″

_________ **7.** The foundation plan is drawn to a scale of ___″ = 1′-0″.
A. ⅛
B. ¼
C. ⅜
D. ½

_________ **8.** Vertical and horizontal rebar are placed ___″ OC in the foundation walls.
A. 12
B. 16
C. 18
D. 24

_________ **9.** Anchor bolts are placed ___″ OC.
A. 16
B. 24
C. 32
D. 48

_________ **10.** A ___ film vapor barrier is placed over the soil in the crawl space area.
A. polyurethane
B. polyethylene
C. rubberized
D. none of the above

_________ **11.** The concrete piers are ___ tall.
A. 4″
B. 1′-4″
C. 2′-0″
D. 2′-4″

_________ **12.** The center foundation wall is ___ high.
A. 8″
B. 3′-0″
C. 7′-4″
D. 9′-6″

__________ **13.** The divider between the two door pockets in the garage is ___″ thick.
A. 9
B. 12
C. 15
D. 18

__________ **14.** The front entrance wall extends ___ in front of the south foundation wall.
A. 14′-4″
B. 24′-3″
C. 29′-6″
D. 30′-0″

__________ **15.** The south foundation wall extends ___″ above the finished grade level.
A. 4
B. 6
C. 8
D. 10

__________ **16.** A 7″ ___ I beam extends under the floor in the crawl space area.
A. wood
B. steel
C. concrete
D. none of the above

__________ **17.** The section views are drawn to a scale of ___″ = 1′-0″.
A. ¼
B. ⅜
C. ½
D. ¾

__________ **18.** The I beam is supported by ___ piers.
A. steel
B. wood
C. concrete
D. masonry

Section 2

Slab-on-Grade Foundation Printreading Exercise

CHAPTER 8

Name ______________________________ **Date** ______________

Slab-on-grade foundations combine foundation walls with a concrete floor slab. The top of the floor slab is at the same elevation as the top of the foundation walls. The concrete floor slab receives its main support from the ground directly below. The plan view of a slab-on-grade foundation provides overall dimensions of the building and dimensions and locations of footings.

Section view drawings indicate the thickness and height of foundation footings, and thickness and elevation of the concrete floor slab. Section view drawings also include the type, size, and spacing of reinforcement in the footings and floor slab.

Refer to the Slab-on-Grade Foundation print on pages 281 and 282 to answer the following question.

Completion

__________ **1.** The south foundation wall is ___ long.

__________ **2.** The foundation walls extend ___″ above the finished grade line.

__________ **3.** The concrete slab is ___″ thick.

__________ **4.** The outside of the east garage wall is ___ from the centerline of the footing under the west garage wall.

__________ **5.** Two-inch ___ insulation is used around the perimeter of the slab.

__________ **6.** The footings under the interior walls are ___″ wide at the lowest point.

__________ **7.** The east foundation wall is ___ long.

__________ **8.** The footings under the interior walls are ___″ high.

__________ **9.** Anchor bolts along the exterior foundation walls are placed ___ OC.

__________ **10.** The exterior foundation walls are ___ high.

__________ **11.** The welded wire reinforcement in the concrete slab is designated as ___.

__________ **12.** The difference in elevation between the garage floor and the house is ___″.

__________ **13.** Two No. ___ rebar are used in the interior wall footings.

__________ **14.** The exterior foundation walls are ___″ thick.

__________ **15.** A ___ mil vapor barrier is placed under the concrete slab.

__________ **16.** The fireplace foundation is ___ long × ___ wide.

Multiple Choice

__________ **1.** Studs for the interior walls are placed ___″ OC.
A. 12
B. 14
C. 16
D. 18

__________ **2.** The outside of the north foundation wall is ___ from the fireplace foundation.
A. 9′-8″
B. 12′-2″
C. 12′-6″
D. 26′-7″

__________ **3.** No. 4 rebar are placed ___″ OC in the exterior foundation walls.
A. 16
B. 20
C. 24
D. 28

__________ **4.** The porch slab is ___ wide.
A. 3′-0″
B. 3′-5″
C. 5′-0″
D. 6′-5″

__________ **5.** The interior wall footings are detailed in Sections ___ and ___.
A. A; B
B. B; D
C. A; E
D. D; E

__________ **6.** A(n) ___″ corbel is used to support the front porch slab.
A. 1
B. 2
C. 4
D. 8

__________ **7.** The finished grade line is ___ above the frost line.
A. 8″
B. 1′-4″
C. 2′-0″
D. 2′-4″

__________ **8.** The footing below the south side of the porch slab is ___″ wide.
A. 5
B. 8
C. 10
D. 20

Section 2

Heavy Construction Foundation Printreading Exercise

8 CHAPTER

Name ______________________ **Date** ______________

A heavy construction foundation print includes a plan view, schedules, and section view drawings. The plan view provides overall dimensions and locations of foundation walls and footings, columns, and pier footings. Specific dimensions for stairways, door and window openings, and areaways are also included in the plan view. Elevations for various floor levels and tops of foundation footings and floor slabs are also indicated.

Footing schedules are commonly included in a heavy construction foundation print. Footing schedules provide the dimensions of the footings and indicate the size and location of reinforcement.

Section view drawings indicate the thickness and heights of foundation walls and footings and the size and location of reinforcement. Stair details or other details are commonly used to show complex structural features.

Refer to the Heavy Construction Foundation print on pages 283 to 286 to answer the following questions.

Completion

__________ **1.** The overall length of the building from east to west is ___.

__________ **2.** The elevation at the top of the footing in the southwest corner of the building is ___.

__________ **3.** The center of the first row of the column footings is ___ from the outside of the north foundation wall.

__________ **4.** The radius of the curved southwest wall is ___.

__________ **5.** The two openings in the areaway in the east wall are ___ × ___.

__________ **6.** The elevation at the top of the F8 footings is ___.

__________ **7.** The center-to-center distance between F8 footings is ___ from east to west.

__________ **8.** The elevation at the top of the first floor slab is ___.

__________ **9.** The elevation at the top of the F4 footing along the interior stairway is ___.

__________ **10.** The elevation at the top of the basement slab along the curved southwest wall is ___.

__________ **11.** ___″ of gravel is required beneath the floor slab.

__________ **12.** A(n) ___″ clearance must be maintained between the bottom footing rebar and soil.

__________ **13.** No. ___ rebar are required in the F6 footing.

__________ **14.** The dimensions of an F10 footing are ___ × ___ × ___.

__________ **15.** ___ No. 8 rebar are required for long way reinforcement for F8 footings.

______ **16.** Typical stepped footings require No. 5 rebar spaced ___″ vertically.

______ **17.** A(n) ___° angle is the maximum slope for the bottom of a stepped footing.

______ **18.** The north and south foundation footings are ___ wide × ___ deep.

______ **19.** No. 4 rebar are placed horizontally at ___″ intervals along the interior of the foundation wall in Section 8.

______ **20.** The top of the floor slab in Section 1 is ___ above the top of the foundation footing.

______ **21.** Exterior foundation walls are ___″ thick.

______ **22.** The total run of the exterior stairway is ___.

______ **23.** No. ___ nosing rebar are embedded in the stairway.

______ **24.** The total rise of the exterior stairway is ___.

______ **25.** The concrete landing at the foot of the exterior stairway is ___ wide × ___ long.

______ **26.** The elevation of the basement slab adjacent to the exterior stairway is ___.

______ **27.** The elevation at the top of the wall footings next to the exterior stairway is ___.

______ **28.** The areaway along the east wall is ___ wide.

______ **29.** The exterior foundation footing along the east wall is ___ wide × ___ deep.

______ **30.** The exterior wall of the areaway along the east foundation wall is ___″ thick.

______ **31.** The dimensions of the sump pit are ___ × ___ × ___.

______ **32.** The footing beneath the southwest curved wall is ___ wide × ___ deep.

______ **33.** Section 7 is drawn at a scale of ___″ = 1′-0″.

______ **34.** The design of the foundation footings is based on a soil bearing capacity of ___ psf.

______ **35.** Footing F___ is the smallest foundation footing shown on the basement and foundation plan.

Multiple Choice

______ **1.** The basement slab along the areaway is ___″ thick.

A. 4
B. 5
C. 6
D. 7

______ **2.** The wall adjacent to the sump pit is ___″ thick.

A. 8
B. 10
C. 12
D. 16

_______ **3.** The foundation footing beneath the southwest curved wall projects ___.
A. 8¾″
B. 1′-0″
C. 1′-4½″
D. 1′-8″

_______ **4.** The vertical rebar placed along the inside of the southwest curved wall are spaced at ___″ intervals.
A. 4
B. 5
C. 8
D. 12

_______ **5.** The foundation footing beneath the interior wall projects ___″.
A. 8
B. 10
C. 12
D. 16

_______ **6.** F6 footings are ___ wide.
A. 5′-0″
B. 8′-6″
C. 10′-0″
D. 13′-6″

_______ **7.** The step height for the stepped foundation walls is ___.
A. 1′-3″
B. 2′-0″
C. 4′-6″
D. 5′-0″

_______ **8.** The first floor slab along the interior wall is ___″ thick.
A. 5½
B. 7
C. 8½
D. 10

_______ **9.** The landing for the exterior stairway is ___″ below the basement floor slab.
A. 6
B. 8
C. 10
D. 12

_______ **10.** The outside foundation wall along the exterior stairway is ___ thick.
A. 6″
B. 8″
C. 11″
D. 1′-0″

__________ **11.** The wall footing beneath the north end of the exterior stairway projects ___ to the north.

A. 6″
B. 1′-0″
C. 2′-8″
D. 4′-8″

__________ **12.** The elevation at the top of the exterior foundation wall along the areaway is ___.

A. –12′-0″
B. –0′-3″
C. +0′-0″
D. +0′-3″

__________ **13.** The basement slab is reinforced with No. ___ rebar spaced at 18″ intervals.

A. 4
B. 5
C. 6
D. 8

__________ **14.** The footing for the elevator shaft (area southeast of the interior stairway) is ___ wide.

A. 2′-6″
B. 6′-11″
C. 8′-9″
D. 11′-9″

__________ **15.** The sump pit is on the ___ side of the elevator shaft.

A. north
B. south
C. east
D. west

SECTION 3

FORM MATERIALS AND CONCRETE QUANTITY TAKEOFF

Estimating form materials and concrete is a rough calculation of the amount of form materials and volume of concrete required for a specific construction project. Professional estimators commonly estimate the form materials and concrete for heavy construction projects. On small construction projects, such as the construction of foundation footings and walls for a residence, the estimating is performed by the contractor or job supervisor.

FORM MATERIALS QUANTITY TAKEOFF

Form materials are estimated separately for each section of the concrete work. When estimating, dimensions of the form materials are rounded to the next highest foot increment before calculations are performed. For example, the dimension of a wall section measuring 5′-4½″ is rounded to 6′-0″.

When estimating plywood form components, the total surface area of the forms is determined by multiplying the length of the forms by the height. When estimating dimensional lumber such as planks, studs, walers, braces, and stakes, the total length of the lumber is calculated.

Waste occurs when form components are cut from standard sizes of form materials. Estimators add from 5% to 15% of the total amount of form materials to compensate for waste. Underestimation of form materials results in a delay in form construction.

Sheathing

Sheathing is the form material in direct contact with the concrete. Plywood or 2″ thick members are used as sheathing for foundation and pier footing forms. Foundation walls are sheathed with plywood reinforced with studs and/or walers, or 2″ thick members reinforced with cleats and strongbacks.

The following examples describe a fundamental approach to estimating form materials and are based on the Foundation Plan and Section A-A on page 287.

Foundation Footing Forms. Foundation footing forms consist of outside and inside form walls. Foundation footing forms are constructed with 2″ thick members or plywood reinforced with stakes and braces. When using 2″ thick members as the sheathing, the total length of the form walls must be determined. The lengths of the outside form walls are calculated by adding the length of the foundation wall and the footing projections at the ends of the foundation walls. The lengths of the inside form walls are calculated by adding the adjacent wall thicknesses to the footing projections and subtracting the sum from the length of the foundation wall.

Example

Determine the total length of 2 × 10 members required to sheath the foundation footing forms.

Solution

1. Calculate the lengths of outside footing form walls A and B.

   Lengths of outside footing form walls A and B = lengths of foundation walls A and B + (footing projections)

   length = 48′-0″ + (5″ + 5″)
   length = 48′-0″ + 10″
   length = 48′-10″

   Lengths of outside footing form walls are rounded to 49′-0″.

2. Calculate the lengths of outside footing form walls C and D.

   Lengths of outside footing form walls C and D = lengths of foundation walls C and D + (footing projections)

   length = 26′-0″ + (5″ + 5″)
   length = 26′-0″ + 10″
   length = 26′-10″

   Lengths of outside footing form walls C and D are rounded to 27′-0″.

3. Calculate the lengths of inside footing form walls A and B.

   Lengths of inside footing form walls A and B = lengths of foundation walls A and B – [(adjacent wall thicknesses) + (footing projections)]

   length = 48′-0″ – [(10″ + 10″) + (5″ + 5″)]
   length = 48′-0″ – (20″ + 10″)
   length = 48′-0″ – 30″
   length = 48′-0″ – 2′-6″
   length = 45′-6″

   Lengths of inside footing form walls A and B are rounded to 46′-0″.

4. Calculate the lengths of inside footing form walls C and D.

Lengths of inside footing form walls C and D = lengths of foundation walls C and D – [(adjacent wall thicknesses) + (footing projections)]

length = 26′-0″ – [(10″ + 10″) + (5″ + 5″)]
length = 26′-0″ – [20″ + 10″]
length = 26′-0″ – 30″
length = 26′-0″ – 2′-6″
length = 23′-6″

Lengths of inside footing form walls C and D are rounded to 24′-0″.

5. Calculate the total length of 2 × 10 members required.

total length of 2 × 10 members = sum of lengths of individual footing form members
total length = 49′ + 49′ + 27′ + 27′ + 46′ + 46′ + 24′ + 24′

total length of 2 × 10 members = 292′-0″

When using plywood as sheathing for foundation footing forms, the surface area of individual footing forms is calculated by multiplying the foundation footing form lengths by the height of the forms. The individual surface areas are added together to determine the total surface area. The total surface area is divided by the area of a plywood panel to determine the number of panels required to sheath the foundation footing forms.

Example

Determine the number of 4 × 8 plywood panels required to sheath the foundation footing forms.

Solution

1. Determine the lengths of the individual footing forms. (See previous example for individual footing form lengths.)

Footing Form	Length
Outside footing form wall A	49′-0″
Outside footing form wall B	49′-0″
Outside footing form wall C	27′-0″
Outside footing form wall D	27′-0″
Inside footing form wall A	46′-0″
Inside footing form wall B	46′-0″
Inside footing form wall C	24′-0″
Inside footing form wall D	24′-0″

2. Calculate the surface area of the individual footing forms. Round the footing height to the next highest foot.

Footing Form	Length	Height	Surface Area (sq ft)
Outside footing form wall A	49′-0″	1′-0″	49
Outside footing form wall B	49′-0″	1′-0″	49
Outside footing form wall C	27′-0″	1′-0″	27
Outside footing form wall D	27′-0″	1′-0″	27
Inside footing form wall A	46′-0″	1′-0″	46
Inside footing form wall B	46′-0″	1′-0″	46
Inside footing form wall C	24′-0″	1′-0″	24
Inside footing form wall D	24′-0″	1′-0″	24

3. Calculate the total surface area of the foundation footing forms.

Total surface area = sum of individual surface areas
Total surface area = 49 + 49 + 27 + 27 + 46 + 46 + 24 + 24
Total surface area = 292 sq ft

4. Calculate the number of 4 × 8 plywood panels required to sheathe the foundation footing forms.

no. of plywood panels = total surface area ÷ area of panel
no. of plywood panels = 292 ÷ (4 × 8)
no. of plywood panels = 292 ÷ 32
no. of plywood panels = 9.125

The number of 4 × 8 plywood panels is rounded to 10.

Pier Footing Forms. Pier footings support steel columns, wood posts, and masonry or concrete piers. Pier footing design determines the type of form to be constructed. Rectangular and tapered pier footings require forms around the perimeter of the pier footing. Stepped pier footings require a base form and a step form for each additional step. Circular pier footings are formed using the required length of tubular fiber form.

The surface area of a rectangular or tapered pier footing form is determined by multiplying the combined length of all sides of the pier footing form by the height of the form. If multiple pier footings with the same dimensions are constructed, the individual surface area is multiplied by the number of piers.

Example

Determine the number of 4 × 8 plywood panels required to sheathe the pier footings.

Solution

1. Calculate the combined length of the four sides of one pier footing form.
   combined length = *sum of all side lengths*
   combined length = 2′-0″ + 2′-0″ + 2′-0″ + 2′-0″
   combined length = 8′-0″
2. Calculate the surface area of one pier footing form. Round the pier footing height to the next highest foot increment.
   Surface area of one pier footing form = *combined length* × *height* = 8′-0″ × 1′-0″
   Surface area of one pier footing form = 8 sq ft
3. Calculate the total surface area of two pier footing forms.
   Total surface area = *surface area of one pier footing form* × *no. of pier footings* = 8 sq ft × 2
   Total surface area of two pier footing forms = 16 sq ft
4. Calculate the number of 4 × 8 plywood panels required.
   no. of 4 × 8 plywood panels = *total surface area* ÷ *area of panel*
   no. of 4 × 8 plywood panels = 16 sq ft ÷ (4 × 8)
   no. of 4 × 8 plywood panels = 16 sq ft ÷ 32
   no. of 4 × 8 plywood panels = .5
   The number of plywood panels required to sheathe the pier footing forms is rounded to 1.

Foundation Wall Forms. Foundation wall forms consist of the inside and outside form walls. The foundation wall forms are constructed with plywood and reinforced with studs, walers, and/or strongbacks. The surface areas of the outside form walls are determined by multiplying the length of the form wall by the height. The surface areas of the inside form walls are determined by subtracting the adjacent wall thicknesses from the outside form wall lengths and multiplying by the height.

Example

Determine the number of 4 × 8 plywood panels required to sheathe the foundation walls.

Solution

1. Calculate the lengths of the outside form walls.
   Lengths of outside form walls A and B = 48′-0″
   Lengths of outside form walls C and D = 26′-0″
2. Calculate the lengths of the inside form walls A and B.
   Lengths of inside form walls A and B = *lengths of outside form walls A and B* – *(adjacent wall thicknesses)*
   length = 48′-0″ – (10″ + 10″)
   length = 48′-0″ – 20″
   length = 48′-0″ – 1′-8″
   length = 46′-4″
   The lengths of inside form walls A and B are rounded to 47′-0″.
3. Calculate the lengths of inside form walls C and D.
   Lengths of inside form walls C and D = *lengths of outside form walls C and D* – *(adjacent wall thicknesses)*
   length = 26′-0″ – (10″ + 10″)
   length = 26′-0″ – 20″
   length = 26′-0″ – 1′-8″
   length = 24′-4″
   The lengths of inside form walls C and D are rounded to 25′-0″.
4. Determine the surface areas of the individual form walls.

Form Wall	Length	Height	Surface Area (sq ft)
Outside form wall A	48′-0″	8′-0″	384
Outside form wall B	48′-0″	8′-0″	384
Outside form wall C	26′-0″	8′-0″	208
Outside form wall D	26′-0″	8′-0″	208
Inside form wall A	47′-0″	8′-0″	376
Inside form wall B	47′-0″	8′-0″	376
Inside form wall C	25′-0″	8′-0″	200
Inside form wall D	25′-0″	8′-0″	200

5. Calculate the total surface area of the foundation wall forms.

   Total surface area = sum of individual surface areas
   total surface area = 384 + 384 + 208 + 208 + 376 + 376 + 200 + 200
   total surface area = 2336 sq ft

6. Calculate the number of 4 × 8 plywood panels required.
   no. of 4 × 8 plywood panels = total surface area ÷ area of panel
   no. of 4 × 8 plywood panels = 2336 ÷ (4 × 8)
   no. of 4 × 8 plywood panels = 2336 ÷ 32
   The total number of 4 × 8 plywood panels required to sheathe the form walls is 73.

Low form walls are often sheathed with members instead of plywood. When form walls are sheathed with members, divide the total surface area by the width of the members (in feet) to determine the total length of members required.

Example

Determine the total length of 2 ×10 members required to sheathe the foundation wall forms.

Solution

Total length of 2 × 10 members = total surface area ÷ width of planks (ft)
total length of 2 × 10 members = 2336 ÷ (10″ ÷ 12)
total length of 2 × 10 members = 2336 ÷ 83′
total length of 2 × 10 members = 2814.5
The total length of 2 × 10 planks required to sheathe the foundation wall forms is rounded to 2815′-0″.

Stiffeners and Supports

Stiffeners and supports such as base plates, studs, walers, braces, and stakes are used to reinforce wall form sheathing. Base plates secure the bottom edge of the sheathing and provide a nailing surface for studs. Studs stiffen and directly reinforce wall form sheathing. Walers reinforce the studs and align the wall forms. Braces secure and align the tops of form walls. Stakes secure foundation and pier footing forms and the lower ends of braces.

Base Plates. Base plates are nailed to the foundation footing and secure the outside form wall for most forming operations. Base plates secure the inside form wall when forming a round structure such as a storage tank. The total length of base plate material required is determined by calculating the perimeter (distance around the outside) of the foundation walls.

Example

Determine the total length of base plate material required to form the foundation wall.

Solution

1. Calculate the perimeter of the foundation walls.
   Perimeter of foundation walls = length of foundation wall A + B + C + D = 48′-0″ + 48′-0″ + 26′-0″ + 26′-0″
   perimeter of foundation walls = 148′-0″
   The total length of base plate material required is 148′-0″.

Studs. The number of studs required to reinforce a form wall is determined by the stud spacing. Stud spacing is based on the stud size and type of other stiffeners to be used. The number of studs for outside and inside form walls is determined by dividing the length of the form wall by the recommended spacing, and adding one stud. For example, a 20′-0″ form wall with studs spaced 2′-0″ OC requires 11 studs [(20′-0″ ÷ 2′-0″) + 1 = 11]. The length of odd-length walls is divided by the recommended spacing and the answer is rounded to the next highest whole number. One additional stud is then added. For example, a 23′-0″ form wall with studs spaced 2′-0″ OC requires 13 studs [(23′-0″ ÷ 2′-0″) + 1 = 13]. The total length of stud material is determined by multiplying the total number of studs by the form wall height.

Example

Determine the total length of stud material required to reinforce the outside and inside form walls if the studs are spaced 2′-0″ OC.

Solution

1. Calculate the number of studs required for each form wall.

Outside Form Wall	Length	OC Spacing	Number of Studs
A	48′-0″	2′-0″	25
B	48′-0″	2′-0″	25
C	26′-0″	2′-0″	14
D	26′-0″	2′-0″	14

2. Calculate the total number of studs required for the outside form wall.
   Total number of studs for outside form wall = sum of individual form wall studs
   total number of studs for outside form wall = 25 + 25 + 14 + 14
   The total number of studs required for outside form wall = 78 studs

3. Calculate the number of studs required for the inside form wall.

Inside Form Wall	Length	OC Spacing	Number of Studs
A	47′-0″	2′-0″	25
B	47′-0″	2′-0″	25
C	25′-0″	2′-0″	14
D	25′-0″	2′-0″	14

4. Calculate the total number of studs required for the inside form wall.
   Total number of studs for inside form wall = sum of individual form wall studs
   total number of studs for inside form wall = 25 + 25 + 14 + 14
   The total number of studs required for inside form wall = 78 studs

5. Calculate the total number of studs for the outside and inside form walls.
   Total number of studs for outside and inside form walls = total number of studs for outside form wall + total number of studs for inside form wall
   total number of studs for outside and inside form walls = 78 + 78
   The total number of studs required for outside and inside form walls = 156 studs

6. Calculate the total length of stud material required.
   Length of stud material = total number of studs × form wall height
   length = 156 × 8′-0″
   The total length of stud material = 1248′-0″

Walers. Single or double walers are secured against the studs for reinforcement. When studs are not used, the walers are secured against the sheathing. The spacing of walers determines the number of rows of walers required. The length of waler material for a single row of walers for the outside form wall is determined by calculating the perimeter of the foundation walls and multiplying by the number of rows of walers required. The length of waler material for a double waler system is twice the length of the waler material for a single row of walers.

The length of waler material required for a single row of walers for the inside form wall is determined by calculating the perimeter of the inside form wall and multiplying by the number of rows of walers required. The total length of waler material for the outside and inside form walls is determined by adding the lengths of waler material for the form walls.

Example

Determine the total length of waler material required for a double waler system. The double walers are secured to the outside and inside form walls and are spaced 12″ from the top and bottom of the sheathing with the intervening rows spaced 24″ OC (four rows of double walers).

Solution

1. Calculate the perimeter of the foundation wall.
   Perimeter of foundation wall = length of foundation walls A + B + C + D
   perimeter of foundation wall = 48′-0″ + 48′-0″ + 26′-0″ + 26′-0″
   The perimeter of the foundation wall = 148′-0″

2. Calculate the length of waler material required for a single row of walers for the outside form wall.
   Length of walers = perimeter of foundation wall × no. of rows
   length = 148′-0″ × 4
   The length of waler material for a single row of walers for the outside form wall = 592′-0″

3. Calculate the length of waler material for a double waler system for the outside form wall.
Length of walers = 2 × *length of single waler material*
length = 2 × 592′-0″
The length of waler material for a double waler system for the outside form wall = 1184′-0″

4. Calculate the perimeter of the inside form wall.
Perimeter of inside form wall = *lengths of inside form walls A + B + C + D*
perimeter = 47′-0″ + 47′-0″ + 25′-0″ + 25′-0″
The perimeter of the inside form wall = 144′-0″

5. Calculate the length of waler material required for a single waler system for the inside form wall.
Length of walers = *perimeter of inside form wall* × *no. of rows*
length = 144′-0″ × 4
The length of waler material for a single waler system for the inside form wall = 576′-0″

6. Calculate the length of waler material for a double waler system for the inside form wall.
Length of walers = 2 × *length of single waler material*
length = 2 × 576′-0″
The length of waler material for a double waler system for the inside form wall = 1152′-0″

7. Calculate the total length of waler material required.
Total length of waler material = *length of waler material for outside form wall* + *length of waler material for inside form wall*
total length = 1184′-0″ + 1152′-0″
The total length of water waler material required = 2336′-0″

Braces. Braces support form walls and are secured at the lower ends by stakes. The number of braces required is based on the spacing of the braces. The number of braces is determined by dividing the length of a form wall by the recommended spacing and adding one brace. The numbers of braces for individual form walls are added to obtain the total number of braces required.

The length of brace required is based on the angle that it forms with the form wall. Braces are commonly attached at approximately a 45° angle. The length of braces attached at a 45° angle is determined by multiplying the height from the base of the footing to the brace attachment point by 1.41. The total length of brace material required is determined by multiplying the number of braces by the individual brace length.

Example

Determine the total length of brace material required for the foundation wall. The braces are spaced 6′-0″ OC and are attached at a 45° angle. The distance from the base of the footing to the brace attachment point is 8′-0″.

Solution

1. Calculate the number of braces required for each form wall.

Outside Form Wall	Length	OC Spacing	Number of Braces
A	48′-0″	6′-0″	9
B	48′-0″	6′-0″	9
C	26′-0″	6′-0″	6
D	26′-0″	6′-0″	6

2. Calculate the number of braces required for the foundation wall.
Total number of braces = *sum of individual braces*
total number of braces = 9 + 9 + 6 + 6
The total number of braces required for the foundation wall = 30 braces

3. Calculate the length of an individual brace.
Brace length = *height* × 1.41
brace length = 8′-0″ × 1.41
Individual brace length = 11.28′
Round 11.28′ to 12′-0″

4. Calculate the total length of brace material required.
Total length of brace material = *no. of braces* × *individual brace length*
total length of brace material = 30 × 12′-0″
total length of brace material = 360′-0″
Total length of brace material is 360′-0″

Stakes. Stakes secure foundation footing and pier forms and the lower ends of braces. Foundation footing forms require stakes along the outer surface of both form walls. The number of stakes for individual outside form walls is determined by dividing the length of the outside form wall by the recommended spacing and adding one stake. The number of stakes required for the outside form wall of an entire foundation footing is determined by adding the number of stakes for the individual form walls.

The number of stakes required for individual inside form walls is determined by dividing the length of the inside form wall by the recommended spacing and adding one stake. The number of stakes required for the inside form wall of an entire foundation is determined by adding the number of stakes for the individual form walls. The total length of stake material required for the outside and inside form walls is determined by adding the number of stakes for the outside and inside form walls and multiplying by the individual stake length.

Example

Determine the total length of stake material required for the foundation footing forms. The stakes are 2′-0″ long and are spaced 2′-0″ OC.

Solution

1. Calculate the number of stakes required for the outside form wall of each foundation footing.

Form Wall	Length	OC Spacing	Number of Stakes
A	49′-0″	2′-0″	26
B	49′-0″	2′-0″	26
C	27′-0″	2′-0″	15
D	27′-0″	2′-0″	15

2. Calculate the number of stakes required for the outside form wall of the entire foundation footing.
   Number of stakes = sum of individual footing stakes
   number of stakes = 26 + 26 + 15 + 15
   The total number of stakes required for the outside form wall = 82 stakes

3. Calculate the number of stakes required for the inside form wall of each foundation footing.

Form Wall	Length	OC Spacing	Number of Stakes
A	46′-0″	2′-0″	24
B	46′-0″	2′-0″	24
C	24′-0″	2′-0″	13
D	24′-0″	2′-0″	13

4. Calculate the number of stakes required for the inside form wall of the entire foundation footing.
   Number of stakes = sum of individual footing stakes
   number of stakes = 24 + 24 + 13 + 13
   The total number of stakes for the inside form wall = 74 stakes

5. Calculate the total number of stakes for the inside and outside form walls.
   Total number of stakes for outside and inside form walls = sum of footing stakes for outside and inside form walls = 82 + 74
   The total number of stakes for outside and inside form walls = 156 stakes

6. Calculate the total length of stake material required for the foundation footing forms.
   Total length of stake material = no. of stakes × individual stake length
   total length of stake material = 156 × 2′-0″
   The total length of stake material for the foundation footing forms = 312′-0″

Pier footing forms require stakes only along the perimeter of the pier box. Small pier footing forms require four corner stakes. Large pier footing forms require corner stakes and intermediate stakes. The total length of stake material required is determined by multiplying the number of stakes by the individual stake length.

Example

Determine the total length of stake material required for the pier footing forms. The stakes are 2′-0″ long and are placed at the four corners of the pier box.

Solution

1. Calculate the length of stake material required for one pier box.
   Length of stake material = no. of stakes × individual stake length = 4 × 2′-0″
   The length of stake material for one pier box = 8′-0″

2. Calculate the total length of stake material required.

   Total length of stake material = no. of pier boxes × length of stake material for one pier box = 2 × 8′-0″
   The total length of stake material for two pier boxes = 16′-0″

Stakes secure the lower ends of braces to the ground. The total length of stake material required to secure the braces is determined by multiplying the number of braces by the individual stake length.

Example

Determine the total length of stake material required for the form wall. The stakes are 2′-0″ long and placed at the lower end of every brace.

Solution

1. Calculate the total length of stake material.
   Total length of stake material = no. of braces × stake length = 30 × 2′-0″
   Total length of stake material = 60′-0″

CONCRETE QUANTITY TAKEOFF

Concrete is estimated by volume and is expressed in cubic yards (cu yd). The volume of each section of a structure is calculated separately and added to other sections to obtain the total volume of concrete required. The volume of a horizontal structural member such as a floor slab is determined by multiplying the thickness, width, and height. The volume of a vertical member such as a foundation wall is determined by multiplying thickness, length, and height. Since dimensions for structural members are expressed in feet, the volume of concrete required is initially expressed in cubic feet. The volume is divided by 27 to obtain the volume of concrete in cubic yards. (One cubic yard equals 27 cubic feet.)

The following examples are based on the Foundation Plan and Section A-A on page 287.

Foundation Footings

Foundation footings support foundation walls. Since foundation footings commonly project beyond foundation walls, the lengths of the footing projections must be added to the length of the foundation walls to obtain a total foundation footing length. The volume is then determined by multiplying the height, width, and length.

Example

Determine the volume of concrete required for the foundation footings.

Solution

1. Calculate the lengths of foundation footings A and B.
   Lengths of footings A and B = lengths of foundation footings A and B + (footing projections)
   length = 48′-0″ + (5″ + 5″)
   length = 48′-0″ + 10″
   The lengths of foundation footings A and B = 48′-10″
2. Calculate the lengths of foundation footings C and D. To avoid calculating the corners of the footings twice, subtract the sum of the wall thicknesses and footing projections of adjacent walls A and B from the length of foundation walls C and D.
   Lengths of footings C and D = lengths of foundation walls C and D – [(wall A and B thicknesses) + (footing projections)]
   length = 26′-0″ – [(10″ + 10″) + (5″ + 5″)]
   length = 26′-0″ – (20″ + 10″)
   length = 26′-0″ – 30″
   length = 26′-0″ – 2′-6″
   The lengths of foundation footings C and D = 23′-6″
3. Calculate the volume of concrete required for foundation footings A and B.
   Volume of footings A and B = height × width × length
   V = 10″ × 1′-8″ × 48′-10″
   V = .83′ × 1.67′ × 48.83′
   The total volume of footings A and B = 67.68 cu ft

4. Calculate the volume of concrete required for foundation footings C and D.
Volume of footings C and D = height × width × length
V = 10″ × 1′-8″ × 23′-6″
V = .83′ × 1.67′ × 23.5′
The volume of footings C and D = 32.57 cu ft

5. Calculate the total volume of concrete required for the foundation footings in cubic feet.
Total volume (cu ft) = sum of individual volumes = 67.68 + 67.68 + 32.57 + 32.57
Total volume (cu ft) = 200.5 cu ft

6. Calculate the total volume of the concrete required for the foundation footings in cubic yards.
Total volume (cu yd) = total volume (cu ft) ÷ 27 = 200.5 ÷ 27
The total volume of concrete required for the foundation footings (in cu yd) = 7.43 cu yd

Pier Footings

Pier footings support steel columns, wood posts, and concrete or masonry piers. The volume of a pier footing is determined by multiplying the height, width, and length.

Example

Determine the volume of concrete required for the pier footings.

Solution

1. Calculate the total volume of concrete required for the pier footings in cubic feet.
Total volume (cu ft) = no. of pier footings × (height × width × length)
total volume = 2 × (10″ × 2′-0″ × 21′ -0″)
total volume = 2 × (.83′ × 2.0′ × 2.0′)
total volume = 2 × 3.32
The total volume of concrete required for the pier footings (in cu ft) = 6.64 cu ft

2. Calculate the volume of concrete required for the pier footings in cubic yards.
Volume of pier footings (cu yd) = total volume (cu ft) ÷ 27
V = 6.64 ÷ 27
The volume of concrete required for the pier footings (in cu yd) = .25 cu yd

Foundation Walls

Foundation walls support the superstructure. Foundation walls for crawl space foundations are shorter than foundation walls for full basement foundations. The volume of foundation walls is determined by multiplying the thickness, length, and height.

Example

Determine the volume of concrete required for the foundation walls.

Solution

1. Calculate the volume of concrete required for foundation walls A and B.
Volume of foundation walls A and B = thickness × length × height
V = 10″ × 48′-0″ × 8′-0″ = .83′ × 48.0′ × 8.0′
The volume of concrete required for foundation walls A and B = 318.72 cu ft

2. Calculate the lengths of foundation walls C and D.
Lengths of foundation walls C and D = lengths of foundation walls C and D – (thicknesses of foundation walls A and B)
length = 26′-0″ – (10″ + 10″)
length = 26′-0″ – 20″
length = 26′-0″ – 1′-8″
The lengths of foundation walls C and D = 24′-4″

3. Calculate the volume of concrete required for foundation walls C and D.
Volume of foundation walls C and D = thickness × length × height
V = 10″ × 24′-4″ × 8′-0″ = .83′ × 24.33′ × 8.0′
The volume of concrete required for foundation walls C and D = 161.55 cu ft

4. Calculate the total volume of the concrete required for the foundation walls in cubic feet.
   Total volume (cu ft) = sum of individual volumes
   V = 318.72 + 318.72 + 161.55 + 161.55
   The total volume of the concrete required for the foundation walls (in cu ft) = 960.54 cu ft
5. Calculate the total volume of concrete required for the foundation walls in cubic yards.
   Total volume (cu yd) = total volume (cu ft) ÷ 27 = 960.54 ÷ 27
   Total volume (cu yd) = 35.58 cu yd

Complete Foundation

The complete foundation consists of foundation and pier footings, and the foundation walls. The volume of concrete required for the complete foundation is determined by adding the volumes of all the individual sections of the foundation.

Example

Determine the volume of concrete required for the complete foundation.

Solution

1. Calculate the total volume of concrete for the complete foundation.
   Total volume = sum of individual volumes
   V = foundation footings + pier footings + foundation walls
   V = 7.43 + .25 + 35.58
   V = 43.26 cu yd
   The total volume of concrete is rounded to 44 cu yd.

Section 3

Review Questions

CHAPTER 8

Name ______________________ **Date** ______________

Estimate the total amount of form material required to construct the foundation forms for the full basement foundation on pages 275 and 276. Use the encircled boldface letters on the foundation plan to identify the foundation footings, walls, and pier footings on the quantity takeoffs.

Foundation wall G extends from the outside of the west foundation wall to the outside of the foundation wall between the garage and the basement. When estimating, round the height of foundation wall G to 4′-0″. Foundation wall A extends 39′-4″ from the east foundation wall. Foundation wall J extends 20′-8″ from the west foundation wall. The perimeters of walls E and K are determined by adding the lengths of the sides not adjacent to the foundation walls.

Foundation Footing Forms. Foundation footing forms are constructed with 2 × 8 and 2 × 10 members. The footing forms are secured with 2 × 4 × 2′-0″ long stakes spaced 2′-0″ OC.

Foundation Wall Forms. Foundation wall forms are sheathed with ¾″ × 4′ × 8′ Plyform® Class 1 panels with the face grain running vertically. The Plyform panels are reinforced with 2 × 4 studs spaced 2′-0″ OC and 2 × 4 base plates. The stud length is equal to the height of the finished concrete wall. Four rows of 2 × 4 double walers stiffen the inside and outside form walls of foundation walls F, D, B, and A. The remaining foundation walls are stiffened with two rows of 2 × 4 double walers along the inside and outside form walls. Braces are attached to the outside form walls. Braces for foundation walls F, D, B, and A are 2 × 4 × 12′-0″ long and are spaced 8′-0″ OC. Braces for the remaining foundation walls are 2 × 4 × 6′-0″ long and spaced 8′-0″ OC. All braces are secured at the lower ends with 2 × 4 × 1′-6″ long stakes.

Pier Footing Forms. Pier footing forms are constructed with 2 × 12 planks reinforced with four 2 × 4 × 2′-0″ long stakes. Additional bracing is not required.

Foundation Footings

1.

QUANTITY TAKEOFF FOUNDATION FOOTINGS—OUTSIDE FORM WALL				
Wall	Planks		Stakes	
	Thickness × Width	Length	Quantity	Total Length
A				
B				
C				
D				
E	FOOTINGS NOT REQUIRED			
F				
G				
H				
I				
J				
K	FOOTINGS NOT REQUIRED			

2.

QUANTITY TAKEOFF FOUNDATION FOOTINGS—INSIDE FORM WALLS				
Wall	Planks		Stakes	
	Thickness × Width	Length	Quantity	Total Length
A				
B				
C	INSIDE FORM WALL NOT REQUIRED			
D				
E	FOOTINGS NOT REQUIRED			
F				
G				
H				
I				
J				
K	FOOTINGS NOT REQUIRED			

__________ **3.** Length of 2 × 4 stakes (in ft)

__________ **4.** Length of 2 × 8 members (in ft)

__________ **5.** Length of 2 × 10 members (in ft)

__________ **6.** Length of 2 × 12 members (in ft)

Foundation Walls

1.

QUANTITY TAKEOFF FOUNDATION WALLS—OUTSIDE FORM WALL											
Wall	Length	Height	Area (sq ft)	Studs		Base Plates	Walers	Braces		Stakes	
				Quantity	Total Length	Total Length	Total Length	Quantity	Total Length	Quantity	Total Length
A											
B											
C						FORM WALL NOT REQUIRED					
D											
E											
F											
G											
H											
I											
J											
K											

2.

QUANTITY TAKEOFF FOUNDATION WALLS—INSIDE FORM WALL												
Wall	Length	Height	Area (sq ft)	Studs		Base Plates	Walers	Braces		Stakes	
				Quantity	Total Length	Total Length	Total Length	Quantity	Total Length	Quantity	Total Length
A						NOT REQ'D FOR INSIDE FORM WALL		NOT REQ'D FOR INSIDE FORM WALL			
B											
C		FORM WALL NOT REQ'D									
D											
E											
F											
G											
H											
I											
J											
K											

_________ **3.** Total surface area of outside and inside form walls (in sq ft)

_________ **4.** Number of ¾″ × 4 × 8 Plyform® Class 1 panels

_________ **5.** Length of 2 × 4 studs (in ft)

_________ **6.** Length of 2 × 4 base plates (in ft)

_________ **7.** Length of 2 × 4 double walers (in ft)

_________ **8.** Length of 2 × 4 braces (in ft)

_________ **9.** Length of 2 × 4 stakes (in ft)

_________ **10.** Total length of 2 × 4 material (in ft)

Estimate the total volume of concrete required to construct the foundation footings, walls, and pier footings for the full basement foundation on pages 275 and 276. Use the encircled boldface letters on the foundation plan to identify the foundation footings, walls, and pier footings on the quantity takeoffs. Decimal equivalents and volumes are rounded to two places after the decimal point.

When determining the volume of concrete required for the foundation, first calculate the volumes of foundation walls and footings G, D, A, and J. When calculating the volume of foundation wall G, use the height shown in Section 6. Next, calculate the volumes of foundation walls and footings F, B, H, and I to eliminate calculating the volume of concrete for the corners twice. Finally, calculate the volumes of stoops E and K, and pier footing C.

Foundation Footings and Walls

1.

QUANTITY TAKEOFF FOUNDATION FOOTINGS							
Footing	Height		Width		Length		Volume
	Inches	Decimal Foot	Inches	Decimal Foot	ft and in.	Decimal Foot	cu ft
A							
B							
C							
D							
E				FOOTINGS NOT REQUIRED			
F							
G							
H							
I							
J							
K				FOOTINGS NOT REQUIRED			

2.

QUANTITY TAKEOFF FOUNDATION WALLS							
Footing	Thickness		Length		Height		Volume
	Inches	Decimal Foot	ft and in.	Decimal Foot	ft and in.	Decimal Foot	cu ft
A							
B							
C				WALLS NOT REQUIRED			
D							
E							
F							
G							
H							
I							
J							
K							

__________ **3.** Total volume of foundation footings (in cu ft)

__________ **4.** Total volume of foundation footings (in cu yd)

__________ **5.** Total volume of foundation walls (in cu ft)

__________ **6.** Total volume of foundation walls (in cu yd)

__________ **7.** Total volume of foundation (in cu yd)

__________ **8.** Total volume of concrete to be ordered (in cu yd)

Plot Plan

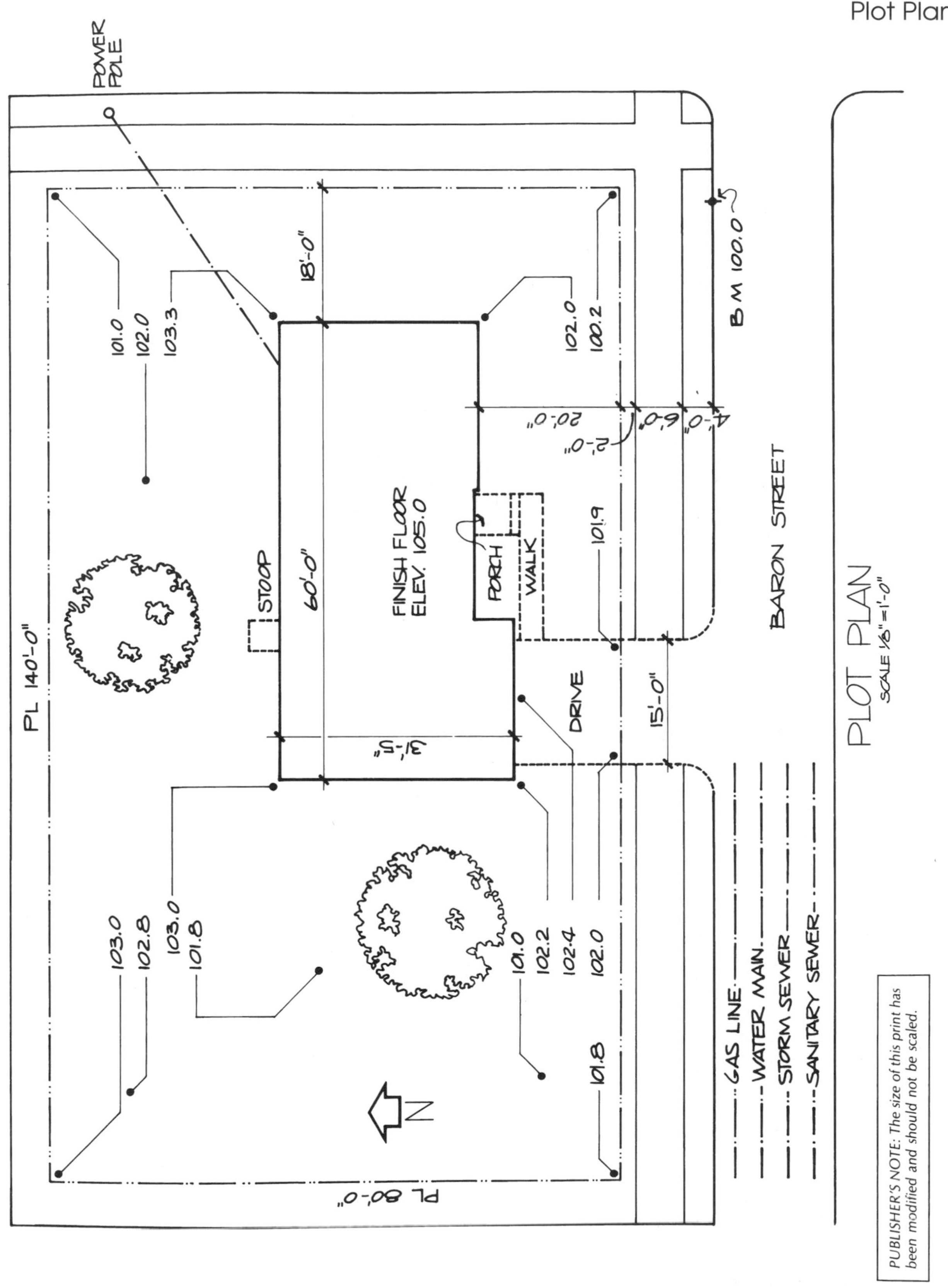

PUBLISHER'S NOTE: The size of this print has been modified and should not be scaled.

Full Basement Foundation

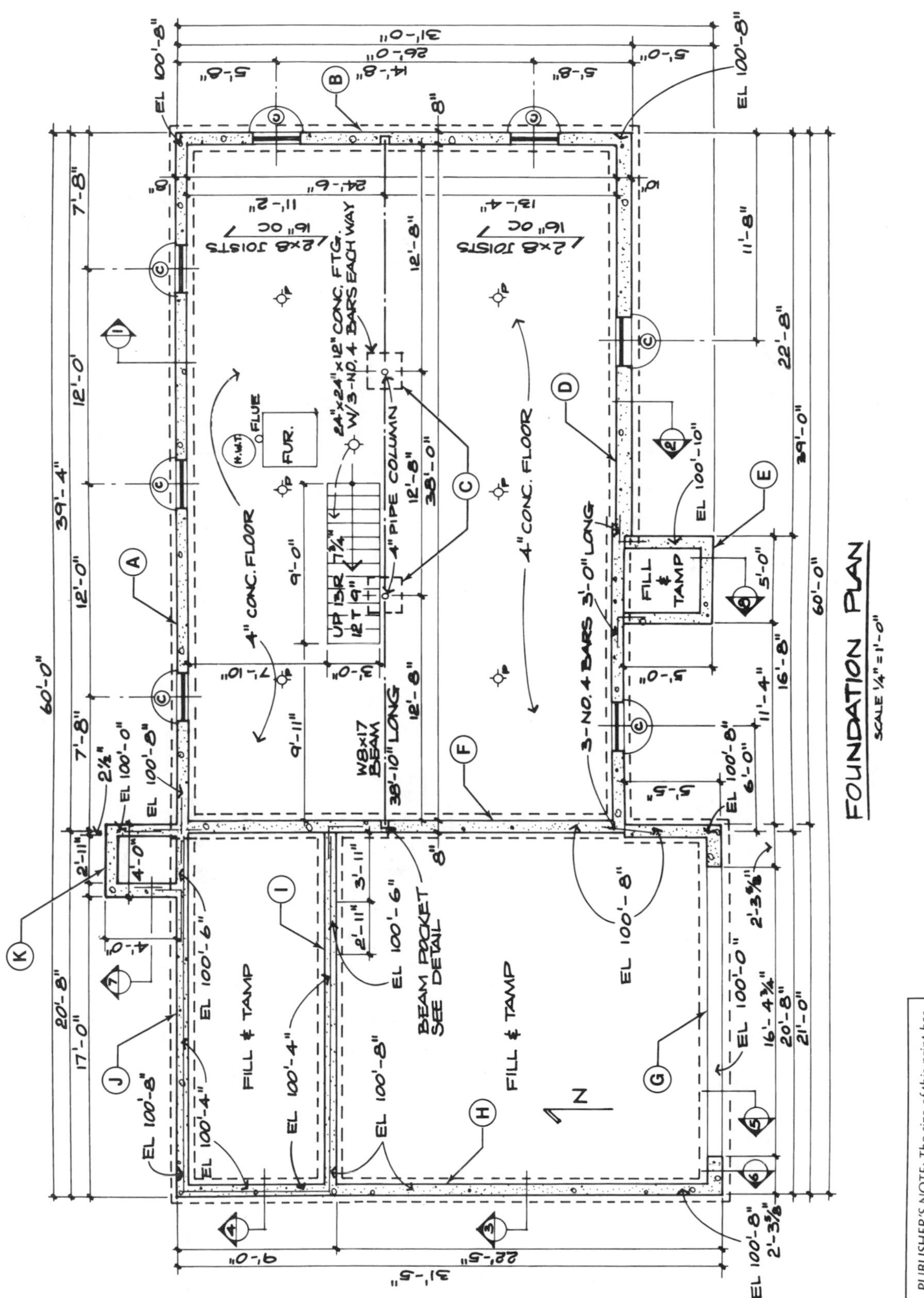

The Carlinghouse Company

PUBLISHER'S NOTE: The size of this print has been modified and should not be scaled.

Full Basement Foundation

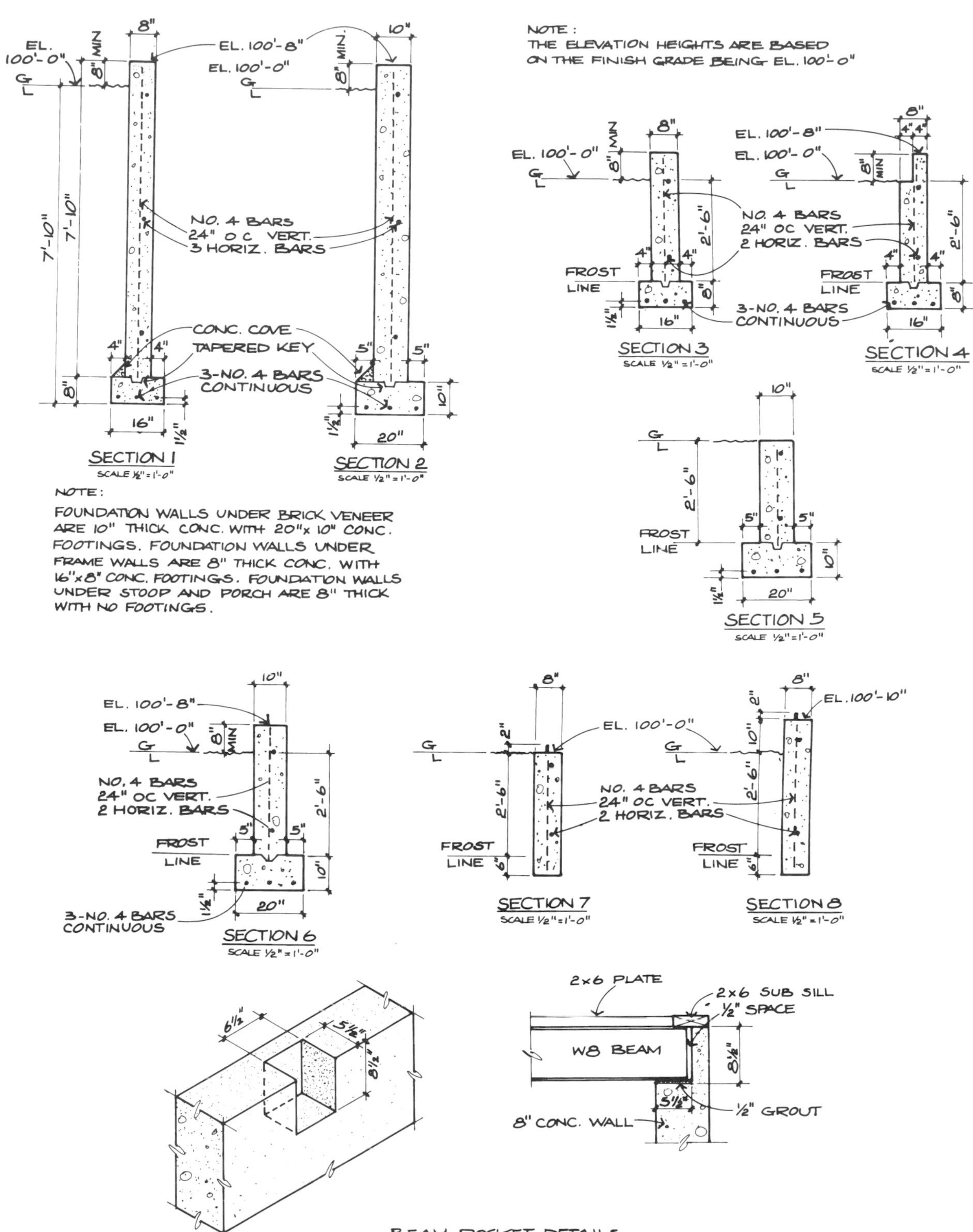

BEAM POCKET DETAILS

PUBLISHER'S NOTE: *The size of this print has been modified and should not be scaled.*

The Garlinghouse Company

Crawl Space Foundation

The Garlinghouse Company

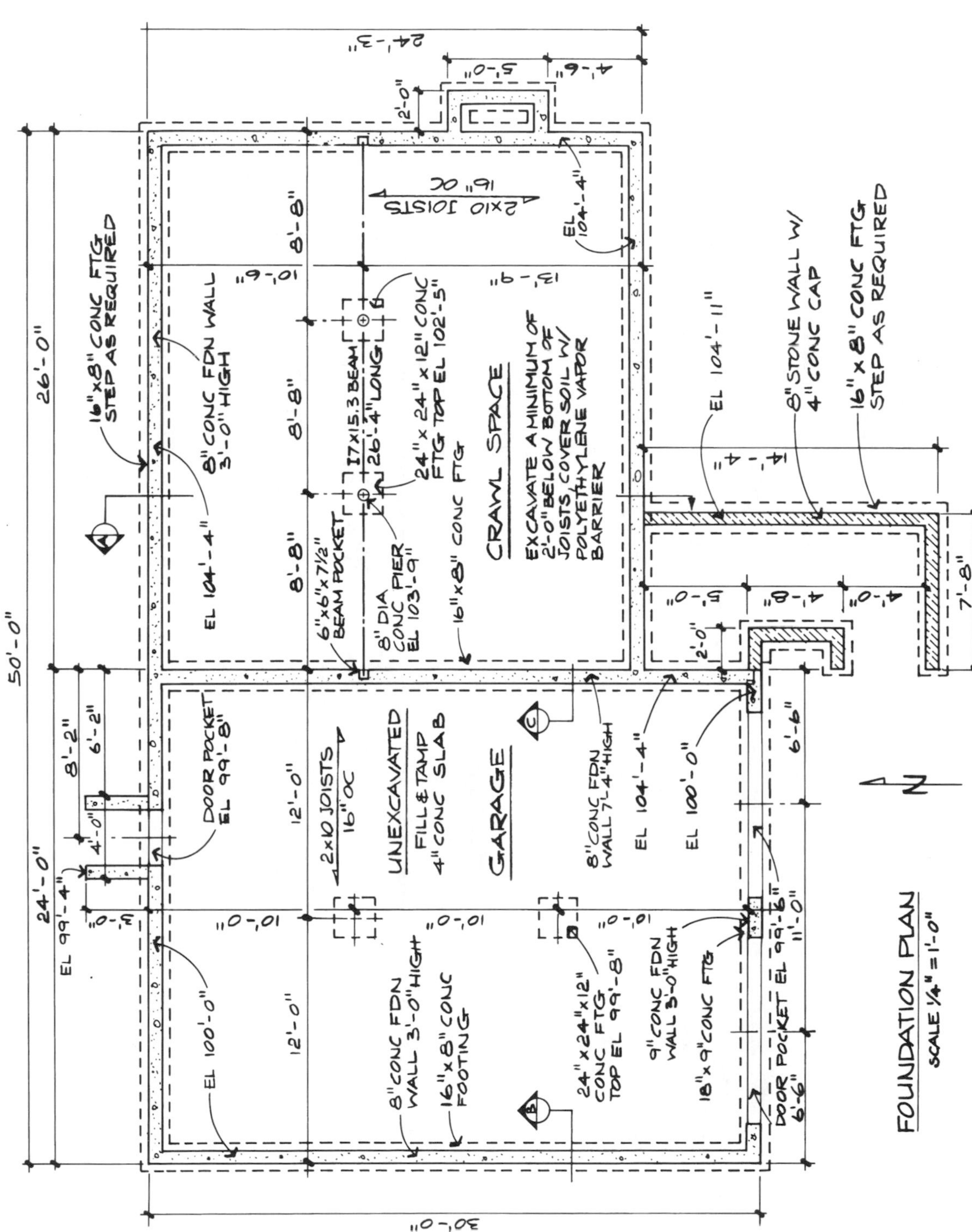

PUBLISHER'S NOTE: *The size of this print has been modified and should not be scaled.*

Crawl Space Foundation

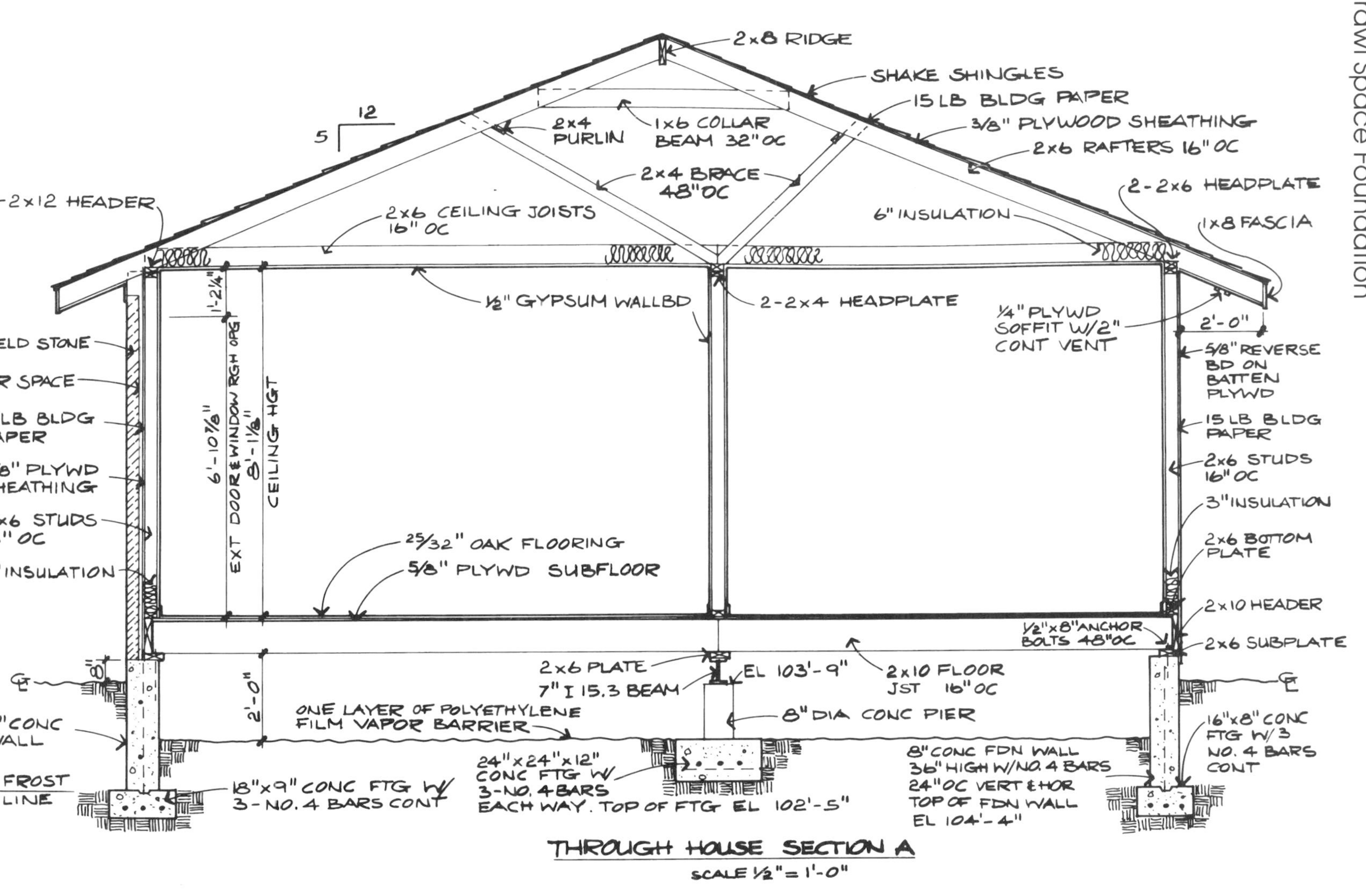

PUBLISHER'S NOTE: The size of this print has been modified and should not be scaled.

The Garlinghouse Company

Crawl Space Foundation

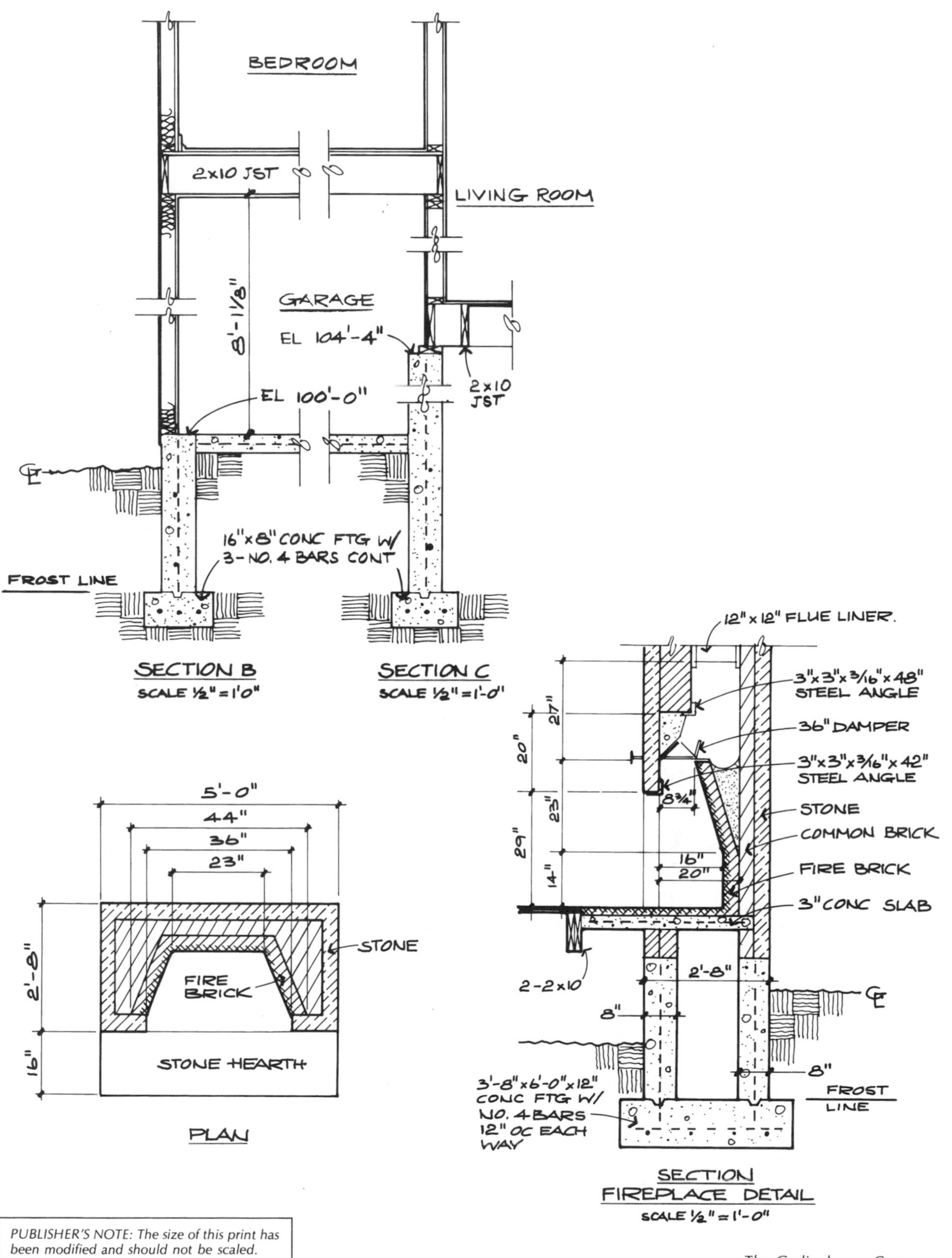

PUBLISHER'S NOTE: The size of this print has been modified and should not be scaled.

The Garlinghouse Company

Slab-on-Grade Foundation

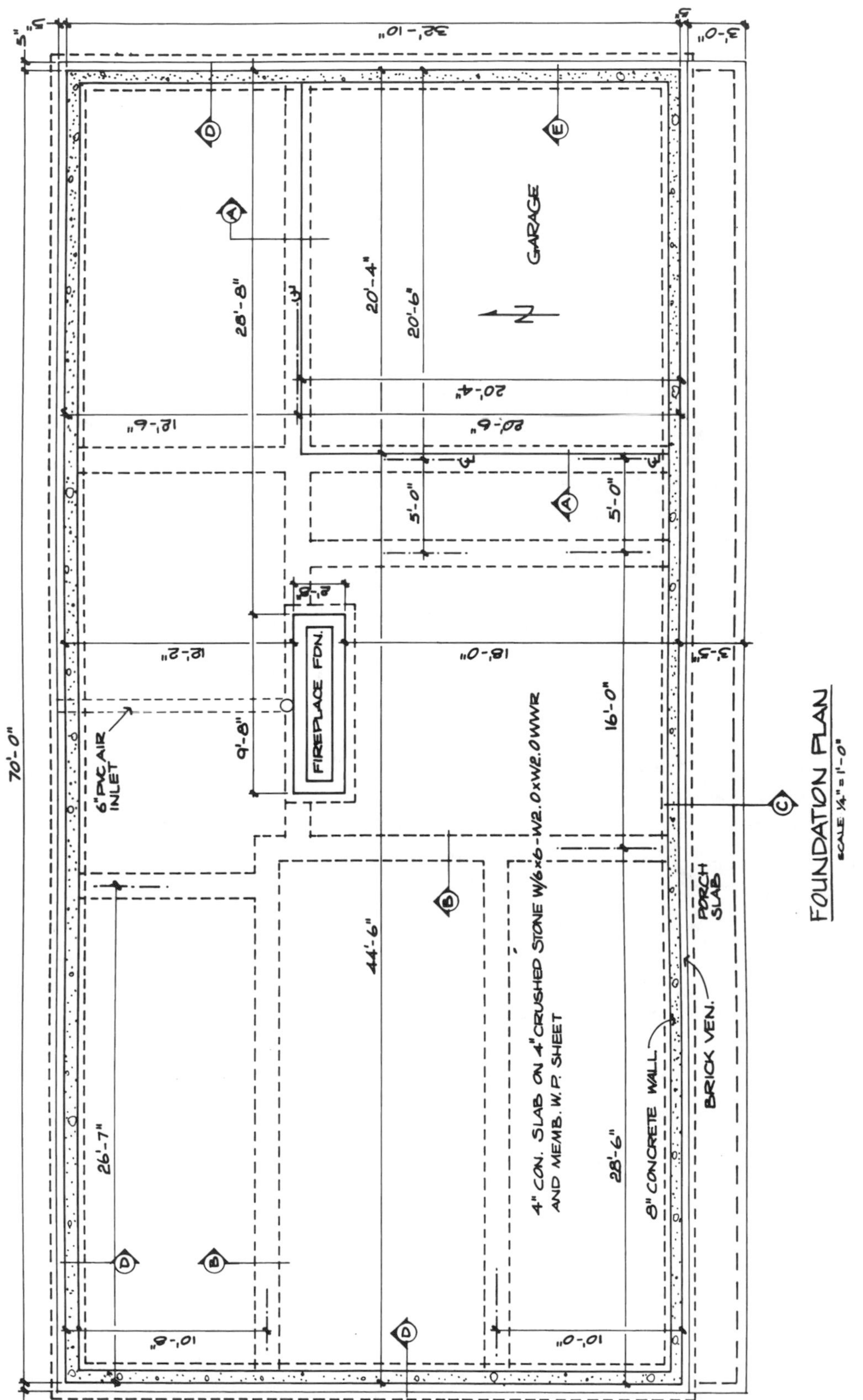

The Garlinghouse Company

PUBLISHER'S NOTE: *The size of this print has been modified and should not be scaled.*

Slab-on-Grade Foundation

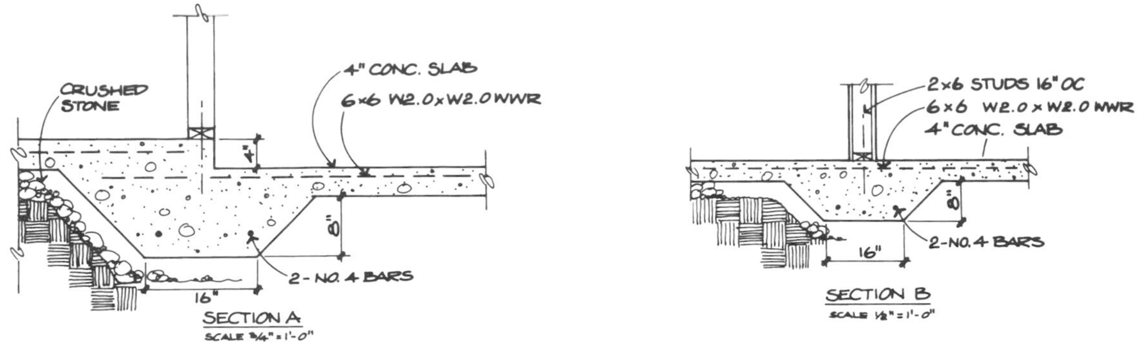

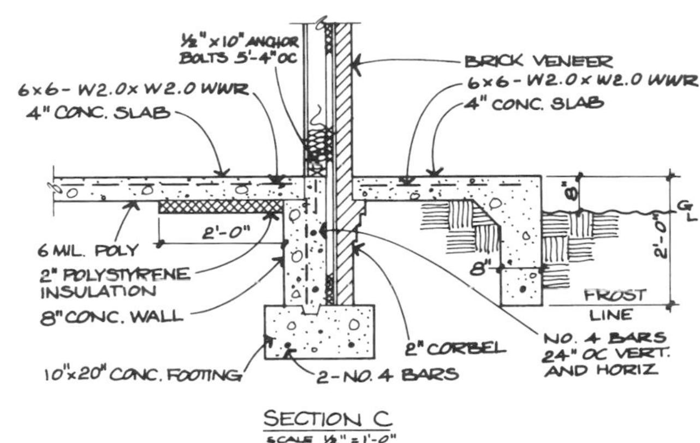

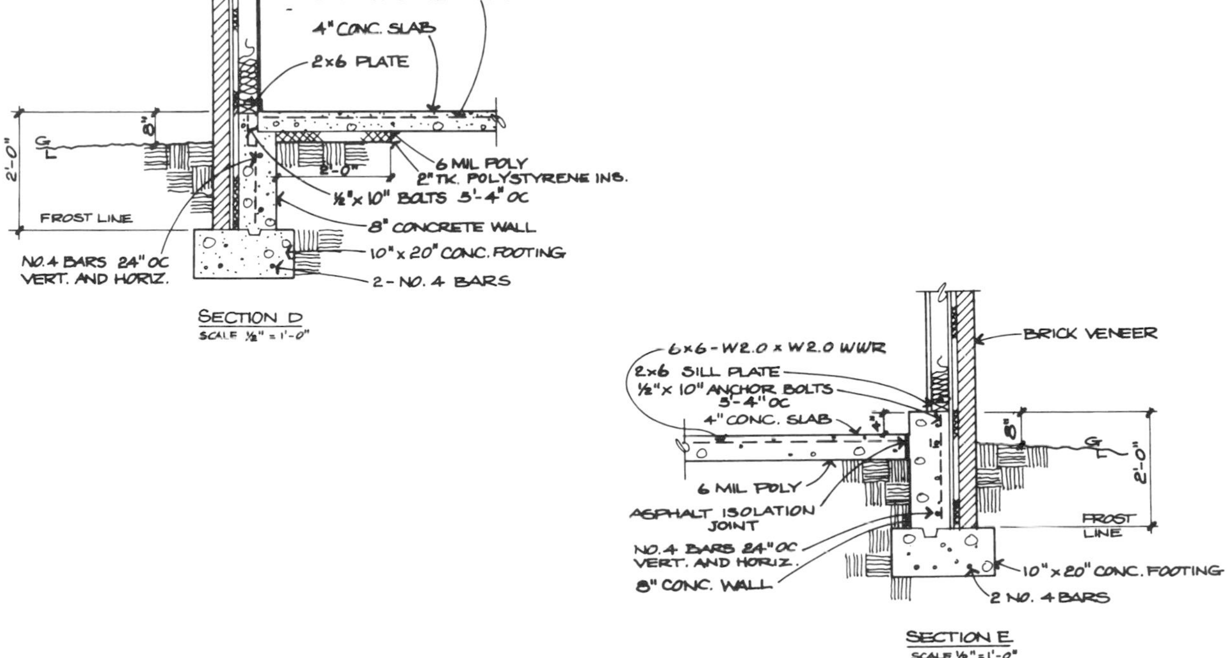

PUBLISHER'S NOTE: The size of this print has been modified and should not be scaled.

The Garlinghouse Company

Heavy Construction Foundation

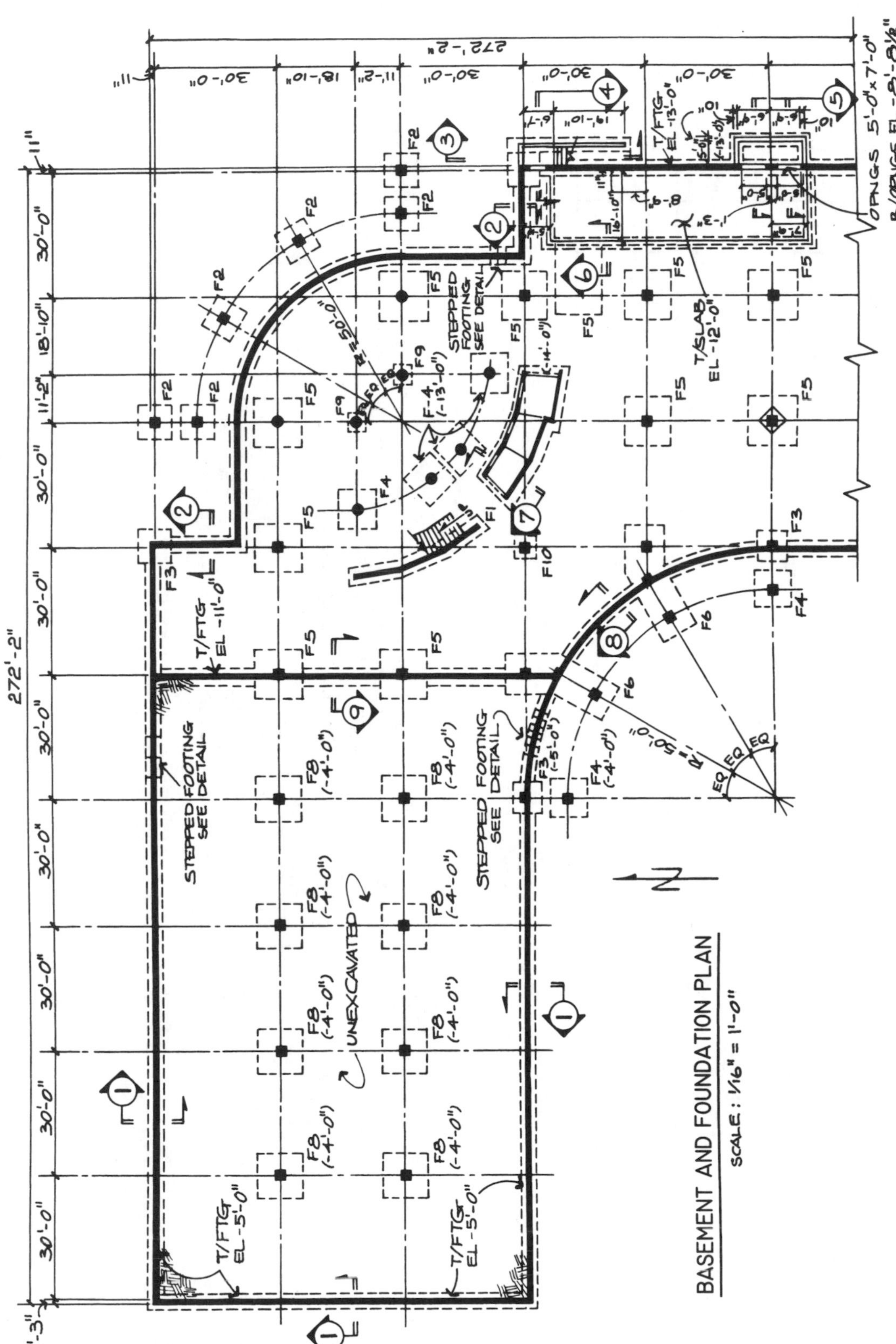

Chris P. Stefanos Associates, Inc.

PUBLISHER'S NOTE: The size of this print has been modified and should not be scaled.

Heavy Construction Foundation

SOIL DATA 5000 psf BEARING	FOOTING SCHEDULE			LC = 4000psi FY = 80,000psi
	SIZE	REINFORCEMENT		
MARK	W × L × D	LONG WAY	SHORT WAY	REMARKS
F1	5′-0″ × 36′-0″ × 14″	8-# 4	34-# 6	
F2	7′-3″ × 7′-3″ × 16″	7-# 7	7-# 7	
F3	8′-3″ × 8′-3″ × 20″	8-# 7	8-# 7	
F4	9′-9″ × 9′-9″ × 26″	9-# 8	9-# 8	
F5	12′-0″ × 12′-0″ × 30″	11-# 9	11-# 9	
F6	SEE DETAIL	11-# 9	30-# 9	
F7	15′-3″ × 15′-3″ × 40″	13-# 9	18-# 9	
F8	11′-0″ × 11′-0″ × 28″	11-# 8	11-# 8	
F9	4′-6″ × 4′6″ × 16″	5-# 5	5-# 5	
F10	5′-6″ × 5′-6″ × 18″	7-# 5	7-# 5	

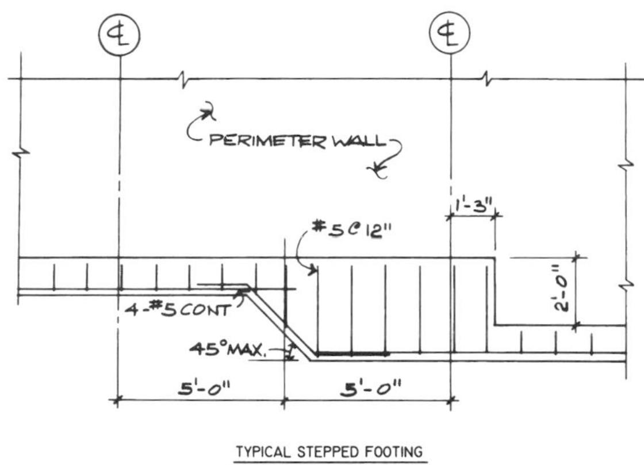

TYPICAL STEPPED FOOTING
PERIMETER BASEMENT WALL
SCALE: 3/8" = 1'-0"

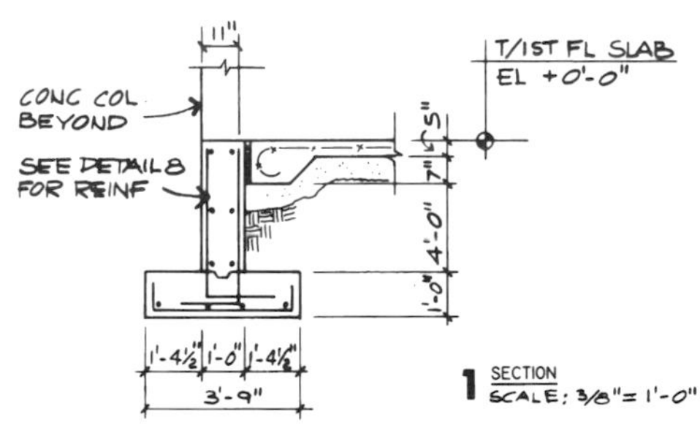

1 SECTION
SCALE: 3/8" = 1'-0"

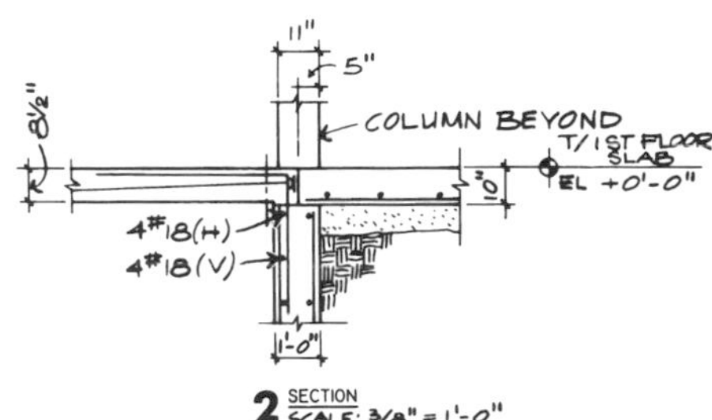

2 SECTION
SCALE: 3/8" = 1'-0"

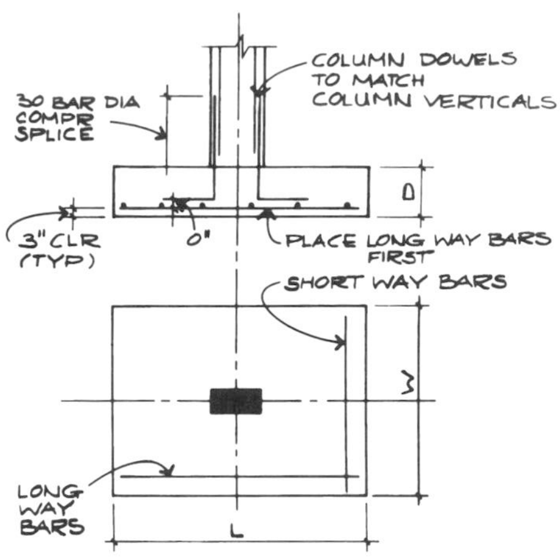

TYPICAL FOOTING DETAIL

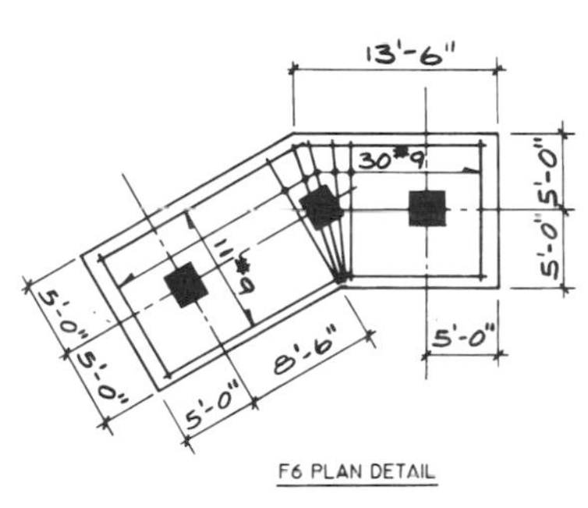

F6 PLAN DETAIL

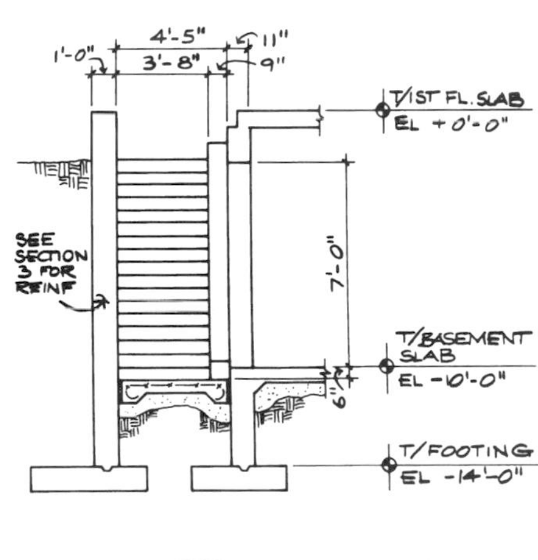

4 SECTION
SCALE: 1/4" = 1'-0"

PUBLISHER'S NOTE: The size of this print has been modified and should not be scaled.

Chris P. Stefanos Associates, Inc.

Heavy Construction Foundation

#4@15"
#4@14"
#5@12"
#4 NOSING BAR TYP.
T/BSMT SLAB
EL -10'-0"
T/FOOTING
EL -14'-0"
4 #4
#4@14"
2'-0"
17 RISERS @6" = 8'-6"
6'-7"
16 TREADS @ 14" = 18'-8"
1'-2"
2'-8"
1'-0"
4'-8"

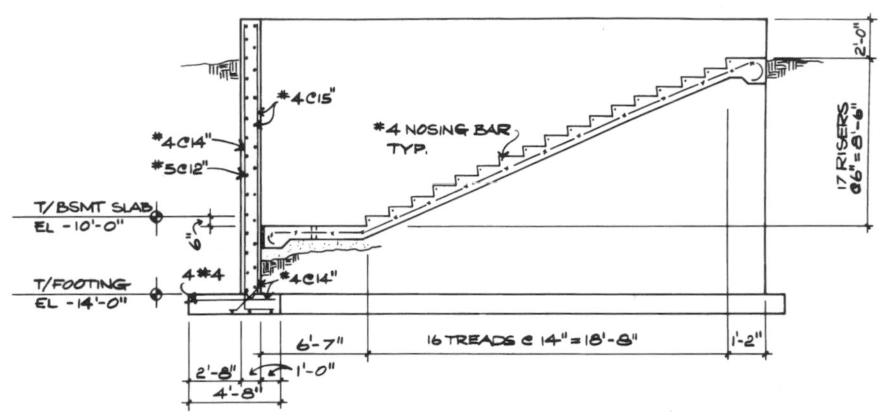

3 SECTION SCALE 1/4" = 1'-0"

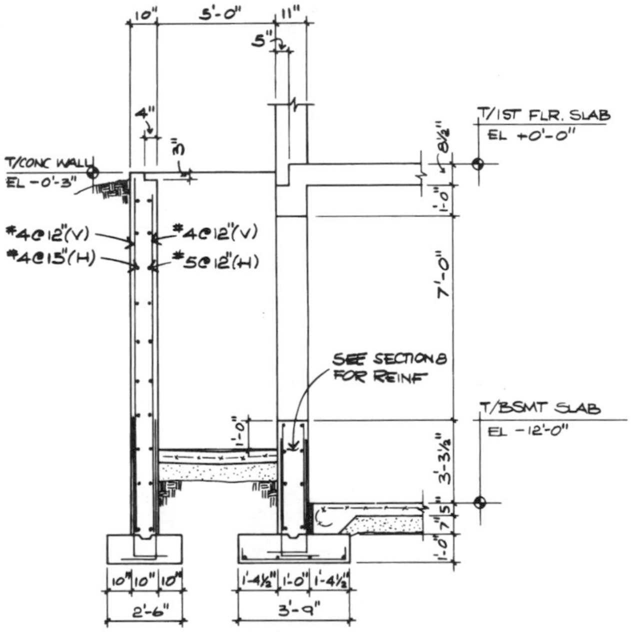

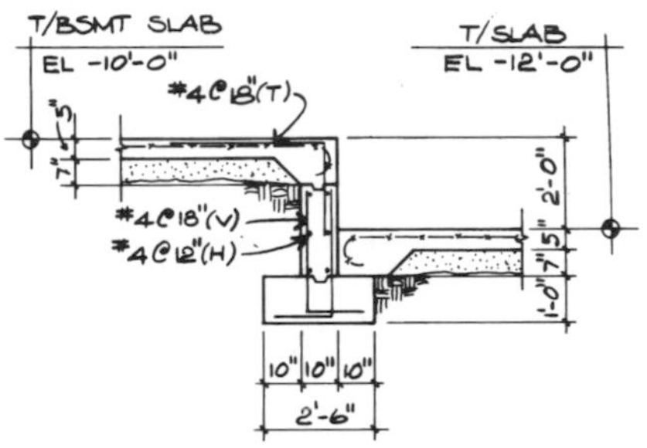

10"
5'-0"
11"
5"
4"
3"
T/CONC WALL
EL -0'-3"
T/1ST FLR. SLAB
EL +0'-0"
8 1/2"
1'-0"
7'-0"
#4@12"(V)
#4@15"(H)
#4@12"(V)
#5@12"(H)
SEE SECTION 8 FOR REINF
T/BSMT SLAB
EL -12'-0"
1'-0"
3'-3 1/2"
5"
7"
1'-0"
10" 10" 10"
2'-6"
1'-4 1/2" 1'-0" 1'-4 1/2"
3'-9"

5 SECTION SCALE: 3/8" = 1'-0"

T/BSMT SLAB
EL -10'-0"
T/SLAB
EL -12'-0"
#4@18"(T)
#4@18"(V)
#4@12"(H)
5"
7"
2'-0"
5"
7"
1'-0"
10" 10" 10"
2'-6"

6 SECTION SCALE: 3/8" = 1'-0"

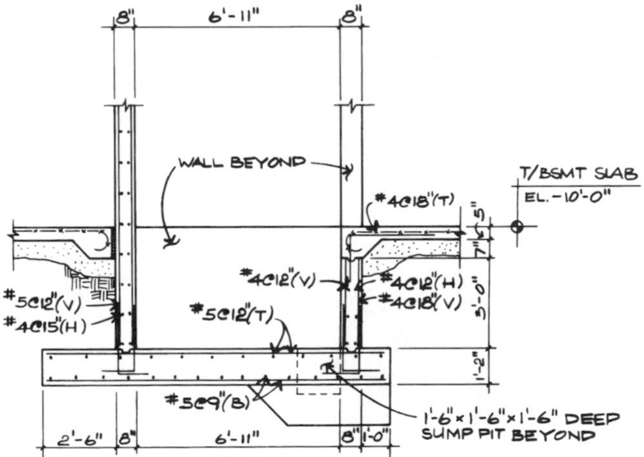

8"
6'-11"
8"
WALL BEYOND
T/BSMT SLAB
EL. -10'-0"
#4@18"(T)
5"
7"
#4@12"(V)
#4@12"(H)
#4@18"(V)
3'-0"
#5@12"(V)
#4@15"(H)
#5@12"(T)
1'-2"
#5@9"(B)
1'-6" x 1'-6" x 1'-6" DEEP SUMP PIT BEYOND
2'-6"
8"
6'-11"
8"
1'-0"
11'-9"

7 SECTION SCALE: 3/8" = 1'-0"

PUBLISHER'S NOTE: The size of this print has been modified and should not be scaled.

Chris P. Stefanos Associates, Inc.

Heavy Construction Foundation

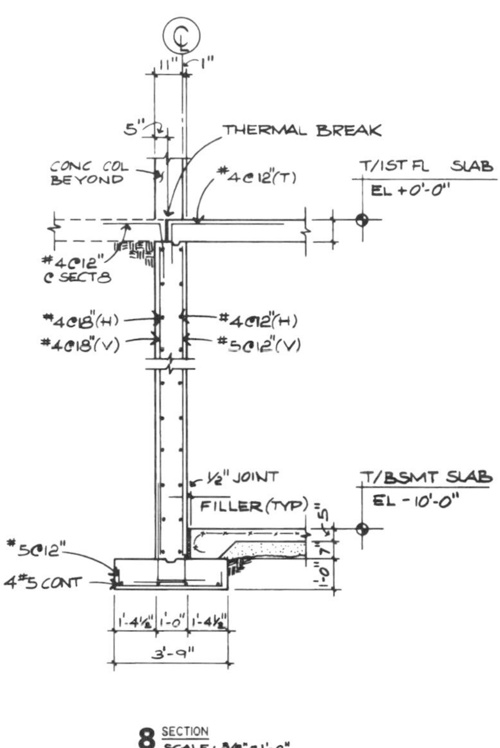

8 SECTION
SCALE: 3/8" = 1'-0"

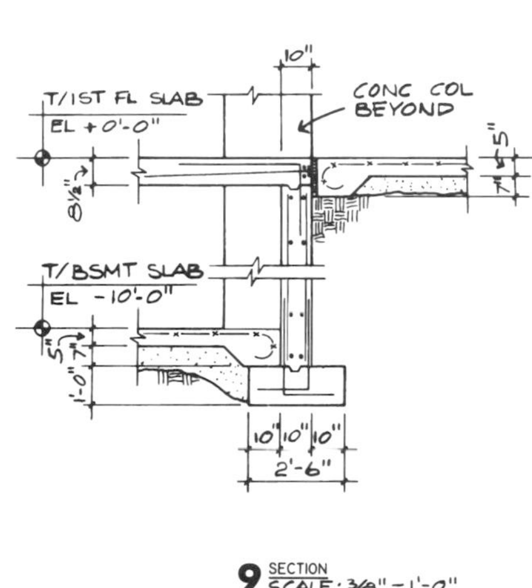

9 SECTION
SCALE: 3/8" = 1'-0"

PUBLISHER'S NOTE: *The size of this print has been modified and should not be scaled.*

Chris P. Stefanos Associates, Inc.

Estimating Print

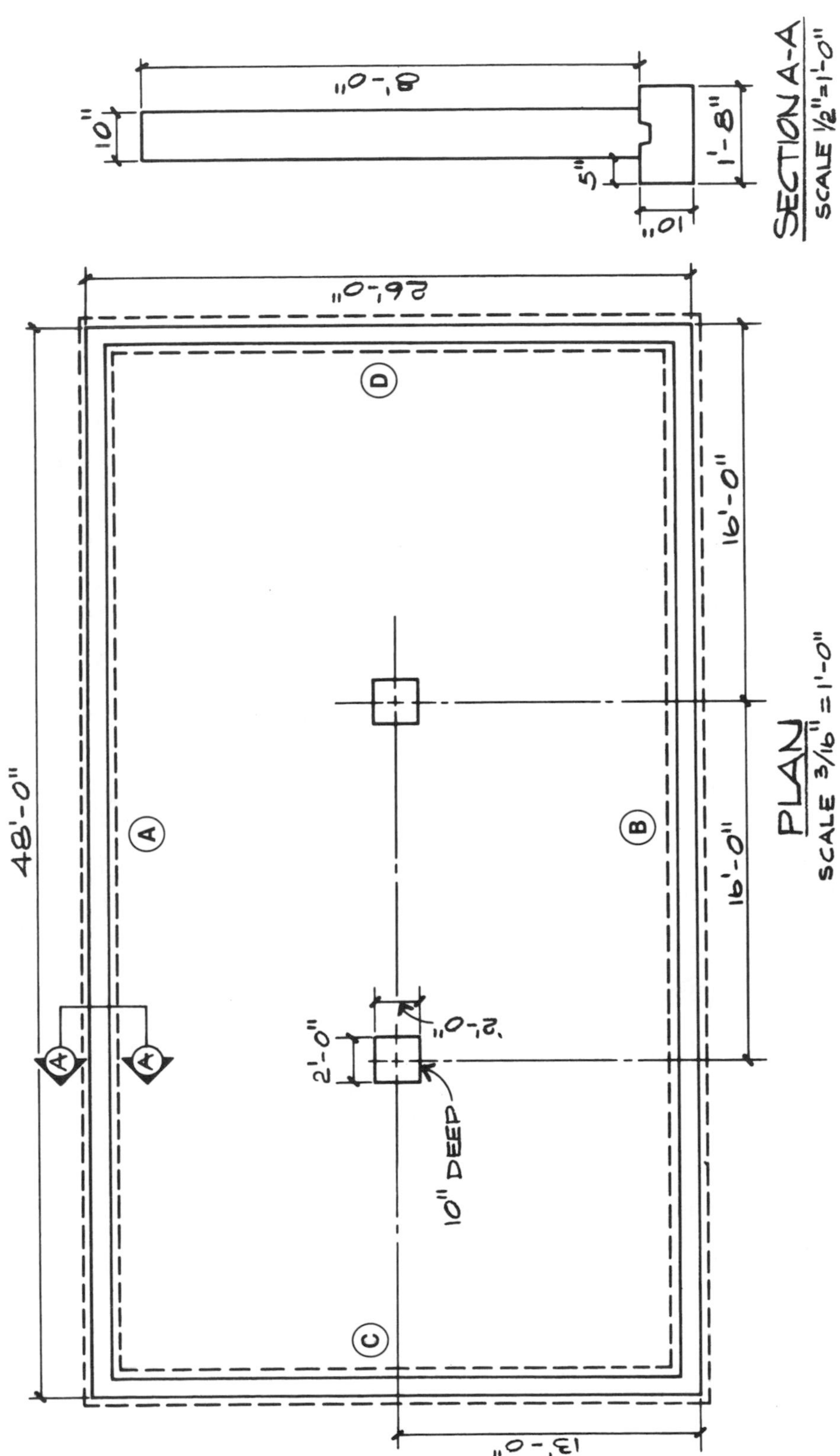

PUBLISHER'S NOTE: The size of this print has been modified and should not be scaled.

Appendix

DECIMAL EQUIVALENTS OF AN INCH

Fraction	Decimal	Fraction	Decimal	Fraction	Decimal	Fraction	Decimal
1/64	0.015625	17/64	0.265625	33/64	0.515625	49/64	0.765625
1/32	0.03125	9/32	0.28125	17/32	0.53125	25/32	0.78125
3/64	0.046875	19/64	0.296875	35/64	0.546875	51/64	0.796875
1/16	0.0625	5/16	0.3125	9/16	0.5625	13/16	0.8125
5/64	0.078125	21/64	0.328125	37/64	0.578125	53/64	0.828125
3/32	0.09375	11/32	0.34375	19/32	0.59375	27/32	0.84375
7/64	0.109375	23/64	0.359375	39/64	0.609375	55/64	0.859375
1/8	0.125	3/8	0.375	5/8	0.625	7/8	0.875
9/64	0.140625	25/64	0.390625	41/64	0.640625	57/64	0.890625
5/32	0.15625	13/32	0.40625	21/32	0.65625	29/32	0.90625
11/64	0.171875	27/64	0.421875	43/64	0.671875	59/64	0.921875
3/16	0.1875	7/16	0.4375	11/16	0.6875	15/16	0.9375
13/64	0.203125	29/64	0.453125	45/64	0.703125	61/64	0.953125
7/32	0.21875	15/32	0.46875	23/32	0.71875	31/32	0.96875
15/64	0.234375	31/64	0.484375	47/64	0.734375	63/64	0.984375
1/4	0.250	1/2	0.500	3/4	0.750	1	1.000

DECIMAL EQUIVALENTS OF A FOOT

Inches	Decimal Foot Equivalent	Inches	Decimal Foot Equivalent	Inches	Decimal Foot Equivalent	Inches	Decimal Foot Equivalent	Inches	Decimal Foot Equivalent	Inches	Decimal Foot Equivalent
1/16	0.0052	2 1/16	0.1719	4 1/16	0.3385	6 1/16	0.5052	8 1/16	0.6719	10 1/16	0.8385
1/8	0.0104	2 1/8	0.1771	4 1/8	0.3438	6 1/8	0.5104	8 1/8	0.6771	10 1/8	0.8438
3/16	0.0156	2 3/16	0.1823	4 3/16	0.3490	6 3/16	0.5156	8 3/16	0.6823	10 3/16	0.8490
1/4	0.0208	2 1/4	0.1875	4 1/4	0.3542	6 1/4	0.5208	8 1/4	0.6875	10 1/4	0.8542
5/16	0.0260	2 5/16	0.1927	4 5/16	0.3594	6 5/16	0.5260	8 5/16	0.6927	10 5/16	0.8594
3/8	0.0313	2 3/8	0.1979	4 3/8	0.3646	6 3/8	0.5313	8 3/8	0.6979	10 3/8	0.8646
7/16	0.0365	2 7/16	0.2031	4 7/16	0.3698	6 7/16	0.5365	8 7/16	0.7031	10 7/16	0.8698
1/2	0.0417	2 1/2	0.2083	4 1/2	0.3750	6 1/2	0.5417	8 1/2	0.7083	10 1/2	0.8750
9/16	0.0469	2 9/16	0.2135	4 9/16	0.3802	6 9/16	0.5469	8 9/16	0.7135	10 9/16	0.8802
5/8	0.0521	2 5/8	0.2188	4 5/8	0.3854	6 5/8	0.5521	8 5/8	0.7188	10 5/8	0.8854
11/16	0.0573	2 11/16	0.2240	4 11/16	0.3906	6 11/16	0.5573	8 11/16	0.7240	10 11/16	0.8906
3/4	0.0625	2 3/4	0.2292	4 3/4	0.3958	6 3/4	0.5625	8 3/4	0.7292	10 3/4	0.8958
13/16	0.0677	2 13/16	0.2344	4 13/16	0.4010	6 13/16	0.5677	8 13/16	0.7344	10 13/16	0.9010
7/8	0.0729	2 7/8	0.2396	4 7/8	0.4063	6 7/8	0.5729	8 7/8	0.7396	10 7/8	0.9063
15/16	0.0781	2 15/16	0.2448	4 15/16	0.4115	6 15/16	0.5781	8 15/16	0.7448	10 15/16	0.9115
1	0.0833	3	0.2500	5	0.4167	7	0.5833	9	0.7500	11	0.9167
1 1/16	0.0885	3 1/16	0.2552	5 1/16	0.4219	7 1/16	0.5885	9 1/16	0.7552	11 1/16	0.9219
1 1/8	0.0938	3 1/8	0.2604	5 1/8	0.4271	7 1/8	0.5938	9 1/8	0.7604	11 1/8	0.9271
1 3/16	0.0990	3 3/16	0.2656	5 3/16	0.4323	7 3/16	0.5990	9 3/16	0.7656	11 3/16	0.9323
1 1/4	0.1042	3 1/4	0.2708	5 1/4	0.4375	7 1/4	0.6042	9 1/4	0.7708	11 1/4	0.9375
1 5/16	0.1094	3 5/16	0.2760	5 5/16	0.4427	7 5/16	0.6094	9 5/16	0.7760	11 5/16	0.9427
1 3/8	0.1146	3 3/8	0.2813	5 3/8	0.4479	7 3/8	0.6146	9 3/8	0.7813	11 3/8	0.9479
1 7/16	0.1198	3 7/16	0.2865	5 7/16	0.4531	7 7/16	0.6198	9 7/16	0.7865	11 7/16	0.9531
1 1/2	0.1250	3 1/2	0.2917	5 1/2	0.4583	7 1/2	0.6250	9 1/2	0.7917	11 1/2	0.9583
1 9/16	0.1302	3 9/16	0.2969	5 9/16	0.4635	7 9/16	0.6302	9 9/16	0.7969	11 9/16	0.9635
1 5/8	0.1354	3 5/8	0.3021	5 5/8	0.4688	7 5/8	0.6354	9 5/8	0.8021	11 5/8	0.9688
1 11/16	0.1406	3 11/16	0.3073	5 11/16	0.4740	7 11/16	0.6406	9 11/16	0.8073	11 11/16	0.9740
1 3/4	0.1458	3 3/4	0.3125	5 3/4	0.4792	7 3/4	0.6458	9 3/4	0.8125	11 3/4	0.9792
1 13/16	0.1510	3 13/16	0.3177	5 13/16	0.4844	7 13/16	0.6510	9 13/16	0.8177	11 13/16	0.9844
1 7/8	0.1563	3 7/8	0.3229	5 7/8	0.4896	7 7/8	0.6563	9 7/8	0.8229	11 7/8	0.9896
1 15/16	0.1615	3 15/16	0.3281	5 15/16	0.4948	7 15/16	0.6615	9 15/16	0.8281	11 15/16	0.9948
2	0.1667	4	0.3333	6	0.5000	8	0.6667	10	0.8333	12	1.0000

CONVERSION TABLE—DECIMAL FEET TO INCHES													
		Inches											
		0	1	2	3	4	5	6	7	8	9	10	11
8th of an Inch	0	.00	.08	.17	.25	.33	.42	.50	.58	.67	.75	.83	.92
	1	.01	.09	.18	.26	.34	.43	.51	.59	.68	.76	.84	.93
	2	.02	.10	.19	.27	.35	.44	.52	.60	.69	.77	.85	.94
	3	.03	.11	.20	.28	.36	.45	.53	.61	.70	.78	.86	.95
	4	.04	.13	.21	.29	.38	.46	.54	.63	.71	.79	.88	.96
	5	.05	.14	.22	.30	.39	.47	.55	.64	.72	.80	.89	.97
	6	.06	.15	.23	.31	.40	.48	.56	.65	.73	.81	.90	.98
	7	.07	.16	.24	.32	.41	.49	.57	.66	.74	.82	.91	.99

AREA

Square, Rectangle, or Parallelogram

*Area = Length × Height**

$A = L \times H$

**Width* may be substituted for *Height*

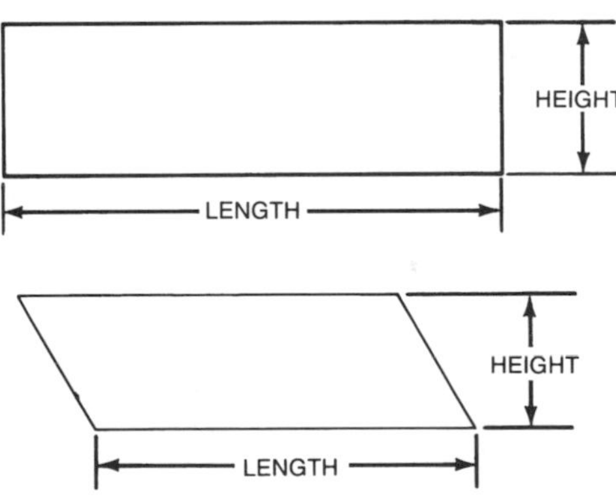

Triangle

$$Area = \frac{base \times altitude}{2}$$

$$Area = \frac{ba}{2}$$

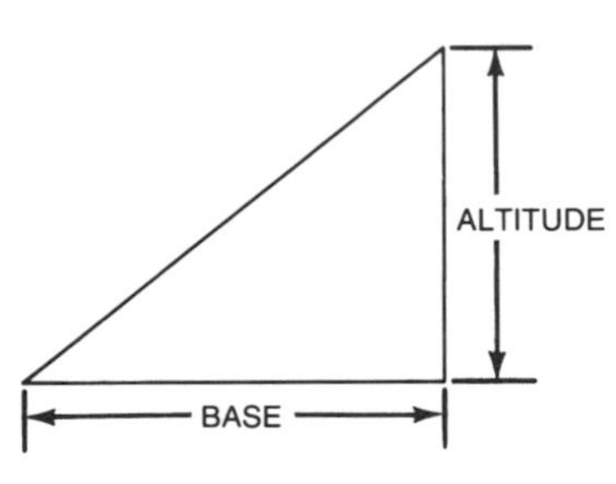

Trapezoid

$$Area = \frac{Height\ (Length\ of\ top + Length\ of\ bottom)}{2}$$

$$Area = \frac{H\ (L_t + L_b)}{2}$$

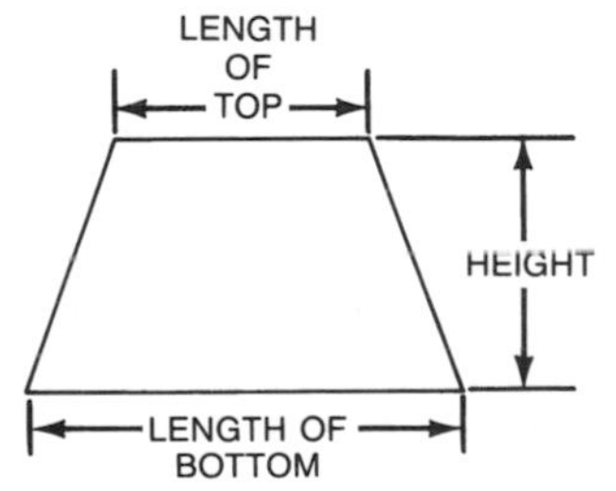

Circle

$Area = \pi \times Radius^2$
$A = \pi R^2$

$Area = .7854 \times Diameter^2$
$A = .7854\ D^2$

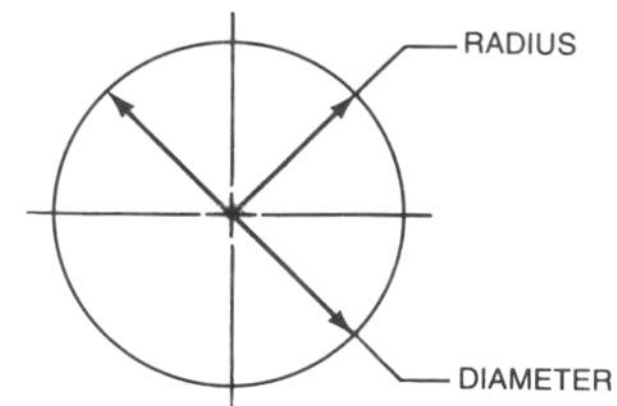

VOLUME

Rectangular Solid

$Volume = Thickness \times Length \times Height$
$V = T \times L \times H$

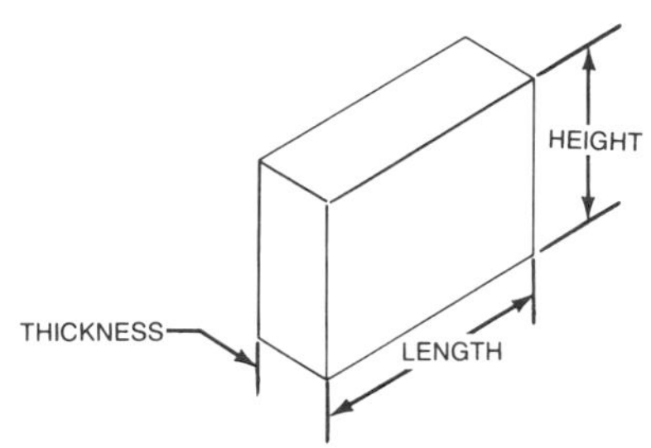

Frustum—One Battered Side

$$Volume = \frac{Top\ Thickness + Bottom\ Thickness}{2} \times Length \times Height$$
$$V = \frac{T_t + T_b}{2} \times L \times H$$

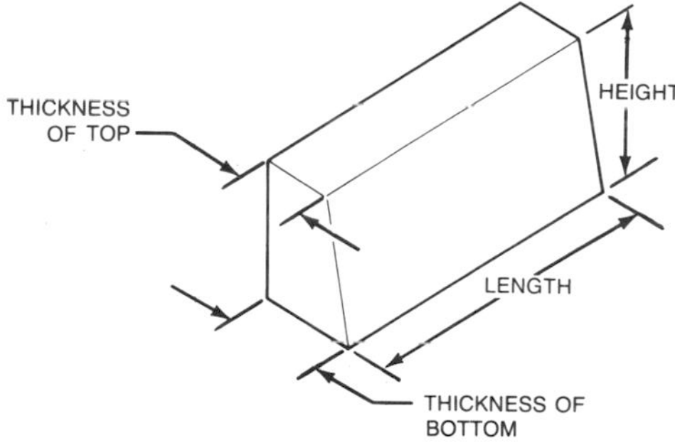

Frustum—Four Battered Sides

$$Volume = \frac{Area\ of\ Top + Area\ of\ Bottom}{2} \times Height$$
$$V = \frac{A_t + A_b}{2} \times Height$$

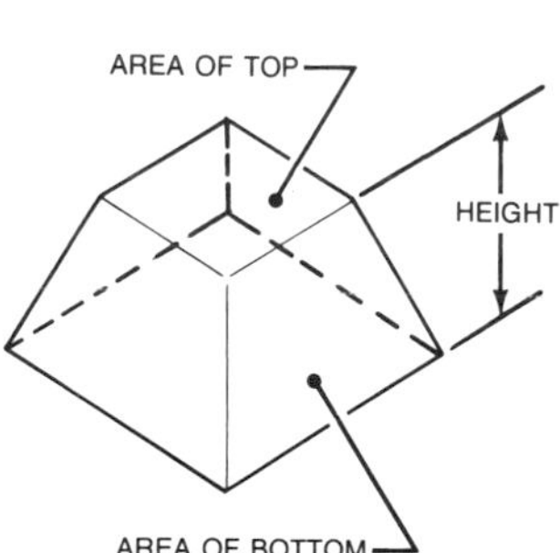

Cylinder

$Volume = \pi \times Radius^2 \times Height$
$A = \pi \times R^2 \times H$

$Volume = .7854 \times Diameter^2 \times Height$
$V = .7854\ D^2 \times H$

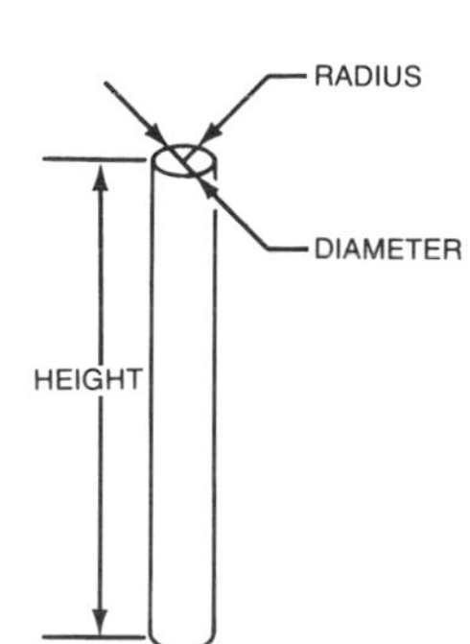

Appendix B: Construction Materials

NOMINAL AND MINIMUM DRESSED SIZES OF BOARDS, DIMENSION LUMBER, AND TIMBERS										
	THICKNESS					WIDTH				
	Nominal Inch	Minimum Dressed				Nominal Inch	Minimum Dressed			
		Dry		Green			Dry		Green	
		inch	mm	inch	mm		inch	mm	inch	mm
Boards	¾	⅝	16	11/16	17	2	1½	38	1 9/16	40
	1	¾	19	25/32	20	3	2½	64	2 9/16	65
	1¼	1	25	1 1/32	26	4	3½	89	3 9/16	90
	1½	1¼	32	1 9/32	33	5	4½	114	4⅝	117
						6	5½	140	5⅝	143
						7	6½	165	6⅝	168
						8	7¼	184	7½	190
						9	8¼	210	8½	216
						10	9¼	235	9½	241
						11	10¼	260	10½	267
						12	11¼	286	11½	292
						14	13¼	337	13½	343
						16	15¼	387	15½	394
Dimension	2	1½	38	1 9/16	40	2	1½	38	1 9/16	40
	2½	2	51	2 1/16	52	2½	2	51	2 1/16	52
	3	2½	64	2 9/16	65	3	2½	64	2 9/16	65
	3½	3	76	3 1/16	78	3½	3	76	3 1/16	78
	4	3½	89	3 9/16	90	4	3½	89	3 9/16	90
	4½	4	102	4 1/16	103	4½	4	102	4 1/16	103
						5	4½	114	4⅝	117
						6	5½	140	5⅝	143
						8	7¼	184	7½	190
						10	9¼	235	9½	241
						12	11¼	286	11½	292
						14	13¼	337	13½	343
						16	15¼	387	15½	394
Timbers	5 & thicker			½ off	13 off	5 & wider			½ off	13 off

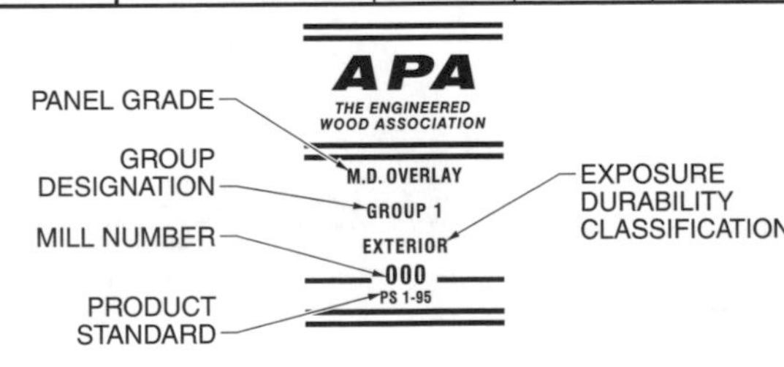

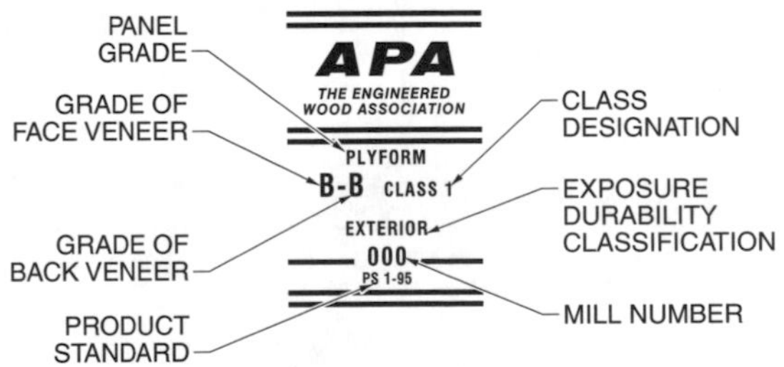

FORM MATERIALS

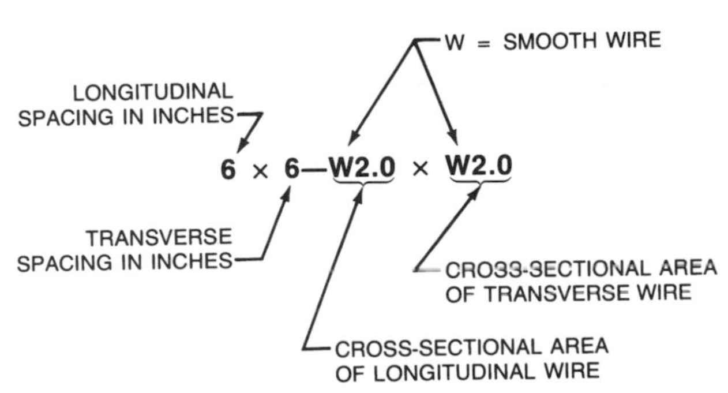

COMMON STOCK SIZES OF WELDED WIRE REINFORCEMENT				
Style Designation		Steel Area sq in. per ft		Weight Approx. lb per 100 sq ft
New Designation (by W-Number)	Old Designation (by Steel Wire Gauge)	Long.	Trans.	
ROLLS				
6 x 6—W1.4 x W1.4	6 x 6—10 x 10	.028	.028	21
6 x 6—W2.0 x W2.0	6 x 6—8 x 8*	.040	.040	29
6 x 6—W2.9 x W2.9	6 x 6—6 x 6	.058	.058	42
SHEETS				
6 x 6—W2.9 x W2.9	6 x 6—6 x 6	.058	.058	42
6 x 6—W4.0 x W4.0	6 x 6—4 x 4	.080	.080	58
6 x 6—W5.5 x W5.5	6 x 6—2 x 2†	.110	.110	80

*Exact W-number size for 8 gauge is W2.1
†Exact W-number size for 2 gauge is W5.4

Wire Reinforcement Institute

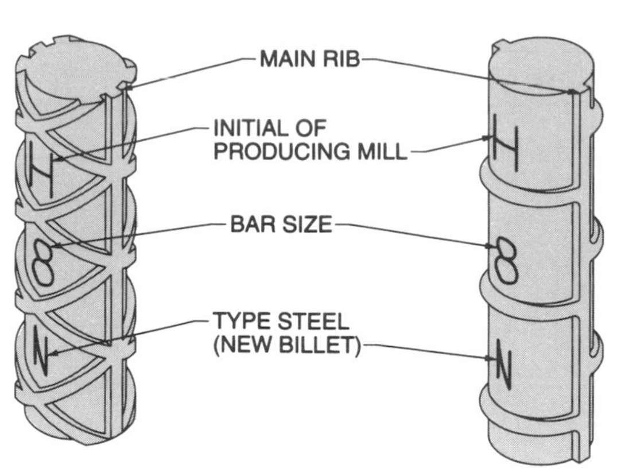

STANDARD REBAR SIZES						
Bar Size Designation	Weight per Foot		Diameter		Cross-Sectional Area Squared	
	lb	kg	in.	cm	in.	cm
#3	0.376	0.171	0.375	0.953	0.11	0.71
#4	0.668	0.303	0.500	1.270	0.20	1.29
#5	1.043	0.473	0.625	1.588	0.31	2.00
#6	1.502	0.681	0.750	1.905	0.44	2.84
#7	2.044	0.927	0.875	2.223	0.60	3.87
#8	2.670	1.211	1.000	2.540	0.79	5.10
#9	3.400	1.542	1.128	2.865	1.00	6.45
#10	4.303	1.952	1.270	3.226	1.27	8.19
#11	5.313	2.410	1.410	3.581	1.56	10.07
#14	7.650	3.470	1.693	4.300	2.25	14.52
#18	13.600	6.169	2.257	5.733	4.00	25.81

American Society for Testing and Materials

REINFORCEMENT

CONCRETE SLAB ESTIMATING TABLE*

Thickness†	Coverage‡	Thickness†	Coverage‡	Thickness†	Coverage‡
1	324	5	65	9	36
1¼	259	5¼	62	9¼	35
1½	216	5½	59	9½	34
1¾	185	5¾	56	9¾	33
2	162	6	54	10	32.5
2¼	144	6¼	52	10¼	31.5
2½	130	6½	50	10½	31
2¾	118	6¾	48	10¾	30
3	108	7	46	11	29.5
3¼	100	7¼	45	11¼	29
3½	93	7½	43	11½	28
3¾	86	7¾	42	11¾	27.5
4	81	8	40	12	27
4¼	76	8¼	39	12¼	21.5
4½	72	8½	38	12½	18
4¾	68	8¾	37	12¾	13.5

* coverage and thickness based on 1 cu yd of concrete
† in in.
‡ in sq ft

CONCRETE FOOTINGS

Width	Height	Volume (cu ft) per Linear Foot
1′-0″	6″	.50
1′-2″		.59
1′-4″		.67
1′-6″		.75
1′-0″	8″	.67
1′-2″		.78
1′-4″		.89
1′-6″		1.00
1′-0″	10″	.83
1′-2″		.97
1′-4″		1.11
1′-6″		1.25
1′-8″		1.39
1′-10″		1.53
2′-0″		1.67
1′-2″	12″	1.17
1′-4″		1.33
1′-6″		1.50
1′-8″		1.67
1′-10″		1.83
2′-0″		2.00

CONCRETE WALLS

Wall Thickness	Volume (cu ft) per 100 sq ft wall
4	33.3
6	50.0
8	66.7
10	83.3
12	100.0

Appendix C: ACI Recommended Practices

Reproduced from Guide to Formwork for Concrete (ACI 347R-03)

Chapter 3—Construction

3.1—Safety precautions

Contractors should follow all state, local, and federal codes, ordinances, and regulations pertaining to forming and shoring. In addition to the very real moral and legal responsibility to maintain safe conditions for workmen and the public, safe construction is, in the final analysis, more economical than any short-term cost savings from cutting corners on safety provisions.

Attention to safety is particularly significant in formwork construction that supports the concrete during its plastic state and until the concrete becomes structurally self-sufficient. Following the design criteria contained in this guide is essential for ensuring safe performance of the forms. All structural members and connections should be carefully planned so that a sound determination of loads may be accurately made and stresses calculated.

In addition to the adequacy of the formwork, special structures, such as multistory buildings, require consideration of the behavior of newly completed beams and slabs that are used to support formwork and other construction loads. It should be kept in mind that the strength of freshly cast slabs or beams is less than that of a mature slab.

Formwork failures can be attributed to substandard materials and equipment, human error, and inadequacy in design. Careful supervision and continuous inspection of formwork during erection, concrete placement, and removal can prevent many accidents.

Construction procedures should be planned in advance to ensure the safety of personnel and the integrity of the finished structure. Some of the safety provisions that should be considered are:

- Erection of safety signs and barricades to keep unauthorized personnel clear of areas in which erection, concrete placing, or stripping is under way;
- Providing experienced form watchers during concrete placement to ensure early recognition of possible form displacement or failure. A supply of extra shores or other material and equipment that might be needed in an emergency should be readily available;
- Provision for adequate illumination of the formwork and work area;
- Inclusion of lifting points in the design and detailing of all forms that will be crane-handled. This is especially important in flying forms or climbing forms. In the case of wall formwork, consideration should be given to an independent work platform bolted to the previous lift;
- Incorporation of scaffolds, working platforms, and guardrails into formwork design and all formwork drawings;
- Incorporation of provisions for anchorage of alternative fall protection devices, such as personal fall arrest systems, safety net systems, and positioning device systems; and
- A program of field safety inspections of formwork.

3.1.1 *Formwork construction deficiencies*—Some common construction deficiencies that can lead to formwork failures are:

- Failure to inspect formwork during and after concrete placement to detect abnormal deflections or other signs of imminent failure that could be corrected;
- Insufficient nailing, bolting, welding, or fastening; Insufficient or improper lateral bracing;
- Failure to comply with manufacturer's recommendations; Failure to construct formwork in accordance with the form drawings;
- Lack of proper field inspection by qualified persons to ensure that form design has been properly interpreted by form builders; and
- Use of damaged or inferior lumber having lower strength than needed.

3.1.1.1 *Examples of deficiencies in vertical formwork*—Construction deficiencies sometimes found in vertical formwork include:

- Failure to control rate of placing concrete vertically without regard to design parameters;
- Inadequately tightened or secured form ties or hardware;
- Form damage in excavations resulting from embankment failure;
- Use of external vibrators on forms not designed for their use;
- Deep vibrator penetration of earlier semihardened lifts; Improper framing of blockouts;

- Improperly located or constructed pouring pockets;
- Inadequate bulkheads;
- Improperly anchored top forms on a sloping face;
- Failure to provide adequate support for lateral pressures on formwork; and
- Failure to provide adequate bracing resulting in attempts to plumb forms against concrete pressure force.

3.1.1.2 *Examples of deficiencies in horizontal formwork*—Construction deficiencies sometimes found in horizontal forms for elevated structures include:

- Failure to properly regulate the rate and sequence of placing concrete horizontally to avoid unanticipated loadings on the formwork;
- Shoring not plumb, thus inducing lateral loading and reducing vertical load capacity;
- Locking devices on metal shoring not locked, inoperative, or missing; Safety nails missing on adjustable two-piece wood shores;
- Failure to account for vibration from adjacent moving loads or load carriers;
- Inadequately tightened or secured shore hardware or wedges;
- Loosening or premature removal of reshores or backshores under floors below;
- Premature removal of supports, especially under cantilevered sections;
- Inadequate bearing area or unsuitable soil under mudsills;
- Mudsills placed on frozen ground subject to thawing;
- Connection of shores to joists, stringers, or wales that are inadequate to resist uplift or torsion at joints;
- Failure to consider effects of load transfer that can occur during post-tensioning;
- Inadequate shoring and bracing of composite construction.

3.2—Construction practices and workmanship

3.2.1—*Fabrication and assembly details*

3.2.1.1—Studs, wales, or shores should be properly spliced.

3.2.1.2—Joints or splices in sheathing, plywood panels, and bracing should be staggered.

3.2.1.3—Shores should be installed plumb and with adequate bearing and bracing.

3.2.1.4—Specified size and capacity of form ties or clamps should be used.

3.2.1.5—All form ties or clamps should be installed and properly tightened as specified. All threads should fully engage the nut or coupling. A double nut may be required to develop the full capacity of the tie.

3.2.1.6—Forms should be sufficiently tight to prevent loss of mortar from the concrete.

3.2.1.7—Access holes may be necessary in wall forms or other high, narrow forms to facilitate concrete placement.

3.2.2—*Joints in the concrete*

3.2.2.1—Contraction joints, expansion joints, control joints, construction joints, and isolation joints should be installed as specified in the contract documents or as requested by the contractor and approved by the engineer/architect.

3.2.2.2—Bulkheads for joints should preferably be made by splitting the bulkhead along the lines of reinforcement passing through the bulkhead. By doing this, each portion can be positioned and removed separately. When required on the engineer/architect's plans, beveled inserts at control joints should be left undisturbed when forms are stripped and removed only after the concrete has been sufficiently cured. Wood strips inserted for architectural treatment should be kerfed to permit swelling without causing pressure on the concrete.

3.2.3 *Sloping surfaces*—Sloped surfaces steeper than 1.5 horizontal to 1 vertical should be provided with a top form to hold the shape of the concrete during placement, unless it can be demonstrated that the top forms can be omitted.

3.2.4 *Inspection*—The inspection should be performed by a person certified as an ACI Concrete Construction Inspector or a person having equivalent formwork training and knowledge.

3.2.4.1—Forms should be inspected and checked before the reinforcing steel is placed to confirm that the dimensions and the location of the concrete members will conform to the structural plans.

3.2.4.2—Blockouts, inserts, sleeves, anchors, and other embedded items should be properly identified, positioned, and secured.

3.2.4.3—Formwork should be checked for camber when specified in the contract documents or shown on the formwork drawings.

3.2.5—*Cleanup and coatings*

3.2.5.1—Forms should be thoroughly cleaned of all dirt, mortar, and foreign matter and coated with a release agent before each use. Where the bottom of the form is inaccessible from within, access panels should be provided to permit thorough removal of extraneous material before placing concrete. If surface appearance is important, forms should not be reused if damage from previous use would cause impairment to concrete surfaces.

3.2.5.2—Form coatings should be applied before placing of reinforcing steel and should not be used in such quantities as to run onto bars or concrete construction joints.

3.2.6—*Construction operations on the formwork*

3.2.6.1—Building materials, including concrete, should not be dropped or piled on the formwork in such a manner as to damage or overload it.

3.2.6.2—Runways for moving equipment should be provided with struts or legs as required and should be supported directly on the formwork or structural member. They should not bear on or be supported by the reinforcing steel unless special bar supports are provided. The formwork should be suitable for the support of such runways without significant deflections, vibrations, or lateral movements.

3.2.7 *Loading new slabs*—Overloading of new slabs by temporary material stockpiling or by early application of permanent loads should be avoided. Loads, such as aggregate, lumber, reinforcing steel, masonry, or machinery should not be placed on new construction in such a manner as to damage or overload it.

3.3—Tolerances

Tolerance is a permissible variation from lines, grades, or dimensions given in contract documents. Suggested tolerances for concrete structures can be found in ACI 117.

The contractor should set and maintain concrete forms, including any specified camber, to ensure completed work is within the tolerance limits.

3.3.1 *Recommendations for engineer/architect and contractor*—Tolerances should be specified by the engineer/architect so that the contractor will know precisely what is required and can design and maintain the formwork accordingly. Specifying tolerances more exacting than needed can increase construction costs.

Contractors should be required to establish and maintain control points and benchmarks in an undisturbed condition until final completion and acceptance of a project. Both should be adequate for the contractor's use and for reference to establish tolerances. This requirement can become even more important for the contractor's protection when tolerances are not specified or shown. The engineer/architect should specify tolerances or require performance appropriate to the type of construction. Specifying tolerances more stringent than commonly obtained for a specific type of construction should be avoided, as this usually results in disputes among the parties involved. For example, specifying permitted irregularities more stringent than those allowed for a Class C surface is incompatible with most concrete one-way joist construction techniques. Where a project involves features sensitive to the cumulative effect of tolerances on individual portions, the engineer/architect should anticipate and provide for this effect by setting a cumulative tolerance. Where a particular situation involves several types of generally accepted tolerances on items such as concrete, location of reinforcement, and fabrication of reinforcement, which become mutually incompatible, the engineer/architect should anticipate the difficulty and specify special tolerances or indicate what governs. The project specifications should clearly state that a permitted variation in one part of the construction or in one section of the specifications should not be construed as permitting violation of the more stringent requirements for any other part of the construction or in any other such specification section.

The engineer/architect should be responsible for coordinating the tolerances for concrete work with the tolerance requirements of other trades whose work adjoins the concrete construction. For example, the connection detail for a building's façade should accommodate the tolerance range for the lateral alignment and elevation of the perimeter concrete member.

3.4—Irregularities in formed surfaces

This section provides a way of evaluating surface variations due to forming quality but is not intended for evaluation of surface defects, such as bugholes (blowholes) and honeycomb, attributable to placing and consolidation deficiencies. The latter are more fully explained by ACI 309.2R. Allowable irregularities are designated either abrupt or gradual. Offsets and fins resulting from displaced, mismatched, or misplaced forms, sheathing, or liners, or from defects in forming materials are considered abrupt irregularities. Irregularities resulting from warping and similar uniform variations from planeness or true curvature are considered gradual irregularities.

Gradual irregularities should be checked with a straightedge for plane surfaces or a shaped template for curved or warped surfaces. In measuring irregularities, the straightedge or template can be placed anywhere on the surface in any direction.

The engineer/architect should indicate which class is required for the work being specified or indicate other irregularity limits where needed, or the concrete surface tolerances as specified in ACI 301 should be followed.

Class A is suggested for surfaces prominently exposed to public view where appearance is of special importance. Class B is intended for coarse-textured, concrete-formed surfaces intended to receive plaster, stucco, or wainscoting. Class C is a general standard for permanently exposed surfaces where other finishes are not specified. Class D is a minimum-quality requirement for surfaces where roughness is not objectionable, usually applied where surfaces will be permanently concealed. Special limits on irregularities can be needed for surfaces continuously exposed to flowing water, drainage, or exposure. If permitted irregularities are different from those given, they should be specified by the engineer/architect.

3.5—Shoring and centering

3.5.1 *Shoring*—Shoring should be supported on satisfactory foundations, such as spread footings, mudsills, or piling.

Shoring resting on intermediate slabs or other construction already in place need not be located directly above shores or reshores below, unless the slab thickness and the location of its reinforcement are inadequate to take the reversal of stresses and punching shear. The reversal of stresses results from the reversal of bending moments in the slab over the shore or reshore below. Where the conditions are questionable, the shoring location should be approved by the engineer/architect. If reshores do not align with the shores above, then calculate for reversal stresses. Generally, the dead load stresses are sufficient to compensate for reversal stresses caused by reshores. Reshores should be prevented from falling.

All members should be straight and true without twists or bends. Special attention should be given to beam and slab or one- and two-way joist construction to prevent local overloading when a heavily loaded shore rests on the thin slab.

Multitier shoring, single-post shoring in two or more tiers, is a dangerous practice and is not recommended.

Where a slab load is supported on one side of the beam only, edge beam forms should be carefully planned to prevent tipping of the beam due to unequal loading.

Vertical shores should be erected so that they cannot tilt and should have a firm bearing. Inclined shores should be braced securely against slipping or sliding. The bearing ends of shores should be square. Connections of shore heads to other framing should be adequate to prevent the shores from falling out when reversed bending causes upward deflection of the forms.

3.5.2 *Centering*—When centering is used, lowering is generally accomplished by the use of sand boxes, jacks, or wedges beneath the supporting members. For the special problems associated with the construction of centering for folded plates, thin shells, and long-span roof structures.

3.5.3 *Shoring for composite action between previously erected steel or concrete framing and cast-in-place concrete—*

3.6—Inspection and adjustment of formwork

Helpful information about forms before, during, and after concreting can be found in Reference 1.3 and ACI 311.1R.

3.6.1—*Before concreting*

3.6.1.1—Telltale devices should be installed on shores or forms to detect formwork movements during concreting.

3.6.1.2—Wedges used for final alignment before concrete placement should be secured in position before the final check.

3.6.1.3—Formwork should be anchored to the shores below so that movement of any part of the formwork system will be prevented during concreting.

3.6.1.4—Additional elevation of formwork should be provided to allow for closure of form joints, settlements of mudsills, shrinkage of lumber, and elastic shortening and dead load deflections of form members.

3.6.1.5—Positive means of adjustment (wedges or jacks) should be provided to permit realignment or readjustment of shores if settlement occurs.

3.6.2 *During and after concreting*—During and after concreting, but before initial set of the concrete, the elevations, camber, and plumbness of formwork systems should be checked using telltale devices.

Formwork should be continuously watched so that any corrective measures found necessary can be promptly made. Form watchers should always work under safe conditions and establish in advance a method of communication with placing crews in case of emergency.

Appendix D: OSHA Concrete and Shoring Regulations

Reproduced from the United States Occupational Safety and Health Administration provisions, U.S. Department of Labor.

Subpart Q—*Concrete and Masonry Construction*

§1926.700—Scope, application, and definitions applicable to this subpart.

(a) *Scope and application.* This subpart sets forth requirements to protect all construction employees from the hazards associated with concrete and masonry construction operations performed in workplaces covered under 29 CFR Part 1926. In addition to the requirements in Subpart Q, other relevant provisions in Parts 1910 and 1926 apply to concrete and masonry construction operations.

(b) *Definitions applicable to this subpart.* In addition to the definitions set forth in §1926.32, the following definitions apply to this subpart.

(b)(1) "Bull float" means a tool used to spread out and smooth concrete.

(b)(2) "Formwork" means the total system of support for freshly placed or partially cured concrete, including the mold or sheeting (form) that is in contact with the concrete as well as all supporting members including shores, reshores, hardware, braces, and related hardware.

(b)(3) "Lift slab" means a method of concrete construction in which floor and roof slabs are cast on or at ground level and, using jacks, lifted into position.

(b)(4) "Limited access zone" means an area alongside a masonry wall, which is under construction, and which is clearly demarcated to limit access by employees.

(b)(5) "Precast concrete" means concrete members (such as walls, panels, slabs, columns, and beams) which have been formed, cast, and cured prior to final placement in a structure.

(b)(6) "Reshoring" means the construction operation in which shoring equipment (also called reshores or reshoring equipment) is placed, as the original forms and shores are removed, in order to support partially cured concrete and construction loads.

(b)(7) "Shore" means a supporting member that resists a compressive force imposed by a load.

(b)(8) "Vertical slip forms" means forms which are jacked vertically during the placement of concrete.

(b)(9) "Jacking operation" means the task of lifting a slab or group of slabs vertically from one location to another (e.g., from the casting location to a temporary (parked) location, or to its final location in the structure), during the construction of a building/structure where the lift-slab process is being used.

§1926.701—General requirements.

(a) *Construction loads.* No construction loads shall be placed on a concrete structure or portion of a concrete structure unless the employer determines, based on information received from a person who is qualified in structural design, that the structure or portion of the structure is capable of supporting the loads.

(b) *Reinforcing steel.* All protruding reinforcing steel, onto and into which employees could fall, shall be guarded to eliminate the hazard of impalement.

(c)(1) *Post-tensioning operations.* No employee (except those essential to the post-tensioning operations) shall be permitted to be behind the jack during tensioning operations.

(c)(2) Signs and barriers shall be erected to limit employee access to the post-tensioning area during tensioning operations.

(d) *Riding concrete buckets.* No employee shall be permitted to ride concrete buckets.

(e)(1) *Working under loads.* No employee shall be permitted to work under concrete buckets while buckets are being elevated or lowered into position.

(e)(2) To the extent practical elevated concrete buckets shall be routed so that no employee, or the fewest number of employees, are exposed to the hazards associated with falling concrete buckets.

(f) *Personal protective equipment.* No employee shall be permitted to apply a cement, sand, and water mixture through a pneumatic hose unless the employee is wearing protective head and face equipment.

§1926.702—Requirements for equipment and tools.

(a)(1) *Bulk cement storage.* Bulk storage bins, containers, and silos shall be equipped with the following:

(a)(1)(i) Conical or tapered bottoms; and

(a)(1)(ii) Mechanical or pneumatic means of starting the flow of material.

(a)(2) No employee shall be permitted to enter storage facilities unless the ejection system has been shut down, locked out, and tagged to indicate that the ejection system is not to be operated.

(b) *Concrete mixers.* Concrete mixers with one cubic yard (.8 m^3) or larger loading skips shall be equipped with the following:

(b)(1) A mechanical device to clear the skip of materials; and (b)(2) Guardrails installed on each side of the skip.

(c) *Power concrete trowels.* Powered and rotating type concrete troweling machines that are manually guided shall be equipped with a control switch that will automatically shut off the power whenever the hands of the operator are removed from the equipment handles.

(d) *Concrete buggies.* Concrete buggy handles shall not extend beyond the wheels on either side of the buggy.

(e)(1) *Concrete pumping systems.* Concrete pumping systems using discharge pipes shall be provided with pipe supports designed for 100 percent overload.

(e)(2) Compressed air hoses used on concrete pumping systems shall be provided with positive fail-safe joint connectors to prevent separation of sections when pressurized.

(f)(1) *Concrete buckets.* Concrete buckets equipped with hydraulic or pneumatic gates shall have positive safety latches or similar safety devices installed to prevent premature or accidental dumping.

(f)(2) Concrete buckets shall be designed to prevent concrete from hanging up on top and the sides.

(g) *Tremies.* Sections of tremies and similar concrete conveyances shall be secured with wire rope (or equivalent materials) in addition to the regular couplings or connections.

(h) *Bull floats.* Bull float handles used where they might contact energized electrical conductors, shall be constructed of nonconductive material or insulated with a nonconductive sheath whose electrical and mechanical characteristics provide the equivalent protection of a handle constructed of nonconductive material.

(i)(1) *Masonry saws.* Masonry saws shall be guarded with a semicircular enclosure over the blade.

(i)(2) A method for retaining blade fragments shall be incorporated in the design of the semicircular enclosure.

(j)(1) *Lockout/Tagout procedures.* No employee shall be permitted to perform maintenance or repair activity on equipment (such as compressors, mixers, screens or pumps used for concrete and masonry construction activities) where the inadvertent operation of the equipment could occur and cause injury, unless all potentially hazardous energy sources have been locked out and tagged.

(j)(2) Tags shall read Do Not Start or similar language to indicate that the equipment is not to be operated.

§1926.703—Requirements for cast-in-place concrete.

(a)(1) *General requirements for formwork.* Formwork shall be designed, fabricated, erected, supported, braced and maintained so that it will be capable of supporting without failure all vertical and lateral loads that may reasonably be anticipated to be applied to the formwork. Formwork which is designed, fabricated, erected, supported, braced and maintained in conformance with the Appendix to this section will be deemed to meet the requirements of this paragraph.

(a)(2) Drawings or plans, including all revisions, for the jack layout, formwork (including shoring equipment), working decks, and scaffolds, shall be available at the jobsite.

(b)(1) *Shoring and reshoring.* All shoring equipment (including equipment used in reshoring operations) shall be inspected prior to erection to determine that the equipment meets the requirements specified in the formwork drawings.

(b)(2) Shoring equipment found to be damaged such that its strength is reduced to less than that required by §1926.703(a)(1) shall not be used for shoring.

(b)(3) Erected shoring equipment shall be inspected immediately prior to, during, and immediately after concrete placement.

(b)(4) Shoring equipment that is found to be damaged or weakened after erection, such that its strength is reduced to less than that required by §1926.703(a)(1), shall be immediately reinforced.

(b)(5) The sills for shoring shall be sound, rigid, and capable of carrying the maximum intended load.

(b)(6) All base plates, shore heads, extension devices, and adjustment screws shall be in firm contact, and secured when necessary, with the foundation and the form.

(b)(7) Eccentric loads on shore heads and similar members shall be prohibited unless these members have been designed for such loading.

(b)(8) Whenever single post shores are used one on top of another (tiered), the employer shall comply with the following specific requirements in addition to the general requirements for formwork:

(b)(8)(i) The design of the shoring shall be prepared by a qualified designer and the erected shoring shall be inspected by an engineer qualified in structural design.

(b)(8)(ii) The single post shores shall be vertically aligned.

(b)(8)(iii) The single post shores shall be spliced to prevent misalignment.

(b)(8)(iv) The single post shores shall be adequately braced in two mutually perpendicular directions at the splice level. Each tier shall also be diagonally braced in the same two directions.

(b)(9) Adjustment of single post shores to raise formwork shall not be made after the placement of concrete.

(b)(10) Reshoring shall be erected, as the original forms and shores are removed, whenever the concrete is required to support loads in excess of its capacity.

(c)(1) *Vertical slip forms.* The steel rods or pipes on which jacks climb or by which the forms are lifted shall be—

(c)(1)(i) Specifically designed for that purpose; and

(c)(1)(ii) Adequately braced where not encased in concrete.

(c)(2) Forms shall be designed to prevent excessive distortion of the structure during the jacking operation.

(c)(3) All vertical slip forms shall be provided with scaffolds or work platforms where employees are required to work or pass.

(c)(4) Jacks and vertical supports shall be positioned in such a manner that the loads do not exceed the rated capacity of the jacks.

(c)(5) The jacks or other lifting devices shall be provided with mechanical dogs or other automatic holding devices to support the slip forms whenever failure of the power supply or lifting mechanism occurs.

(c)(6) The form structure shall be maintained within all design tolerances specified for plumbness during the jacking operation.

(c)(7) The predetermined safe rate of lift shall not be exceeded.

(d)(1) *Reinforcing steel.* Reinforcing steel for walls, piers, columns, and similar vertical structures shall be adequately supported to prevent overturning and to prevent collapse.

(d)(2) Employers shall take measures to prevent unrolled wire mesh from recoiling. Such measures may include, but are not limited to, securing each end of the roll or turning over the roll.

(e)(1) *Removal of formwork.* Forms and shores (except those used for slabs on grade and slip forms) shall not be removed until the employer determines that the concrete has gained sufficient strength to support its weight and superimposed loads. Such determination shall be based on compliance with one of the following:

(e)(1)(i) The plans and specifications stipulate conditions for removal of forms and shores, and such conditions have been followed, or

(e)(1)(ii) The concrete has been properly tested with an appropriate ASTM standard test method designed to indicate the concrete compressive strength, and the test results indicate that the concrete has gained sufficient strength to support its weight and superimposed loads.

(e)(2) Reshoring shall not be removed until the concrete being supported has attained adequate strength to support its weight and all loads in place upon it.

§1926.704—Requirements for precast concrete.

(a) Precast concrete wall units, structural framing, and tilt-up wall panels shall be adequately supported to prevent overturning and to prevent collapse until permanent connections are completed.

(b) Lifting inserts which are embedded or otherwise attached to tilt-up precast concrete members shall be capable of supporting at least two times the maximum intended load applied or transmitted to them.

(c) Lifting inserts which are embedded or otherwise attached to precast concrete members, other than the tilt-up members, shall be capable of supporting at least four times the maximum intended load applied or transmitted to them.

(d) Lifting hardware shall be capable of supporting at least five times the maximum intended load applied transmitted to the lifting hardware.

(e) No employee shall be permitted under precast concrete members being lifted or tilted into position except those employees required for the erection of those members.

§1926.705—Requirements for lift-slab operations.

(a) Lift-slab operations shall be designed and planned by a registered professional engineer who has experience in lift-slab construction. Such plans and designs shall be implemented by the employer and shall include detailed instructions and sketches indicating the prescribed method of erection. These plans and designs shall also include provisions for ensuring lateral stability of the building/structure during construction.

(b) Jacks/lifting units shall be marked to indicate their rated capacity as established by the manufacturer.

(c) Jacks/lifting units shall not be loaded beyond their rated capacity as established by the manufacturer.

(d) Jacking equipment shall be capable of supporting at least two and one-half times the load being lifted during jacking operations and the equipment shall not be overloaded. For the purpose of this provision, jacking equipment includes any load bearing component which is used to carry out the lifting operation(s). Such equipment includes, but is not limited to, the following: threaded rods, lifting attachments, lifting nuts, hook-up collars, T-caps, shearheads, columns, and footings.

(e) Jacks/lifting units shall be designed and installed so that they will neither lift nor continue to lift when they are loaded in excess of their rated capacity.

(f) Jacks/lifting units shall have a safety device installed which will cause the jacks/lifting units to support the load in any position in the event any jacklifting unit malfunctions or loses its lifting ability.

(g) Jacking operations shall be synchronized in such a manner to ensure even and uniform lifting of the slab. During lifting, all points at which the slab is supported shall be kept within ½ inch of that needed to maintain the slab in a level position.

(h) If leveling is automatically controlled, a device shall be installed that will stop the operation when the ½ inch tolerance set forth in paragraph (g) of this section is exceeded or where there is a malfunction in the jacking (lifting) system.

(i) If leveling is maintained by manual controls, such controls shall be located in a central location and attended by a competent person while lifting is in progress. In addition to meeting the definition in §1926.32(f), the competent person must be experienced in the lifting operation and with the lifting equipment being used.

(j) The maximum number of manually controlled jacks/lifting units on one slab shall be limited to a number that will permit the operator to maintain the slab level within specified tolerances of paragraph (g) of this section, but in no case shall that number exceed 14.

(k)(1) No employee, except those essential to the jacking operation, shall be permitted in the building/structure while any jacking operation is taking place unless the building/structure has been reinforced sufficiently to ensure its integrity during erection. The phrase "reinforced sufficiently to ensure its integrity" used in this paragraph means that a registered professional engineer, independent of the engineer who designed and planned the lifting operation, has determined from the plans that if there is a loss of support at any jack location, that loss will be confined to that location and the structure as a whole will remain stable.

(k)(2) Under no circumstances, shall any employee who is not essential to the jacking operation be permitted immediately beneath a slab while it is being lifted.

(k)(3) For the purpose of paragraph (k) of this section, a jacking operation begins when a slab or group of slabs is lifted and ends when such slabs are secured (with either temporary connections or permanent connections).

(k)(4) Employers who comply with appendix A to §1926.705 shall be considered to be in compliance with the provisions of paragraphs (k)(1) through (k)(3) of this section.

(l) When making temporary connections to support slabs, wedges shall be secured by tack welding, or an equivalent method of securing the wedges to prevent them from falling out of position. Lifting rods may not be released until the wedges at that column have been secured.

(m) All welding on temporary and permanent connections shall be performed by a certified welder, familiar with the welding requirements specified in the plans and specifications for the lift-slab operation.

(n) Load transfer from jacks/lifting units to building columns shall not be executed until the welds on the column shear plates (weld blocks) are cooled to air temperature.

(o) Jacks/lifting units shall be positively secured to building columns so that they do not become dislodged or dislocated.

(p) Equipment shall be designed and installed so that the lifting rods cannot slip out of position or the employer shall institute other measures, such as the use of locking or blocking devices, which will provide positive connection between the lifting rods and attachments and will prevent components from disengaging during lifting operations.

Appendix E: Leveling Instruments

Preliminary site work must be done accurately to ensure that the building is in the proper location and at the appropriate elevation. Building lines are established and the locations of columns, piers, and other structural components are determined from the building lines. The builder's level, transit-level, and laser transit-level are commonly used to lay out and establish building lines and other reference points on the job site. Each of the leveling instruments rotates horizontally and can be used for horizontal measurements. In addition, the transit-level rotates vertically to plumb and align vertical surfaces.

THE BUILDER'S LEVEL

A builder's level is used to establish and verify grades and elevations and set up reference points over long distances. It is used extensively for grading operations and general foundation layout.

The main parts of a builder's level are the telescope, leveling vial, and leveling screws. The telescope is used to sight objects and contains lenses, an eyepiece, a detachable sunshade, and a focusing knob. The lenses magnify the object being sighted. The eyepiece is rotated to bring the crosshairs into focus. The detachable sunshade protects the objective lens from damage and reduces glare. The focusing knob is used to adjust the telescope until the object being viewed appears sharp and clear. A leveling vial is a sensitive device used to indicate the levelness of the telescope. The leveling vial is located above or below the telescope. If the leveling vial indicates that the telescope is out-of-level, the leveling screws are used to level the telescope in all directions. **See Figure E-1.**

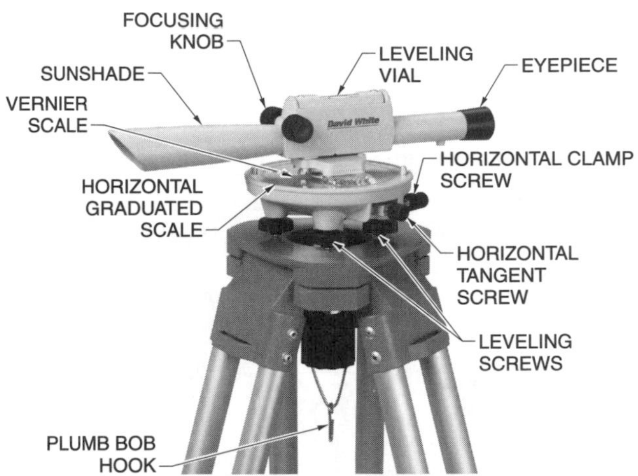

Figure E-1. A builder's level is used to establish grade levels and elevations on a job site. The telescope can only be moved horizontally.

A horizontal clamp screw holds the builder's level in a fixed horizontal position. A horizontal tangent screw is used to make slight adjustments to the telescope in a horizontal direction after the horizontal clamp screw has been tightened.

Some builder's levels and transit-levels have an adjoining horizontal circle and a horizontal vernier scale to measure horizontal angles. The horizontal circle is moved manually, but does not move as the telescope is rotated. The horizontal circle is divided into four quadrants; each quadrant indicating 0° to 90°. The horizontal vernier scale is attached to the leveling instrument's frame and moves along the inside of the horizontal circle as the telescope is turned. The horizontal vernier scale is commonly graduated in 15 minute increments, but 5 minute increments are used for more precise readings. **See Figure E-2.**

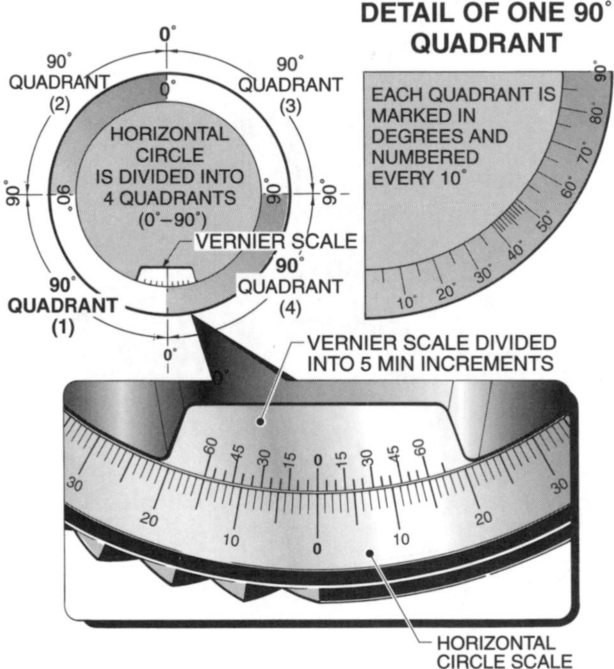

Figure E-2. The horizontal circle and vernier scale are used to measure horizontal angles.

THE TRANSIT-LEVEL

A transit-level can perform all of the functions of a builder's level. In addition, the telescope of the transit-level can be tilted vertically to plumb and align vertical surfaces. Many of the parts of a transit-level, such as the telescope, leveling vial, leveling screws, and horizontal adjustments, are similar to a builder's level. In addition to these parts, a telescope lock lever, vertical clamp screw, and vertical tangent screw are used to make and maintain vertical adjustments. The telescope lock lever holds the telescope in the correct position for horizontal leveling. When the telescope lock lever is released, the telescope can be adjusted vertically. When the telescope is adjusted to its desired position, the vertical clamp screw is tightened. The vertical tangent screw is then used to make fine vertical adjustments. **See Figure E-3.**

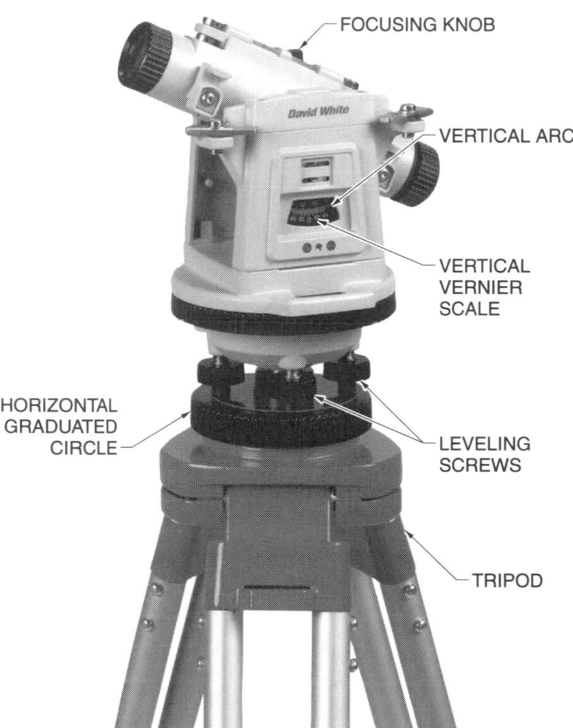

Figure E-3. A transit-level is used to make vertical and horizontal angular measurements.

Most transit-levels are also equipped with a vertical arc and vertical vernier scale. The vertical arc is used to measure vertical angles and is graduated from 0° to 45° in two directions. The vertical arc moves as the telescope is adjusted vertically. The vertical vernier scale is attached to the transit-level's frame and allows the transit-level to be adjusted in 5 minute increments. Other models of transit-levels have a fixed pointer that indicates the angle to which the telescope has been set. The fixed pointer gives the measurement to the nearest whole degree. **See Figure E-4.**

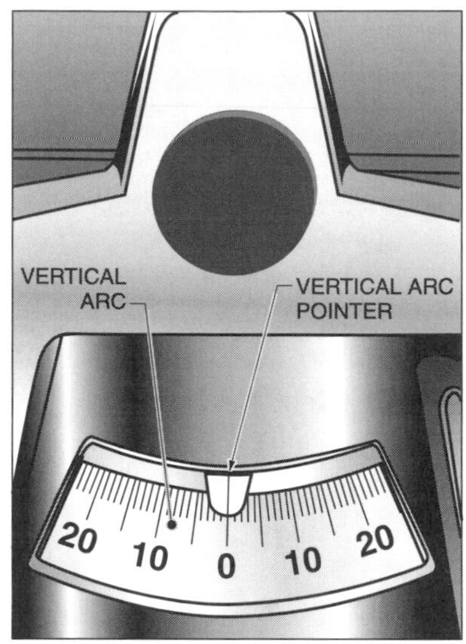

Figure E-4. A fixed pointer (vertical arc pointer) may be used to indicate an angular measurement on a vertical arc. Measurements are made to the nearest degree.

Tripods

A builder's level or transit-level must be mounted on a tripod. **See Figure E-5.** Threaded or cup assemblies are commonly used to fasten the leveling instrument to the tripod head. A leveling instrument with a threaded base screws into a threaded tripod head. If the tripod head has a cup assembly, a threaded mounting stud at the base of the leveling instrument is screwed into the cup assembly.

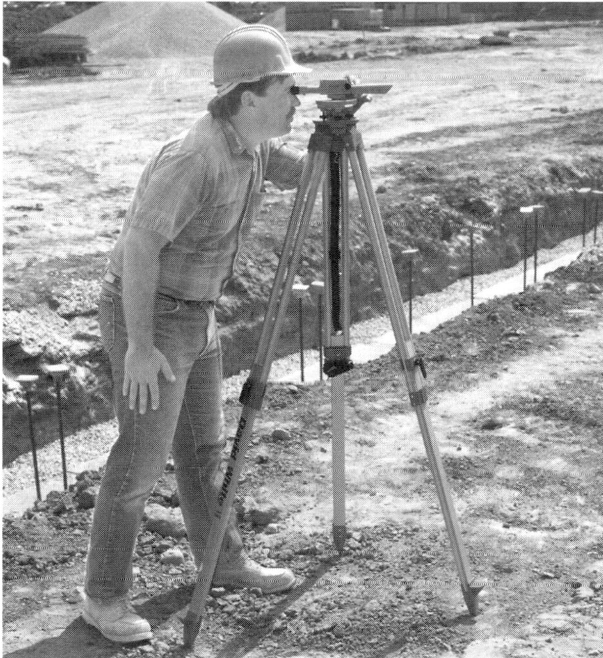

David White Instruments

Figure E-5. A leveling instrument is mounted on a tripod. The leveling head must be kept as level as possible.

When setting up the tripod, position the legs about 3′ apart and keep the tripod head as level as possible. Push the legs firmly into the ground and tighten the wing nuts. On sloping ground, force the uphill leg of the tripod into the slope. A triangular wood base frame helps prevent movement when placing a tripod over concrete or other smooth, hard surfaces. **See Figure E-6.**

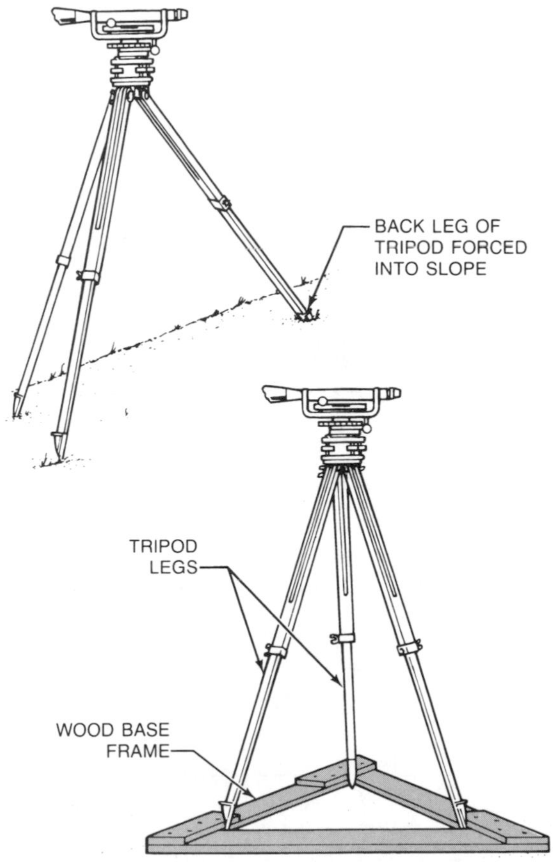

Figure E-6. The back leg of a tripod should be forced into the ground when setting up a leveling instrument on a slope. A triangular wood base frame can be fabricated when setting up a leveling instrument on a hard, smooth surface.

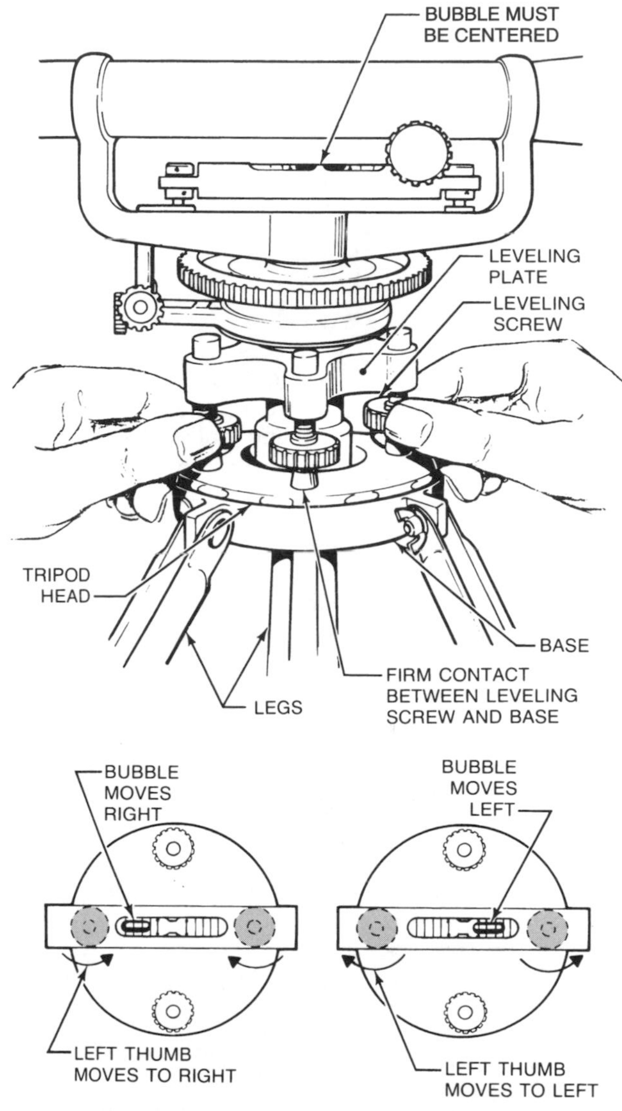

Figure E-7. Leveling screws must be adjusted to level the instrument. The leveling vial bubble moves in the direction the left thumb moves.

Setting Up Builder's Levels and Transit-Levels

The builder's level and transit-level must be leveled in all positions over the base to ensure accurate measurements. When leveling a transit-level, the telescope lock lever must be in the closed position. The leveling screws are adjusted so that the telescope is level when it rotates on top of the base. **See Figure E-7.** Firm contact must be maintained between the leveling screws and the base.

When adjusting the leveling screws, turn the leveling screws equal amounts at the same time and in opposite directions. The direction in which the left thumb moves when turning the screws is the direction that the bubble moves. The leveling screws should not be overtightened. Repeated overtightening will damage the leveling instrument. **See Figure E-8.**

Focusing and Sighting. Leveling instruments focus on a very small area or field of vision. The field of vision is the total magnified area seen through the telescope after it has been focused.

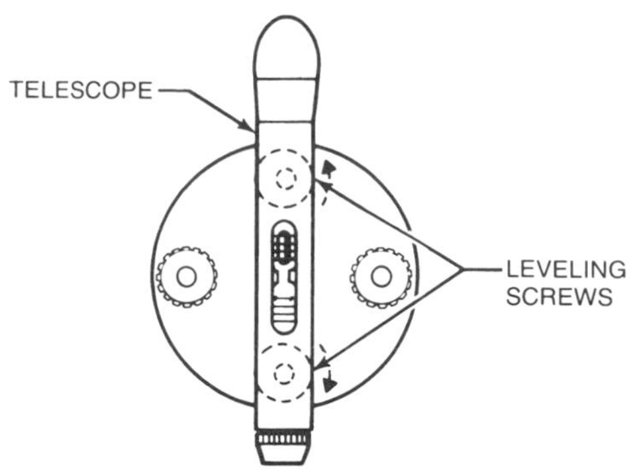

1. Position the telescope directly over a pair of leveling screws. Turn the two screws at the same time and in opposite directions until the leveling vial bubble is centered.

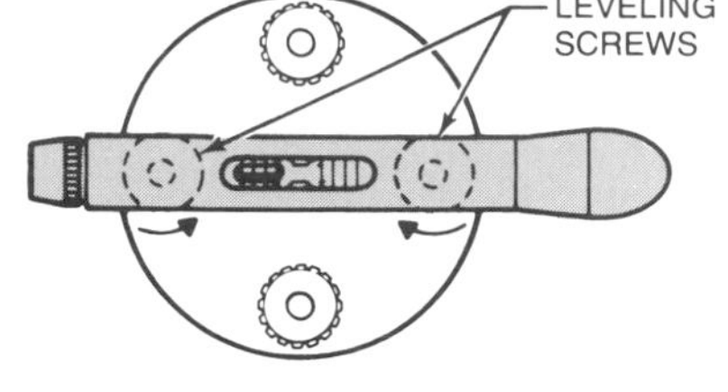

2. Rotate the telescope 90° and level the telescope.

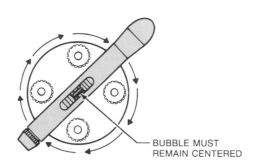

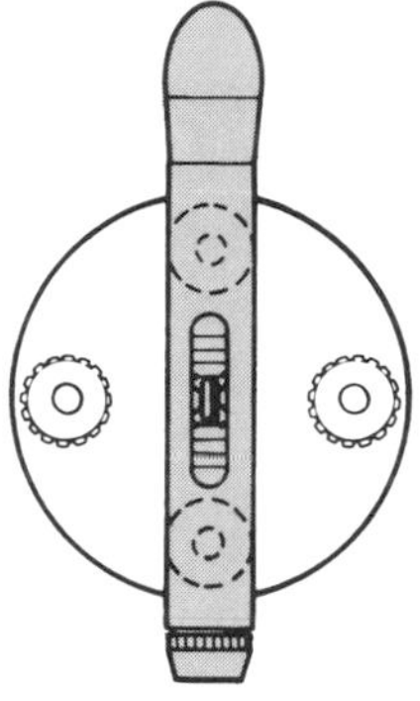

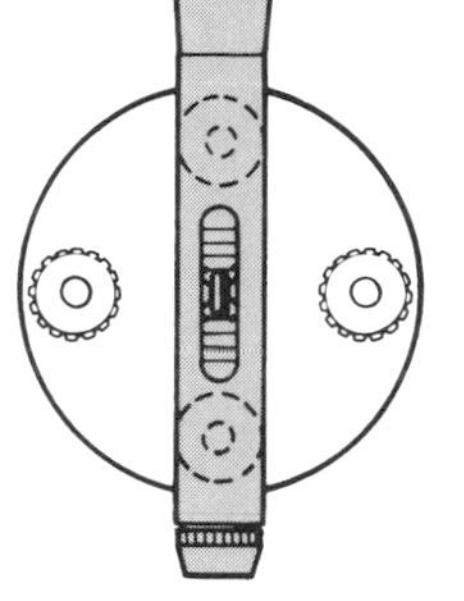

3. Rotate the telescope to the original position. Make adjustments as necessary.

BUBBLE MUST REMAIN CENTERED

4. Rotate the telescope over all four leveling screws, making sure the leveling vial bubble remains centered in all positions.

Figure E-8. Leveling screws are adjusted while the telescope is positioned over them. The telescope is rotated 360° to check that the telescope is level.

The telescope contains crosshairs (fine vertical and horizontal lines) that indicate the center of the field of vision. When viewing a target through a telescope, the crosshairs appear to be projected on the target. An imaginary straight line extending from the intersection of the crosshairs to the target is the line of sight. **See Figure E-9.**

When sighting and focusing on a target using a builder's level, look across the top of the telescope barrel and aim the telescope at the target. Some leveling instruments have devices similar to gun sights to assist in aiming the telescope. Look through the telescope and turn the focusing knob until the target is sharp and clear. Move the telescope until the crosshairs are aligned with the target. Tighten the horizontal clamp screw and make final adjustments to the field of vision using the horizontal tangent screw.

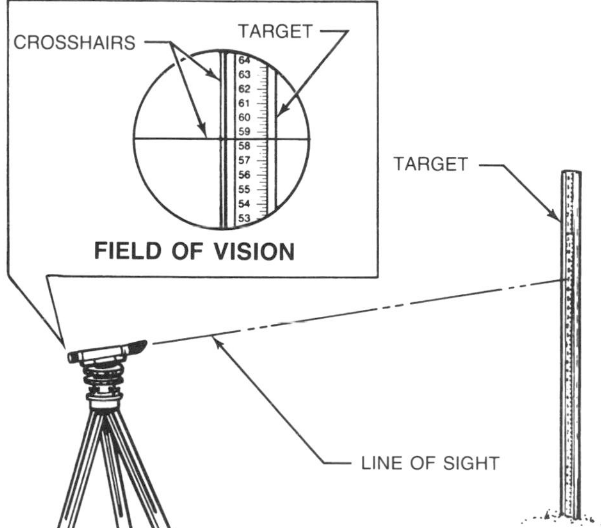

Figure E-9. Leveling instruments focus on a small field of vision. The line of sight extends from the intersection of the crosshairs to the target.

When adjusting the telescope of a transit-level vertically, release the telescope lock lever that secures the telescope. Sight and focus the telescope on the target. Move the telescope until the crosshairs align with the target, and tighten the vertical clamp screw. Make final adjustments to the field of vision by turning the vertical tangent screw.

Leveling Rods

Leveling rods are made of wood, plastic, or aluminum. They commonly consist of two or three sections and are adjustable from 8′ to 14′ in length. When using a builder's level or transit-level, one person sights through the telescope while a second person holds the leveling rod at the desired location. **See Figure E-10.**

Leica Geosystems

Figure E-10. A leveling rod is held vertically when being sighted. The hands should be positioned so as not to obstruct the view.

Large numbers and graduation marks on the leveling rod facilitate reading measurements on the rod. The foot measurements are red and the graduation marks and other figures are usually black. Movable metal targets may be used when sighting the telescope from a long distance. The center of the metal target aligns with the desired position on the leveling rod. **See Figure E-11.**

An architect's rod or engineer's rod is commonly used in construction. The architect's rod is graduated in feet, inches, and eighths of an inch and is used by carpenters and other construction workers. The engineer's rod is graduated in feet and tenths and hundredths of a foot, and is commonly used by surveyors and engineers. **See Figure E-12.**

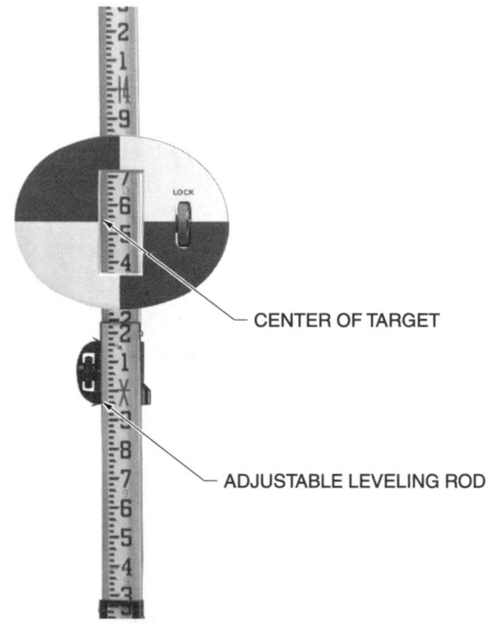

Figure E-11. A movable metal target is used to facilitate sighting over a long distance.

A measuring tape or wood rule held against an unmarked wood rod may be used as a leveling rod when sighting over short distances. In some cases, an unmarked wood rod can be used as a leveling rod without using a measuring tape or wood rule. The line of sight is marked on the unmarked wood rod while it is positioned over an established reference point. The wood rod is moved to another location and the line of sight is marked on the rod. The distance between the two marks is the difference in elevation.

Verifying Grades and Elevations. A builder's level or transit-level is used to verify grade differences on a building site. The leveling instrument is positioned where all desired grade levels can be conveniently read through the telescope. The various grade levels are read and recorded. The higher the grade level reading, the lower the actual grade level. The difference in grade level between the highest and lowest grade levels is calculated by subtracting the lowest grade level reading from the highest grade level reading. **See Figure E-13.**

The builder's level or transit-level is used to establish and verify grade differences on steeply sloped lots. Stakes are positioned along the slope and at the top and bottom of the slope and driven flush with the surface. The leveling instrument is positioned between the first and second stakes and a rod reading is recorded at the first stake. The rod is positioned over the second stake and a rod reading is recorded. The leveling instrument is then positioned between the second and third stakes, and rod readings are recorded for the second and third stakes. This procedure is repeated until a rod reading is recorded at the top of the slope. Calculations are then performed to determine the total grade difference between the top and bottom of the slope. **See Figure E-14.**

Elevations are established for foundation footings and walls using two methods: using an unmarked wood rod and rule or using just an unmarked wood rod. When using an unmarked wood rod and rule, the difference in elevation is determined by subtracting an initial reading from a second reading. For example, if the top of a footing form stake is to be located 6″ below the benchmark or established reference point, an initial reading is taken at the benchmark or established reference point. The unmarked wood rod and rule are positioned on top of the footing form stake and a second reading is taken. The footing form stake is driven until the difference between the initial reading and the second reading is equal to the desired difference in elevation. **See Figure E-15.**

When using just an unmarked wood rod, an initial reading is marked on the rod. The difference in elevation (6″) is then measured and marked on the wood rod. The wood rod is positioned on top of the footing form stake and the stake is driven until the second mark aligns with the line of sight. **See Figure E-16.**

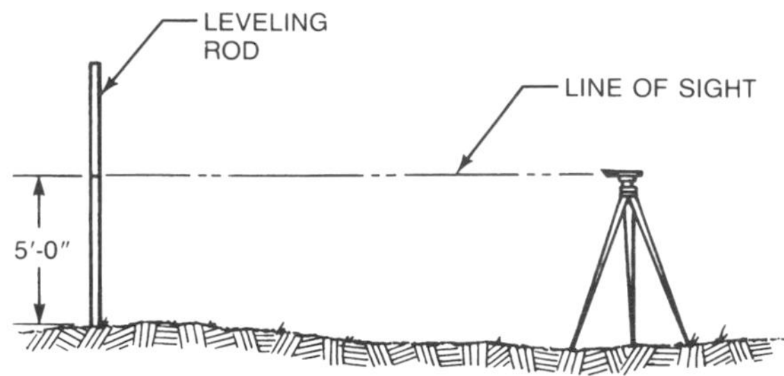

1. Set the leveling instrument up at a convenient location on the building site. Take a line of sight reading of the leveling rod held at a specific point. In this example, the line of sight reading is 5′-0″.

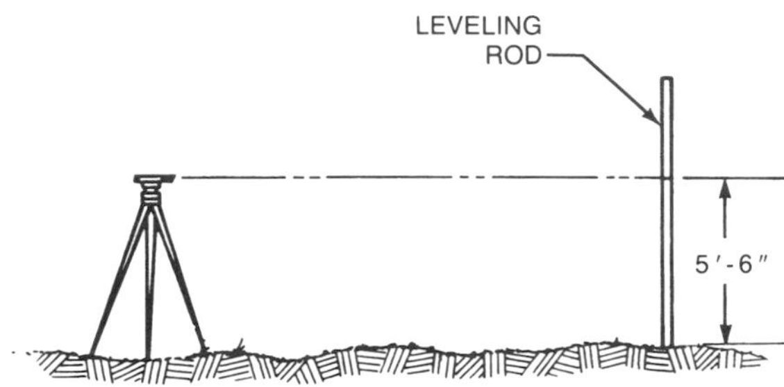

2. Move the leveling rod to another point and take a line of sight reading (5′-6″). Subtract the smaller reading from the larger reading to determine the grade difference (5′-6″ – 5′-0″ = 6″).

Figure E-13. Grade differences are determined by subtracting the lowest grade level from the highest grade level.

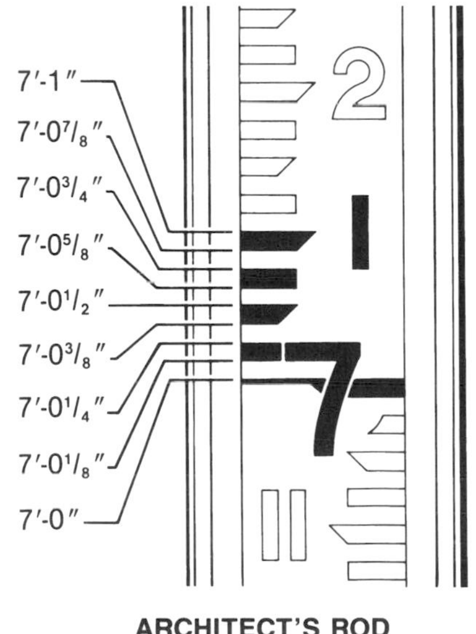

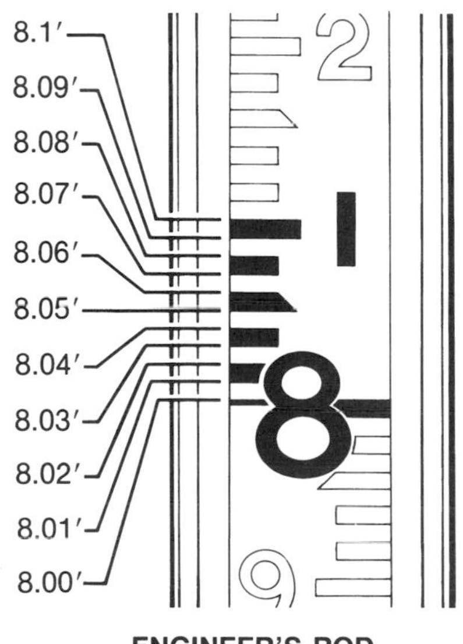

Figure E-12. An architect's rod or an engineer's rod is commonly used with leveling instruments. The architect's rod is graduated in eighths of an inch and the engineer's rod is graduated in hundredths of an inch.

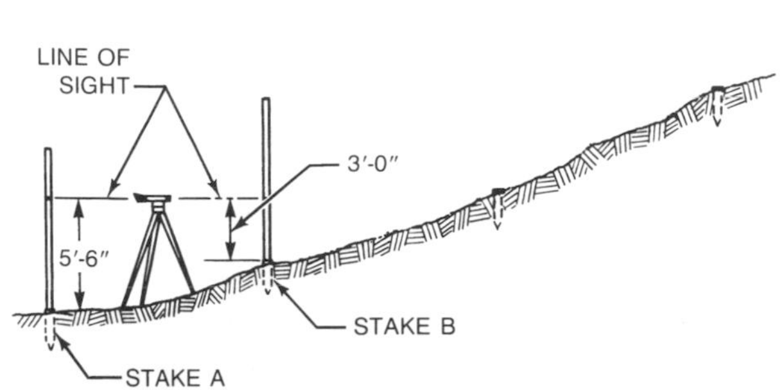

1. Drive stakes at equal intervals along the slope. Set up a leveling instrument and take a line of sight reading at stake A. (In this example the line of sight reading is 5'-6".) Take a line of sight reading at stake B (3'-0").

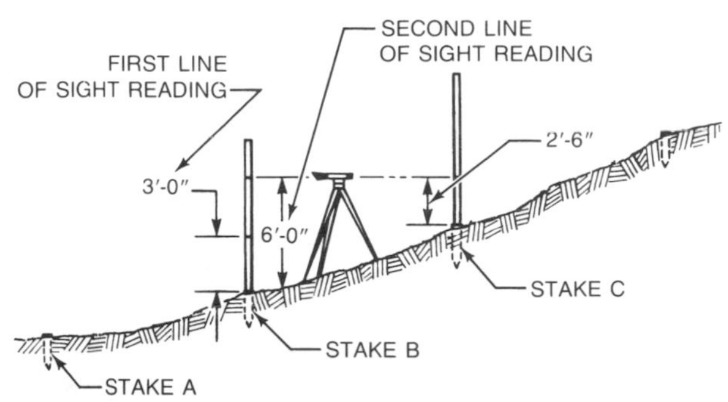

2. Move the leveling instrument to a point above stake B. Take a second line of sight reading at stake B (6'-0"). Subtract the first line of sight reading from the second reading (6'-0" − 3'-0" = 3'-0"). Add this figure to the line of sight reading at stake A (5'-6" + 3'-0" = 8'-6"). Take a line of sight reading at stake C (2'-6").

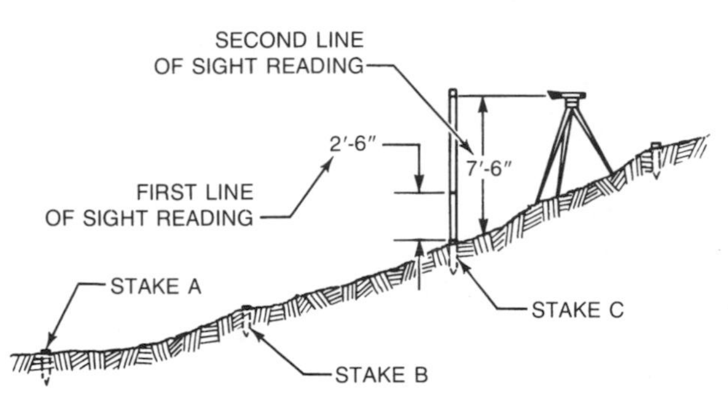

3. Move the leveling instrument to a point above stake C. Take a second line of sight reading at stake C (7'-6"). Subtract the first line of sight reading at stake C from the second reading (7'-6" − 2'-6" = 5'-0"). Add this figure to the sum of the line of sight readings in step 2 (5'-0" + 8'-6" = 13'-6").

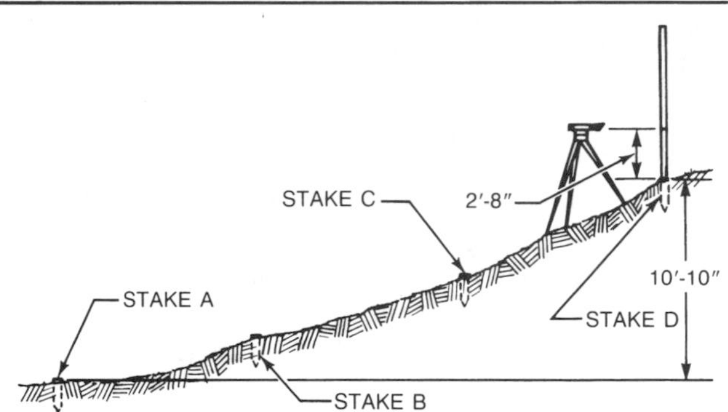

4. Take a line of sight reading at stake D (2'-8"). Subtract this figure from the sum in step 3 to determine the difference in grade level between stakes A and D (13'-6" − 2'-8" = 10'-10").

Figure E-14. The builder's level or transit-level is used to establish elevations on a steeply sloped lot.

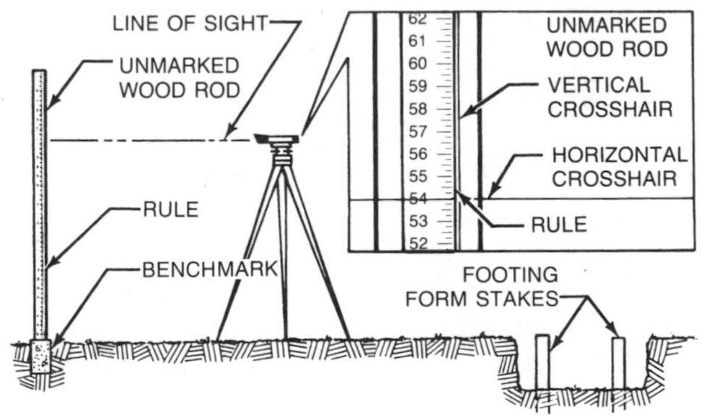

1. Hold an unmarked wood rod over a benchmark and take a line of sight reading on the rule.

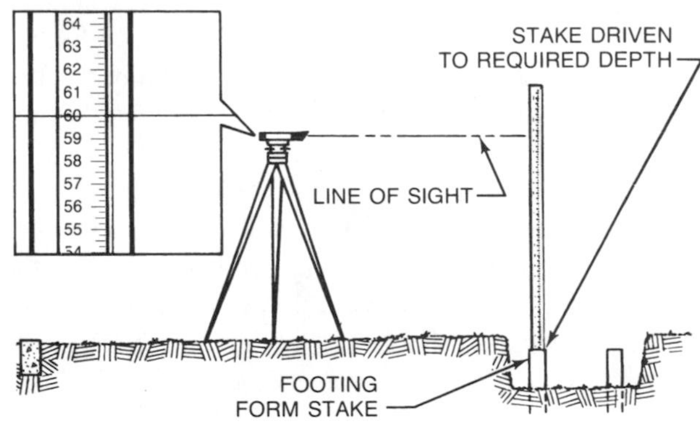

2. Hold an unmarked wood rod on top of a stake. Drive the stake until the desired reading aligns with the horizontal crosshair. In this example, the top of the footing form stake is 6" lower than the benchmark (60" − 54" = 6").

Figure E-15. When using an unmarked wood rod and rule to determine elevations, the initial reading is subtracted from the second reading to obtain the difference in elevation.

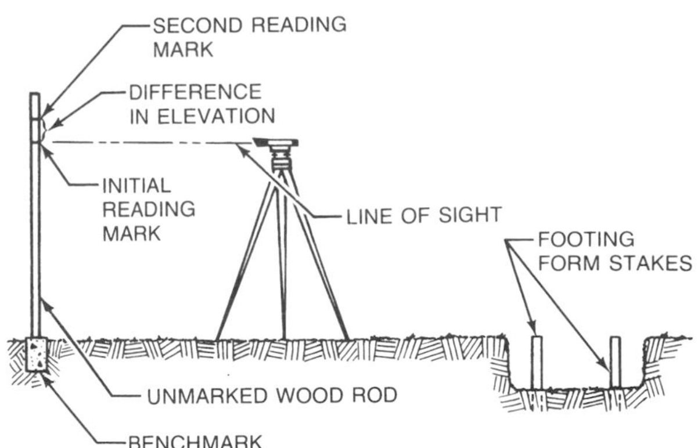

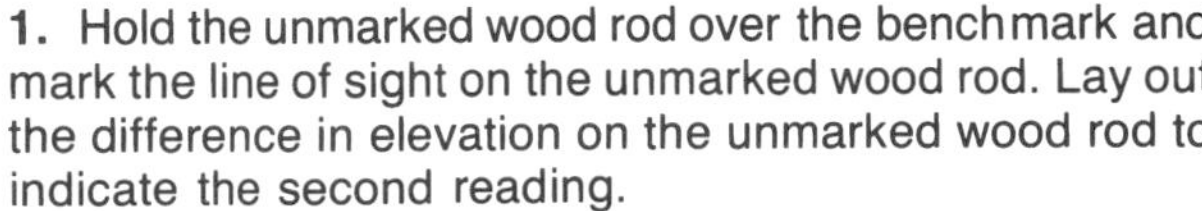

1. Hold the unmarked wood rod over the benchmark and mark the line of sight on the unmarked wood rod. Lay out the difference in elevation on the unmarked wood rod to indicate the second reading.

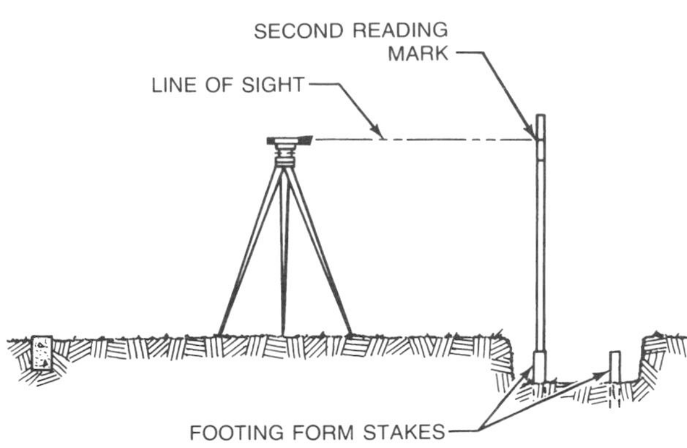

2. Place the unmarked wood rod on top of a stake. Drive the stake until the second reading mark aligns with the line of sight.

Figure E-16. When using an unmarked wood rod as a leveling rod, the elevations are measured and marked. The difference between the two marks is the difference in elevation.

A leveling instrument and an unmarked wood rod are also used to establish level points over a long distance, such as for corner stakes for foundation footing forms. After the desired elevation is established, an unmarked wood rod is positioned against the stake with the bottom of the wood rod even with the elevation mark. The wood rod is moved to another stake, and the elevation mark is aligned with the line of sight. The second stake is then marked at the bottom of the wood rod to indicate the same elevation as the initial stake. **See Figure E-17.**

Laying Out Right Angles. Buildings are commonly constructed in the shape of a rectangle or variation of a rectangle. The sides of a rectangle are at right angles (90°) to each other. A transit-level is the most efficient instrument to use to lay out the building lines for foundations of rectangular, L-, or T-shaped buildings. **See Figure E-18.** When squaring building lines with a transit-level, the instrument is set up and leveled over an established reference point, such as a nail driven into the top of a stake indicating the corner of a lot or building. A plumb bob is attached to a line that is hooked to the bottom of the leveling instrument. The plumb bob ensures that the transit-level is in its exact location when laying out the building lines. **See Figure E-19.** Following is a possible procedure for setting up a transit-level:

1. Spread the tripod legs. Position the tripod so the tripod head is directly over an established reference point.
2. Attach the transit-level to the tripod head. Secure a line to the plumb bob hook at the bottom of the instrument. Adjust the line length so the plumb bob is approximately ¼″ above the reference point.
3. Roughly level the transit-level. Do not tighten the leveling screws.
4. Move the transit-level on the shifting center until the plumb bob aligns exactly with the reference point. If the transit-level does not have a shifting center, carefully shift the tripod until the plumb bob is in the correct position.
5. Adjust and tighten the leveling screws so the transit-level is level in all directions. A right angle is laid out by sighting back to another established reference point, such as the second corner of a lot or building. The instrument is then rotated 90° to establish a line at a right angle to the first line. **See Figure E-20.**

Plumbing with a Transit-Level. The transit-level is used to plumb high walls or columns. It should be positioned at a convenient distance from the object being plumbed. If possible, the distance should be greater than the height of the object being plumbed. After setting up and leveling the transit-level, release the telescope lock lever and aim the instrument at the bottom edge of the member being plumbed. Align the vertical crosshair with the bottom edge of the object and tighten the vertical clamp screw to hold the telescope in position. Tighten the horizontal clamp screw and make final adjustments to align the vertical crosshair by turning the horizontal tangent screw. Loosen the vertical clamp screw and aim the telescope toward the top edge of the member. After it is in position, tighten the vertical clamp screw and sight through the telescope. The member is plumb when the top edge of the member aligns with the vertical crosshair.

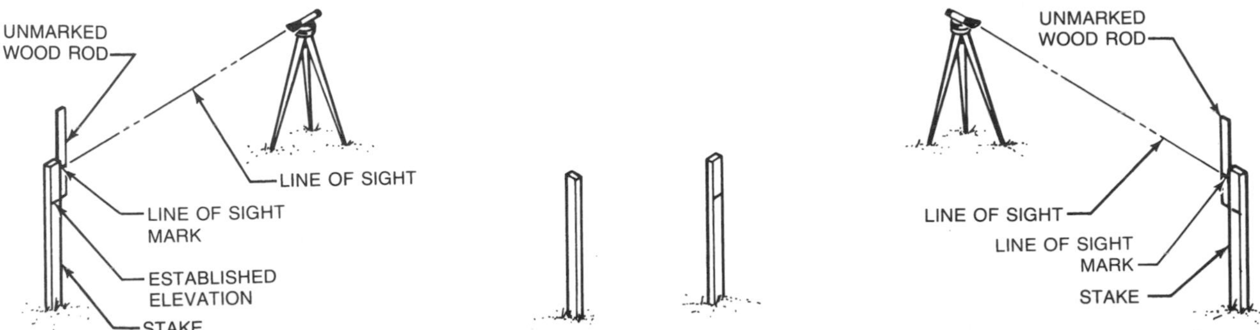

1. Establish the desired elevation on a stake. Position an unmarked wood rod along the stake, aligning the bottom with the mark on the stake. Mark the line of sight reading on the unmarked wood rod.

2. Move the unmarked wood rod to another stake and position it so the line of sight mark aligns with the line of sight of the leveling instrument. Mark the stake at the bottom of the unmarked wood rod.

Figure E-17. An unmarked wood rod and a leveling instrument are used to establish elevations over a long distance.

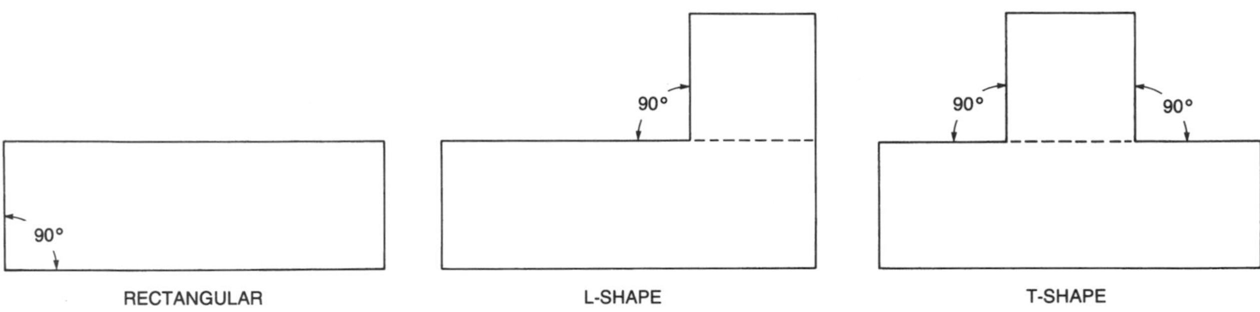

Figure E-18. Building shapes commonly consist of lines that are at right angles to each other.

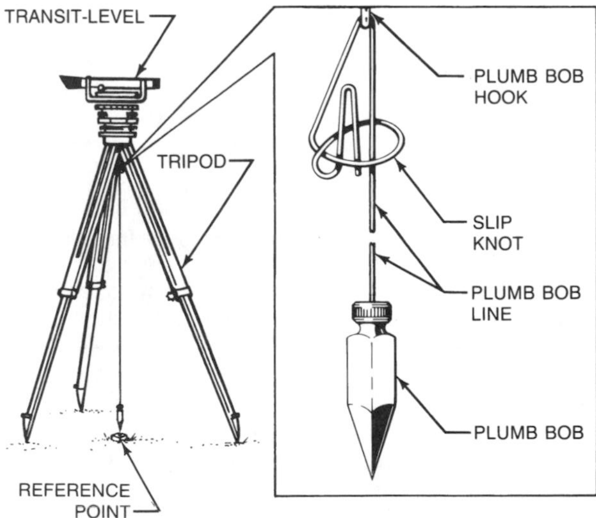

Figure E-19. A transit-level is placed directly over a reference point when establishing elevations. A slipknot is used to attach the plumb bob to the plumb bob hook to allow it to be raised or lowered.

Establishing Points in a Straight Line. The transit-level is used to establish points in a straight line, such as a row of form stakes or piers. The transit-level is set up over one of the end stakes and aimed toward another end stake. The telescope is sighted and focused, and the vertical crosshair is aligned with the second end stake. Intermediate stakes are positioned and aligned with the vertical crosshair. The stakes are driven to the required depth and checked again for accurate layout. **See Figure E-21.**

LASER TRANSIT-LEVEL

The laser transit-level is a leveling instrument increasingly used in construction work for leveling and plumbing. It performs most of the operations of a transit-level, but only requires one person to set up and operate. A laser transit-level is used for establishing grades and elevations over long distances.

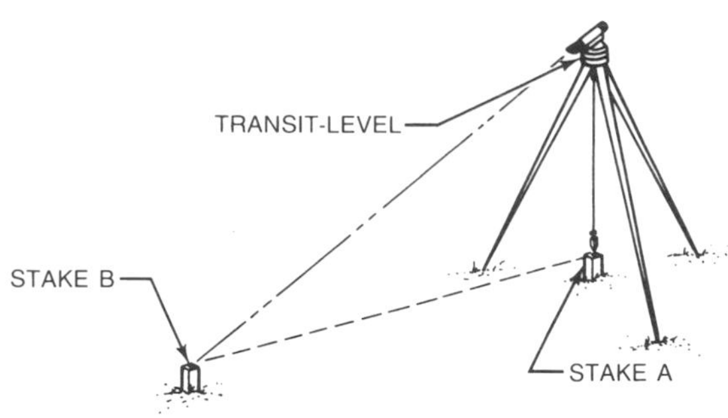

1. Set up the transit-level over stake A. Sight through the telescope and align the vertical crosshair with stake B.

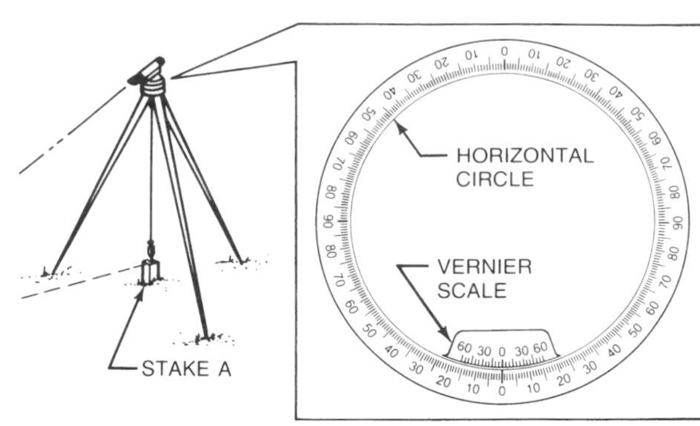

2. Turn the horizontal circle of the transit-level to align one of the "0" readings on the circle with the "0" on the vernier scale.

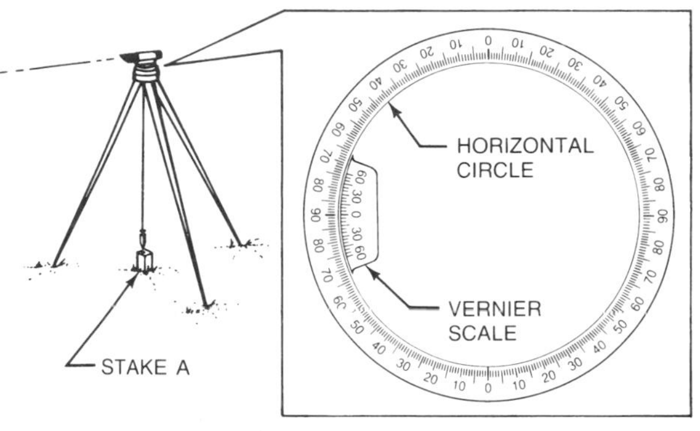

3. Rotate the transit-level until the "0" index on the vernier scale aligns with the 90° index on the horizontal circle.

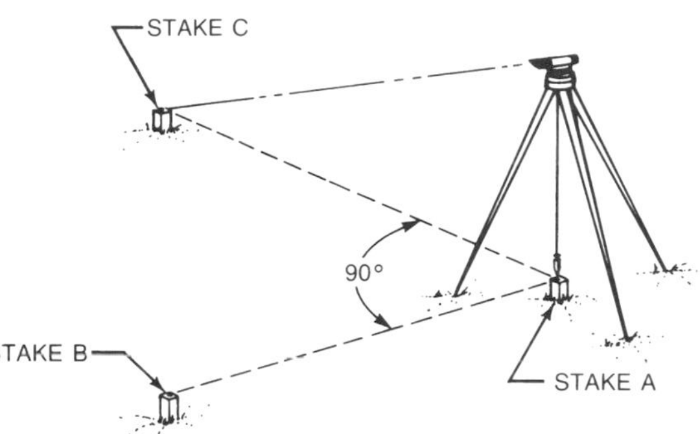

4. Measure the required distance and drive stake C. Aim the telescope at the top of stake C and drive a nail where the vertical crosshair and the exact measurement intersect. A right angle is formed between lines A-B and A-C.

Figure E-20. A transit-level is used to lay out right angles on a job site.

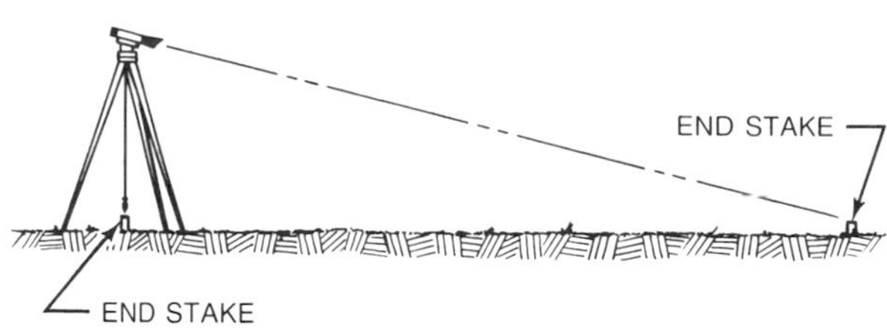

1. Set up and level the transit-level over one of the end stakes. Aim the telescope toward the other end stake and adjust the horizontal tangent screw until the vertical crosshair aligns exactly with the second end stake.

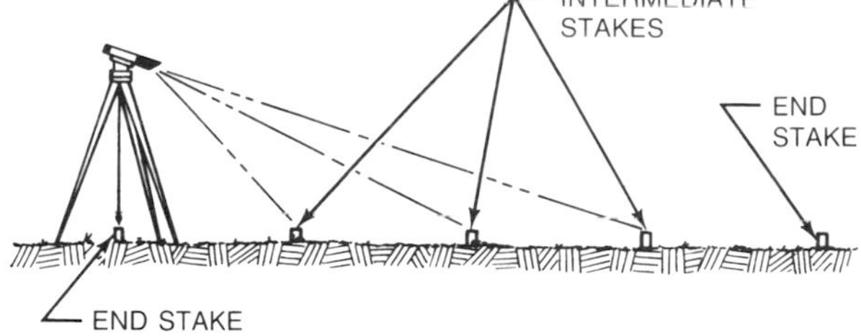

2. Position an intermediate stake and focus the telescope on the stake. Move the stake until the edge aligns with the vertical crosshair. Drive the stake to the required depth. Repeat the process until all stakes are driven.

Figure E-21. A transit-level is used to establish reference points in a straight line.

The main parts of the laser transit-level are the rotating beacon, out-of-level indicator, and leveling screws. The rotating beacon revolves at a maximum rate of 360 revolutions per minute (rpm), causing a beam projection. The laser barrel contains a helium-neon sealed-in tube that activates the laser beam. **See Figure E-22.** The laser beam is a concentrated red beam of light approximately 3/8″ in diameter. The instrument can be mounted on a tripod or stationary column with gimbals. The leveling screws level the laser transit-level for accurate measurements. Some models of laser transit-levels have a self-leveling mechanism that keeps the instrument at its original setting despite changes in thermal conditions or minor jolts.

Figure E-22. The rotating head of a laser transit-level revolves at a maximum rate of 360 rpm. The rate that the head rotates is adjusted using the variable speed control.

The laser beam is directed toward a sensor when grade levels or elevations are measured. A sensor is a light-sensitive target attached to a leveling rod that lights up when properly aligned with the laser beam. **See Figure E-23.** A laser transit-level uses various targets or sensors for different operations. The sensor is attached to a leveling rod with brackets. **See Figure E-24.** Grade readings are made at any point where sensors are positioned on the job site. **See Figure E-25.**

Figure E-24. A sensor is attached to a leveling rod. When the laser beam is directly aligned with the sensor's eye, the indicator light comes on.

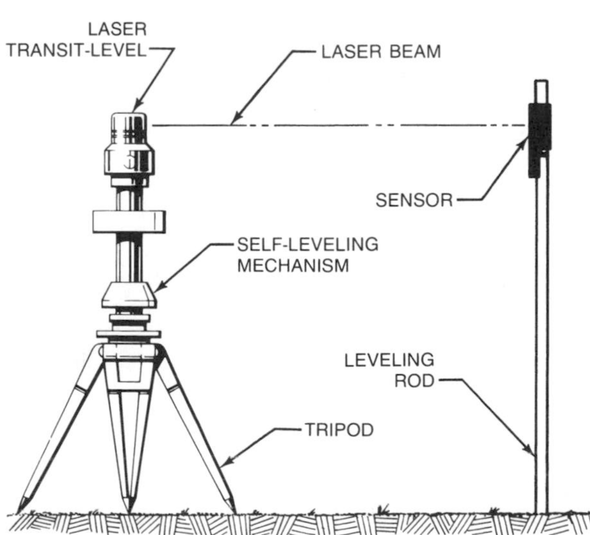

Figure E-23. A laser transit-level emits a concentrated beam of light.

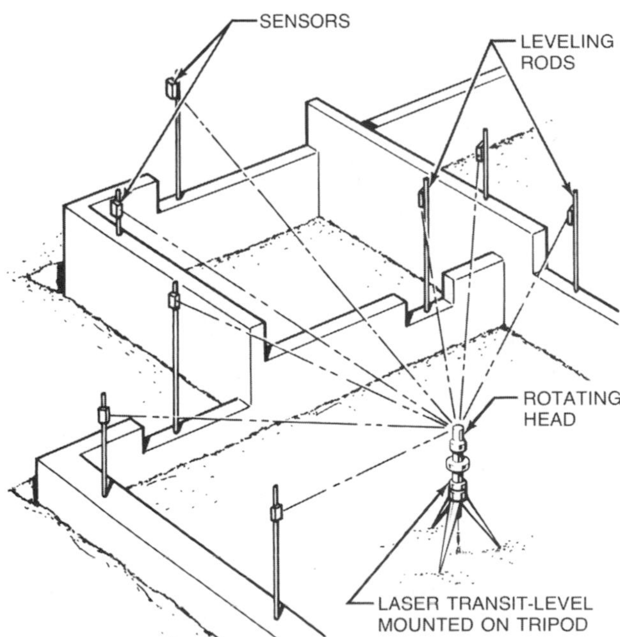

Figure E-25. A laser transit-level is used to establish the elevations at various points on the job site.

A laser transit-level may be used to plumb walls and columns by removing the rotating beacon and plumbing the instrument. With the rotating beacon removed, the laser beam is projected upward. The instrument is plumbed over a point on the ground with a plumb bob and aimed at a target or sensor attached to the wall or column being plumbed. A measurement is taken from the wall or column to the point of the plumb bob and from the wall or column to the sensor. When the two measurements are equal, the wall or column is plumb. **See Figure E-26.**

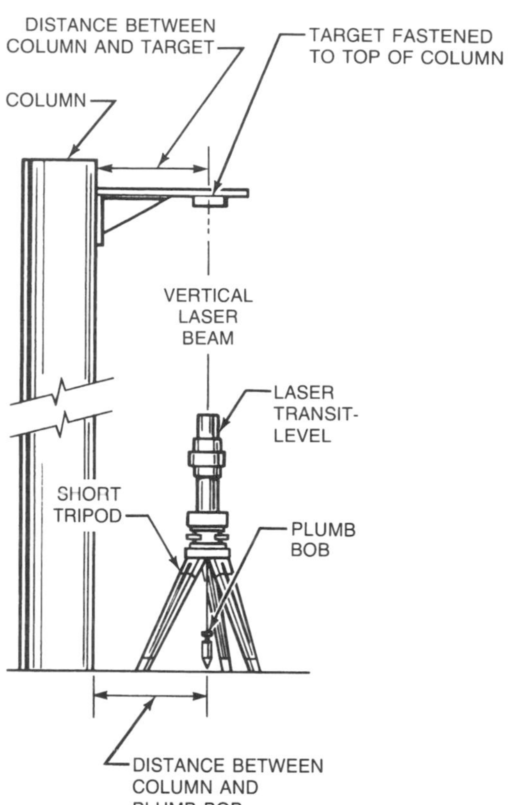

Figure E-26. A laser transit-level is used for plumbing walls and columns. When the distance between the column and target is equal to the distance between the column and plumb bob, the column is plumb.

TOTAL STATION INSTRUMENTS

Total station instruments were first used as survey instruments operated by surveyors for land terrain surveys, cartography (mapmaking), bridge building, and mining operations. Currently, total station instruments are used by carpenters and other building tradesworkers for job site layout.

Total stations perform the functions of traditional transit-levels including leveling, plumbing, and horizontal and vertical measurements. In addition, total stations electronically measure distances, record and store data using an integral computer, and perform mathematical calculations. Although there are many models of total stations available, the basic features and operating procedures are similar. **See Figure E-27.**

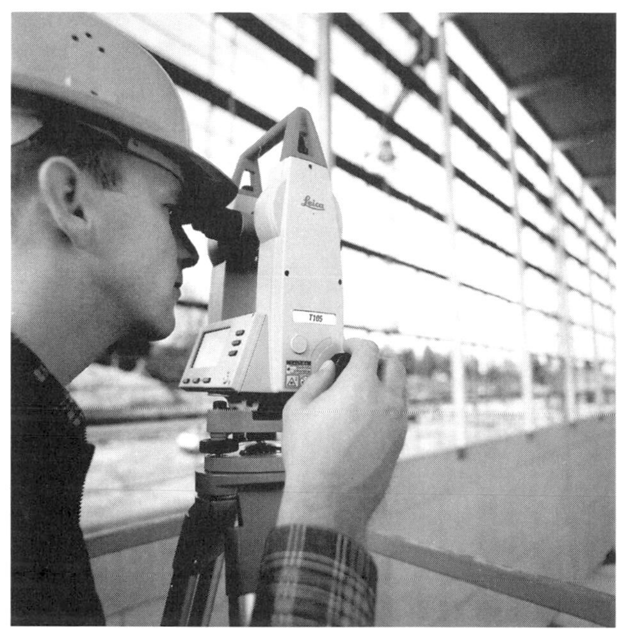

Leica Geosystems

Figure E-27. Total stations perform the functions of traditional transit-levels including leveling, plumbing, and horizontal and vertical measurements.

Setting Up Total Stations

As with other survey instruments, the tripod must first be set up prior to mounting the total station instrument. Place the tripod directly over an established ground mark (station point) so the tripod head is roughly level. Press the tripod legs firmly into the ground so the tripod is in a stable position.

The total station instrument is mounted on the tripod and the instrument is leveled using the tribrach leveling screws. The total station instrument must be directly over the ground mark. Some total station instruments may require the use of a plumb bob to precisely align the total station with the ground mark; newer models are equipped with an optical laser plummet that projects a small-diameter laser beam downward. The laser beam is aligned with the ground mark. **See Figure E-28.**

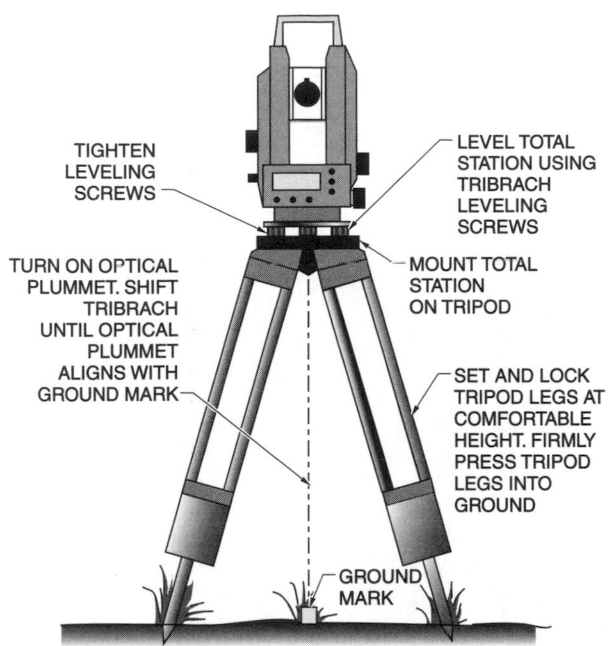
TIGHTEN LEVELING SCREWS
LEVEL TOTAL STATION USING TRIBRACH LEVELING SCREWS
TURN ON OPTICAL PLUMMET. SHIFT TRIBRACH UNTIL OPTICAL PLUMMET ALIGNS WITH GROUND MARK
MOUNT TOTAL STATION ON TRIPOD
SET AND LOCK TRIPOD LEGS AT COMFORTABLE HEIGHT. FIRMLY PRESS TRIPOD LEGS INTO GROUND
GROUND MARK

Figure E-28. The total station instrument is mounted on the tripod and the instrument is leveled using the tribrach leveling screws.

Total station instruments are equipped with an integral computer that not only records survey data, but can also run other software programs. The computer can be used to calculate sines, cosines, and tangents, as well as to determine angles and measure distances.

Electronic Distance Measurement. An integral feature of a total station instrument is electronic distance measurement (EDM). EDM ensures a high degree of accuracy (within .001) when measuring distances between points, without the use of a measuring tape.

The total station is focused on reflective tape adhered to a surface or on a prism mounted on a pole or stake. An infrared beam is then directed from the total station instrument toward the tape or prism. The beam bounces back to the total station and the distance is calculated from the time elapsed. The measurement is recorded by the integral data collector or by a data-collection unit attached to the total station. The information can also be transmitted to a computer, often located in the contractor's office, where computer-aided design and drafting (CADD) software can develop precise field drawings and contour maps.

abutment: End structure that supports the beams, girders, and deck of a bridge or arch.

accelerating admixture (accelerator): Substance added to a concrete mixture to reduce setting time and improve the early strength of concrete.

accident report: Document that details facts about an accident.

actual size: Thickness and width of lumber after shrinkage resulting from surfacing and seasoning.

adjustable flat tie: Patented wall tie device that consists of a flat piece of metal set on edge between metal side rails.

adjustable wood shore: Two-piece shore consisting of overlapping wood members that are secured in place with post clamps.

admixture: Material other than cement, aggregate, and water that is added to a batch of concrete immediately before or during the mixing process.

aggregate: Granular material, such as sand and gravel, used with cement to produce mortar or concrete.

agitating truck: Truck with an agitator to transport freshly mixed concrete from batch plant to job site.

air detrainer: Admixture that decreases air content in concrete mixtures so hardeners may be cast on a wet slab and incorporated into the surface.

air-entrained concrete: Concrete containing an admixture that produces microscopic air bubbles in the concrete. Used to improve workability and freeze resistance of concrete.

air-entraining admixture: A foaming substance used to add microscopic air bubbles to concrete.

American Concrete Institute (ACI): Association that sets standards for concrete construction.

anchor bolt: Bolt used to secure sill plates, columns, and beams to concrete or masonry.

anchor clip: Strap-like device embedded in the top of a foundation wall and used to secure sill plates.

architectural concrete: Permanently exposed concrete surface that features special designs or patterns such as textured finishes, and ribbed and fluted surfaces.

aspect ratio: Length of a fiber divided by its diameter.

auger: Earth-boring device attached to rig. Used to bore holes in the soil for deep piers or piles.

backer rod: Foam material used to prevent moisture from seeping between the wall panel and the footing.

backfilling: Replacing of soil around the outside foundation walls after the walls have been completed.

backhoe: Excavating equipment used for small loading jobs and digging trenches for foundations.

base course: Layer of selected material, usually gravel, placed beneath a concrete slab. Its main purpose is to control the capillary rise of water to the slab bed.

batch: Quantity of mortar or concrete mixed at one time.

batching: Measuring and proportioning of the concrete mix.

batch plant: Location where concrete is mixed according to specifications.

batterboard: A 1″ or 2″ level piece of dimensional lumber formed to hold the building lines in position and to show the exact boundaries of a building.

battered foundation: Monolithic structural support consisting of a wall with a vertical exterior face and a sloping interior face.

beam: Horizontal member that supports a bending load over a span, such as from column to column.

beam bottom form: Bottom soffit of a beam form resting directly over the shore system.

beam pocket: Space in a foundation wall that receives a beam.

beam side form: Vertical member of a beam form. Beam side is nailed against or rests on top of the beam bottom.

bearing capacity: The ability of soil to support weight.

bearing pile: Pile that penetrates through layers of unstable soil until it reaches firm bearing soil.

belled caisson: Concrete caisson flared at its bottom to provide a greater bearing area.

benching: Excavation method in which a series of steps is carved with vertical surfaces between the levels.

benchmark: Point of reference for grades and elevation on a construction site.

bleeding: Segregation in a concrete mix in which water rises to the surface of freshly placed concrete.

bleedwater: Excess water that collects on the surface of concrete as aggregate material sinks in the concrete mixture.

bored caisson: Caisson constructed by placing a metal casing into a hole that is bored into the earth and filled with concrete after it is in position.

brace: Diagonal or horizontal wood or metal member used to stiffen and support various parts of a form.

breakback: Grooved section between spreader cones.

bridge deck: Slab of the bridge superstructure that supports the traffic load.

buck: Frame placed inside a form to provide an opening for a door or window after the concrete has set.

bucket: Large metal container into which concrete is discharged. The bucket is then raised by crane to the placement area.

buggy: Manual or motor-driven cart used to move small amounts of concrete from hoppers or mixers to the placement area.

builder's level: Telescope-like instrument used for leveling operations over long distances and establishing grades and elevations.

building site: Location where construction occurs.

built-in-place form: Form assembly built entirely in place.

built-ins: The frames (bucks) for the door and window openings, and beam pockets where required.

bulkhead: Wood member installed inside or at the end of a concrete form to prevent the concrete from flowing into a section or out of the end of the form.

bulldozer: Earth-moving equipment used to start excavations and strip rocks and topsoil at the surface of the building site.

butt: Large upper portion of a pile.

caisson: Cast-in-place pile formed by drilling a hole, inserting reinforcement, and filling the hole with concrete.

calcium chloride: Crystalline solid accelerating admixture.

capillary action: Physical process in soil that causes water and vapor to rise from the water table and move up toward the surface of the ground.

capital: Flared section at the top of a concrete column that supports the floor slab above.

casing: Metal cylindrical shell that is driven into the ground to restrain uncompacted soil near the surface.

casting bed: Base and support for casting precast structural members.

cast-in-place concrete: Concrete placed in forms where it is required to set as part of the structure.

catch basin: Area excavated and filled with gravel that receives surface water runoff. Also called *dry well.*

cement: Ingredient that binds the sand and aggregate together in concrete after water is added.

chair: Support structure made from metal, plastic, or precast concrete used to provide an accurate, consistent spacing between welded wire reinforcement or rebar and subgrade.

chamfer strip: Narrow strip of wood ripped at a 45° angle.

chute: Trough-like device used to move concrete from a higher point to a lower point.

clay: Fine-grained, natural mineral that is plastic when moist and hard and brittle when dry.

cleanout: Opening in the bottom of a form wall or column form that allows debris to be removed before the concrete is placed.

climbing form: Large panel or ganged panel form that is lifted vertically for succeeding lifts.

closed stairway: Stairway enclosed by walls.

clutch-type insert: Insert that consists of a T-bar anchor and a recess former supported by a base.

coarse aggregate: Crushed stone retained on a U.S. Standard #4 (4.75 mm) sieve.

cofferdam: Large, rectangular, watertight enclosure constructed of interlocking sheet piling.

cohesive (fine-grain) soil: Soil that consists mostly of silt and clay with particles that usually can be seen only with a microscope.

coil insert: Anchoring device that allows bolts to be inserted after concrete has set.

coil tie: Internal disconnecting tie with external bolts that screw into an internal device consisting of metal struts with helical coils at each end.

cold joint: Joint formed when concrete for a wall is placed over a concrete footing that has already set.

coloring admixture: Substance that imparts a desired color to concrete.

column: Vertical member supporting beams, girders, and/or floor slabs.

column clamp: Device to hold column form sides together.

common nail: Flat-headed nail used where appearance is not important, such as in form construction.

compression: Stress caused by pushing together or a crushing force.

compression test: Test that measures the compressive strength of concrete.

compressive strength: Maximum resistance of a concrete or mortar specimen to vertical loads.

concrete: Material consisting of aggregate combined with a binding medium of portland cement and water.

concrete admixture: Material other than cement, aggregate, and water that is added to a batch of concrete immediately before or during the mixing process to modify the concrete's properties.

concrete joist: Narrow, closely spaced beam that supports a floor or roof slab.

concrete mix: Proportion of cement, sand, and aggregate that is combined with water to produce concrete.

confined space: Space large enough and configured so an employee can physically enter and perform assigned work, has limited or restricted means for entry and exit, and is not designed for continuous employee occupancy.

consistency: Ability of concrete, in its plastic form, to flow as it is being placed into the form.

consolidation: Process of working fresh concrete so that a closer arrangement of particles is created and the number of voids is decreased or eliminated.

construction joint: Joint used where two successive placements of concrete meet, across which a bond is maintained between the placements.

contour line: Line that shows the slope of the ground with lines extending from identified grade levels.

control joint: Groove made in a horizontal or vertical concrete surface to create a weakened plane and control the location of cracking.

corner tie: Wood kicker or metal device used to brace the corners of forms.

course: Horizontal layer of concrete. Several courses of concrete make up a lift.

crawl space: Space between the bottom of the floor joists resting on top of the foundation walls and the ground below.

crawl space foundation: Low foundation featuring a narrow accessible space between the first floor joist and the ground.

crystalline silica (quartz): Natural compound found in the crust of the Earth; basic component of sand and granite.

curing: Process of maintaining concrete moisture content and temperature in its early stages to allow desired properties to develop.

curtain wall: Light, non-load-bearing section of wall made of metal or precast lightweight concrete that is attached to the exterior framework of a building.

dampproofing admixture: Substance added to a concrete mixture to improve the impermeability (resistance to water penetration) of hardened concrete.

dead load: Constant weight of an entire superstructure.

decibel (dB): Unit used to express the relative intensity of sound.

deck: Concrete surface that supports a traffic load.

deck form: Form upon which concrete for a floor or roof slab is placed.

decking: Sheathing material used for a deck form.

deflection: Amount of bending or distortion of a material due to direct pressure.

dome pan: Square prefabricated pan form nailed in position through holes in the flanges.

double-headed nail: Nail with two heads that permits easy removal. Commonly used in temporary construction. Also called *duplex nail.*

double-post shore: Wood shore consisting of a head supported by two vertical posts.

doubling up: Placing of the second or opposite form wall.

dowel: Deformed or plain round steel bar extending into adjoining portions of separately placed sections of concrete.

drain pipe: Plastic or clay pipe used to convey water away from an area, such as a foundation wall.

drain tile: Pipe constructed of clay, plastic, or concrete sections. Commonly placed alongside the foundation footing to carry water away from the foundation.

drop panel: Thickened structural area over a column or capital that supports a flat slab floor.

driving head: Metal device placed on top of the pile head to receive the pile driver's blows and protect it from damage.

dry batching: Procedure in which the dry concrete ingredients are placed in the truck and then mixed with water on the way to the job.

early strength: Concrete or mortar strength during the first 72 hours after placement.

ear muff: Ear protection device worn over the ears.

earplug: Ear protection device inserted into the ear canal and made of moldable rubber, foam, or plastic.

earth-formed footing: Footing formed by digging a trench to the dimensions of the footing and filling it with concrete.

edge form: Low wall form positioned around the perimeter of the placement area to contain fresh concrete. Used in flatwork, precast, and tilt-up construction.

electrical shock: Condition that results when a body becomes part of an electrical circuit.

elephant trunk: Flexible tube extending from the bottom of a hopper.

elevation: Grade level established to indicate vertical distance above or below a reference point.

encasement ball insert: One-piece insert that rests on a base.

encasement ball lifting unit: Lifting unit that consists of a shaft containing encasement balls and an adjusting mechanism, two spring-loaded plungers, and a shackle.

entrance platform: Low stoop or porch located at an entrance to a building.

excavation: Removal of earth to allow for construction of a foundation.

expansion joint: Joint that separates adjoining sections of concrete to allow for movement caused by expansion and contraction of the slabs. Also called *isolation joint.*

external vibrator: Vibrator that generates and transmits vibration waves from the exterior to the interior of the concrete.

fiber-reinforced concrete (FRC): Concrete mixture that uses glass, metal, or plastic fibers mixed with concrete to provide extra strength.

fill: Soil or other material brought in from another location and deposited at the building site to raise the grade level.

fine-grained soil: Soil composed of fine particles such as silt or clay.

flat plate floor: Concrete floor slab system supported by columns tied directly to the floor.

flat slab floor: Concrete floor slab system supported by columns and reinforced in two or more directions without beams or girders.

flatwork: Construction of floor slabs, patios, sidewalks, or other horizontal surfaces.

floating foundation: Thickened, reinforced slab placed monolithically with walls and/or footings that transmits the load over a large area. See *mat* and *raft foundations.*

foot: Lower section of a pile.

footing: Part of a concrete foundation that spreads and transmits the load over soil or piles.

form: Temporary structure constructed to contain concrete while it sets.

form anchor: Device embedded in concrete during placement to fasten formwork that will be constructed later.

form hanger: Metal device used to support formwork suspended from structural framework such as steel or precast beams and girders.

form liner: Wood or plastic material placed against the inside of a form wall to produce a textured or patterned finish or to absorb moisture.

form tie: Device used to space and tie opposite form walls and prevent them from spreading or shifting while concrete is being placed.

formwork: System of supporting freshly placed concrete.

foundation: Part of a structure that rests on and extends into the ground and provides support for the load of the superstructure.

foundation footing: Part of a foundation that rests directly on the soil, acts as a base for the foundation wall, and distributes the entire foundation load over a wide soil area.

foundation wall: Load-bearing wall that extends above and below the ground level.

free fall: Distance of descent of freshly placed concrete into forms without using a drop chute or other means of control.

friction pile: Pile that receives its support from friction between the surrounding soil and exterior surface of the pile.

front setback: Distance from a building to the front property line.

front walk: Walk that extends from the front entrance of a building to a driveway or sidewalk.

frost line: Depth to which soil freezes in a particular area.

full basement foundation: Foundation that provides living and/or storage areas below the superstructure.

ganged panel forms: Wall forms constructed of many small panels bolted together.

gas former: Admixture that facilitates expansion setting and is used in nonshrink grouts.

girder: Large horizontal member that supports a bending load over a span, such as from column to column.

goggles: Form of eye protection with a flexible frame and secured on the face with an elastic headband.

grade: Existing or proposed ground level of a building site.

grade beam: Reinforced concrete beam used as a foundation for superstructures. Main support is received from piers or piles extending into the ground.

granular (coarse-grained) soil: Soil that consists mostly of sand and gravel with large, visible particles.

green concrete: Concrete that has been placed, but has not yet reached full strength.

ground beam: Reinforced beam running along the surface of the ground that does not rest on supporting piers or piles.

groundwater: Water beneath the Earth's surface that is primarily affected by the water table.

groundwork: Preliminary grading, excavating, trenching, and backfilling required at a job site.

grouped pile: Numerous piles driven in close arrangement

grout: Mixture obtained by combining cement, sand, and water.

heave: Upward thrust due to frost or moisture absorption and expansion of soil.

heavy construction: Concrete or heavy timber construction methods used to erect structures such as factories, bridges, freeways, and dams.

honeycomb: Void left in concrete due to mortar not effectively filling the spaces among the coarse aggregate.

hook points: Locations in a precast member providing access for crane attachment.

hopper: Funnel-shaped box used to place concrete in a form.

hub: Stake used to indicate a corner of a property.

hydration: Chemical reaction between cement and water that produces hardened concrete.

immersion vibrator (thermal vibrator): Tool that consists of a motor, a flexible shaft, and an electrically or pneumatically powered metal vibrating head that is dipped into and pulled through concrete.

impact force compaction: Compaction using a machine that alternately strikes and leaves the ground at high speed to increase soil density.

insert: Anchoring device placed in precast concrete members that provides a crane attachment point.

insulating concrete forms (ICFs): Specialized forming system that consists of a layer of concrete sandwiched between layers of insulating foam material on each side.

internal disconnecting tie: Form tie consisting of two external sections that screw into an internally threaded bolt and remain in place after the forms are stripped. Eliminates the use of spreader cones and reduces the size of holes remaining in the concrete.

isolation joint: Separation created between adjoining parts of a concrete structure to allow for movement of the parts. Usually filled with caulking compound or asphalt-impregnated material.

isolation strip: Piece of ½″ thick premolded asphalt-impregnated material placed before the concrete is placed.

J-bolt: Type of anchor bolt embedded into the concrete at the time of placement. The threaded end projects from the concrete to allow for the attachment of a sill plate or other structural member.

kerf: Cut or groove made by a saw blade.

key strip: Chamfered piece pressed into the concrete immediately after placement; used to form a keyway.

keyway: Tapered groove formed in concrete at the top surface of a spread footing.

kicker: A wood block or board that reinforces another form member against an outward thrust.

knee pads: Rubber, plastic, or leather pads strapped onto the knees for protection.

Lally column: Steel pipe column that rests on a concrete pier footing. Used to support wood or steel beams.

large panel forms: Wall forms constructed in large prefabricated units.

lateral pressure: Horizontal pressure such as the force of soil against the side of a high foundation wall.

ledger: Horizontal member that supports permanent or temporary structural members.

leveling rod: Graduated rod used in conjunction with a builder's level or transit-level.

L-foundation: Foundation that has a footing on only one side of the foundation wall.

L-head shore: Shore formed with the horizontal member projecting from one side. Commonly used to support forms for spandrel beams.

lift: Layer of concrete placed in a wall and separated by horizontal construction joints.

lift bar: Horizontal bar used with cranes that are equipped with cables threaded over pulleys. Cables are attached to precast panels to be raised into position.

lift plate: Metal plate that is bolted to inserts embedded in precast concrete members.

line of sight: Imaginary straight line extending from a builder's level or transit-level to object being sighted.

live load: Varying load supported by a structure.

material safety data sheet (MSDS): Written document used to relay hazardous information from the manufacturer, importer, or distributor to the worker.

mat foundation: Thickened reinforced slab that transmits the load of the structure as one unit over the surface of the soil.

minimum bending radius: Smallest bending radius that the plywood can be subjected to without structural damage.

mix: Mixture of aggregate, cement, water, and required admixtures.

monolithic concrete: Concrete placed in forms without construction joints.

motor grader: Earth-moving machine used for final grading operations.

mudsill: **1.** Plank or timber placed on top of soil to support shores. **2.** Wood member fastened to the top of foundation walls to which joists or studs are nailed.

nominal lumber size: Dimension of sawed lumber before it is surfaced and seasoned.

Occupational Safety and Health Administration (OSHA): U.S. government agency that establishes safety regulations for the construction trade and other industries.

open stairway: Unenclosed stairway running between two levels.

pan: Metal or plastic prefabricated form unit used in construction of concrete floor joist systems.

panel: Form section consisting of sheathing and stiffeners that can be erected and stripped as a unit.

panel form: Prebuilt form section made up of panel sheathing, studs, and top and bottom plates.

parapet: Short walls that act as a safety barrier along the edge of a superstructure.

patented tie: Patented device used to secure and space opposite walls of forms during concrete placement.

personal protective equipment (PPE): Safety equipment worn by a construction worker for protection against safety hazards on a job site.

pier box: Form for pier footings.

pier footing: Part of the foundation system that supports wood or metal posts bearing girders.

pier foundation: Foundation in which the exterior and interior walls of a building are supported by beams, posts, and pier footings.

pigtail bolt: Anchoring device that has curved and angular shapes that increase holding power.

pilaster: Rectangular column incorporated with a concrete wall to strengthen the wall and provide support for the end of a beam.

pile: Long structural member that penetrates deep into the soil.

pile cutoff: Portion of the pile head that is removed after the pile is in its desired position.

pile head: Upper surface of a precast pile in its final position.

pile shoe: Metal cone placed over the tip of the pile to protect it from damage while the pile is being driven.

pipe pile: Round steel pile. Hollow interior is filled with concrete after pile has been driven into ground.

placement: Process of placing and consolidating concrete. Pour is a term used interchangeably with placement.

plank: Lumber over 1″ thick and 6″ or more in width.

plasticity: Property and state of freshly mixed concrete that determines its shaping qualities and workability.

plasticizer: Type of water-reducing admixture that provides concrete with increased workability with less water.

plate: Flat horizontal member placed at the top and bottom of panel studs.

plot plan: Drawing that provides information regarding preliminary sitework. Also called *site plan.*

plumb: 1. Vertical. **2.** To make vertical.

Plyform®: Plywood product specifically designed for concrete formwork.

plywood: Manufactured panel product consisting of veneers that are glued together under intense heat and pressure. Used extensively as form sheathing.

polyethylene film: Thin sheet of plastic frequently used as a vapor barrier.

portland cement: Product obtained by pulverizing and mixing limestone with other products to produce cement required for concrete mix.

post base: Metal device embedded in concrete piers or walls. Used to secure the bottom of wood posts.

post-tensioning: Method of prestressing in which the steel cables are tensioned after the concrete has been placed.

powder-actuated fastener: Special concrete nail driven with a powder-actuated fastening tool.

power shovel: Large earth-moving equipment used for excavation.

pozzolan: Fine substance that chemically reacts with calcium hydroxide, which is produced in the hydration process of cement.

precast concrete: Concrete structural member formed somewhere other than its final position.

prefabricated forms: Forms constructed from prebuilt panel sections.

pressure treating: Process in which chemical preservatives are forced into the wood under intense pressure.

prestressed concrete: Concrete in which internal stresses are introduced to such a degree that tensile stresses resulting from service loads are counteracted to the desired degree.

pretensioning: Method of prestressing in which the steel cables are tensioned before the concrete is placed in the casting bed.

property line: Line that defines the boundaries of a building lot.

public walk: Walk that runs along a street bordering the building lot. Also called *sidewalk.*

raft foundation: Thickened reinforced slab placed monolithically with the walls.

rammer: Soil compaction tool that alternately strikes and leaves the ground at high speed to increase soil density.

ready-mixed concrete: Concrete manufactured at batch plants and delivered by truck to the job site in a plastic state.

rebar: Steel reinforcing bar with deformations on the surface to allow the bar to interlock with concrete.

rectangular foundation: Monolithically placed structural support consisting of two vertical faces with no dimensional changes.

reshore: Temporary shore firmly placed under concrete beams, girders, or slabs after form shores have been removed. Used to avoid deflection of the shored member or damage to concrete that is partially cured.

ribbed floor slab: Thin floor slab integrated with concrete joists that tie into supporting girders and columns. Also called *one-way joist system.*

ribbon: Narrow strip of material, usually wood, used in formwork.

riser: Vertical member between two stair steps.

rock pocket: Porous void in hardened concrete that consists primarily of coarse aggregate and open voids with little or no mortar.

rodding: Compaction of concrete with tamping rod.

roller: Soil compaction tool that uses weight, or weight and vibration, to increase soil density.

safety glasses: Glasses with impact-resistant lenses, reinforced frames, and side shields.

scaffold shoring: Shoring that consists of tubular steel frames assembled to support beam and floor forms.

screed: System used to level and strike off the concrete when placing concrete for slabs.

segregation: Separation of sand and cement ingredients from the coarse aggregate in a concrete mix.

seismic risk zone: Area where conditions for earthquakes exist.

separation: Coarse aggregate separating from the rest of the mix during placement occurring because of improper placement and consolidation.

service walk: Walk that is located between a driveway or public sidewalk and rear entrance of building.

set: The stiffening of concrete after it has been placed. *Initial set* refers to first stiffening. *Final set* occurs when concrete has attained significant hardness and rigidity.

set-retarding admixture (retarder): Substance added to concrete to extend its setting time.

shear wall: Panel-sheathed wall used to withstand severe seismic activity and heavy wind loads.

sheathing: Material used to form the face of wall forms or the deck of floor forms.

she bolt: See *internal disconnecting tie.*

sheet piling: Interlocking metal piles designed to resist lateral pressure. Frequently used in deep excavations.

shore: Temporary support for formwork and fresh concrete that has not developed full strength.

shore jack: Metal device fastened to the bottom of a wood shore to allow height adjustment to be made.

shore post: Vertical member used in shoring systems.

shoring: System used to prevent sliding or collapse of earth banks around an excavation.

shrink mixing: Mixing all the concrete ingredients (including water) for approximately 30 sec at the batch plant, depositing the mix in the drum of the ready-mixed truck, then mixing en route to the job.

shut-off: Vertical bulkhead placed at the end of wall forms to contain fresh concrete.

silicosis: Disease of the lungs caused by inhaling dust containing crystalline silica particles.

sill plate: Wood member placed on top of a wall.

silt: Granular material consisting of fine mineral particles that range from 2 to 50 μm in diameter.

single-post shore: Shore consisting of a single post placed under stringers supporting floor forms.

single wall form: Wall form system consisting of one constructed form wall and the other wall composed of solid rock, extremely hard soil, or an existing foundation of an adjoining building.

site cast: Precast structural concrete members cast in casting beds on the job site.

site work: Preliminary layout, excavation, and other preparations required before construction can begin.

slab: Flat horizontal layer of concrete supported by ground, beams, columns, or walls.

slab-on-grade foundation: System that combines concrete foundation walls with a concrete floor slab resting directly on a bed of gravel.

sleeve: Metal or fiber cylinder set inside a form to shape a passage for pipes or other objects through the finished concrete wall.

sloping: An excavation method that slants the sides away from the bottom of an excavation.

slump cone: Cone made of galvanized metal that is 8″ in diameter at the bottom, 4″ in diameter at the top, and 12″ high and is used to perform slump tests.

slump test: Test that measures the consistency, or slump, of concrete.

snap tie: Patented wall tie device with cones acting as form spreaders.

soil compaction: Process of applying energy to loose soil to increase its density and load-bearing capacity through consolidation and removal of voids.

soil mechanics: Study of soil types and the resulting effect upon the behavior of the soil.

soil moisture: Water in the soil that affects soil conditions.

soldier piles: H-shaped piles that are driven into the ground with a pile-driving rig and spaced approximately 8″ apart. Also called *soldier beams.*

spading: Consolidation of concrete with a narrow wood rod or spading tool.

spandrel beam: Beam located in outer wall of a building usually to support floors or roof.

spandrel tie: Type of snap tie with a hooked end.

spread footing: Base for the wall above that distributes the load of the building over a wider area.

stack cast: Precast method in which a series of panels are cast on top of each other.

stake: Wood or metal piece sharpened at its lower end and driven into the ground to either anchor the lower ends of braces or to hold the sides of footing forms in place.

static force compaction: Compaction using a heavy machine that squeezes soil particles together without vibratory motion to increase soil density.

steel reinforcing bar: Deformed steel bar placed in concrete to increase its ability to withstand lateral pressure and tie adjoining concrete members together. Also called *rebar.*

stepped foundation: Foundation shaped like a series of long steps. Usually constructed on sloping lots.

stepped pier: Pier consisting of two or more rectangular piers decreasing in size and placed on top of one another.

strike board: Wood or metal straightedge used for screeding concrete.

strike off: To level concrete to its correct finish grade.

stringer: Horizontal timber placed on top of shores. Commonly used as part of a floor form system.

strip: To remove forms from set concrete surfaces.

strongback: Vertical member attached to the back of forms or precast members to reinforce or stiffen them.

stud: Vertical member used to stiffen and support form sheathing.

substructure: Footings, piers, pier caps, and abutments that support bridges, ramps, or overpass decks.

superstructure: Bridge deck, sidewalks, and parapets (low walls formed along the edges of the deck) of a highway system.

suspended formwork: Formwork suspended from a structural member and supported with U-shaped snap ties or coil hangers that are positioned over the beam or girder.

tapered pier: Pier footing with inclined sides.

template: **1.** Frame used in positioning formwork members. **2.** Wood piece used to lay out and secure anchor bolts during concrete placement.

temporary spoil: Soil material and other waste that is dug out of an excavation.

tensile strength: Resistance of a material to forces attempting to pull it apart.

tension: Pulling or stretching force.

tension crack: Crack in soil caused by increases and decreases in moisture content and by earth movement.

test pit: Shallow excavation dug to examine soil conditions on the job site.

textured plywood: Form sheathing producing special surface effects such as wood grain, boards, and other designs.

T-foundation: Foundation consisting of a wall placed above a spread footing that extends on both sides of the wall.

T-head shore: Shore formed with the horizontal member projecting equally on both sides of post.

3-4-5 method: Layout method used to establish right angles for the building corners.

tilt-up construction: Concrete construction method in which wall panels of a building are cast horizontally at a location adjacent to its eventual position and lifted into position by crane.

tip: Small lower end of a pile.

toenail: **1.** Nail driven at an angle. **2.** To drive a nail at an angle.

total rise: Vertical distance from one floor to the floor above.

total run: Horizontal length of a stairway measured from the foot of the stairway to a point plumbed down from where the stairway ends at a floor or landing.

tower crane: Crane consisting of a high tower and gib. Sections are added to the tower to achieve greater heights.

transit-level: Surveying instrument used for leveling over long distances as well as establishing grades and elevations. Telescope can also be moved vertically for plumbing and other layout work.

transit-mixer truck: Truck equipped with a large drum to mix concrete for delivery to job site.

tread: Horizontal surface of the step in a stairway.

trench: Narrow excavation that is deeper than it is wide.

tubular fiber form: Round column forms constructed of spirally wound fiber plies.

unit rise: Vertical height of riser calculated by dividing the total rise of the stairway by the total number of risers.

unit run: Width of the stair tread calculated by dividing the total run of the stairway by the total number of treads.

vapor barrier: Waterproof membrane placed under slabs-on-grade to contain ground surface dampness.

veneer: One of an odd number of thin layers of wood that are glued together under intense heat and pressure.

vibration compaction: Compaction using a machine to apply high-frequency vibration to the soil to increase soil density.

vibratory plate: Soil compaction tool that applies high-frequency vibration to the ground to increase the soil density.

virgin soil: Soil over which most construction takes place such as gravel, sand, silt, and clay.

voids: Air spaces in set concrete resulting from segregation and improper consolidation during placement.

waffle slab: Thin floor slab integrated with concrete joists that, at right angles to each other, produce a waffle-like appearance. Also called *two-way joist system.*

waler: Horizontal member attached to the outside of form walls to strengthen and stiffen walls. Also called *wale.*

waler rod: Form tie assembly consisting of an inner rod threaded at each end that screws into two outside rods. Outside rods are tightened against walers with large nut washers.

water-cement ratio (w/c): Amount of water used in a concrete mix in relation to the amount of cement.

water-reducing admixture: Substance added to a concrete mixture to reduce the amount of water needed to produce a desired mixture.

water-reducing, set-retarding admixture: Substance that allows less mix water to be used to produce concrete of a desired slump while retarding the set of concrete.

waterstop: Thin piece of rubber, metal, or plastic placed across a construction joint to prevent water leakage.

water table: Highest point below the Earth's surface in a given area that is normally saturated with water.

wedge: **1.** Device driven against the walers at the ends of form ties to hold opposite form walls in position. **2.** Tapered pieces of wood placed beneath shore posts.

welded wire fabric: Heavy steel wire welded together in a grid pattern. Used to reinforce concrete slabs. Also called *wire mesh.*

welding plate: Metal plate embedded in a precast concrete member to attach plates or rebars of adjoining members.

wing wall: Short section of wall at an angle to the abutment used as a retaining wall and to stabilize the abutment.

working scaffold: Temporary elevated structure providing a working platform supporting workers and materials.

yield strength: Maximum load that a material will bend or stretch to accommodate a load and still return to its original size or shape.

A

B

C

J

K

L

M

N

O

P

Q

R

S